Biomaterial-Assisted 3D In Vitro Tumor Models: From Organoid towards Cancer Tissue Engineering Approaches

Biomaterial-Assisted 3D In Vitro Tumor Models: From Organoid towards Cancer Tissue Engineering Approaches

Editors

Serena Danti
Nicola Contessi Negrini
Alessandro Franchi

MDPI • Basel • Beijing • Wuhan • Barcelona • Belgrade • Manchester • Tokyo • Cluj • Tianjin

Editors

Serena Danti
University of Pisa
Italy

Nicola Contessi Negrini
Imperial College London
UK

Alessandro Franchi
University of Pisa
Italy

Editorial Office
MDPI
St. Alban-Anlage 66
4052 Basel, Switzerland

This is a reprint of articles from the Special Issue published online in the open access journal *Cancers* (ISSN 2072-6694) (available at: https://www.mdpi.com/journal/cancers/special_issues/3D_In_Vitro).

For citation purposes, cite each article independently as indicated on the article page online and as indicated below:

LastName, A.A.; LastName, B.B.; LastName, C.C. Article Title. *Journal Name* **Year**, *Volume Number*, Page Range.

ISBN 978-3-0365-6842-3 (Hbk)
ISBN 978-3-0365-6843-0 (PDF)

Cover image courtesy of Contessi Negroni.

Contents

About the Editors

Serena Danti

Serena Danti is senior researcher at the Dept. of Civil and Industrial Engineering, University of Pisa, Pisa, Italy. She received her degree in Chemical Engineering from the University of Pisa in 2003. In 2006, she was a scholar research associate at the Centre for Excellence in Tissue Engineering, Rice University, Houston, TX. Dr. Danti obtained her Ph.D. in "Health Technologies: Evaluation and Management of Innovations in the Biomedical Field" in 2007 from the Centre for Excellence in Computer Assisted Surgery (ENDOCAS), University of Pisa, Italy. Her postdoctoral training was performed at the Centre for the Clinical Use of Stem Cells (CUCCS), and as a senior post-oc at the Otorhinolaryngology Unit, formerally the Dept. of Neurosciences, currently the Dept. of Surgical, Medical and Molecular Pathology, University of Pisa, Italy. She received two research fellowships from Pisa Hospital, ENT, Audiology and Phoniatrics Unit. She has been an affiliate researcher at Massachusetts Institute of Technology (MIT), Cambridge, MA, USA, since 2017 and she has worked at the BioRobotics Institute, Scuola Superiore Sant'Anna, Pontedera (Pisa), Italy, since 2014. She is a member of the 3R's center (Italy). Serena Danti has been involved in several national and international projects in the field of in vitro approaches to understand and treat different pathologies. She has published 120 articles in international journals and contributed about 15 chapters to international books. She has participated in more than 150 congresses, presenting oral and poster communications, and she has been invited to give about 25 plenary lectures at international symposia and at several universities. Dr. Danti is also named as the inventor in five patents. Her research interest involves smart biomaterial approaches for pathology resolution, including ear, bone, and cancer.

Nicola Contessi Negrini

Nicola Contessi Negrini is a research associate at the Department of Bioengineering, Imperial College London (UK). Nicola graduated with a degree in Biomedical Engineering from Politecnico di Milano (IT) and received his European Ph.D. in Bioengineering from Politecnico di Milano, working on the development of hydrogel-based scaffolds for adipose tissue regeneration via advanced fabrication technologies. He has been a visiting researcher at the University of Brighton (UK), Warsaw University of Technology (PL), and Massachusetts Institute of Technology (USA). After completing his Ph.D., Nicola joined the Celiz Lab at Imperial College London to investigate the application of bio-orthogonal chemistries for the development of tissue engineering scaffolds and in vitro models and to engineer innovative translational biofabrication technologies and bioinks. Nicola is a representative of the early career forums of the European Society for Biomaterials (ESB) and the UK Society for Biomaterials (UKSB), has received several awards for his research including the Julia Polak European Doctorate Award, has presented his work at more than 15 international conferences, and has published more than 25 peer-reviewed articles on biomaterials and biofabrication. Nicola's research interest is focused on the development of biomimetic hydrogels and additive manufacturing technologies to obtain scaffolds and in vitro models, with a particular interest in natural derived polymers for soft tissue regeneration and modelling.

Alessandro Franchi

Alessandro Franchi is a full professor of pathology at the University of Pisa, Italy, and a medical director at Azienda Ospedaliera-Universitaria Pisana (AOUP), Pisa, Italy. Since the beginning of his career, he has focused his attention on the histological and molecular characterization of sinonasal tumors. He is also an expert in head and neck pathology, sarcomas and rare tumors, bone and soft tissue pathology, dermatopathology, and ultrastructural pathology. The importance of his research contributions in this field has led to his participation in several research projects funded by the Italian Ministry of University and by the Istituto Toscano Tumori. Prof. Franchi is a member of national and international scientific societies in the field of anatomic pathology and head and neck diseases. He is the principal investigator of a Tuscany region research project focused on the study of rare sinonasal tumors. He has contributed to the 4th Edition of the WHO Classification of Head and Neck Tumours published by IARC Press in 2017, with several chapters dedicated to sinonasal tumors. In 2017–18, he chaired the Dataset Authoring Committee "Nasal cavity and paranasal sin uses" within the project "Dataset Otolaryngology/ENT" of the International Collaboration for Cancer Reporting (ICCR). Prof. Franchi collaborates with the Italian Registry of Sinonasal Tumors RENATuNS as an expert pathologist and with several European and American groups focused on the research of sinonasal tumors. He has authored about 220 publications in peer-reviewed scientific journals. He is the co-editor of the second edition of the textbook "Pathology of the Head and Neck" published by Springer. His research activity is focused on head and neck pathology, with special reference to sinonasal cancers.

Preface to "Biomaterial-Assisted 3D In Vitro Tumor Models: From Organoid towards Cancer Tissue Engineering Approaches"

The merging of disciplines is generating new knowledge at the cutting edge of science. This collection has been envisaged to bring bioengineering, materials science, and cancer research a step closer by explaining, designing and investigating new tools to release complex, more reliable yet ethically sustainable tumor in vitro models. Our aim is to comprehensively foster the scientific community working in this field to collaborate even more interactively and ultimately accelerate the achievement of more powerful and precise therapies to fight cancer than those currently available. Here, we have collected excellent contributions from researchers working in different countries (Italy, Spain, Germany, UK, USA, Japan, Australia) and from four continents, including general overviews and research papers facing diverse tumor types (e.g., pancreatic ductal adenocarcinoma, ovarian cancer, breast cancer, colorectal cancer, and bone cancers), which avail themselves of biomaterial scaffolds to study tumorigenesis, metastasis and screen different therapeutic options (chemo-, radio-, and immuno-therapy) up to personalized approaches via patient-derived 3D models.

Serena Danti , Nicola Contessi Negrini, and Alessandro Franchi
Editors

cancers

MDPI

Editorial

Biomaterial-Assisted 3D In Vitro Tumor Models: From Organoid towards Cancer Tissue Engineering Approaches

Nicola Contessi Negrini [1], Alessandro Franchi [2] and Serena Danti [3,*]

1 Department of Bioengineering, Imperial College London, London SW7 2BX, UK
2 Department of Translational Research and of New Technologies in Medicine and Surgery, University of Pisa, 56126 Pisa, Italy
3 Department of Civil and Industrial Engineering, University of Pisa, 56126 Pisa, Italy
* Correspondence: serena.danti@unipi.it

Cancers are a leading cause of death around the world, accounting for nearly 10 million deaths yearly. Scientists across the globe are intensifying more and more of their efforts to understand tumor development, to develop early-stage diagnostic (and theragnostic) methods, as well as to identify new and personalized biomarkers to obtain novel and more efficient drugs. The connecting link of all these aspects is the availability of reliable, reproducible, cost-effective, and ethically sustainable models. Currently used in vivo animal models and in vitro bi-dimensional (2D) models are affected by drawbacks that are hindering the understanding of tumor structural and biological complexity and the development of efficient treatments. In recent years, biomaterial-assisted three-dimensional (3D) in vitro modelling has gained tremendous momentum as a more reliable and biomimetic alternative to better understand tumor biology and develop new therapeutics or screen drugs and drug combinations for personalized medicine approaches. The growing knowledge produced in the field of tissue engineering has lately opened the way for cancer tissue engineering [1]. The multi- and inter-disciplinary nature of this topic needs to fill a methodology gap, which involves bringing the cancer biology community closer to the field of biomaterials science and *vice versa*. This is the objective of our Special Issue, in which we collected a series of 13 papers (7 research papers, 4 reviews, 1 systematic review, and 1 commentary) published by international leaders in relation to the development and application of 3D in vitro tumor models using biomaterials as instructive elements.

Citation: Contessi Negrini, N.; Franchi, A.; Danti, S. Biomaterial-Assisted 3D In Vitro Tumor Models: From Organoid towards Cancer Tissue Engineering Approaches. *Cancers* **2023**, *15*, 1201. https://doi.org/10.3390/cancers15041201

Received: 6 February 2023
Accepted: 9 February 2023
Published: 14 February 2023

Furthermore, 3D biomaterial-assisted in vitro models can be used to culture multiple cell populations using biomaterials to recapitulate the 3D structure of the tissue in vivo, not achievable by culturing cells on traditional 2D plastic substrates. Several biomaterials or biomaterial combinations can be used to culture different cells to in vitro mimic the complexity of specific cancers, replicating cell–cell interactions, cell–extracellular matrix (ECM) interactions, and vascularization, which is not largely possible using traditional 2D culture plastics as substrates [2,3]. The different structural and biological complexity obtained with 3D in vitro models not only emulates the in vivo morphology, but it also increases the functional similarity of the pathophysiological in vivo tissue/organ. For instance, the drug sensitivity of tissues in vivo, such as the response to innovative immunotherapy, is better mimicked and replicated by 3D in vitro models compared to 2D in vitro models [4].

Betriu et al. [5] studied the internalization and degradation of epithelial growth factor receptors (EGFRs) of pancreatic ductal adenocarcinoma after treatment with Erlotinib and compared the response of cells cultured on 2D plastics and 3D self-assembling peptide scaffolds. The authors demonstrated how Erlotinib treatment promoted epidermal growth factor (EGFR) degradation by cells cultured in 3D cultures, but not on 2D cultures, which proves how the 3D environment allowed for an improved mimicking of in vivo cellular morphology, matrix dimensionality ad stiffness, molecular gradient, and response to therapeutics. In order to prepare 3D in vitro models of soft and hard cancers, Tomar et al. [6]

used polyhydroxyalkanoates synthesized by microorganisms to fabricate 3D porous scaffolds via particulate leaching. The authors cultured breast and colon cancer cells and demonstrated a nutrient diffusion across the scaffolds and cell penetration in the biomaterial 3D structure, proving the suitability of these scaffolds as 3D biomaterial-assisted in vitro cancer models as more biomimetic alternatives to conventional cell cultures. The use of these in vitro models is particularly promising when patient-derived cancer cells are cultured in biomaterial scaffolds and used to develop a patient-specific 3D in vitro model, enabling the possibility of developing personalized and patient-specific cancer treatments [7]. However, the use of 3D models imposes new technical challenges that still need to be fully addressed; namely, biomaterial-assisted in vitro models that use highly reproducible, cost-effective, and adequate technologies for analysis are yet to be optimized. In this context, for example, Shembrey et al. developed a method to monitor intra-tumoral cell heterogeneity using the optical barcoding of patient-derived cancer cells in a 3D in vitro model of colorectal cancer [8].

Biomaterial-assisted 3D in vitro models of cancers have the great potential of overcoming the limitations of the currently used 2D in vitro model. However, developing biomaterial-assisted 3D models comes with several challenges that need to be carefully tackled to achieve an efficient model. Design parameters that need to be addressed include morphological and topographical cues, structural and mechanical cues, and chemical/biological cues to correctly mimic the in vivo tumor microenvironment found [9]. Wieland et al. prepared 3D aligned microfiber scaffolds via melt electrowriting to mimic brain structures and demonstrated the role of topography in determining cell behavior in a metastatic brain model [10]. The authors showed the importance of aligned microfibers to guide the formation of an in vitro relevant biomimetic model and applied the model to validate specific genotypes and their involvement in controlling cell morphology, durotaxis, adhesion, plasticity, and migration in a brain metastasis in vitro model. Mechanical cues are also fundamental to obtaining a relevant biomaterial-assisted in vitro cancer model for both soft and hard tissues, as in the case of bone metastasis and bone cancer [11]. For instance, we developed an in vitro osteosarcoma model using 3D-printed polyurethane and we tuned the porosity and mechanical properties of the scaffolds to achieve an optimal scaffold to promote cell colonization in the scaffold pores and osteogenic differentiation, which could be a platform for osteosarcoma study [12]. After optimizing the physico-mechanical properties of the scaffolds, we introduced biomimetic cues by pre-generating a bone-like ECM on the prepared scaffolds with osteo-induced mesenchymal stem cells [13]. The presence of biomimetic chemical/biological cues is indeed critical for an efficient biomaterial-assisted 3D in vitro model. In the context of the 3D in vitro model of ovarian cell cancer [14], Gupta et al. developed and compared different polymeric scaffolds and highlighted the importance of mimicking ECM components in vitro to accurately mimic the in vivo tumor [15]. Finally, Wishart et al. demonstrated the importance of a fibronectin coating on their polyurethane scaffolds to develop an in vitro model capable of recapitulating hypoxic conditions in 3D pancreatic cancer models to study the effects of radiotherapy treatment [16].

Biomaterial-assisted in vitro models have the great potential of allowing the biological and structural complexity of tumors to be recapitulated in vitro, thus opening new frontiers never investigated before with traditionally used 2D in vitro models and in vivo animal models. The continuous innovation in biomaterials science and new biomolecular tools for the characterization of in vitro models will critically contribute to improving our understanding of cancer and develop new therapeutics to treat such dismal conditions. We hope that this Special Issue will motivate cancer biology scientists and biomaterials and tissue engineers to join forces to tackle this challenging research field, in order to make advances in cancer understanding and treatment.

Conflicts of Interest: The authors declare no conflict of interest.

References

1. Ricci, C.; Moroni, L.; Danti, S. Cancer Tissue Engineering-New Perspectives in Understanding the Biology of Solid Tumours—A Critical Review. *OA Tissue Eng.* **2013**, *1*, 1–7. [CrossRef]
2. Nii, T.; Makino, K.; Tabata, Y. Three-Dimensional Culture System of Cancer Cells Combined with Biomaterials for Drug Screening. *Cancers* **2020**, *12*, 2754. [CrossRef] [PubMed]
3. Negrini, N.C.; Volponi, A.A.; Higgins, C.A.; Sharpe, P.T.; Celiz, A.D. Scaffold-Based Developmental Tissue Engineering Strategies for Ectodermal Organ Regeneration. *Mater. Today Bio* **2021**, *10*, 100107. [CrossRef] [PubMed]
4. Vitale, C.; Marzagalli, M.; Scaglione, S.; Dondero, A.; Bottino, C.; Castriconi, R. Tumor Microenvironment and Hydrogel-Based 3D Cancer Models for In Vitro Testing Immunotherapies. *Cancers* **2022**, *14*, 1013. [CrossRef] [PubMed]
5. Betriu, N.; Andreeva, A.; Semino, C.E. Erlotinib Promotes Ligand-Induced EGFR Degradation in 3D but Not 2D Cultures of Pancreatic Ductal Adenocarcinoma Cells. *Cancers* **2021**, *13*, 4504. [CrossRef] [PubMed]
6. Tomar, A.; Uysal-Onganer, P.; Basnett, P.; Pati, U.; Roy, I. 3D Disease Modelling of Hard and Soft Cancer Using PHA-Based Scaffolds. *Cancers* **2022**, *14*, 3549. [CrossRef] [PubMed]
7. Hasbum, A.; Karabulut, O.; Reyes, R.E.; Ricci, C.; Franchi, A.; Danti, S.; Chew, S.A. Combined Application of Patient-Derived Cells and Biomaterials as 3D In Vitro Tumor Models. *Cancers* **2022**, *14*, 2503. [CrossRef] [PubMed]
8. Shembrey, C.; Smith, J.; Grandin, M.; Williams, N.; Cho, H.J.; Mølck, C.; Behrenbruch, C.; Thomson, B.N.J.; Heriot, A.G.; Merino, D.; et al. Longitudinal Monitoring of Intra-Tumoural Heterogeneity Using Optical Barcoding of Patient-Derived Colorectal Tumour Models. *Cancers* **2022**, *14*, 581. [CrossRef] [PubMed]
9. Fischetti, T.; di Pompo, G.; Baldini, N.; Avnet, S.; Graziani, G. 3d Printing and Bioprinting to Model Bone Cancer: The Role of Materials and Nanoscale Cues in Directing Cell Behavior. *Cancers* **2021**, *13*, 4065. [CrossRef] [PubMed]
10. Wieland, A.; Strissel, P.L.; Schorle, H.; Bakirci, E.; Janzen, D.; Beckmann, M.W.; Eckstein, M.; Dalton, P.D.; Strick, R. Brain and Breast Cancer Cells with Pten Loss of Function Reveal Enhanced Durotaxis and Rhob Dependent Amoeboid Migration Utilizing 3d Scaffolds and Aligned Microfiber Tracts. *Cancers* **2021**, *13*, 5144. [CrossRef] [PubMed]
11. Hughes, A.M.; Kolb, A.D.; Shupp, A.B.; Shine, K.M.; Bussard, K.M. Printing the Pathway Forward in Bone Metastatic Cancer Research: Applications of 3d Engineered Models and Bioprinted Scaffolds to Recapitulate the Bone–Tumor Niche. *Cancers* **2021**, *13*, 507. [CrossRef] [PubMed]
12. Negrini, N.C.; Ricci, C.; Bongiorni, F.; Trombi, L.; D'alessandro, D.; Danti, S.; Farè, S. An Osteosarcoma Model by 3D Printed Polyurethane Scaffold and In Vitro Generated Bone Extracellular Matrix. *Cancers* **2022**, *14*, 2003. [CrossRef] [PubMed]
13. Pham, Q.P.; Kurtis Kasper, F.; Scott Baggett, L.; Raphael, R.M.; Jansen, J.A.; Mikos, A.G. The Influence of an in Vitro Generated Bone-like Extracellular Matrix on Osteoblastic Gene Expression of Marrow Stromal Cells. *Biomaterials* **2008**, *29*, 2729–2739. [CrossRef]
14. Sevinyan, L.; Gupta, P.; Velliou, E.; Madhuri, T.K. The Development of a Three-Dimensional Platform for Patient-Derived Ovarian Cancer Tissue Models: A Systematic Literature Review. *Cancers* **2022**, *14*, 5628. [CrossRef] [PubMed]
15. Gupta, P.; Miller, A.; Olayanju, A.; Madhuri, T.K.; Velliou, E. A Systematic Comparative Assessment of the Response of Ovarian Cancer Cells to the Chemotherapeutic Cisplatin in 3D Models of Various Structural and Biochemical Configurations—Does One Model Type Fit All? *Cancers* **2022**, *14*, 1274. [CrossRef] [PubMed]
16. Wishart, G.; Gupta, P.; Nisbet, A.; Schettino, G.; Velliou, E. On the Evaluation of a Novel Hypoxic 3d Pancreatic Cancer Model as a Tool for Radiotherapy Treatment Screening. *Cancers* **2021**, *13*, 6080. [CrossRef] [PubMed]

cancers

MDPI

Article

An Osteosarcoma Model by 3D Printed Polyurethane Scaffold and In Vitro Generated Bone Extracellular Matrix

Nicola Contessi Negrini [1,*], Claudio Ricci [2], Federica Bongiorni [1], Luisa Trombi [3], Delfo D'Alessandro [3], Serena Danti [2,*] and Silvia Farè [1]

1 Department of Chemistry, Materials and Chemical Engineering "G. Natta", Politecnico di Milano, 20131 Milan, Italy; federica.bongiorni@mail.polimi.it (F.B.); silvia.fare@polimi.it (S.F.)
2 Department of Civil and Industrial Engineering, University of Pisa, 56122 Pisa, Italy; claudio.ricci@med.unipi.it
3 Department of Surgical, Medical, Molecular Pathology, University of Pisa, 56126 Pisa, Italy; l.trombi@yahoo.it (L.T.); delfo.dalessandro@unipi.it (D.D.)
* Correspondence: nicola.contessi@polimi.it (N.C.N.); serena.danti@unipi.it (S.D.)

Simple Summary: Development of new therapeutics to treat osteosarcoma is fundamental to decreasing its current health impact. 3D in vitro models are gaining tremendous momentum as, compared to traditional 2D in vitro models and in vivo models, can speed up new treatment discovery and provide clarification of the pathology development, by ultimately offering a reproducible and biomimetic tool. However, engineering a 3D osteosarcoma in vitro model is challenging, since the reliability of the models strictly depends on their ability to correctly mimic the physical, mechanical, and biological properties of the pathological tissue to be replicated. Here, we designed 3D printed polyurethane scaffolds enriched by in vitro pre-generated bone extracellular matrix, synthesized by osteo-differentiated human mesenchymal stromal cells, to replicate in vitro an osteosarcoma model, which can be potentially used to study tumor progression and to assess new treatments.

Citation: Contessi Negrini, N.; Ricci, C.; Bongiorni, F.; Trombi, L.; D'Alessandro, D.; Danti, S.; Farè, S. An Osteosarcoma Model by 3D Printed Polyurethane Scaffold and In Vitro Generated Bone Extracellular Matrix. *Cancers* **2022**, *14*, 2003. https://doi.org/10.3390/cancers14082003

Academic Editors: Marilena Loizidou and Roberto Benelli

Received: 12 February 2022
Accepted: 11 April 2022
Published: 15 April 2022

Abstract: Osteosarcoma is a primary bone tumor characterized by a dismal prognosis, especially in the case of recurrent disease or metastases. Therefore, tools to understand in-depth osteosarcoma progression and ultimately develop new therapeutics are urgently required. 3D in vitro models can provide an optimal option, as they are highly reproducible, yet sufficiently complex, thus reliable alternatives to 2D in vitro and in vivo models. Here, we describe 3D in vitro osteosarcoma models prepared by printing polyurethane (PU) by fused deposition modeling, further enriched with human mesenchymal stromal cell (hMSC) secreted biomolecules. We printed scaffolds with different morphologies by changing their design (i.e., the distance between printed filaments and printed patterns) to obtain different pore geometry, size, and distribution. The printed PU scaffolds were stable during in vitro cultures, showed adequate porosity (55–67%) and tunable mechanical properties (Young's modulus ranging in 0.5–4.0 MPa), and resulted in cytocompatible. We developed the in vitro model by seeding SAOS-2 cells on the optimal PU scaffold (i.e., 0.7 mm inter-filament distance, 60° pattern), by testing different pre-conditioning factors: none, undifferentiated hMSC-secreted, and osteo-differentiated hMSC-secreted extracellular matrix (ECM), which were obtained by cell lysis before SAOS-2 seeding. Scaffolds pre-cultured with osteo-differentiated hMSCs, subsequently lysed, and seeded with SAOS-2 cells showed optimal colonization, thus disclosing a suitable biomimetic microenvironment for osteosarcoma cells, which can be useful both in tumor biology study and, possibly, treatment.

Keywords: fused deposition modeling; cancer tissue engineering; in vitro model; mechanical properties; mesenchymal stromal cell; bone matrix; bone cancer; tumor microenvironment

1. Introduction

The development of three-dimensional (3D) tissue or organ biomimetic models has become a topic of major interest in the field of biomaterials and tissue engineering. In vitro bioartificial 3D models are unique tools to study the development of pathologies and test potential new drugs [1], to better understand tissue and organ morphogenesis [2], and to recapitulate tumorigenesis and cancer progression to potentially develop new personalized treatments and therapeutics for precision medicine [3]. In the latter case, cancer tissue engineering has emerged in the bioengineering field aiming at fabricating 3D in vitro models to mimic the multi-dimensional structure, organization, and complex relationship in diverse tumor microenvironments (TMEs) [4]. Conventional studies to assess therapeutic efficacy on tumor cells have usually been performed using traditional two-dimensional (2D) in vitro models and small animal in vivo models. However, a wide consensus has now been reached on the evidence that 2D models lack mimicking structural, mechanical, and biochemical cues of the tissue microenvironment, thus oversimplifying cell-cell and cell-extracellular matrix (ECM) interactions [5]. Moreover, animal models are affected by several limitations, including low reproducibility, the impossibility of performing longitudinal studies due to animal sacrifice at the end of most tests, associated high costs, time-consuming studies, as well as ethical concerns and the necessity of ethical approvement by dedicated bodies [3]. 3D in vitro models represent an optimal compromise between the use of 2D in vitro models and in vivo models, as they mimic the complex 3D architecture of tissues and they can recapitulate the cell-cell and cell-ECM interactions, thus leading to improved predictions [6,7]. Despite undeniable advantages, the fabrication of 3D in vitro models is still challenging for bioengineers, as prepared models should possess adequate properties to mimic the specific tissue and pathology to be bioartificially recreated [8].

Osteosarcoma is an osteoid-producing primary malignant bone tumor of mesenchymal origin, primarily affecting long bones of the extremities, with the highest incidence in the distal femur, proximal humerus, and proximal tibia [9]. Despite being relatively rare, race and age have been identified as risk factors, with those aged 10–14 years old and over 65 years old being the most affected [10]. Current treatments for osteosarcoma include surgical resection of the diseased portion, coupled with systemic chemotherapy. However, differently from carcinomas (epithelial origin), the safe resection margin in osteosarcoma resection is still unknown due to the unconfined structure of this tumor. As such, the 5-year event-free survival for patients with early diagnosed localized osteosarcoma is 65–70%, whereas patients with recurrent disease or metastases, most commonly in the lungs (85–90%), have a very poor prognosis of approximatively 20–30% survival rate [11,12]. The development of tools to allow a deep understanding of the tumor evolution and its interaction with the tissue microenvironment is thus impellent to identify molecular targets, anticancer drugs, and patient-specific treatments, which would improve the current pathological outcomes.

Typical steps used to study the efficiency of drugs and treatments avail themselves of 2D in vitro cell cultures, followed by in vivo animal models, and finally pre-clinical and clinical trials. However, most of the drugs/treatments demonstrated to be efficient on 2D in vitro models fail in replicating this high efficiency in vivo, given the oversimplification of the 2D models, thus leading to a waste of resources [13]. 3D in vitro osteosarcoma models are required to speed up new therapeutic evaluation and they are the objective of several studies. Such advanced osteosarcoma models can be distinguished in: *(i)* scaffold-free systems (e.g., multicellular tumor spheroids and organoids cultured in suspension), and *(ii)* scaffold-based cancer tissue engineering approaches. In the latter case, the fabrication of biomaterial scaffolds to support the 3D tumor development allows reproducible and easy to handle culture systems to be obtained, provided with tunable biochemical/mechanical properties, which are in turn capable of mimicking those of the native tumor microenvironment [14,15]. To date, the scientific literature mostly reports on 3D scaffolds to study secondary tumors in bone. Several studies describe scaffolds designed to mimic in vitro the development of breast cancer bone metastasis, including polyurethane (PU) foams [16], porous chitosan scaffolds loaded with hydroxyapatite [17], silk fibroin scaffolds [18,19], poly(lactide-co-glycolide) (PLGA) porous scaffolds mineralized with hydroxyapatite [20], 3D hyaluronate-based hydrogels [21], and polyethylene glycol (PEG) hydrogels coupled with printed polycaprolactone (PCL) [22]. Only a few works describe the development of scaffolds addressing in vitro models of primary bone tumors. Electrospun poly(ε-caprolactone) scaffolds were used to prepare Ewing osteosarcoma in vitro models [23,24], porous silk sponges were fabricated via freeze-drying, seeded with osteosarcoma cells, and the prepared model was compared to traditional 2D in vitro cultures and in vivo mice models [25], and methacryloyl platelet lysate was used to sustain spheroid growth and invasion [26]. Nevertheless, replicating in vitro a biomimetic microenvironment of bone cancer in terms of morphological, mechanical, chemical, and biological cues represents a challenge [27]. Among the plethora of technologies used for scaffold fabrication, 3D printing by fused deposition modeling is a subset of additive manufacturing techniques, with high control of geometric parameters, such as pore and filament size, fundamental to tailoring biomaterials properties. PU is a class of polymers entitled with diverse properties. PUs are widely known in the biomedical field for their applications in medical devices and have recently become attractive as 3D printing biomaterials [28]. Moreover, previous evidence highlighted the importance of biochemical signaling, often provided by artificial hydroxyapatite coatings of the polymeric surfaces [13,16]. A possible strategy, previously used in bone tissue engineering, relies on the application of bone cells in a first culture step to produce bone ECM and then to remove these cells by lysis to have a biohybrid scaffold in which to seed the stem cells of interest [29,30]. The advantage of this strategy is that a biomimetic microenvironment is generated, not only containing hydroxyapatite, but also bone proteins and growth factors secreted by non-pathogenic cells [31].

Here, we developed a biomimetic osteosarcoma in vitro model via 3D printed polyether-urethane (hereafter referred to as PU) scaffolds, post-decorated with human stromal cell (hMSC)-synthesized ECM. In the first phase of this study, we tested different printed geometries to achieve optimal morphological, physical, and mechanical properties to replicate a bone-like model and allow hMSC colonization and differentiation towards osteoblastic lineage. After selecting the most promising scaffolds, we developed different biomimetic models by the deposition of biomolecules synthesized by undifferentiated or osteo-differentiated hMSC. The constructs were then lysed to remove live cells and seeded with SAOS-2 osteosarcoma cells to investigate the potential use of the prepared scaffold as in vitro osteosarcoma model.

2. Materials and Methods

2.1. Materials

Thermoplastic PU (hardness: 80 Shore A) 3D printing filaments were purchased from Sharebot S.r.l. (Nibionno, Italy). Lymphoprep was purchased by GE healthcare (Hatfield, UK). Low-glucose Dulbecco's Modified Eagles Medium (DMEM), L-glutamine, penicillin, streptomycin heat-inactivated fetal bovine serum, trypsin, $CaCl_2$, human AB plasma, *p*-nitrophenol standards, alkaline buffer solution (1.5 M 2-amino-2-methyl-1-propanol at pH 10.3), 4-nitrophenyl phosphate disodium salt hexahydrate capsules, NaOH, acetic acid, phosphate-buffered saline (PBS), methanol, 36 vol-H_2O_2, bovine serum albumin (BSA), Mayer's hematoxylin, dibutyl-phthalate polystyrene xylene (DPX), silver nitrate, pyrogallol, and sodium thiosulfate were obtained from Sigma-Aldrich, (Milan, Italy). The osteogenic differentiation medium was bought from Lonza (Basel, Switzerland). AlamarBlue™, PicoGreen kit, and Paraffin Histoplast LP were supplied by Thermo Fisher Scientific (Waltham, MA, USA). Arsenazo III was bought from Diagnostic Chemicals Ltd. (Oxford, CT, USA). Neutral buffered formalin, absolute ethanol, and xylene were purchased from Bio-Optica (Milan, Italy). Aluminum sulfate was obtained from Carlo Erba (Milan, Italy). Goat serum was obtained from Vector Lab (Burlingame, CA, USA). Primary antibody for collagen type I (COLL-1; ab34710) was bought from AbCam (Cambridge, UK). Primary antibodies for osteopontin (OPN; sc-166261), fibronectin (FN; sc-59826), transforming growth factor beta 1 (TGF-β1; sc-146), and alkaline phosphatase (ALP; sc-166261) were supplied by Santa Cruz Biotechnology, (Santa Cruz, CA, USA). Goat anti-rabbit and goat anti-mouse biotinylated secondary antibodies and streptavidin (Vectastain Elite ABC Kit Standard) were obtained from Vector Laboratories (Burlingame, CA, USA). SAOS-2 cell line (89050205) was bought from ATCC-LG standards (Milan, Italy). L929 (85011425) cells were obtained by ECACC (Salisbury, UK). McCoy's 5A medium was bought from Gibco (Rodano, Italy). HMSCs were supplied by Merck Millipore S.A.S. (Burlington, MA, USA).

2.2. 3D Printing of PU Scaffolds

Computer-aided design (CAD) models were prepared by SolidWorks®software (Dossault Systems Solidworks Corp., Waltham, MA, USA). The CAD models were then processed by Slic3r software (open source) to obtain stl files. These files were used to print PU scaffolds by processing PU 3D printing filaments via fused deposition modeling (Sharebot 42, Sharebot s.r.l., Nibionno, Italy). Each scaffold was printed by subsequent deposition of 12 parallel layers (thickness = 4.8 mm) and scaffold specimens were subsequently cut with a scalpel (square shape, side = 6 mm). The angle shift between subsequent layers and the distance between the axis of the printed filaments were varied (i.e., 90°, 60°, and 45° angle shift between subsequent layers of parallel filaments printed at 1.0 mm or 0.7 mm distance) to obtain different scaffold types, as summarized in Table 1. The surface roughness of printed filaments (n = 6) was measured according to DIN 4768 (R_a, R_zIso, UBM MICROFOCUS).

Table 1. 3D printed polyurethane samples fabricated by varying the angle shift between subsequent layers (i.e., 90°, 60°, or 45°) and the distance between the center of filament printed in parallel on the same layer (i.e., 1.0 mm or 0.7 mm).

Filament Distance [mm]	Angle Shift [°]		
	90°	60°	45°
0.7	90_0.7	60_0.7	45_0.7
1.0	90_1	60_1	45_1

2.3. Morphological, Physical, and Mechanical Characterization

The morphological characterization of the printed PU scaffolds was performed by stereomicroscope (Leica DFC290, Leica Microsystems). Top-view images of the scaffolds (n = 6 images per n = 3 scaffolds per type) were acquired and analyzed by ImageJ software (version 1.52e). The average pore area for each scaffold type was calculated by measuring the area of pores in binary images. The density of the scaffolds was calculated as a mass-to-volume ratio of the samples (n = 5). The porosity percentage of each scaffold formulation was calculated according to EN ISO 4590, following Equation (1):

$$\text{Porosity } [\%] = \left(1 - \frac{\rho_{\text{scaffold}}}{\rho_{\text{material}}}\right) \times 100 \quad (1)$$

where $\rho_{scaffold}$ is the density of the scaffolds, while $\rho_{material}$ is the density of the PU samples printed with 100% infill.

The weight variation of the printed scaffold in a culture medium, tested to simulate their potential use as in vitro model, was evaluated after scaffold sterilization (immersion in absolute ethanol, exposure under UV light on both sides for 20 min). Dry samples (n = 3) were weighted (w_0), immersed in a culture medium, and stored in an incubator to simulate the cell culture condition (37 °C, 5% CO_2). At established time points, samples were removed from the culture medium and weighted (w_t). The culture medium was refreshed every three days. The percentage weight variation in time (ΔW%) was then calculated as Equation (2) [32]:

$$\Delta W\,[\%] = \frac{w_t - w_0}{w_0} \times 100 \tag{2}$$

Static compression tests were performed by Dynamic Mechanical Analyzer (DMA Q800, TA Instruments) to investigate the mechanical properties of the scaffolds. Hydrated samples (i.e., immersed until weight variation reached the plateau) were tested at 37 °C by applying a load ramp at 0.5 $N{\cdot}min^{-1}$ (0.05 N preload) up to 18 N (i.e., load limit of the machine). The Young modulus E was calculated as the slope of the curve in the 0–1% strain range ($R^2 > 0.9$).

2.4. *In Vitro Cytotoxicity Tests on PU Scaffolds*

Samples printed with a 0.7 mm distance between filaments were selected for the in vitro tests. Samples were sterilized by immersion in absolute ethanol and by exposure to UV light (20 min per side).

In vitro, indirect cytotoxicity tests (according to UNI EN ISO 10993) were performed by using L929 and SAOS-2 cells. Scaffold samples (n = 3) were lodged in 24-multiwell tissue culture polystyrene (TCPS), immersed in a 1.2 mL culture medium, and stored in an incubator (37 °C, 5% CO_2) for 1, 3, and 7 days. At each time point, a culture medium without samples was placed in the incubator as a control. L929 and SAOS-2 cells were seeded in 96-multiwell TCPS at 1×10^4 cells/well density and cultured until 70% confluency was reached. Then, the culture medium was replaced with the scaffold supernatant to continue cell cultures for 24 h, using a non-conditioned culture medium as control. The cell metabolic activity was evaluated using the AlamarBlue™ assay, as Equation (3) [33]:

$$\text{Cell Viability}\,[\%] = \frac{f_{eluate} - f_{AlamarBlue}}{f_{control} - f_{AlamarBlue}} \times 100 \tag{3}$$

where f_{eluate} is the fluorescence intensity of the supernatant of the cells cultured with culture medium eluates, $f_{control}$ the intensity of the supernatant measured for cells incubated with culture medium controls, and $f_{AlamarBlue}$ the background fluorescence of the unreacted AlamarBlue solution ($\lambda_{excitation}$ = 540 nm, $\lambda_{emission}$ = 595 nm, using a UV spectrophotometer (Tecan GENius Plus plate reader, Tecan Trading AG, Männedorf, Switzerland).

2.5. *3D Bone Model Preparation*

The suitability of the printed scaffolds as 3D in vitro culture systems for bone was first evaluated by seeding and then osteo-differentiating hMSCs on scaffolds printed with 90°, 60°, and 45° angle shift patterns. HMSC was expanded in a culture medium consisting of low-glucose DMEM added with 2 mM L-glutamine, 100 IU/mL penicillin, 100 mg/mL streptomycin, and 10% heat-inactivated FBS until 70–80% confluence was reached. HMSCs at 500.000 cells/sample were resuspended in 50 µL of human AB plasma with 20 µL $CaCl_2$ 25 mM, seeded on 90_0.7, 60_0.7, and 45_0.7 printed PU scaffolds, and incubated for 30 min at 37 °C in 24-multiwell TCPS. Subsequently, 1 mL of culture medium was added to each sample. The following day, the expansion medium was changed to an osteogenic differentiation medium, and the culture was continued for 2 weeks by replacing the culture medium every 2–3 days. The metabolic activity of hMSC/PU scaffold constructs was measured by AlamarBlue assay. On days 1, 7, and 14 of culture, the samples (n = 3) were incubated for 3 h at 37 °C with the dye diluted in the culture medium as per the manufacturer's instruction. For each test, 100 µL of supernatant from samples or controls were loaded in 96-multiwell TCPS and the absorbance was measured by spectrophotometry (Victor 3; PerkinElmer, Waltham, MA, USA) under double wavelength reading (i.e., at 570 nm and 600 nm). The AlamarBlue percentage (%AB_{red}) was calculated by the Equation (4), as for the manufacturer's protocol, in which: λ = absorbance, s = sample, and c = control:

$$\%AB_{red} = \frac{\left(117,216 \cdot \lambda_{s(570\ nm)} - 80,586 \cdot \lambda_{s(600\ nm)}\right)}{\left(155,677 \cdot \lambda_{c(600\ nm)} - 14,652 \cdot \lambda_{c(570\ nm)}\right)} \times 100 \quad (4)$$

Cellularity (i.e., DNA content) in construct lysates was quantified by PicoGreen assay. Constructs (n = 3) previously frozen in 2 mL of double-distilled (dd)-water underwent three freeze/thaw cycles followed by vortexing to allow DNA extraction from the samples. DNA standards (range: 0–6 μg/mL) were prepared and 50 μg/mL of standard or sample was poured into individual wells of a 96-multiwell plate. Working buffer and PicoGreen dye solution, prepared following the manufacturer's instructions, were added at 100 μg/mL and 150 μg/mL per well, respectively. After 10 min incubation in the dark at room temperature (RT), the fluorescence intensity was measured at $\lambda_{excitation}$ = 480 nm, $\lambda_{emission}$ = 520 nm, using Victor 3 plate reader.

On the residual samples not used for cellularity (n = 3), intracellular ALP was quantified via a colorimetric end-point assay using p-nitrophenol phosphate as a reagent. The rate of p-nitrophenol formation, catalyzed by ALP, is directly related to the enzyme activity in the sample. p-nitrophenol standards (range: 0–250 μM) were prepared, and 80 μL of standard or sample was inserted into individual wells of a 96-multiwell plate. The alkaline buffer solution was added at 20 μL/well. Substrate solution was prepared by dissolving the capsules into dd-water with a final concentration of 4 mg/mL and added at 100 μL/well. The microplate was incubated at 37 °C for 1 h and the reaction was stopped by adding 0.3 M NaOH at 100 μL/well. The absorbance was measured at 405 nm using Victor 3 plate reader.

Finally, the amount of mineralized ECM present, in terms of calcium content, was quantified using the Arsenazo III reagent. 1 N acetic acid was added to equal volumes of residual lysate solutions to obtain 0.5 M acetic acid sample solutions. Samples (n = 3) were incubated overnight to dissolve calcium deposits. Ca^{2+} standards (range: 0–100 μg/mL) were prepared from dilutions of a 1 mg/mL stock solution of $CaCl_2$ and 20 μL of standard or sample was added to individual wells of a 96-multiwell plate. MilliQ water solution at 300 μL/well of Arsenazo III > 0.15 M was added and the microplate was incubated for 10 min at RT. The absorbance at 650 nm was measured on a Victor 3 plate reader. Cell distribution and morphology on the scaffolds were investigated via histological analysis. The constructs were fixed in neutral buffered formalin (4% w/v) at 4 °C overnight, and washed in PBS. Dehydration was done using a graded series of ethanol aqueous solutions at 40 °C. The samples were clarified in xylene overnight at RT and rinsed in liquid paraffin in a vacuum oven at 60 °C for 24 h, placed in a wax embedding box, and air-dried. Paraffin blocks were sectioned by a microtome into 6 μm-thick sections that were mounted on glass slides. The sections were deparaffinized in xylene, rehydrated in absolute ethanol, washed in distilled water, and PBS×1. The specimens were incubated in a methanolic solution containing 0.6% v/v 36-volume H_2O_2, in the dark for 15 min to quench endogenous peroxidases. After washings in PBS, specimens were incubated with 5% v/v goat serum for 20 min at 37 °C to block a-specific binding sites of the secondary antibodies. Sections were then incubated with primary antibodies diluted in 0.1% w/v bovine BSA in a moist chamber overnight at 4 °C. The following antibodies were used: rabbit polyclonal anti-human collagen type I 1:1200 mouse monoclonal anti-human Osteopontin 1:2000 and mouse monoclonal anti-human Alkaline Phosphatase 1:100. Negative controls were obtained by incubating sections with only BSA/PBS solution. Specimens were then incubated with goat anti-rabbit or goat anti-mouse biotinylated secondary antibodies diluted 1:200 in 1.5% v/v goat serum-1x PBS solution for 60 min, then with streptavidin solution for 30 min, prepared according to the manufacturer's instructions. Sections were incubated in the substrate-chromogen solution (0.5 mg/mL of 3,3′-diaminobenzidine tetrahydrochloride) and added with H_2O_2, for 5 min in the dark, then counterstained with Mayer's hematoxylin for 2 min and washed in tap water for 2 min. Sections were dehydrated in absolute ethanol, clarified, and mounted with a coverslip DPX mounting medium. Stained sections were observed with a Nikon Eclipse Ci microscope (Nikon Instruments, Amsterdam, The Netherlands) equipped with a digital camera.

2.6. 3D Osteosarcoma Model Preparation and Validation

In vitro osteosarcoma models were prepared by seeding SAOS-2 cells on PU scaffolds printed with a 60° angle shift pattern. SAOS-2 cells (1×10^6 cells) were resuspended in 50 µL of human AB plasma with 20 µL $CaCl_2$ 25 mM and seeded on differently pre-treated PU scaffolds for 30 min at 37 °C in 24-multiwell TCPS. The different pre-treatments included *(i)* printed PU scaffolds (i.e., no pre-treatment), *(ii)* PU scaffolds containing non-mineralized ECM produced by pre-seeded undifferentiated hMSCs (PU_ECM), which were further lysed in sterile H_2O, and *(iii)* PU scaffolds containing mineralized ECM produced by osteodifferentiated hMSCs, which were further lysed in sterile H_2O (PU_bECM). The PU_ECM constructs were obtained as described in Section 2.5, either using an hMSC growth medium or an osteo-differentiating culture medium for 3 weeks. After cell seeding, 1 mL of McCoy's 5A culture medium was added, and the constructs were cultured for 1 week. The different prepared models are summarized in Table 2.

Table 2. 3D in vitro osteosarcoma models prepared using PU 60° pattern printed scaffolds.

3D Osteosarcoma Model Acronym	Scaffold Pre-Treatment (1st Step Culture, 3 Weeks)	Substrate Used for Osteosarcoma Cells	2nd Step Culture (1 Week)
PU	None	Plain scaffold	SAOS-2 cells
PU_ECM	Undifferentiated hMSC culture followed by cell lysis	Scaffold containing immature bone ECM	SAOS-2 cells
PU_bECM	Osteo-differentiated hMSC culture followed by cell lysis	Scaffold containing mature (mineralized) bone ECM	SAOS-2 cells

The in vitro models were then processed for histological analysis, as previously described. H&E staining was performed to study cell morphology and scaffold colonization, as reported in Section 2.6. In addition, von Kossa staining was performed to reveal deposits of calcium phosphate, typical of bone mineralized ECM. After deparaffination and rehydration, sections were washed in PBS, incubated with 1% *w/v* silver nitrate exposed to light, 0.5% *w/v* pyrogallol, and 5% *w/v* sodium thiosulfate. The counterstaining was performed by incubating cells with 0,1% *w/v* nuclear fast red diluted in a distilled water solution containing 5% *w/v* aluminum sulfate, followed by washing in tap water to visualize the staining. Sections were dehydrated, clarified, mounted, and observed as previously described.

2.7. Statistical Analysis

Data were analyzed by GraphPad Prism software. All data are represented as mean ± standard deviation. Statistical analysis was performed by one-way ANOVA test and Tukey's multiple comparisons, considering $p < 0.05$ as statistically significant.

3. Results

In vitro osteosarcoma, 3D models were developed by first optimizing the physicomechanical properties of the printed PU scaffolds for a bone 3D model. Different layouts of the PU scaffolds were obtained by changing the angle shift printing pattern and the distance between the filaments. The cytocompatibility of the produced scaffolds was assessed using SAOS-2 and L929. Thereafter, the most suitable scaffold allowing a 3D in vitro model of bone tissue was selected using osteo-differentiated hMSCs. The selected PU scaffold was then used to generate osteosarcoma 3D models by seeding SAOS-2 cells. Different treatments were preliminarily applied to the polymeric scaffold to recapitulate in vitro the development of osteosarcoma, aiming at understanding the role of biological molecules.

3.1. Morphological, Physical, and Mechanical Properties of 3D Printed Polyurethane Scaffolds

Optimization of the fabrication parameters and procedure was required to 3D print defect-less scaffolds with reproducible geometry. The initial printing trials achieved scaffolds characterized by morphological irregularities, given by insufficient extrusion of material and non-uniform deposition of the printed filaments. Thus, the printing parameters were adjusted to achieve a uniformly printed structure. The optimized printing parameters are summarized in Table 3. Afterward, the Gcode files used to print the scaffolds were modified by Repetier-Host software to guide the movements of the printing head, thus avoiding defects in the deposited material during the printing process. Optimization of the printing and design parameters and modification of the printing movements allowed scaffolds with desired and reproducible geometries to be finally obtained (Figure S1). These parameters were then applied to print scaffolds with different geometries and subsequently tested as potential in vitro osteosarcoma models.

Table 3. Optimized 3D printing parameters used for the fabrication of 3D printed PU scaffolds as in vitro osteosarcoma models.

<table>
<tr><th rowspan="2">Printing Parameters</th><th>Printing Speed [mm·s^{-1}]</th><th>Extrusion Temperature [°C]</th><th>Plate Temperature [°C]</th><th>Extrusion Multiplier</th></tr>
<tr><td>12</td><td>240</td><td>30</td><td>1.5</td></tr>
<tr><th rowspan="2">Design parameters</th><th>Layer height [mm]</th><th colspan="2">Filament thickness [mm]</th><th>Filament diameter [mm]</th></tr>
<tr><td>0.4</td><td colspan="2">0.4</td><td>1.7</td></tr>
</table>

The printed PU filaments were characterized by roughness values $R_a = 0.60 \pm 0.20$ µm and R_zIso = 4.09 ± 1.22 µm. Six different scaffold geometries (i.e., 90_1, 60_1, 45_1, 90_0.7, 60_0.7, and 45_0.7) were printed by varying the printing pattern (i.e., 90°, 60°, and 45° angle shift between subsequent printed layers) and distance between filaments (i.e., 1.0 mm and 0.7 mm). Representative images of the printed patterns compared to the computer-assisted design (CAD) drawing set for the printing process are shown in Figure 1.

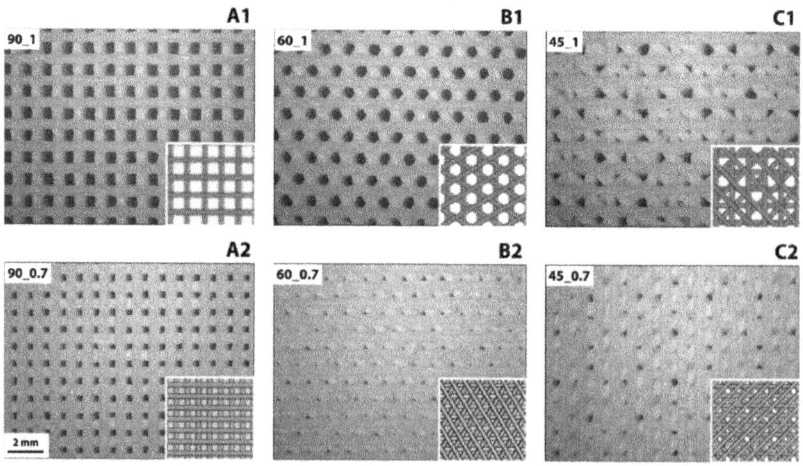

Figure 1. Morphology of 3D printed PU scaffolds imaged via stereomicroscope. Columns (**A1**,**A2**,**B1**,**B2**,**C1**,**C2**) show different printing patterns: 90°, 60°, and 45° angle shifts between subsequent printed layers, respectively. The top row (**A1**,**B1**,**C1**) shows a 1.0 mm distance between filaments, whereas the bottom row (**A2**,**B2**,**C2**) 0.7 mm distance between filaments. Scale bar = 2 mm for all. Inset images depict computer-assisted design (CAD) files used to print the scaffolds.

All the printed geometries reproduced the CAD designs used to print the scaffolds. The measured distance between the center of printed filaments corresponded to the one set in the different CAD models (i.e., for all the printed geometries, error < 1%), proving the correct deposition of filaments following the CAD model. The average diameter of the printed filaments was 0.57 ± 0.07 mm, approximatively 40% higher than the set filament thickness. This increase in thickness was attributed to the printing process that melted the filament, thus causing it's partial spreading upon deposition. The latter is a well-studied phenomenon occurring during fused deposition modeling printing procedures [34]. Different pore architectures were achieved by varying the angle shift between printed layers (Figure 1(A1,A2,B1,B2,C1,C2)) and the distance between adjacent filaments (Figure 1(A1,A2,B1,B2,C1,C2)). SEM micrographs displaying the printed geometries and surface of the printed filaments are displayed in Figure S2. Scaffolds printed with 90° shift angle patterns showed squared pores, for both 1.0 mm and 0.7 mm inter-filament distances. Scaffolds printed with 60° shift angle patterns had hexagonal pores in case of 1 mm inter-filament distance, whereas triangular pores in case of 0.7 mm inter-filament distance. Scaffolds printed with 45° shift angle patterns showed pores with a more complex, rose-like shape. For all the considered geometries, no structure collapse was observed, as confirmed by the porous structure observed for the cross-sections of the 3D printed PU scaffolds (Figure S3).

The physicomechanical properties of the scaffolds were successfully modulated by changing the printing patterns. The average pore area approximatively ranged from 0.43–0.08 mm^2 (Figure 2A). The 90_1 and 60_1 scaffolds were characterized by the highest pore area (i.e., 0.43 ± 0.04 mm^2 and 0.42 ± 0.06 mm^2, respectively, $p > 0.05$, and $p < 0.05$ vs. other scaffolds), as qualitatively proved by the previously shown morphology. The 45_1 and 90_0.7 type scaffolds had comparable pore sizes (i.e., 0.15 ± 0.12 mm^2 and 0.16 ± 0.03 mm^2, respectively; $p > 0.05$), which were higher than 60_0.7 and 45_0.7 type scaffolds ($p < 0.05$). The 60_0.7 and 45_0.7 type scaffolds showed the smallest pore areas (i.e., 0.08 ± 0.03 mm^2 and 0.08 ± 0.06 mm^2, respectively; $p > 0.05$). The high standard deviation in pore size of the scaffolds printed with 45° angle shift was given by the different shapes of pores obtained by printing with this angle shit, compared to the repeated motifs (i.e., squared, hexagonal, or triangular) observed in 90° and 60° angle shifts. Percent porosity values ranged from 55 to 67% (Figure 2B). Scaffolds printed with a 1.0 mm distance between filaments resulted to possess higher porosity than the scaffolds printed with a 0.7 mm distance ($p < 0.05$), while no differences were detected comparing the scaffolds printed with the same distance between filaments and different angle shifts ($p > 0.05$). The apparent density of the scaffolds also varied by changing the distance between the filaments, being the scaffolds printed with 0.7 mm distance were characterized by higher density values than scaffolds printed with 1.0 mm distance (Figure S4).

Sterile scaffolds were then immersed in a culture medium and incubated at 37 °C, 5% CO_2, to investigate the absorption of fluids inside the porous structure in simulated cell culture conditions. The percent weight variation of the scaffolds up to 3 weeks and the weight variation values at plateau are shown in Figure 2C. All the scaffolds showed a quick fluid absorption, namely, increasing weight variation values in the first 3 h after immersion in the culture medium, thus good hydrophilicity. At plateau, the scaffolds printed with 1.0 mm inter-filament distance demonstrated higher weight variation compared to that of the scaffolds printed with 0.7 mm inter-filament distance (Figure 2D; $p < 0.05$).

Representative stress-strain (σ-ε) curves obtained by compression tests for PU scaffolds printed with 1.0 and 0.7 mm distance are shown in Figure 3A,B, respectively.

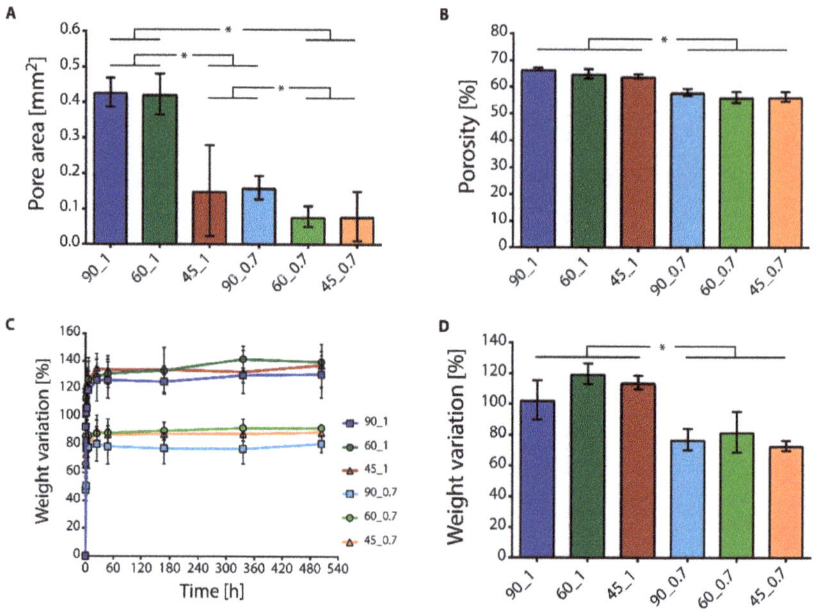

Figure 2. Physical characterization of 3D polyurethane (PU) scaffolds. (**A**) Average pore area determined by top-view images of the printed scaffolds (* $p < 0.05$), and (**B**) percent porosity (* $p < 0.05$). (**C**) Weight variation of the printed PU scaffolds after immersion in culture medium, and (**D**) weight variation at plateau (* $p < 0.05$).

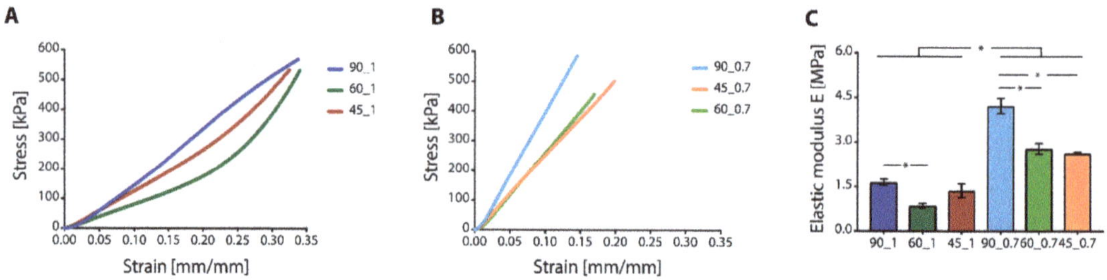

Figure 3. Compressive mechanical properties of 3D printed PU scaffolds. (**A**,**B**) Representative stress-strain curves of (**A**) samples printed with 1.0 mm distance, and (**B**) 0.7 mm distance between fibers, with 90°, 60°, and 45° angle shifts between subsequent printed layers. (**C**) Elastic modulus E was calculated from the compressive curves (* $p < 0.05$).

Scaffolds printed with 0.7 mm inter-filament distance showed a linear response in the range of the applied stress, while scaffolds printed with 1.0 mm inter-filament distance showed a first linear region (i.e., $\varepsilon < 0.2$ mm·mm^{-1}) and subsequent non-linear response (i.e., $\varepsilon > 0.2$ mm·mm^{-1}). All the scaffolds proved to be able to support the applied loads with neither failure nor disruption up to 18 N. Elastic modulus (E) values of the scaffolds under compression ranged from 0.5–4.0 MPa, depending on the printed geometry and distance between filaments (Figure 3C). The scaffolds printed with 0.7 mm inter-filament distance resulted in higher E values, compared to those printed with 1.0 mm inter-filament distance between filaments ($p < 0.05$). This difference was given by the different number of intersection points between the printed filaments that sustained the load during compression. If the same surface area was considered, the scaffolds printed with filaments closer one to the other were characterized by a higher number of intersections between the filaments that were able to sustain the load, thus resulting in higher E values.

3.2. In Vitro Cytotoxicity of PU Scaffolds

In vitro tests were conducted on scaffolds printed with a 0.7 mm inter-filament distance between filaments. Scaffolds printed with 1.0 mm inter-filament distance were excluded from the in vitro tests as they showed low E values (i.e., E < 1.5 MPa). In addition, high pore size due to the larger inter-filament distance hindered the pore colonization from seeded SAOS-2 and L929 cells (data not shown). In vitro indirect cytotoxicity showed that SAOS-2 cells (Figure 4A) and L929 cells (Figure 4B), used as osteosarcoma and connective tissue cell line models, respectively, displayed cell metabolic activity higher than 90%, thus proving the non-cytotoxicity of the printed PU scaffolds.

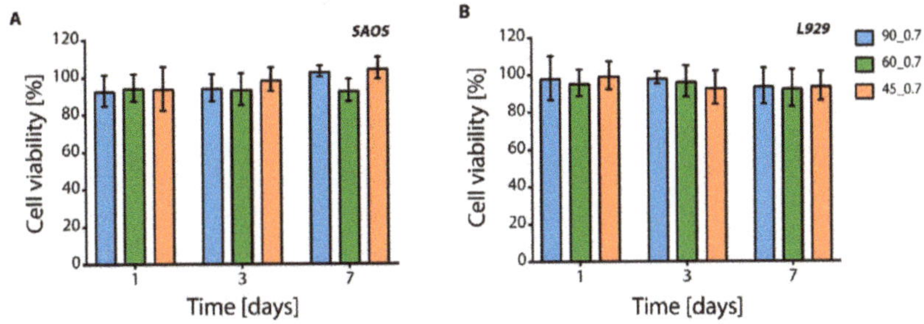

Figure 4. In vitro indirect cytotoxicity tests. Cell viability was intended as metabolic activity, measured for (**A**) SAOS-2 and (**B**) L929 cells cultured in a culture medium extracted by contact with PU samples with 0.7 inter-filament distance for 1, 3, and 7 days.

3.3. 3D In Vitro Bone Model

After a preliminary cytotoxicity assessment was performed by using two different cell lines, the ability of the printed PU scaffolds in supporting primary bone cell adhesion and colonization to different scaffold inner geometries was tested by using hMSCs. All the scaffolds supported hMSC adhesion and an increase in metabolic activity was observed in time (Figure 5A).

No statistical differences were observed comparing the cell viability (Figure 5A) and cellularity, namely the cell number in the scaffolds (Figure 5B), of hMSCs cultured on scaffolds with different geometries ($p > 0.05$). After 2 weeks of culture, ALP activity (Figure 5C) was significantly reduced for cells cultured on scaffolds with 90° and 60° angle shifts, compared to cells cultured on scaffolds with 45° ($p < 0.05$). Calcium content was higher for scaffolds printed with 90° geometry, compared to scaffolds printed with 45° geometry (Figure 5D).

Histological analysis confirmed the presence of osteodifferentiated hMSCs colonizing the printed PU scaffolds. Figure 6 shows the H&E and von Kossa staining in the three scaffold types after being cultured with osteo-differentiated hMSCs, the latter revealing presence of mineralization nodules. It is important to outline that, unlike in 60° Figure 6(A2,B2) and 45° Figure 6(A3,B3), the cells cultured in 90° Figure 6(A1,B1) shift angle geometry scaffolds adhered only to the PU fibers and did not grow inside the pores.

Bone markers were also revealed via immunohistochemistry to confirm hMSC osteo-differentiation (Figure 7). The early markers ALP Figure 7(A1–A3), osteopontin Figure 7(B1–B3), and osteocalcin Figure 7(C1–C3) were expressed on the printed PU scaffolds with different intensities. A good but not uniform positivity of intracellular ALP antigen was observed for 45° and 60° angle shift scaffolds, which was less but more uniformly observed for 90° angle shift scaffolds (Figure 7A). The expression levels confirmed the highest hMSC osteo-differentiation, in terms of intensity of late-stage marker expression, on 90° angle shift scaffolds Figure 7(A1,B1,C1).

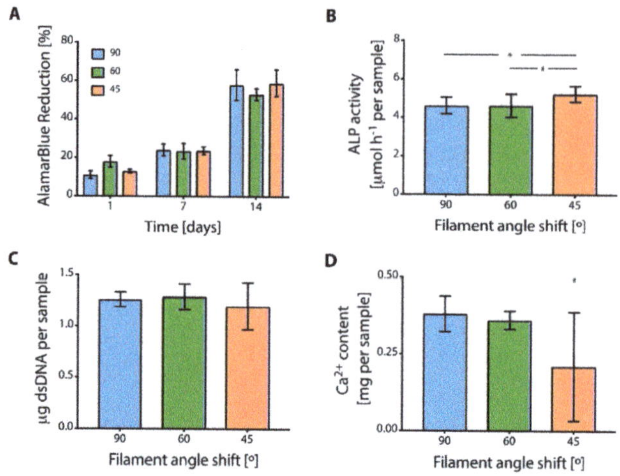

Figure 5. Characterization of the in vitro response of hMSCs osteo-differentiated on 90°, 60°, and 45° angle shift printed geometries with 0.7 inter-filament distance: (**A**) metabolic activity of cells over 2 weeks of culture, (**B**) content of DNA (cellularity) per sample (* $p < 0.05$), (**C**) alkaline phosphatase (ALP) activity, and (**D**) Ca^{2+} content (* $p < 0.05$).

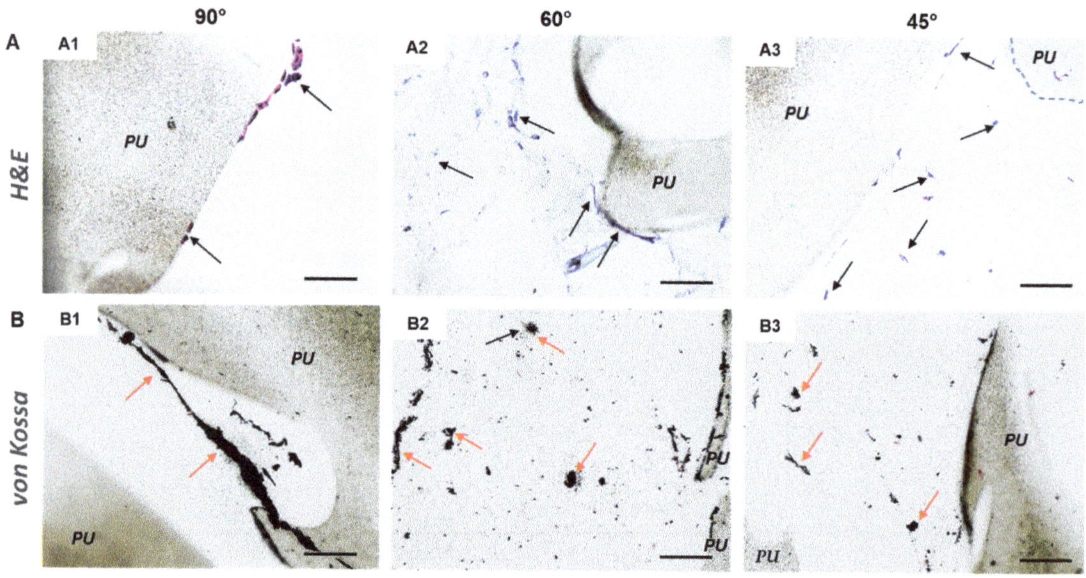

Figure 6. Histological staining on osteodifferentiated hMSC/scaffold constructs. 3D printed scaffolds consisted of PU fibers oriented at 90° (**A1**,**B1**), 60° (**A2**,**B2**), and 45° (**A3**,**B3**) angle shift printed geometries with 0.7 inter-filament distance. (**A**) H&E staining: cell nuclei are in blue-violet, cytoplasm in pink, black arrows = some representative cells; (**B**) von Kossa staining: mineralized matrix granules are in black, cell nuclei in red, red arrows = mineralized ECM. "PU" indicates polyurethane scaffolds; scale bar = 100 μm for all micrographs.

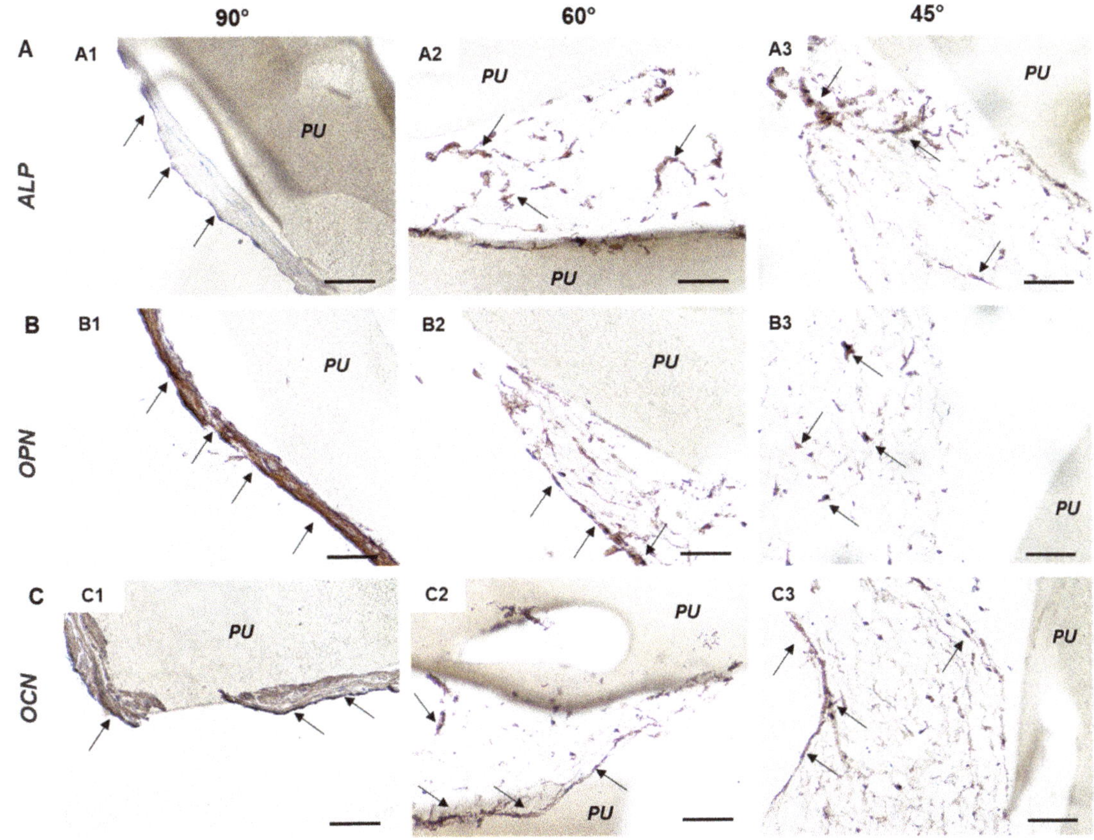

Figure 7. Immunohistochemistry of hMSCs osteo-differentiated on 3D PU scaffolds at 90° (**A1**,**B1**,**C1**), 60° (**A2**,**B2**,**C2**), and 45° (**A3**,**B3**,**C3**) angle shift printed geometries with 0.7 inter-filament distance. Representative micrographs revealing bone markers in brown and specifically the immunopositivity to (**A**) alkaline phosphatase (ALP), (**B**) osteopontin (OPN), (**C**) osteocalcin (OCN). "PU" indicates polyurethane scaffolds; arrows point to well visible cells; scale bar = 100 µm for all micrographs.

Figure 8 shows the results of the immunohistochemical analysis in the hMSC/PU scaffold constructs for some key ECM molecules, such as collagen type I (Figure 8A) and fibronectin (Figure 8C), as well as a fundamental signaling molecule, TGF-β1 (Figure 8B), present in bone, which is considered to play a role in cancer progression and metastasis. Immunohistochemical analysis confirmed the absence of pore colonization in 90° angle shift PU scaffolds in Figure 7(A1,B1,C1) and Figure 8(A1,B1,C1). The cells adhered to the scaffold's inner surfaces without being able to fill the scaffold pores.

Collagen I, TGF-β1, and fibronectin were expressed in the three scaffold types. Collagen type I was revealed in all the three scaffold types, in particular in 60° and 45° angle shift PU scaffolds Figure 8(A2,A3). TGF-β1 was most intensely expressed in 90° angle shift PU scaffolds Figure 8(B1). Fibronectin was very intensely revealed in 90° angle shift PU scaffolds Figure 8(C1) and at the adhesion sites between cells and scaffold surfaces in 60° and 45° angle shift PU scaffolds Figure 8(C2,C3). All these findings supported the fact that the three PU scaffold types were cytocompatible and allowed for hMSC osteo-differentiation. The cells differentiated in 90° angle shift PU scaffolds showed higher expression of late bone markers than others, even though the scaffold colonization was not effective. The 60° angle shift PU scaffolds were thus used to generate the osteosarcoma model.

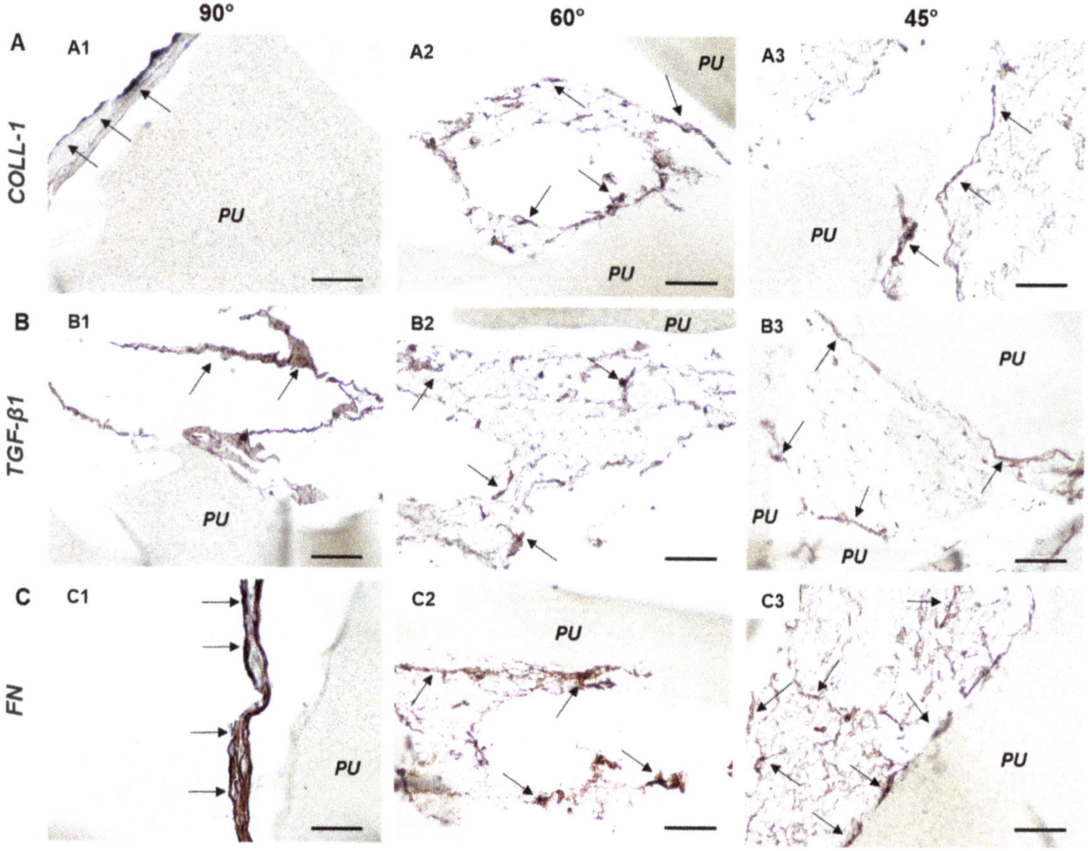

Figure 8. Immunohistochemistry of hMSCs osteo-differentiated on 3D PU scaffolds at 90° (**A1**,**B1**,**C1**), 60° (**A2**,**B2**,**C2**), and 45° (**A3**,**B3**,**C3**) angle shift printed geometries with 0.7 inter-filament distance. Representative micrographs revealing some ECM and signaling molecules in brown and specifically the immunopositivity to (**A**) collagen type I (COLL-1), (**B**) transforming growth factor beta 1 (TGF-β1), (**C**) fibronectin (FN). "PU" indicates polyurethane scaffolds; arrows point to well visible cells; scale bar = 100 μm for all micrographs.

3.4. 3D In Vitro Osteosarcoma Model

The PU scaffolds printed with a 60° angle shift were then selected for the development of the in vitro osteosarcoma model (Table 2, see Discussion). The PU filaments arranged in these geometric features provided an optimal compromise between efficient pore colonization by hMSCs, and the highest calcium production. SAOS-2 cells were seeded on PU_bone ECM scaffolds, as well as on plain PU and PU_ECM scaffolds as controls (Figure 9). H&E staining revealed the presence of rounded cells, i.e., SAOS-2 cells, keeping contact with scaffold in all the analyzed samples (Figure 9A), particularly in the pores of scaffolds with pre-generated bone ECM Figure 9(A3).

Von Kossa staining showed the presence of mineralized ECM only in the constructs constituted by scaffolds with pre-generated bone ECM and SAOS-2 cells Figure 9(B3). Different from the two controls in which SAOS-2 cells were layered on the polymeric surface Figure 9(B1,B2), SAOS-2 cells were detected in contact with the calcium nodules in the poral spaces Figure 9(B3).

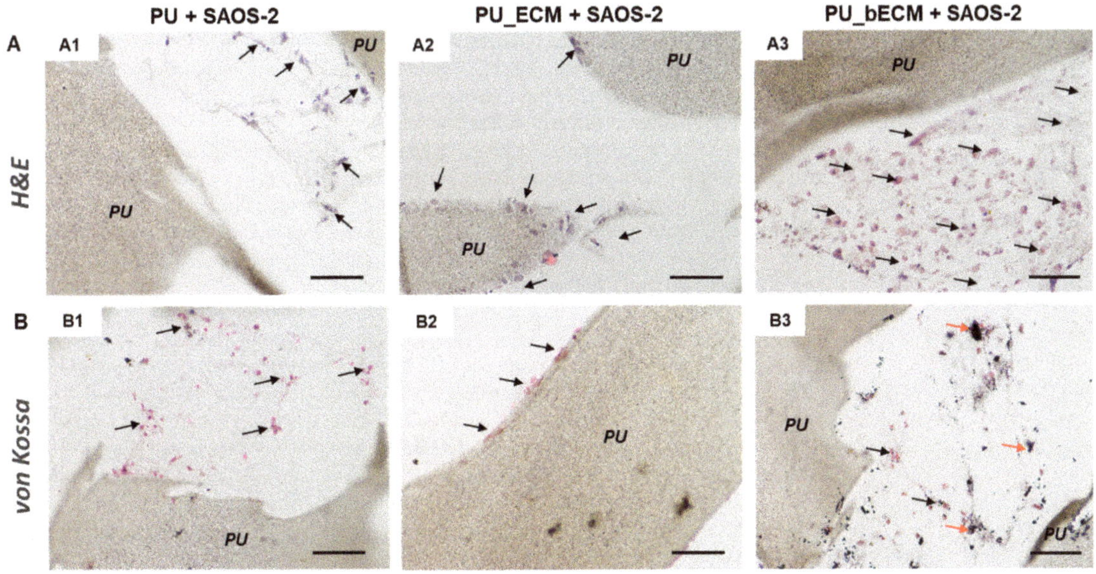

Figure 9. 3D PU scaffolds with 60° shift angle geometry with 0.7 inter-filament distance as in vitro osteosarcoma model. Histological sections showing: (**A**) H&E staining; and (**B**) von Kossa staining of: (**A1**,**B1**) 3D printed PU scaffolds cultured with SAOS-2 cells; (**A2**,**B2**) 3D printed scaffolds containing immature (i.e., hMSC pre-generated) ECM cultured with SAOS-2 cells, and (**A3**,**B3**) 3D printed PU scaffolds containing mature (i.e., osteodifferentiated hMSC pre-generated) bone ECM cultured with SAOS-2 cells. Black arrows indicate SAOS-2 cells, red arrows indicate mineralized ECM granules, "PU" indicates polyurethane scaffolds; scale bar = 100 μm for all the micrographs.

4. Discussion

3D in vitro osteosarcoma models are of paramount importance to deepen the knowledge on the development and progression of this cancer, as well as to ease and optimize the discovery of new treatments and drugs [3]. A reliable 3D in vitro model would allow the current limitations of 2D in vitro models to be overcome, as they oversimplify the physiological features of the pathological tissue [35]. On the other hand, such a 3D in vitro model would allow efficient management of the complex in vivo studies, without raising ethical concerns [6]. Developing reliable 3D in vitro cell/biomaterial constructs strictly depends on the fabrication of scaffolds with morphology (e.g., pore size and porosity) and mechanical properties suitable for the selected tissue, thus resulting in an adequately biomimetic microenvironment for cancer cell growth. Additive manufacturing technologies have demonstrated a unique potential to target these features [36]. In our study, the control of the morphological and mechanical features of the fabricated scaffold is achieved by using fused deposition modeling [37], while a chemically biomimetic microenvironment is artificially replicated in vitro by seeding hMSCs on the printed PU scaffolds, promoting their osteogenic differentiation and culturing osteosarcoma cells on the developed biomimetic 3D printed scaffold after hMSCs were removed by lysis. Under these circumstances, the osteosarcoma cells are cultured on a morphologically and mechanically suitable scaffold that contains the organic and mineral components of bone ECM produced by normal (i.e., non-cancerous) cells.

PU is a class of polymers widely used in the biomedical field, both for clinical uses (e.g., cardiovascular applications [38]), which has been investigated to prepare scaffolds for bone regeneration [39], as well as in vitro bone models [16]. We selected PU as a polymer for the fabrication of our models due to its versatility, possibility of controlling its physicomechanical properties, non-cytotoxicity, and biostability, the latter being useful when long-term in vitro cultures are required since the same characteristics will be maintained on the long run. For these reasons, PU was considered a valuable polymer to prepare an in vitro osteosarcoma model. The commercial PU filaments were processed by fused deposition modeling, after optimizing the printing parameters, to obtain scaffolds with defined and reproducible architectures. Compared to other traditional fabrication technologies employed to prepare scaffolds for bone tissue engineering (e.g., particulate leaching [20], gas foaming [16], lyophilization [17]), the 3D printing technology used in this study allowed scaffolds with controlled pore size, distribution, and geometry to be obtained in a relatively short time [37,40]. Three patterns (i.e., 90°, 60°, and 45° angle shifts) and two different distances between adjacent printed filaments (i.e., 0.7 mm and 1.0 mm) were applied to fabricate scaffolds with different pore sizes and morphology, as these design parameters also served to modulate the physicomechanical properties of the scaffolds [41]. Tuning the design of the pores allowed the scaffold porosity to be modulated, which in turn influenced the mechanical properties and, eventually, the in vitro cell response. The distance between the printed filaments that we set was in the 700–1000 μm range, which led to pore size resembling the optimal pore dimension described in the literature for in vitro bone tumor models, namely, ranging from 200–300 μm [25], up to 1100 μm [42]. Different distances between filaments combined with diverse geometrical patterns resulted in distinct pore areas, that approximatively ranged from 0.40–0.08 mm^2. Changes in the pore areas gave rise to a variation of the scaffold density, which, for all the prepared scaffolds, was in the trabecular bone density range. The porosity of the prepared scaffolds (55–67%) was comparable to the one achieved for scaffolds fabricated by additive manufacturing and used for bone regeneration [43], as well as for in vitro bone models [44].

The first step of this study addressed the identification of the most suitable scaffold type able to support normal bone tissue growth. We considered that osteosarcoma, as a primary bone tumor, develops within bone tissue. Preliminary in vitro tests using SAOS-2 and L929 cells showed that large pores of scaffolds printed with a 1.0 mm distance between filaments, resulted in poor cellularity and lack of pore occupation by cells and ECM molecules. In addition, those scaffolds had low rigidity. Thus, scaffolds printed with 0.7 mm inter-filament distance were further investigated as in vitro osteosarcoma models. Moreover, 1.0 mm inter-filament scaffolds demonstrated relatively low mechanical properties (i.e., $E < 2$ MPa), compared to 0.7 mm inter-filament distance scaffolds (i.e., $E > 2$ MPa). The mechanical properties of the scaffolds must be tuned to achieve a suitable biomimetic model, as they contribute to guiding cell fate and morphology [45,46], as well as the in vitro mimicry of the TME [47]. We tuned the mechanical properties of the scaffolds by designing their pore geometry and size [41]. Our scaffolds had Young's modulus values in the order of MPa, which are superior [16,48], or comparable [49,50] to those described in literature able to mimic metastatic and primary bone tumors. Using the 0.7 inter-filament distance, we obtained bone-like constructs with similar cellularity using osteo-differentiated hMSCs. However, only in the 45° and 60° filament angle shift, the cells were able to colonize the poral spaces. In the 90° type, the cells resulted were laid onto the PU filaments, efficiently differentiated, but the pores were not populated. In the 45° type, we observed a less mature bone ECM generation, as the ALP activity was higher and calcium content lower. Therefore, the 60° filament angle shift was chosen as the optimal scaffold type. The selected material and fabrication technology used to design the morphology, physical, and mechanical properties, allowed us to define the most promising scaffold type (i.e., 60_0.7), which was subsequently used to prepare biomimetic in vitro models by incorporating different cell-secreted molecules.

The TME is made of the cellular and extracellular components in and by which the tumor exists and develops. TME is thus the result of dynamic and complex cell-cell and cell-ECM interactions taking place between normal and cancerous cells and their secreted molecules [4,13]. The TME consists of cancer cells, tumor stromal cells (e.g., stromal fibroblasts, endothelial cells, and immune cells), and ECM molecules (e.g., collagenic and non-collagenic molecules, hyaluronan, laminin). Among others, the role of collagen and its interaction with fibronectin is conditioned by tumor cells to enable signaling pathways and receptors that support cancer progression and metastasis. It is also considered that these phenomena are stimulated by the TME, as it promotes cancer cell heterogeneity and clonal evolution [51]. For instance, during tumor progression, cancer cells modify and divert the surrounding ECM towards the enhancement of anchorage-dependent molecules and the storage of pro-tumorigenic factors [52]. Polymeric scaffolds incorporating biomolecules, like gelatin, were able to stimulate fibroblast proliferation [53], as well as cancer cell interaction by metalloproteinase production [8], and integrin expression [54], which is associated with tumor progression [52]. Another signaling molecule involved in carcinogenesis is TGF-β, as in normal conditions, it possesses tumor-suppressing activity. Dysregulation of TGF-β has been associated with cancer invasiveness and metastasis [55]. Therefore, multidrug resistance often occurs because of the TME, which, by also recruiting normal cells to the scope, is revealed to be a protective environment for the tumor. To better target, the tumors, including osteosarcoma, drugs able to interact with the TME along with cell proliferation are a subject of study [56]. In this view, the mechanical consistency of scaffold-based in vitro models, particularly relevant for hard tissue tumors, maybe not be exhaustive.

To better recapitulate the complexity of the TME, and specifically, the cancer cell-ECM cross-talk, we recreated some biological traits of the bone ECM within our 3D printed synthetic polymeric substrate. By cultivating hMSCs embedded in fibrin clots in the 60_0.7 scaffold type, followed by cell lysis, we generated the biomolecules proper of bone ECM, to be further explored as a microenvironment for osteosarcoma cell colonization. An immature (i.e., not mineralized, to be used as control) bone ECM was obtained by culturing undifferentiated hMSCs, whereas a mature (i.e., mineralized) bone ECM was obtained by differentiating the hMSCs towards the osteoblastic lineage for 3 weeks inside the PU scaffolds. This pre-generated bone ECM, as obtained in 3D printed PU 60_0.7, was analyzed via immunohistochemistry, resulting in rich collagen type I fibers, was positive for osteopontin, osteocalcin, fibronectin, TGF-β, and ALP, and via histochemistry, revealing calcium nodules. Quantitative tests corroborated the success of osteogenic differentiation by the presence of calcium [57]. The used procedure for cell lysis is known to preserve ECM molecules and growth factors [29]. In this way, we supported a suitable physicomechanical microenvironment, as given by the PU scaffolds, with a biological microenvironment, as given by cell-deposited bone-ECM. At this stage of the investigation, we aimed to have a simple model of SAOS-2 cells interacting with a bone ECM like scaffold. As such, we did not co-culture normal (i.e., osteodifferentiated hMSCs) and cancerous (i.e., SAOS-2) cells together; however, this will be possible by removing the lysis step to save the generated osteoblasts [53]. Moreover, immune and/or vascular cells could be added [54,58]. The obtained findings revealed a remarkable interaction of SAOS-2 cells with the biohybrid scaffold. Indeed, SAOS-2 cells were able to colonize the pores of the scaffolds by interacting with the deposited bone ECM. This study aims to provide a model of improved complexity for osteosarcoma, enabling the study of drugs that target the TME. Combined drugs may disclose effective therapeutic options to improve the prognosis of osteosarcoma.

5. Conclusions

We 3D-printed PU scaffolds with reproducible patterns, geometries, and properties. The selected scaffolds (i.e., 0.7 mm distance between filaments, 60° angle shift) showed optimal porosity, density, cytocompatibility, and mechanical properties to be used for in vitro bone regeneration. The developed pre-culture condition (i.e., pre-seeding with osteo-differentiated hMSCs, subsequently lysed) was a suitable procedure to obtain a biohybrid scaffold, which combined the optimal physicomechanical properties of the 3D printed synthetic polymer, with the relevant biological properties of an in vitro pre-generated bone ECM. The developed bone-biomimetic 3D microenvironment was used for culturing SAOS-2 cells to create a TME recapitulating the main ECM features for the osteosarcoma study. The 3D in vitro model here developed could be used, in the future, to screen new therapeutics and to in-depth investigate osteosarcoma progression.

Supplementary Materials: The following supporting information can be downloaded at: https://www.mdpi.com/article/10.3390/cancers14082003/s1, Figure S1: optimization of printing parameters; Figure S2: Representative SEM micrographs of printed PU scaffolds; Figure S3: representative microscopy images of the section of the scaffolds; Figure S4: apparent density of the scaffolds.

Author Contributions: Conceptualization, N.C.N., S.F. and S.D.; methodology, N.C.N., F.B., C.R., D.D., L.T. and S.D.; investigation, F.B., L.T., D.D. and C.R.; resources, S.F. and S.D.; data curation, N.C.N., F.B., C.R. and D.D.; writing—original draft preparation, N.C.N. and S.D.; writing—review and editing, N.C.N., S.F. and S.D.; supervision, N.C.N., S.F. and S.D.; project administration, N.C.N., S.F. and S.D.; validation, D.D. and S.D.; visualization, C.R. and D.D.; formal analysis, S.F., D.D. and S.D. funding acquisition, S.F. and S.D. All authors have read and agreed to the published version of the manuscript.

Funding: This research received no external funding.

Institutional Review Board Statement: Not applicable.

Informed Consent Statement: Not applicable.

Data Availability Statement: Data are available upon request to the corresponding authors.

Conflicts of Interest: The authors declare no conflict of interest.

References

1. Caddeo, S.; Boffito, M.; Sartori, S. Tissue Engineering Approaches in the Design of Healthy and Pathological in Vitro Tissue Models. *Front. Bioeng. Biotechnol.* **2017**, *5*, 40. [CrossRef]
2. Negrini, N.C.; Volponi, A.A.; Higgins, C.A.; Sharpe, P.T.; Celiz, A.D. Scaffold-Based Developmental Tissue Engineering Strategies for Ectodermal Organ Regeneration. *Mater. Today Bio* **2021**, *10*, 100107. [CrossRef]
3. Nii, T.; Makino, K.; Tabata, Y. Three-Dimensional Culture System of Cancer Cells Combined with Biomaterials for Drug Screening. *Cancers* **2020**, *12*, 2754. [CrossRef]
4. Bregenzer, M.E.; Horst, E.N.; Mehta, P.; Novak, C.M.; Raghavan, S.; Snyder, C.S.; Mehta, G. Integrated Cancer Tissue Engineering Models for Precision Medicine. *PLoS ONE* **2019**, *14*, e0216564. [CrossRef]
5. Wang, C.; Tang, Z.; Zhao, Y.; Yao, R.; Li, L. Three-Dimensional in Vitro Cancer Models: A Short Review. *Biofabrication* **2014**, *14*, 02201. [CrossRef]
6. Ricci, C.; Moroni, L.; Danti, S. Cancer Tissue Engineering-New Perspectives in Understanding the Biology of Solid Tumours-a Critical Review. *OA Tissue Eng.* **2013**, *1*, 4. [CrossRef]
7. Hughes, A.M.; Kolb, A.D.; Shupp, A.B.; Shine, K.M.; Bussard, K.M. Printing the Pathway Forward in Bone Metastatic Cancer Research: Applications of 3D Engineered Models and Bioprinted Scaffolds to Recapitulate the Bone—Tumor Niche. *Cancers* **2021**, *13*, 507. [CrossRef]
8. Ricci, C.; Mota, C.; Moscato, S.; D'Alessandro, D.; Ugel, S.; Sartoris, S.; Bronte, V.; Boggi, U.; Campani, D.; Funel, N.; et al. Interfacing Polymeric Scaffolds with Primary Pancreatic Ductal Adenocarcinoma Cells to Develop 3D Cancer Models. *Biomatter* **2014**, *4*, e955386. [CrossRef]
9. Lindsey, B.A.; Markel, J.E.; Kleinerman, E.S. Osteosarcoma Overview. *Rheumatol. Ther.* **2017**, *4*, 25–43. [CrossRef]
10. Sadykova, L.R.; Ntekim, A.I.; Muyangwa-Semenova, M.; Rutland, C.S.; Jeyapalan, J.N.; Blatt, N.; Rizvanov, A.A. Epidemiology and Risk Factors of Osteosarcoma. *Cancer Investig.* **2020**, *38*, 259–269. [CrossRef]
11. Harrison, D.J.; Geller, D.S.; Gill, J.D.; Lewis, V.O.; Gorlick, R. Current and Future Therapeutic Approaches for Osteosarcoma. *Expert Rev. Anticancer Ther.* **2018**, *18*, 39–50. [CrossRef]
12. Meazza, C.; Scanagatta, P. Metastatic Osteosarcoma: A Challenging Multidisciplinary Treatment. *Expert Rev. Anticancer Ther.* **2016**, *16*, 543–556. [CrossRef]
13. Rodrigues, J.; Heinrich, M.A.; Teixeira, L.M.; Prakash, J. 3D In Vitro Model (R)Evolution: Unveiling Tumor–Stroma Interactions. *Trends Cancer* **2021**, *7*, 249–264. [CrossRef]
14. Narkhede, A.A.; Crenshaw, J.H.; Crossman, D.K.; Shevde, L.A.; Rao, S.S. An in Vitro Hyaluronic Acid Hydrogel Based Platform to Model Dormancy in Brain Metastatic Breast Cancer Cells. *Acta Biomater.* **2020**, *107*, 65–77. [CrossRef]
15. Lee, H.J.; Mun, S.; Pham, D.M.; Kim, P. Extracellular Matrix-Based Hydrogels to Tailoring Tumor Organoids. *ACS Biomater. Sci. Eng.* **2021**, *7*, 4128–4135. [CrossRef]
16. Angeloni, V.; Contessi, N.; de Marco, C.; Bertoldi, S.; Tanzi, M.C.; Daidone, M.G.; Farè, S. Polyurethane Foam Scaffold as in Vitro Model for Breast Cancer Bone Metastasis. *Acta Biomater.* **2017**, *63*, 306–316. [CrossRef]

17. Zhu, W.; Wang, M.; Fu, Y.; Castro, N.J.; Fu, S.W.; Zhang, L.G. Engineering a Biomimetic Three-Dimensional Nanostructured Bone Model for Breast Cancer Bone Metastasis Study. *Acta Biomater.* **2015**, *14*, 164–174. [CrossRef]
18. Talukdar, S.; Kundu, S.C. Engineered 3D Silk-Based Metastasis Models: Interactions between Human Breast Adenocarcinoma, Mesenchymal Stem Cells and Osteoblast-like Cells. *Adv. Funct. Mater.* **2013**, *23*, 5249–5260. [CrossRef]
19. Subia, B.; Dey, T.; Sharma, S.; Kundu, S.C. Target Specific Delivery of Anticancer Drug in Silk Fibroin Based 3D Distribution Model of Bone-Breast Cancer Cells. *ACS Appl. Mater. Interfaces* **2015**, *7*, 2269–2279. [CrossRef]
20. Pathi, S.P.; Kowalczewski, C.; Tadipatri, R.; Fischbach, C. A Novel 3-D Mineralized Tumor Model to Study Breast Cancer Bone Metastasis. *PLoS ONE* **2010**, *5*, e8849. [CrossRef]
21. Pan, T.; Fong, E.L.S.; Martinez, M.; Harrington, D.A.; Lin, S.H.; Farach-Carson, M.C.; Satcher, R.L. Three-Dimensional (3D) Culture of Bone-Derived Human 786-O Renal Cell Carcinoma Retains Relevant Clinical Characteristics of Bone Metastases. *Cancer Lett.* **2015**, *365*, 89–95. [CrossRef] [PubMed]
22. Sieh, S.; Taubenberger, A.V.; Lehman, M.L.; Clements, J.A.; Nelson, C.C.; Hutmacher, D.W. Paracrine Interactions between LNCaP Prostate Cancer Cells and Bioengineered Bone in 3D in Vitro Culture Reflect Molecular Changes during Bone Metastasis. *Bone* **2014**, *63*, 121–131. [CrossRef] [PubMed]
23. Santoro, M.; Lamhamedi-Cherradi, S.E.; Menegaz, B.A.; Ludwig, J.A.; Mikos, A.G. Flow Perfusion Effects on Three-Dimensional Culture and Drug Sensitivity of Ewing Sarcoma. *Proc. Natl. Acad. Sci. USA* **2015**, *112*, 10304–10309. [CrossRef] [PubMed]
24. Fong, E.L.S.; Lamhamedi-Cherradi, S.E.; Burdett, E.; Ramamoorthy, V.; Lazar, A.J.; Kasper, F.K.; Farach-Carson, M.C.; Vishwamitra, D.; Demicco, E.G.; Menegaz, B.A.; et al. Modeling Ewing Sarcoma Tumors in Vitro with 3D Scaffolds. *Proc. Natl. Acad. Sci. USA* **2013**, *110*, 6500–6505. [CrossRef] [PubMed]
25. Tan, P.H.S.; Aung, K.Z.; Toh, S.L.; Goh, J.C.H.; Nathan, S.S. Three-Dimensional Porous Silk Tumor Constructs in the Approximation of in Vivo Osteosarcoma Physiology. *Biomaterials* **2011**, *32*, 6131–6137. [CrossRef]
26. Monteiro, C.F.; Custódio, C.A.; Mano, J.F. Bioengineering a Humanized 3D Tri-Culture Osteosarcoma Model to Assess Tumor Invasiveness and Therapy Response. *Acta Biomater.* **2021**, *134*, 204–214. [CrossRef]
27. Costard, L.S.; Hosn, R.R.; Ramanayake, H.; O'Brien, F.J.; Curtin, C.M. Influences of the 3D Microenvironment on Cancer Cell Behaviour and Treatment Responsiveness: A Recent Update on Lung, Breast and Prostate Cancer Models. *Acta Biomater.* **2021**, *132*, 360–378. [CrossRef]
28. Griffin, M.; Castro, N.; Bas, O.; Saifzadeh, S.; Butler, P.; Hutmacher, D.W. The Current Versatility of Polyurethane Three-Dimensional Printing for Biomedical Applications. *Tissue Eng. Part B Rev.* **2020**, *26*, 272–283. [CrossRef]
29. Datta, N.; Holtorf, H.L.; Sikavitsas, V.I.; Jansen, J.A.; Mikos, A.G. Effect of Bone Extracellular Matrix Synthesized in Vitro on the Osteoblastic Differentiation of Marrow Stromal Cells. *Biomaterials* **2005**, *26*, 971–977. [CrossRef]
30. Datta, N.; Pham, Q.P.; Sharma, U.; Sikavitsas, V.I.; Jansen, J.A.; Mikos, A.G. In Vitro Generated Extracellular Matrix and Fluid Shear Stress Synergistically Enhance 3D Osteoblastic Differentiation. *Proc. Natl. Acad. Sci. USA* **2006**, *103*, 2488–2493. [CrossRef]
31. Danti, S.; Stefanini, C.; D'Alessandro, D.; Moscato, S.; Pietrabissa, A.; Petrini, M.; Berrettini, S. Novel Biological/Biohybrid Prostheses for the Ossicular Chain: Fabrication Feasibility and Preliminary Functional Characterization. *Biomed. Microdevices* **2009**, *11*, 783–793. [CrossRef] [PubMed]
32. Campiglio, C.E.; Ponzini, S.; de Stefano, P.; Ortoleva, G.; Vignati, L.; Draghi, L. Cross-Linking Optimization for Electrospun Gelatin: Challenge of Preserving Fiber Topography. *Polymers* **2020**, *12*, 2472. [CrossRef] [PubMed]
33. Negrini, N.C.; Celikkin, N.; Tarsini, P.; Farè, S.; Święszkowski, W. Three-Dimensional Printing of Chemically Crosslinked Gelatin Hydrogels for Adipose Tissue Engineering. *Biofabrication* **2020**, *12*, 025001. [CrossRef] [PubMed]
34. Agassant, J.F.; Pigeonneau, F.; Sardo, L.; Vincent, M. Flow Analysis of the Polymer Spreading during Extrusion Additive Manufacturing. *Addit. Manuf.* **2019**, *29*, 100794. [CrossRef]
35. Betriu, N.; Andreeva, A.; Semino, C.E. Erlotinib Promotes Ligand-Induced EGFR Degradation in 3D but Not 2D Cultures of Pancreatic Ductal Adenocarcinoma Cells. *Cancers* **2021**, *13*, 4504. [CrossRef]
36. Fischetti, T.; Di Pompo, G.; Baldini, N.; Avnet, S.; Graziani, G. 3d Printing and Bioprinting to Model Bone Cancer: The Role of Materials and Nanoscale Cues in Directing Cell Behavior. *Cancers* **2021**, *13*, 4065. [CrossRef]
37. Milazzo, M.; Contessi Negrini, N.; Scialla, S.; Marelli, B.; Farè, S.; Danti, S.; Buehler, M.J. Additive Manufacturing Approaches for Hydroxyapatite-Reinforced Composites. *Adv. Funct. Mater.* **2019**, *29*, 1903055. [CrossRef]
38. Meskinfam, M.; Bertoldi, S.; Albanese, N.; Cerri, A.; Tanzi, M.C.; Imani, R.; Baheiraei, N.; Farokhi, M.; Farè, S. Polyurethane Foam/Nano Hydroxyapatite Composite as a Suitable Scaffold for Bone Tissue Regeneration. *Mater. Sci. Eng. C* **2018**, *82*, 130–140. [CrossRef]
39. Pitton, M.; Fiorati, A.; Buscemi, S.; Melone, L.; Farè, S.; Contessi Negrini, N. 3D Bioprinting of Pectin-Cellulose Nanofibers Multicomponent Bioinks. *Front. Bioeng. Biotechnol.* **2021**, *9*, 732689. [CrossRef]
40. Moroni, L.; de Wijn, J.R.; van Blitterswijk, C.A. 3D Fiber-Deposited Scaffolds for Tissue Engineering: Influence of Pores Geometry and Architecture on Dynamic Mechanical Properties. *Biomaterials* **2006**, *27*, 974–985. [CrossRef]
41. Trachtenberg, J.E.; Santoro, M.; Williams, C.; Piard, C.M.; Smith, B.T.; Placone, J.K.; Menegaz, B.A.; Molina, E.R.; Lamhamedi-Cherradi, S.E.; Ludwig, J.A.; et al. Effects of Shear Stress Gradients on Ewing Sarcoma Cells Using 3D Printed Scaffolds and Flow Perfusion. *ACS Biomater. Sci. Eng.* **2018**, *4*, 347–356. [CrossRef] [PubMed]

42. Domingos, M.; Chiellini, F.; Cometa, S.; de Giglio, E.; Grillo-Fernandes, E.; Bártolo, P.; Chiellini, E. Evaluation of in Vitro Degradation of Pcl Scaffolds Fabricated via Bioextrusion. Part 1: Influence of the Degradation Environment. *Virtual Phys. Prototyp.* **2010**, *5*, 65–73. [CrossRef]
43. Lynch, M.E.; Chiou, A.E.; Lee, M.J.; Marcott, S.C.; Polamraju, P.V.; Lee, Y.; Fischbach, C. Three-Dimensional Mechanical Loading Modulates the Osteogenic Response of Mesenchymal Stem Cells to Tumor-Derived Soluble Signals. *Tissue Eng. Part A* **2016**, *22*, 1006–1015. [CrossRef] [PubMed]
44. Liu, G.; Pastakia, M.; Fenn, M.B.; Kishore, V. Saos-2 Cell-Mediated Mineralization on Collagen Gels: Effect of Densification and Bioglass Incorporation. *J. Biomed. Mater. Res. Part A* **2016**, *104*, 1121–1134. [CrossRef] [PubMed]
45. Contessi Negrini, N.; Angelova Volponi, A.; Sharpe, P.T.; Celiz, A.D. Tunable Cross-Linking and Adhesion of Gelatin Hydrogels via Bioorthogonal Click Chemistry. *ACS Biomater. Sci. Eng.* **2021**, *7*, 4330–4346. [CrossRef]
46. Kostic, A.; Lynch, C.D.; Sheetz, M.P. Differential Matrix Rigidity Response in Breast Cancer Cell Lines Correlates with the Tissue Tropism. *PLoS ONE* **2009**, *4*, e6361. [CrossRef]
47. Al-Munajjed, A.A.; Plunkett, N.A.; Gleeson, J.P.; Weber, T.; Jungreuthmayer, C.; Levingstone, T.; Hammer, J.; O'Brien, F.J. Development of a Biomimetic Collagen-Hydroxyapatite Scaffold for Bone Tissue Engineering Using a SBF Immersion Technique. *J. Biomed. Mater. Res. Part B Appl. Biomater.* **2009**, *90*, 584–591. [CrossRef]
48. Zhu, W.; Castro, N.J.; Cui, H.; Zhou, X.; Boualam, B.; McGrane, R.; Glazer, R.I.; Zhang, L.G. A 3D Printed Nano Bone Matrix for Characterization of Breast Cancer Cell and Osteoblast Interactions. *Nanotechnology* **2016**, *27*, 315103. [CrossRef]
49. Miao, H.; Shen, R.; Zhang, W.; Lin, Z.; Wang, H.; Yang, L.; Liu, X.Y.; Lin, N. Near-Infrared Light Triggered Silk Fibroin Scaffold for Photothermal Therapy and Tissue Repair of Bone Tumors. *Adv. Funct. Mater.* **2021**, *31*, 2007188. [CrossRef]
50. Baghban, R.; Roshangar, L.; Jahanban-Esfahlan, R.; Seidi, K.; Ebrahimi-Kalan, A.; Jaymand, M.; Kolahian, S.; Javaheri, T.; Zare, P. Tumor Microenvironment Complexity and Therapeutic Implications at a Glance. *Cell Commun. Signal.* **2020**, *18*, 59. [CrossRef]
51. Lorusso, G.; Rüegg, C.; Kuonen, F. Targeting the Extra-Cellular Matrix—Tumor Cell Crosstalk for Anti-Cancer Therapy: Emerging Alternatives to Integrin Inhibitors. *Front. Oncol.* **2020**, *10*, 01231. [CrossRef] [PubMed]
52. Lazzeri, L.; Cascone, M.G.; Danti, S.; Serino, L.P.; Moscato, S.; Bernardini, N. Gelatine/PLLA Sponge-like Scaffolds: Morphological and Biological Characterization. *J. Mater. Sci. Mater. Med.* **2007**, *18*, 1399–1405. [CrossRef] [PubMed]
53. Moscato, S.; Ronca, F.; Campani, D.; Danti, S. Poly(Vinyl Alcohol)/Gelatin Hydrogels Cultured with HepG2 Cells as a 3D Model of Hepatocellular Carcinoma: A Morphological Study. *J. Funct. Biomater.* **2015**, *6*, 16–32. [CrossRef] [PubMed]
54. Syed, V. TGF-β Signaling in Cancer. *J. Cell. Biochem.* **2016**, *117*, 1279–1287. [CrossRef]
55. Liu, J.; Wu, J.; Zhou, L.; Pan, C.; Zhou, Y.; Du, W.; Chen, J.-M.; Zhu, X.; Shen, J.; Chen, S.; et al. ZD6474, a New Treatment Strategy for Human Osteosarcoma, and Its Potential Synergistic Effect with Celecoxib. *Oncotarget* **2015**, *6*, 21341. [CrossRef]
56. Viti, F.; Landini, M.; Mezzelani, A.; Petecchia, L.; Milanesi, L.; Scaglione, S. Osteogenic Differentiation of MSC through Calcium Signaling Activation: Transcriptomics and Functional Analysis. *PLoS ONE* **2016**, *11*, e0148173. [CrossRef]
57. de la Ossa, J.G.; Trombi, L.; D'Alessandro, D.; Coltelli, M.B.; Serino, L.P.; Pini, R.; Lazzeri, A.; Petrini, M.; Danti, S. Pore Size Distribution and Blend Composition Affect In Vitro Prevascularized Bone Matrix Formation on Poly(Vinyl Alcohol)/Gelatin Sponges. *Macromol. Mater. Eng.* **2017**, *302*, 1700300. [CrossRef]
58. Grolman, J.M.; Zhang, D.; Smith, A.M.; Moore, J.S.; Kilian, K.A. Rapid 3D Extrusion of Synthetic Tumor Microenvironments. *Adv. Mater.* **2015**, *27*, 5512–5517. [CrossRef]

cancers

MDPI

Article

3D Disease Modelling of Hard and Soft Cancer Using PHA-Based Scaffolds

Akanksha Tomar [1], Pinar Uysal-Onganer [2], Pooja Basnett [3], Uttam Pati [1,*] and Ipsita Roy [4,*]

1 School of Biotechnology, Jawaharlal Nehru University, New Delhi 110067, India; aakansha.tomar103@gmail.com
2 Cancer Research Group, School of Life Sciences, College of Liberal Arts and Sciences, University of Westminster, London W1W 6UW, UK; p.onganer@westminster.ac.uk
3 School of Life Sciences, College of Liberal Arts and Sciences, University of Westminster, London W1W 6XH, UK; p.basnett@westminster.ac.uk
4 Department of Materials Science and Engineering, Faculty of Engineering, University of Sheffield, Sheffield S10 2TN, UK
* Correspondence: uttampati472@gmail.com (U.P.); i.roy@sheffield.ac.uk (I.R.)

Simple Summary: Tumour progression in vivo was able to be well mimicked in 3D culture by utilizing biodegradable 10 mm × 10 mm × 8 mm P(3HO-*co*-3HD) and P(3HB)-based 3D scaffolds with a pore size of 30 to 300 μm. Both hard (MCF7 and MDA-MB-231) and soft (HCT116) tumour-related cells were successfully grown on the scaffolds, and their growth patterns were studied for 5 days. MDA-MB-231 tend to grow in clusters, and MCF7 cells form an evenly dispersed layer, which covered most of the 3D PHA scaffolds, while HCT116 formed large colonies within the pockets of the 3D PHA scaffold. Epithelial mesenchymal transition (EMT) marker genes, including *Wnt-11*, *E-cadherin*, *Vim* and *Snail* expression profiles, were like those seen in real tumour samples, which confirmed that the cancer models were exhibiting real tumour-like characteristics with high fidelity. These models are important in mimicking hypoxic tumours and in studying gene expression, cellular signalling, angiogenesis and drug response for translational research.

Citation: Tomar, A.; Uysal-Onganer, P.; Basnett, P.; Pati, U.; Roy, I. 3D Disease Modelling of Hard and Soft Cancer Using PHA-Based Scaffolds. *Cancers* **2022**, *14*, 3549. https://doi.org/10.3390/cancers14143549

Academic Editor: Christian D. Muller

Received: 28 June 2022
Accepted: 18 July 2022
Published: 21 July 2022

Abstract: Tumour cells are shown to change shape and lose polarity when they are cultured in 3D, a feature typically associated with tumour progression in vivo, thus making it significant to study cancer cells in an environment that mimics the in vivo milieu. In this study we established hard (MCF7 and MDA-MB-231, breast cancer) and soft (HCT116, colon cancer) 3D cancer tumour models utilizing a blend of P(3HO-*co*-3HD) and P(3HB). P(3HO-*co*-3HD) and P(3HB) belong to a group of natural biodegradable polyesters, PHAs, that are synthesised by microorganisms. The 3D PHA scaffolds produced, with a pore size of 30 to 300 μm, allow for nutrients to diffuse within the scaffold and provide the cells with the flexibility to distribute evenly within the scaffold and grow within the pores. Interestingly, by Day 5, MDA-MB-231 showed dispersed growth in clusters, and MCF7 cells formed an evenly dispersed dense layer, while HCT116 formed large colonies within the pockets of the 3D PHA scaffolds. Our results show Epithelial Mesenchymal Transition (EMT) marker gene expression profiles in the hard tumour cancer models. In the 3D-based PHA scaffolds, MDA-MB-231 cells expressed higher levels of *Wnt-11* and mesenchymal markers, such as *Snail* and its downstream gene *Vim* mRNAs, while MCF7 cells exhibited no change in their expression. On the other hand, MCF7 cells exhibited a significantly increased *E-Cadherin* expression as compared to MDA-MB-231 cells. The expression levels of EMT markers were comparative to their expression reported in the tumour samples, making them good representative of cancer models. In future these models will be helpful in mimicking hypoxic tumours, in studying gene expression, cellular signalling, angiogenesis and drug response more accurately than 2D and perhaps other 3D models.

Keywords: tumour modelling; polyhydroxyalkanoates (PHAs); scaffold; breast cancer; colon cancer; epithelial-mesenchymal transition (EMT)

1. Introduction

Cancer remains a major cause of death worldwide with 19.3 million new cases and 10 million deaths worldwide in 2020. The most prevalent types of diagnosed cancers include breast (11% of the total cases), lung (11.4%) and colorectal (10%) cancer, followed by prostrate (7.3%) cancer [1]. Cancers that begin in epithelial cells of glandular tissues and produce fluids and mucous are termed as adenocarcinomas. Most cancers of the breast, colon and prostate are adenocarcinomas. Invasive ductal carcinoma is the most common breast adenocarcinoma. Tissue engineered models are required to provide the cells with a mimic of their native environment while preserving their phenotypes, genotypes and behaviour, in order to provide a high-throughput analysis and cost-effective drug screening [2,3].

Two-dimensional (2D) monolayer cultures have been used to study tumour biology where cells are cultured on rigid materials like glass and polystyrene with no contribution from the extracellular matrix. A growing number of studies recognise the limitations of such 2D cultures for in vitro studies [4–9]. Apart from the simplicity and low cost, this method does not entirely reflect the essential physiology of real tissues. The cells are forced into polarity and a flattened shape, and modified mechanical and biochemical signals affect cell–cell communication [7,10,11]. As a result, most 2D studies of cellular network functions do not translate to the in vivo models, thereby hindering the development of effective therapies that can be successfully translated into the clinic. The development of 3D cell culture provides better models for drug screening, translational research and cancer prediction. In animal models, cancer is induced by genetic modification or is surgically implanted into mice. However, the success rates of establishing and propagating theses human solid tumours range from 20% to 50% [12]; in genetically engineered mouse models, the tumours fail to grow synchronously and make comparisons of drug responses difficult, and human versus murine differences further add up to the dissimilarities from the human tumours [13].

In vivo cancer-associated stroma is a three-dimensional (3D) structure consisting of neighbouring cells, extracellular matrix (ECM), blood vessels, immune cells and cytokines. Cancer cells interact with their microenvironment during proliferation, metastasis and during chemo- or radiotherapy [14–22]. Therefore, it is important to study cancer cells in an environment that mimics the in vivo milieu. The tumour microenvironment has been investigated extensively [23–29], nevertheless tumour cell biology in a three-dimensional (3D) environment remains poorly understood. Still, 3D polymer-based cancer models can provide several advantages when compared to animal models, such as reproducibility, complexity in terms of cell types, substrate chemistry, topography, tailored mechanical properties, engineered gas diffusion gradients and ethical sustainability [30].

Several 3D in vitro culture models have been proposed to mimic physiological conditions in the tumour, recreating cell-to-cell contact, tumour cell microenvironment and generating hypoxic-necrotic areas [31]. Studies reveal that various cancer cell lines grown in 2D and 3D cell cultures show differences in cancer-related pathways like mTOR-AKT-S6K (mechanistic target of rapamycin (mTOR)-ribosomal S6 kinase (S6K) pathway), and also vary in their drug response, thus making it important to switch to 3D models [32–34]. mTOR is an essential regulator of cell homeostasis including protein translation, glucose and lipid metabolism, as well as cell survival and autophagy, and it is a central player that senses and responds to various extracellular growth signals [35]. Scaffold-free cell culture models do not utilise exogenous artificial platforms for promoting cell growth. Such cellular aggregates, termed as multicellular tumour spheroids (MCTS), are the most popular 3D cell culture method to mimic the tumour microenvironment and the cells produce their own ECM. These only partially recapitulate the microenvironment cues and are difficult to optimise for inconsistencies in their formation. They are often applied to mimic structures of breast cancer, epithelial cancer and endothelial cell angiogenesis. Breast and ovarian cancer cell lines have been studied using Matrigel in multi-well or transwell plates for gene expression, cellular signal pathways, angiogenesis and chemotherapy response [22,36–39]. In addition, different types of scaffolds, ranging from non-woven fibre ECM-derived materials to polymers in the form of foams and hydrogels, have been investigated. For developing effective scaffolds it is important that the biomaterial should meet the requirement of the physical properties and biocompatibility. Natural biodegradable polymers are attractive because they are highly biocompatible and hence may be used to support cell growth in vitro and tissue growth in vivo. They are composed of polysaccharides (amylose, cellulose, alginate, chitosan or hyaluronic acid), proteins (collagen or gelatine), nucleic acids or polyhydroxyalkanoates [40]. Alginate is a popular biomaterial for cellular encapsulation and has been used to produce fibres or scaffolds for the 3D culture of cancer cells. Low concentrations of alginate with other hydrogels including gelatine, agarose and gelatine methacrylate (GelMA) have been used to design networks of moderate stiffness and retain biological characteristics, tumorigenicity, metastatic ability and increased drug resistance [41,42]. Gene expression analysis of tumour cells cultured in 2D versus 3D alginate-based, oxygen-controlled tumour models revealed striking interdependence between culture dimensionality and hypoxia response [43]. Gelatine hydrogel microspheres (GM) incorporated into cell aggregates tackle the problem of low oxygen and nutrient supply and demonstrate longer cell viability [44]. Cancer invasion models based on drug-incorporated gelatine microspheres along with cancer cells and cancer-associated fibroblasts are efficient tools of drug screening [45]. Chitosan is another biomaterial known for its biocompatibility, biodegradability, low immunogenicity and low cost. Deacetylated chitosan scaffolds show better attachment of cancer cell lines, and cells grow as three-dimensional clumps on the chitosan matrix [46]. Chitosan forms electrostatic interactions with amine groups of alginate and blends to form an interconnected and porous 3D structure with mechanical strength and shape maintenance, which is significantly improved as compared to neat chitosan [47]. A blend of the hydrogel form of chitosan and hyaluronic acid has been used as a non-adhesive material for spheroids formation [48].

The formation of a solid tumour is central to the development of most kinds of cancer. Solid tumours involve cancers of the ovary, breast, colon, brain and other tissues [49]. Most solid tumours have a complex 3D architecture with different populations of abnormal cells divided into parenchymal and stromal compartments. The interactions between the tumour and the microenvironment result in complexity and heterogeneity of tumours leading to resistance to chemotherapy [27]. The stiffness and fibrosis increase from healthy to malignant tissues and are accompanied with chemoresistance [50]. The stiffness of fibroadenoma (solid, smooth, firm non-cancerous benign lumps), measured by its Young's modulus value is 11.42 kPa, and that of invasive ductal carcinoma (cancer that begin growing in a milk duct and then invade the fibrous or fatty tissue of the breast outside of the duct) is 22.55 kPa [51]. However, the Young's modulus of colorectal cancer tissue is 7.51 kPa [52]. These values indicate that breast cancer tissues display higher tissue stiffness than colorectal cancer and are hence the former are referred to as hard tumours and the latter as soft tumours.

Hence, a 3D model for cancer with high fidelity needs to meet many criteria including shape, suitable dimensions, adequate interconnected porosity and suitable mechanical properties to mimic the exact tumour environment. Such a model structure thus needs to be made of a material that is processable into porous 3D structures with tuneable stiffness. Polyhydroxyalkanoates (PHAs) comprise such a family of biodegradable polyesters that are produced by bacterial fermentation under nutrient-limiting conditions [53]. There are two types of PHAs, short chain length PHAs (monomer chain length C_4-C_5), or SCL-PHAs, and medium chain length PHAs (monomer chain length C_6-C_{16}), or MCL-PHAs. SCL-PHAs are normally hard and brittle, whereas MCL-PHAs are highly elastomeric in nature. In addition, PHAs are highly biocompatible in nature and exhibit surface properties that allow for the attachment and proliferation of mammalian cells [54–56]. The first PHA to be identified was poly(3-hydroxybutyric acid) (P(3HB)), which is a homopolymer of 3-hydroxybutyrate (HB). P(3HB) is a typical SCL-PHA, i.e., a stiff polymer. Examples of MCL-PHAs, the elastomeric member of this family, include Poly(3-hydroxyoctanoate), or

P(3HO), and Poly(3-hydoxyoctanoate-co-3-hydroxydecanoate), or P(3HO-*co*-3HD) [57]. Blending is an effective way of developing new polymeric material with tailored mechanical properties and in some cases also leads to improved biocompatibility as compared to the parent components. For example, the viability of mouse fibroblasts (cell line L929) on Poly(3-hydroxybutyrate) films increased significantly upon blending with poly(3-hydroxybutyrate-*co*-3-hydroxyhexanoate), P(3HB-*co*-3HHx) [58,59].

Verification of the biomimetic properties of a 3D cancer model can be carried out using a range of methodologies, and quantification of a suitable marker is one of the best strategies. Aberrant Wnt signalling is a hallmark for many cancers and upregulated Wnt-11 expression is reported in breast cancer [60]. Moreover, it has been established that Wnt-11 expression triggers oestrogen receptor alpha and modulates cellular migration in breast cancer [61]. Wnt-11 is downstream of TGF-β—shown previously to be one of the triggers of epithelial-mesenchymal transition (EMT) and chemoresistance [62,63]. EMT is associated with disruption of intracellular tight junctions and loss of cell–cell contact; during cancer progression it is an accepted phenomenon that is due to the loss of epithelial features and gain of mesenchymal morphology [64]. Wnt-11 would thus indeed be a suitable biomarker to validate the 3D cancer models. EMT is a process of cellular reprogramming of epithelial cells into modulating their cell–cell adhesion properties and gaining mesenchymal characteristics such as increased motility and invasiveness linked to metastasis [65]. The EMT programme is orchestrated through transcription factors like Snail, Slug and Zeb1 that are responsible for the gain of mesenchymal properties [66]. Snail is linked to tumour progression and invasiveness due to its ability to alter the expression of the *Vimentin* gene (*vim*) [67]. The latter is one of the mesenchymal markers responsible for maintaining cell shape, cytoplasm integrity and stabilizing cytoskeletal interactions and is found downstream of the Snail gene. An intercellular adhesion protein, E cadherin, displays the gain of epithelial properties and is inversely correlated to invasion of surrounding tissues and metastasis [68]. In the present study, we mainly focus on altered expression of four genes associated with EMT in the human adenocarcinoma cell line.

In this study, we have established, for the first time, 3D cancer models based on the novel family of biocompatible natural polymers, or PHAs. A blend of the MCL-PHA, P(3HO-*co*-3HD), an elastomeric polymer, and a SCL-PHA, P(3HB), a stiff polymer, was used to tailor the mechanical property of the model. Porous 3D PHA scaffolds were generated with variable pore size, allowing for efficient infiltration of the cells and the required nutrients. The growth pattern of the cell lines representing both hard and soft cancer, i.e., breast cancer (MCF7, MDA-MB-231) and colon cancer (HCT116), respectively, have been investigated within these 3D PHA scaffolds. Two different breast cancer cell lines exhibiting variable growth kinetics were used to compare their temporal growth pattern. While MCF-7 are epithelial-like cells associated with a weak invasiveness and good prognosis, MDA-MB-231 are enriched for epithelial to mesenchymal transition (EMT) markers and possess higher phenotypic plasticity and a more invasive behaviour connected to aggressive disease [69]. The size of the 3D tumour models was developed mimicking their real dimensions in patients. Finally, the expression of EMT markers were quantified within these novel tumour models. The unique PHA-based cancer models developed in this work are an excellent step forward in the provision of tailorable 3D models for the in-depth understanding of cancer progression and therapy.

2. Materials and Methods

2.1. Bacterial Strains and Chemicals Used

P(3HO-*co*-3HD) was produced using *Pseudomonas mendocina* CH50, which was obtained from the National Collection of Industrial and Marine Bacteria (NCIMB 10541), Aberdeen, UK. P(3HB) was produced using *Bacillus subtilis* OK2, which was obtained from the University of Westminster culture collection. The chemicals used for the production and characterisation of PHAs were purchased from Sigma-Aldrich or BDH Ltd. (Dorset, UK), VWR (Leicestershire, UK), unless otherwise stated.

2.2. Production and Extraction of P(3HB) and P(3HO-*co*-3HD)

P(3HB) was produced by *Bacillus subtilis* OK2 using glucose at 35 g/L concentration as the sole carbon substrate. The sterile nutrient broth was inoculated with a single colony of *Bacillus subtilis* OK2 and incubated for 16 h at 30 °C, 200 rpm. Then, 10% (*v*/*v*) of the inoculum was used to inoculate the production stage (modified Kannan and Rehacek media), which was incubated at 30 °C, 200 rpm for 48 h. The temperature was controlled at 30 °C, pH was set at 6.8 and 1 vvm air was supplied to the bioreactor. At the end of the fermentation, cells were retrieved by centrifugation. Wet biomass was homogenised and stored at −20 °C overnight, followed by lyophilisation. P(3HO-*co*-3HD) was produced by *P. mendocina* CH50 using glucose at 20 g/L concentration as the sole carbon substrate. Fermentation was carried out in two stages. The seed culture was prepared by inoculating sterile nutrient broth with a single colony of *P. mendocina* CH50. This was incubated for 16 h at 30 °C, 200 rpm. Then, 10% (*v*/*v*) of the inoculum was used to inoculate the second stage seed culture (mineral salt medium—MSM), which was incubated at 30 °C, 200 rpm for 24 h. Next, 10% (*v*/*v*) of the second stage seed culture was used to inoculate the final PHA production media (MSM media). The temperature was controlled at 30 °C, pH was set at 7 and 1 vvm air was supplied to the bioreactor. At the end of the fermentation, cells were retrieved by centrifugation. Wet biomass was homogenised and stored at −20 °C overnight followed by lyophilisation. PHA was extracted from the dried biomass using the Soxhlet extraction method. Methanol was used to remove the impurities from the biomass under reflux for 24 h. Methanol was then replaced with chloroform, and the biomass was subjected to Soxhlet extraction for another 24 h. The chloroform solution containing polymer was concentrated in a rotary evaporator. The PHA was precipitated using ice-cold methanol solution and stored at room temperature [70–72].

2.3. Gas Chromatography Mass Spectrometry (GC-MS)

The monomeric composition of the PHA produced was identified using GC-MS. Prior to the GC-MS analysis, polymer samples were methanolysed. GC-MS analysis was carried out using a Varian GC/MS system consisting of Chrompack CP-3800 gas chromatograph and Saturn 200 MS/MS block as described in Constantinides et al., 2018.

2.4. Production of the 3D PHA Scaffolds

The 3D PHA scaffolds were prepared using the particulate leaching technique. P(3HB) and P(3HO-*co*-3HD) were dissolved in chloroform, in a 50:50 ratio. Sodium chloride (<300 µm) was used as the porogen to create porous 3D PHA scaffolds. It was added to the polymer solution in a 1:9 (polymer:porogen) ratio and stirred for 24 h at room temperature to allow for homogeneous dispersion. The polymer solution containing porogen was poured into a Teflon mould (Dimensions—20 mm × 16 mm × 10 mm) and allowed to dry. The dry 3D PHA scaffolds were removed from the mould using a sterile scalpel and immersed in sterile water to allow porogen leaching. The pH of the sterile water containing 3D PHA scaffolds was measured to ensure complete removal of the porogen. Post leaching, the scaffolds were dried. They were visibly porous and were approximately 20 mm × 15 mm × 8 mm in size.

2.5. Cell Culture on 3D PHA Scaffolds

The 3D PHA scaffolds were cut into 10 mm × 10 mm × 8 mm using a sterile scalpel and placed in 24-well plates. Prior to seeding cells, the 3D PHA scaffolds were sterilised on both sides under high-intensity ultraviolet radiation, concentrated around a wavelength of 253.7 nm for 30 min. The 3D PHA scaffolds were further sterilised by washing them with 70% ethanol for 5 min. This was repeated thrice. Post sterilisation, they were allowed to dry for 12 h in the 37 °C humidified chamber to allow for ethanol to volatilise away. Sterile 3D PHA scaffolds were then rinsed thrice in 1× phosphate buffer saline (PBS) (Sigma Aldrich, St. Louis, MO, USA) and then incubated in 2 mL of Dulbecco's modified Eagle medium

(DMEM) supplemented with 2 mM L-glutamine, 10% FBS and 1% Penicillin-Streptomycin for the next 24 h.

HCT116 (ATCC®-CCL-247), MCF-7 (ATCC®-HTB-22) and MDA-MB-231 (ATCC®-HTB-26) cell lines were obtained from ATCC. The cells were cultured in T75 flasks (Sigma Aldrich) in DMEM (Himedia) containing 10% FBS (Thermo Fisher Scientific, Waltham, MA, USA), 2 mM L-Glutamine (Merck Life Science UK Limited, Dorset, UK) and 1% antibiotic solution (100 U/mL penicillin, 100 μg/mL streptomycin) (Sigma Aldrich), henceforth referred to as complete DMEM. Cultured cells were maintained at 37 °C, 5% CO_2 in a humidified incubator. Static surface seeding method was used to seed cells onto the 3D PHA scaffolds as described in previous papers [73]. The ratio of the cell-seeded and scaffold area was optimised. Different cell seeding densities were used. The higher cell density of 200,000 cells/scaffold of 3 × 10 × 10 or 10 × 10 × 10 mm^3 scaffolds was found to be optimal. Hence, a concentrated cell suspension of 200,000 cells was added to each 3D PHA scaffold and incubated for 30 min to allow for cells to attach to the scaffold. After 30 min, 500 μL (enough to cover the 3D PHA scaffold) of complete DMEM was carefully added from the sides so as to not dislodge any cells seeded on the top of the 3D PHA scaffolds and incubated for 12 to 24 h. After 12 h, 1000 μL of complete DMEM was added from the sides. At the end of incubation, the 3D PHA scaffolds were submerged in complete DMEM and covered with a cell crown 24 (Scaffdex Oy, Tampere, Finland). The cell-seeded 3D PHA scaffolds were incubated for 5 days. After every 24 h, media were aspirated, 3D PHA scaffolds were washed three times with PBS and complete DMEM was added to the 3D PHA scaffolds.

2.6. Cell Proliferation Studies

Proliferation of the cells on the 3D PHA scaffolds at the end of each time point was determined using the Alamar Blue assay following the manufacturer's protocol (Thermo Fisher Scientific, Gloucester, UK). Alamar Blue reagent (10% volume of culture media) was added to the wells. After 3 h of incubation, Alamar Blue solution was transferred to a 96-well plate to obtain absorbance values at 570 nm. Tissue Culture Plastic (TCP) was used as the positive control.

2.7. Scanning Electron Microscopy (SEM)

2.7.1. SEM Characterisation of the 3D PHA Scaffolds

Post leaching, dried 3D PHA scaffolds were observed under the Scanning Electron Microscope (SEM). The 3D PHA scaffolds were cut using a clean scalpel. The samples were vacuum-dried and placed on aluminium stubs. Finally, the 3D PHA scaffolds were gold coated and imaged using SEM Zeiss EVO40. The pore sizes were calculated from the SEM images using the Image J program (Public Domain Image Processing Program, National Institute of Health, Bethesda, MD, USA).

2.7.2. SEM Characterisation of the Cell Laden 3D PHA Scaffolds

SEM was used to view the cross section of the cell laden 3D PHA scaffolds to observe cell morphology within the 3D PHA scaffold. Cell seeded 3D PHA scaffolds were fixed in 2% paraformaldehyde in PBS and kept at 4 °C overnight. They were dehydrated using graded ethanol solution (50%, 70%, 80%, 90% and 100%) and Hexamethyldisilazane (HMDS), and they were gold plated and imaged using SEM Zeiss EVO40 (Public Domain Image Processing Program, National Institute of Health, Bethesda, MD, USA).

2.8. Confocal Microscopy

The cell-seeded 3D PHA scaffolds cultured as described above were stained with CellTrace™ Calcein Green, AM/Ethidium homodimer (Invitrogen™)-1 mix in 1:4 ratio (MDA-MB-231) or 50 mM Image-iT TMRM reagent/10 μM CellTrace™ Calcein Green, AM (Invitrogen™) (HCT116 and MCF-7) for 30 min, cut using a clean scalpel into 1–2 mm slices placed in glass-bottom dishes and viewed using 488 nm and 548 nm wavelength, respectively, using a Nikon AIR microscope. In MDA-MB-231-seeded 3D PHA scaffolds, Calcein Green/Ethidium bromide led to green/red colour for live/dead analysis, whereas in HCT116- and MCF7-seeded 3D PHA scaffolds, TMRM/Calcein Green led to red/green colour, respectively. TMRM is a cell-permeant dye that accumulates in active mitochondria with intact membrane potentials indicating live cells.

2.9. *Total RNA Extraction, cDNA Synthesis and qRT-PCR*

Total RNA was extracted using Trizol (Sigma, Hertfordshire, UK), RNA concentration and purity was measured using the NanoDrop Spectrophotometer using absorbance values at 260 nm and 280 nm. cDNA was generated by the reverse transcriptase reaction and used for qPCR. The following genes were studied (corresponding primer sequences are given in references in parentheses): Wnt-11 [74]; Vim, Snail and E-cadherin [75]. Analysis by real-time qPCR was done by SYBR Green premix (Qiagen, Germantown, UK) using the following conditions: 95 °C for 15 min, 40 cycles at 95 °C for 15 s, 60 °C for 1 min and 72 °C for 15 sec. Relative levels of mRNA expression were calculated according to the CT/2-ΔΔCT method [64]. RNA polymerase II, (RPII) was optimised and used as the reference gene [76,77]. Experiments were performed in triplicate and the standard deviation was calculated as well as the Student's *t*-test using GraphPad Prism 7.00 (La Jolla, CA, USA) software.

2.10. *Data Analysis*

All data were analysed as means ± standard errors. Statistical significance was determined using the student's *t*-test or ANOVA with Newman–Keuls post hoc analysis, as appropriate. Results were considered significant for $p < 0.05$.

3. Results

3.1. *Production and Chemical Characterisation of the Polymer to Make 3D PHA Scaffolds*

Polymer production was carried out using bacterial fermentation. P(3HB) and P(3HO-*co*-3HD) polymers were produced using *Bacillus subtilis* OK2 and *Pseudomonas mendocina* CH50, respectively, with glucose as the sole carbon source as described in Basnett et al., 2021.

Gas Chromatography–Mass Spectrometry was used to confirm the monomeric composition of the PHAs produced. For P(3HB), the gas chromatogram showed one peak (R_t = 4.1 min), originating from the product of polymer methanolysis, and another peak of the internal standard, methyl benzoate (R_t = 6.4 min) (Figure 1a). Mass spectra pattern for the peak at R_t = 4.1 min matched with the methyl ester of 3-hydroxybutyric acid from the NIST Standard Reference Library. For P(3HO-*co*-3HD), the gas chromatogram showed two peaks (R_t = 7.7 min and R_t = 9.3). The mass spectra for the peaks matched with the methyl esters of 3-hydroxyoctanoic acid (3HO) and 3-hydroxydecanoic acid (3HD), respectively, from the NIST Standard Reference Library (Figure 1b).

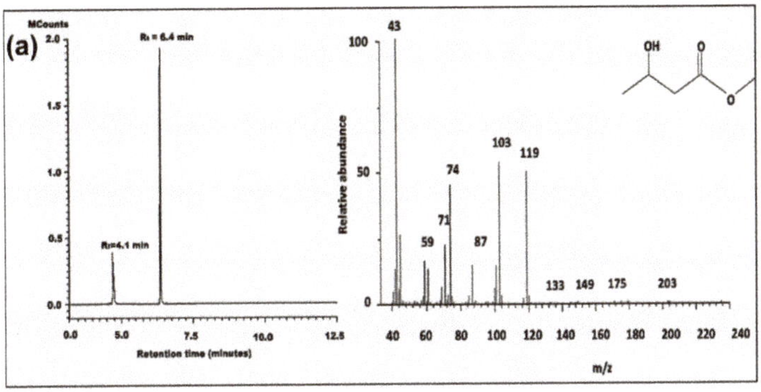

Figure 1. *Cont.*

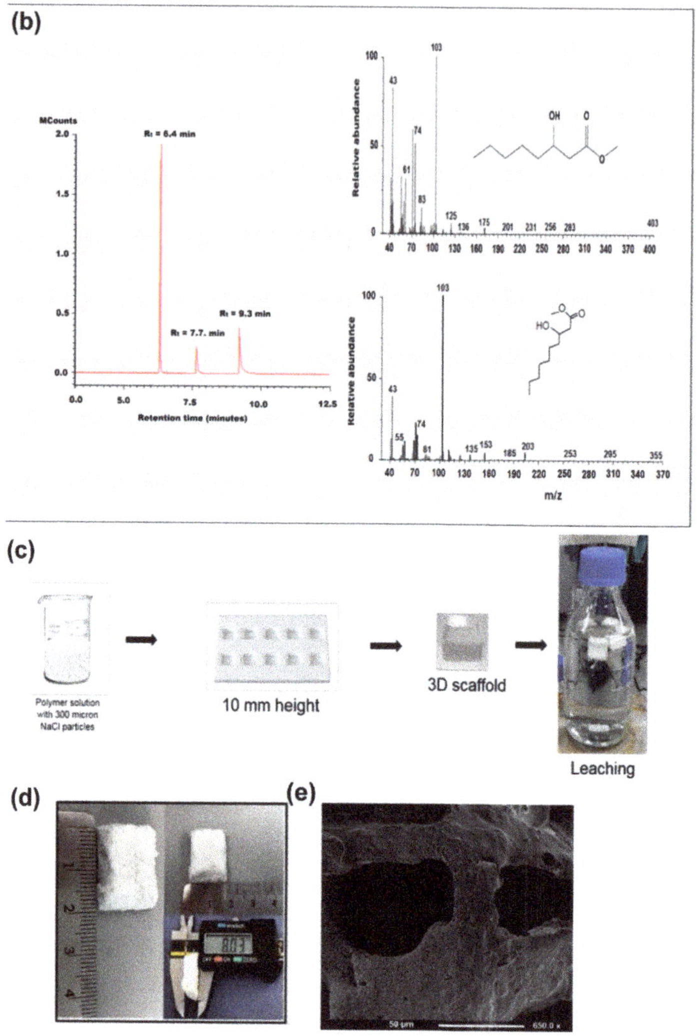

Figure 1. (**a**) GC-MS spectra of P(3HB) and (**b**) P(3HO-*co*-3HD). (**c**) Schematic representation of the preparation of porous 3D PHA scaffolds. (**d**) An optical image of air-dried porous 3D PHA scaffolds 20 mm × 15 mm × 8 mm. (**e**) SEM images of the P(3HB)/P(3HO-*co*-3HD) 50:50 3D PHA scaffold.

In Figure 1a,b, the peak at 6.25 represented the internal standard (methyl benzoate). In the case of P(3HB), the fragment peak at retention time 4.125 represented the 3-hydroxy methyl ester of butyric acid, whereas in the P(3HO-*co*-3HD) spectrum, the fragment peak at retention time of 7.611 and 9.188 represented 3-hydroxyl-methyl ester of octanoic acid and decanoic acid, respectively.

3.2. Production and Characterisation of the 3D PHA Scaffold

The porous 3D PHA scaffolds were prepared by solvent casting–particulate leaching technique [67], which is a standard method to produce polymer-based scaffolds. The NaCl sieved through a 300 μm sieve ensured that the 3D PHA scaffolds have a well-connected variable pore size. The adoption of an appropriate polymer-to-salt ratio, experimentally defined as 1:9 for both P(3HO-*co*-3HD) and P(3HB), resulted in the formation of rich and interconnected porosity, and the choice of sieving salt through a 300 μm sieve allowed for controlling the final pore size. The 3D PHA scaffolds were soaked in distilled water to remove the porogen (Figure 1c). Stable pH of the water into which the porogen was being leached ensured the complete removal of the porogen, i.e., NaCl. The 3D PHA scaffolds appeared to be visibly porous without magnification. The 3D PHA scaffold size was 20 mm × 15 mm × 8 mm (Figure 1d). Larger 3D PHA scaffolds were produced, hoping to induce hypoxic conditions in the core.

Figure 1e shows SEM images of 3D PHA scaffolds, which were analysed further to analyse the structure and calculate the pore size of the 3D PHA scaffolds using Image J. The pore size varied from 30 to 300 μm, which provided a varied level of pore sizes within the 3D PHA scaffolds. The pores of variable sizes were evenly distributed throughout the 3D PHA scaffold, and an interconnected network of pores was observed.

3.3. *Cell Culture on the 3D PHA Scaffolds to Create the Disease Models*

3.3.1. Hard Tumour Disease Models

To show the ability of these 3D PHA scaffolds for development of tumour models two different human cancer cell lines derived from breast epithelium were chosen because of their different invasive properties and stiffness; basal MDA-MB-231 was derived from adenocarcinoma metastatic tumour site that is linked to aggressive disease, and MCF7 was derived from primary breast ductal carcinoma that belongs to the luminal A subtype [2,69]. These cell lines have significantly different stiffness. Several studies report that metastatic cells are softer as compared to their non-invasive counterparts [78,79]. Stiffness values of more aggressive cells (MDA-MB-231) are lower than their non-aggressive counterparts (MCF7) [80]. Each cell line was seeded on pre-soaked 3D PHA scaffolds and allowed to grow for 5 days. High cell seeding density per 3D PHA scaffold sample was chosen for cell viability studies to allow the cells to populate the 3D PHA scaffold, enabling cell–scaffold interaction as shown in Figures 2 and 3.

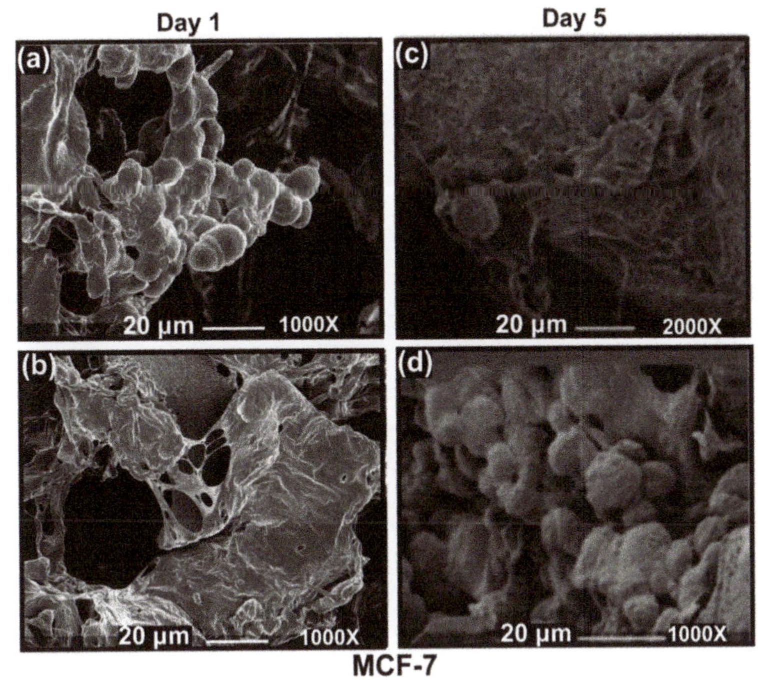

Figure 2. *Cont.*

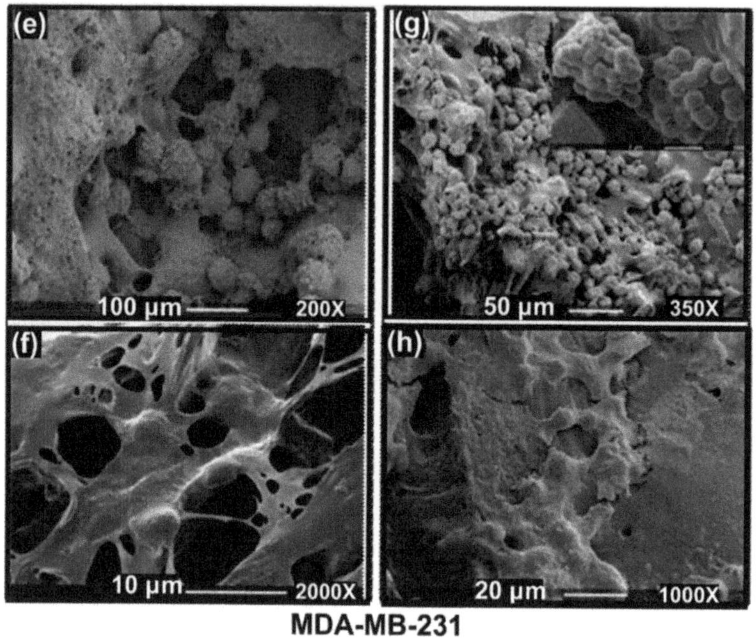

Figure 2. SEM images of cells grown on 10 mm × 10 mm × 8 mm PHA-based 3D scaffolds (**a–d**) MCF7 cells cultured on 3D PHA scaffolds at (**a**,**b**) Day 1 and (**c**,**d**) Day 5. (**b**) MDA-MB-231 cells cultured on 3D PHA scaffolds at (**e**,**f**) Day 1 and (**g**,**h**) Day 5.

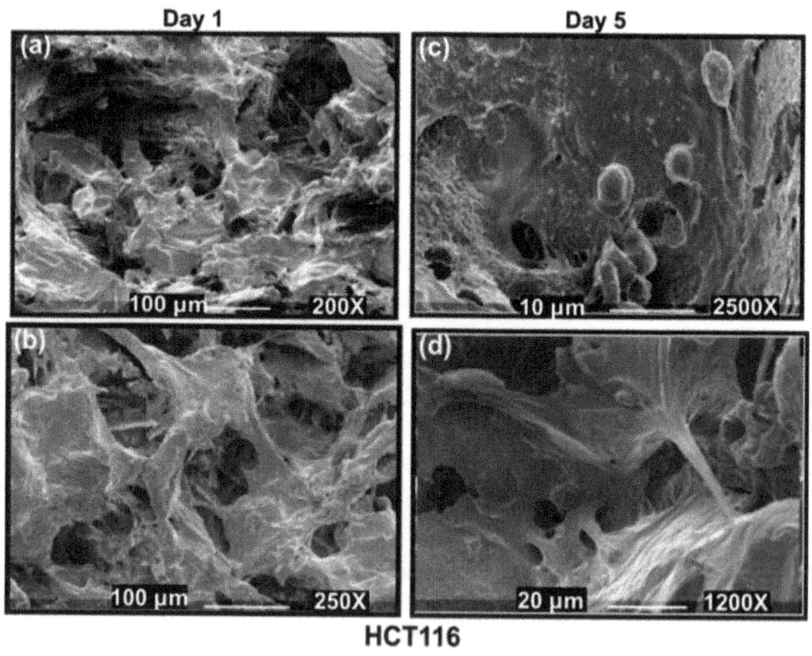

Figure 3. SEM images of HCT116 cells grown on 10 mm × 10 mm × 8 mm 3D PHA scaffolds on Day 1 (**a**,**b**) and Day 5 (**c**,**d**).

3.3.2. Cell Viability of Hard Cancer 3D Models

Viability of the cells on the 3D PHA scaffold was assessed on Day 1, 3 and 7 using the Alamar blue assay that detects metabolically active cells. The MCF-7 cells displayed a significant increase in cell viability with time (Figure 4a). On Day 1, 3 and 7, cell viability was set at 100% for the 2D cell culture on TCP. While at Day 1 on the 3D PHA scaffold, the cell viability was 74%, which significantly increased to 98.8% ($p < 0.05$) at Day 3. Again, on Day 7, there was a further significant increase in the cell viability of the MCF-7 cells to 168% ($p < 0.05$). Similarly, MDA-MB-231 cells displayed a significant increase in cell viability over time (Figure 4b). On Day 1, 3 and 7, the cell viability of the cells cultured on 2D cell culture on TCP were normalised to 100%. While the cells on the 3D PHA scaffolds showed only 59% cell viability on Day 1, which increased significantly to 80.2% on Day 3 ($p < 0.05$), followed by a further increase to 161.45% ($p < 0.05$) on Day 7.

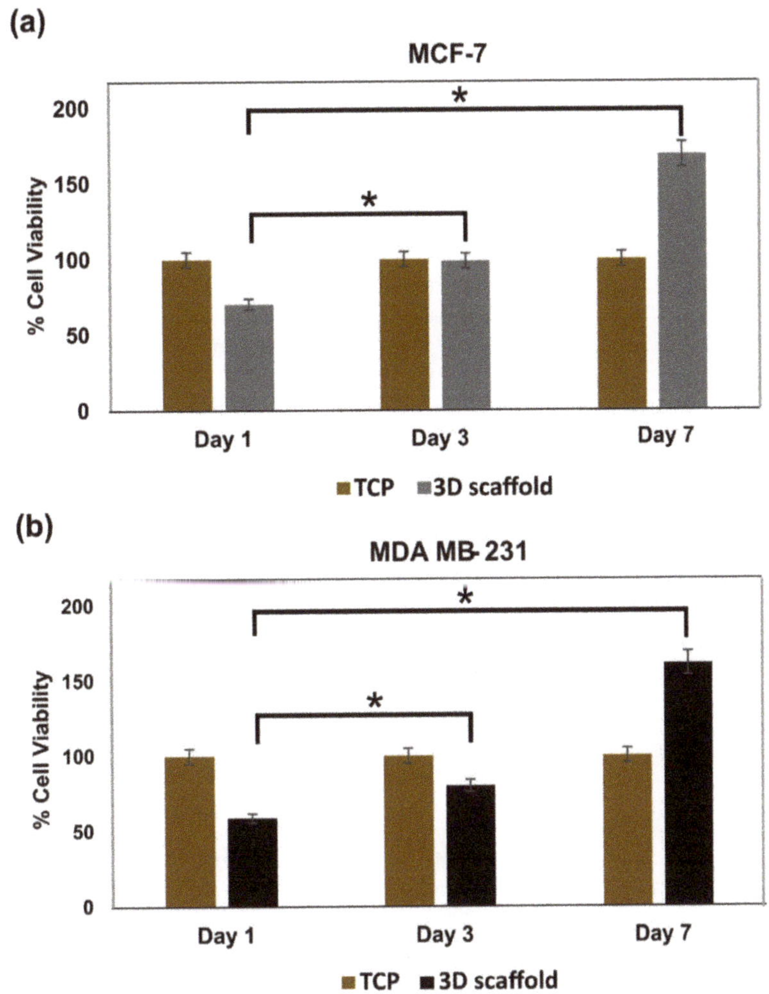

Figure 4. Alamar blue assay of (**a**) MCF-7 cells and (**b**) MDA-MB-231 cells cultured on 2D tissue culture plastic (TCP) and 3D PHA scaffold for 1, 3 and 7 days. Differences were considered statistically significant with $p < 0.05$ (*).

These results show that cells in the 2D cell cultures exhibited higher proliferation rates than those in the 3D cultures initially, but the 3D cultures maintained a longer proliferation phase. This finding was consistent with previous studies [81].

3.3.3. SEM Imaging of the Disease Models: Formation of Hard Cancer 3D Models

After 1 and 5 days of culture, cells were fixed on the 3D PHA scaffolds and observed under the SEM to assess their morphology. SEM images revealed that the MDA-MB-231 cells formed aggregates in contrast to the elongated shape observed when cultured on a 2D surface. The MCF-7 cells displayed diverse morphologies (Figure 2a,b,e,f). They proliferated into extensive layers of cells on the 3D PHA scaffold at Day 5. Hence, it seems like initially, on Day 1, the cells adhere to the 3D PHA scaffold and adjust to the new environment provided by the 3D PHA scaffold. However, by Day 5, the cells proliferate at a high rate, either forming layers covering the 3D PHA scaffold surface or growing in clumps where each cell surface could be viewed.

3.3.4. Live Cell Assessment on the Hard Cancer Disease Models

To investigate the growth pattern of breast cancer cell lines, MCF7 and MDA-MB-231 cells were grown on 10 mm 3D PHA scaffolds for a period of up to 5 days. The cells were stained with TMRM (Red) and Calcein green (green) to check the live cells within the 3D PHA scaffolds. TMRM detects active mitochondrial membrane and Calcein green detects live cells with intact cell membranes. On Day 1, the MCF7 cells were distributed evenly throughout the 3D PHA scaffold and no clumps were observed (Figure 5a,b). The cells spread throughout the centre and infiltrated to the bottom of the 3D PHA scaffolds (Figure 5b). On Day 5, MCF7 cells showed an even distribution throughout the 3D PHA scaffolds and there was no significant cell death, and they appear to be denser than Day 1 (Figure 5c,d). MCF7 cells formed an even dispersed layer on the 3D PHA scaffolds on Day 5, which covered most of the 3D PHA scaffold. The porosity of the 3D PHA scaffolds facilitates the cells to infiltrate to the bottom of the 3D PHA scaffolds, and a consistent distribution of cells were observed throughout the 3D PHA scaffold. Similar to the MCF7 cells, MDA-MB-231 cells also grew well on the 3D PHA scaffolds. Upon staining with Calcein Green (green), a lower density of MDA-MB-231 cells was observed on Day 1 (Figure 5e,f) and Day 5 (Figure 5g,h), and they tended to grow in clusters, as observed in native tumour tissue, throughout the 3D PHA scaffold. In comparison to MCF-7, there was not a marked difference in the growth pattern of MDA-MB-231 between Day 1 and Day 5.

3.3.5. Wnt-11 and E-Cadherin mRNA Expression Profiling

Solid hard tumour models made by culturing breast cancer cell lines MCF7 and MDA-MB-231 were analysed for mRNA expression levels of several EMT marker genes. *Wnt-11*, *E-cadherin*, *Vim* and *Snail* genes were monitored over a period of 0, 7, and 14 days after seeding the cells in 3D PHA scaffolds. MCF7 and MDA-MB-231 cells were cultured in 3D PHA scaffolds, and cells were extracted from 3D PHA scaffolds on Day 0, 7 and 14 and analysed for mRNA levels.

We first analysed *Wnt-11* mRNA levels in both MCF-7 and MDA MB-231 cells. *Wnt-11* mRNA expression in both breast cancer cell lines were quantified after growing cells on 3D PHA scaffolds on Day 0 using qRT-PCR. The MCF-7 cells, which have the most epithelial properties, expressed less *Wnt-11* mRNA than MDA-MB-231 cells (Figure 6a; Day 0; 40-fold $\pm$ 0.5, $n = 3$; $p < 0.01$). Next, MCF-7 cells were seeded onto 3D PHA scaffolds and allowed to grow for 14 days, and *Wnt-11* mRNA levels were analysed in a time dependent manner. qRT-PCR analysis revealed that *Wnt-11* gene expression in MCF7 cells did not significantly change when the cells were grown on the 3D PHA scaffolds for 7 and 14 days. Similarly, MDA-MB-231 cells were seeded on the 3D PHA scaffolds and allowed to proliferate for a period of 14 days. Interestingly, *Wnt-11* mRNA levels were found to be significantly upregulated in MDA-MB-231 cells grown on 3D PHA scaffolds on Day 7 and 14 by 5 $\pm$ 0.2 and 229 $\pm$ 0.8-fold in comparison to Day 0, respectively.

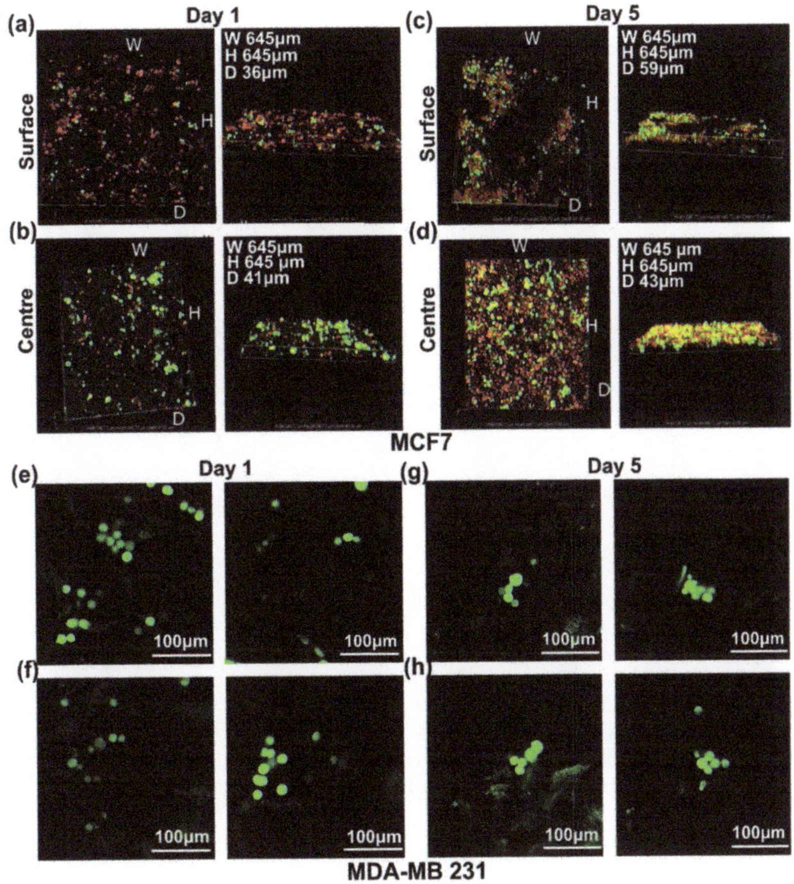

Figure 5. The 3D images of cells grown on 10 mm × 10 mm × 8 mm PHA-based 3D scaffolds (**a–d**). MCF7 cells cultured on 3D PHA scaffolds at (**a**,**b**) Day 1 and (**c**,**d**) Day 5 stained with TMRM (Red) and Calcein Green (Green), both for live cells. (**b**) MDA-MB-231 cells cultured on 3D PHA scaffolds at (**e**,**f**) Day 1 and (**g**,**h**) Day 5 stained with Calcein Green (Green) live cells and Ethidium bromide (Red) dead cells.

E cadherin is another EMT marker that is an intracellular adhesion protein and marker of epithelial characteristics [68]. To see the effects of PHA-based 3D disease models, MCF-7 and MDA-MB-231 cells were seeded on 3D PHA scaffolds and allowed to proliferate for a period of 0, 7 and 14 days. mRNA levels of the *E-Cadherin* gene were studied using qRT-PCR analysis. It was observed that the expression level of *E-cadherin* was significantly higher in MCF7 cells than MDA-MB-231 cells and increased further on the Days 7 and 14, as reported earlier [82]. *Wnt-11* and *E-cadherin* mRNA levels were found to be inversely correlated in MCF-7 and MDA-MB-231 cells (Figure 6d; $n = 3$; $p < 0.05$). Next, mesenchymal markers *Vim* and *Snail* mRNA levels were analysed in MCF-7 and MDA-MB-231 cells cultured on 3D PHA scaffolds. Cells were allowed to proliferate for 0, 7 and 14 days, then subjected to qRT-PCR analysis. It was observed that in MDA-MB-231 cells, the expression levels of *Vim* and *Snail* increased significantly by 429- and 450-fold, respectively, when the cells were grown for 14 days as compared to on Day 0 ($n = 3$; $p < 0.01$). Enhanced *Vim* and *Snail* gene expressions were also found in MCF-7 cells on Day 7 as compared to Day 0 (5- and 8-fold, respectively), but significant upregulation of both *Vim* and *Snail* mesenchymal markers was observed after 14 days with respect to Day 0 (Figure 6e, $n = 3$; $p < 0.01$).

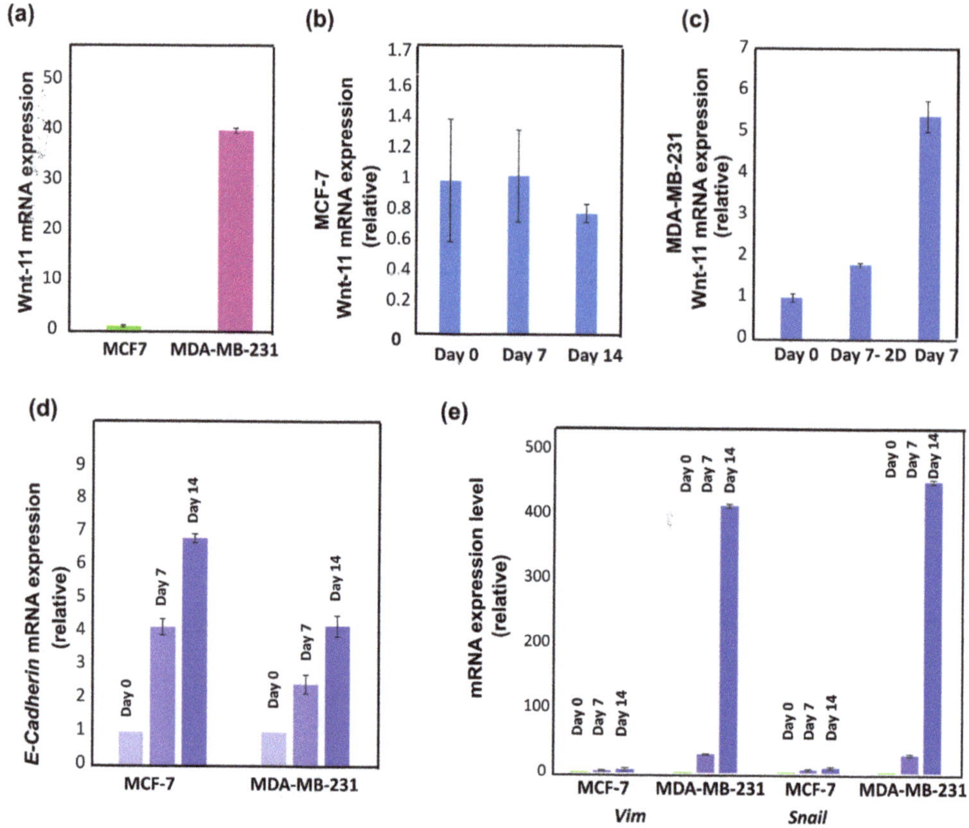

Figure 6. (**a**) qRT-PCR analysis of Wnt-11 expression level in MCF7 and MDA-MB-231 cells cultured in 3D PHA scaffolds. (**b**) Wnt-11 mRNA levels in MCF7 cells cultured in 3D PHA scaffolds for 0, 7 and 14 days. (**c**) Wnt-11 mRNA expression levels in MDA-MB-231 cells cultured on 3D PHA scaffolds for 0, 7 and 14 days. (**d**) E-Cadherin mRNA expression levels in MCF7 and MDA-MB-231 cells cultured in 3D PHA scaffolds for 0, 7 and 14 days. (**e**) mRNA levels of Vim and Snail in MCF7 and MDA-MB-231 cells grown in 3D PHA scaffolds. The column graphic represents the average of three replicates of mRNA isolated from each cell line. The data are normalised according to RPII expression level by fold analysis ($n = 3$; $p < 0.01$).

In summary, all the results obtained are consistent with the fact that 3D cell culture using the 3D PHA scaffolds are superior substrates as mimics of the microenvironment of tumours as compared to 2D cell culture (data not shown) to reliably study gene expression profiles as well as cellular behaviour. The role of increased EMT markers and Wnt-11 in 3D need to be studied further in the future.

3.4. *Cancer Disease Modelling for Soft Tumours*

3.4.1. SEM Imaging of the Disease Models: Formation of Soft Cancer 3D Models

Human adenocarcinoma colorectal cell line HCT116 was seeded on the fabricated 3D PHA scaffolds of 10 ×10 × 8 mm size and was allowed to proliferate for 5 days. The cell line proliferated into flat layered sheets on the 3D PHA scaffold on Day 1. HCT116 cells proliferated further and layers of cells forming colonies spread out within the crevices of the 3D PHA scaffolds by Day 5 (Figure 3c,d). Hence, it seems like initially, on Day 1, the cells adhere and adjust to the new environment provided by the 3D PHA scaffold.

However, by Day 5, the cells proliferated at a high rate, either forming layers covering the 3D PHA scaffold surface or growing in clumps where each cell surface could be viewed.

3.4.2. Live Cell Assessment on the Soft Cancer Disease Models

To make soft cancer models, HCT116 cells were allowed to grow in 3D PHA scaffolds for 5 days and imaged at Day 1 and 5 by staining them with TMRM (Red) and Calcein Green (Green). TMRM detects active mitochondrial membranes, and Calcein green detects live cells with intact cell membranes. HCT116 cells grow in pockets in the 3D PHA scaffolds rather than being dispersed throughout the 3D PHA scaffolds. This could be attributed to the fact that HCT116 grow in colonies in 2D cultures as well. By Day 5, HCT116 cells appeared to form bigger clusters (Figure 7a–c) than Day 1 (Figure 7d–f), and colonies started to appear in the pockets within the 3D PHA scaffolds with no significant cell death because the cells had enough time to adhere and adapt to the 3D PHA scaffold's environment.

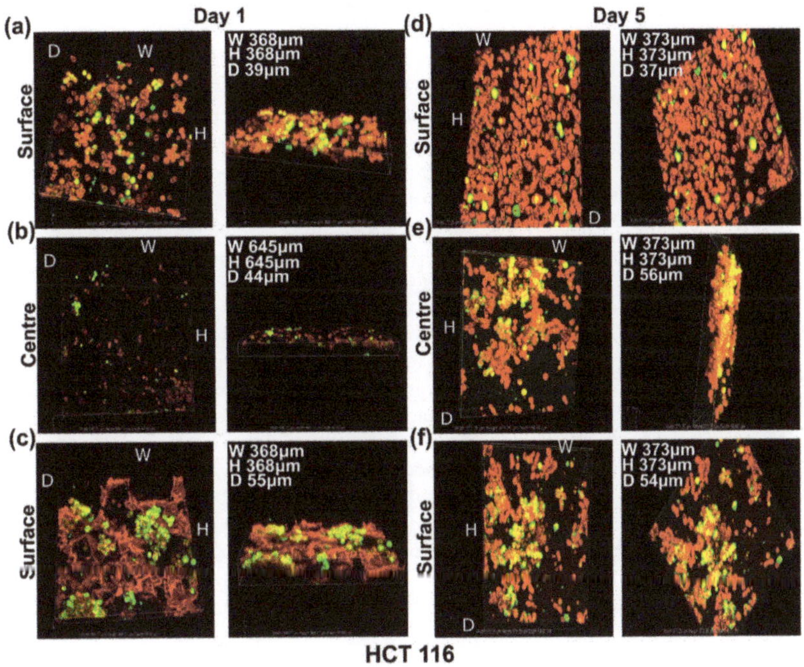

Figure 7. Live cell imaging of HCT116 cells grown on 10 mm × 10 mm × 8 mm 3D PHA scaffolds on Day 1—(**a**) upper side, (**b**) centre and (**c**) lower side—and on Day 5—(**d**) upper side, (**e**) centre and (**f**) lower side—using TMRM (Red) and Calcein Green (Green), both for live cells. The upper side, lower side and centre of the 3D PHA scaffold were imaged using a Nikon confocal microscope.

3.5. Comparison of Growth Patterns in Soft and Hard Cancer Disease Models

As seen in Figure 8, on Day 5, MDA-MB-231(representing hard breast tumour/cancer type) showed dispersed growth in clusters, and MCF7 cells (representing hard breast tumour/cancer type) formed an evenly dispersed dense layer, while HCT116 (representing soft colon tumour/cancer type) formed large colonies within the pockets of the 3D PHA scaffolds.

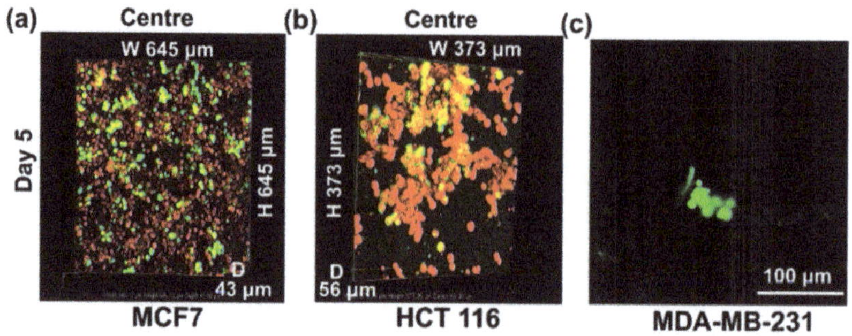

Figure 8. Comparison of cell growth patterns of (**a**) MCF7, (**b**) HCT-116 stained with TMRM (Red) and (**c**) MDA-MB-231 cells stained with Calcein Green (Green), both for live cells.

4. Discussion

In this study, an MCL and SCL-PHA blend (50:50 wt%) was used to fabricate a porous 3D PHA scaffold of size 10 mm × 10 mm × 8 mm and culture breast and colon cancer cells. There are many advantages of using PHAs. One main advantage of using PHAs is the fact that the mechanical properties of PHAs can be modulated to make models of variable stiffness. This is because there are many types of PHAs ranging from C_4-C_{16} units in each monomer unit. By varying the carbon source provided to the bacteria and the bacterial species used, the monomer content of the PHA can be altered. This in turn leads to changes in the mechanical properties. The mechanical properties of the PHA-based 3D scaffolds can be tailored to match the specific tumour type by blending various types of PHAs or producing copolymers with different monomer types. This leads to an enormous range of mechanical properties that are not accessible for scaffolds using alginate/gelatine/hyaluronic acid or chitosan. In addition, PHA-based scaffolds have a slow degradation rate, and the degradation occurs by surface degradation. This results in stable scaffold structures, which can be used for long-term studies as opposed to the other types mentioned above. Additionally, PHAs are thermoplastics in nature and hence can be processed easily using a variety of processing methods, such as 3D printing using fused deposition modelling (FDM) and Selective Laser Sintering (SLS), to produce structurally varied and bespoke models.

Finally, and not the least, the size that can be achieved using PHA-based scaffolds is comparable to that of patient tumours. Since most tumours grow to a size of 1–2 cm when initially diagnosed, cancer cells grown in 3D PHA scaffolds of 1 cm thickness will resemble the oxygen gradient, nutrient and waste removal characteristics of in vivo tumours and emulate a tumour-like microenvironment [83,84]. Most 3D scaffolds that use biomaterials, such as gelatine, alginate and chitosan, have a lower size range, up to 600 μm [41,44,45]. 3D cell culture in the bigger 3D PHA scaffolds resembling tumour sizes is much more relevant to the understanding of cancer cell behaviour, identification of targets for cancer treatments and drug screening.

Hence, novel 3D cancer models were developed, where breast cancer and colon cancer cells were used to mimic hard and soft tumours, respectively. Breast tumours have higher stiffness as compared to colon tumours, which affects the oxygen gradient and nutrient supply in these tumours, and also affects the various cellular mechanisms [85]. PHAs are relatively slow in degradation and undergo surface degradation [86]. For substantial degradation, the scaffolds will need to be maintained in vitro for at least two to three months. In contrast, degradation of hydrogel-based scaffolds, such as for alginate and gelatine, occurs very fast. They undergo bulk degradation in less than 12 h in vitro [87]. It is extremely difficult to match the speed of hydrogel degradation with the pace of tissue formation, which is important in maintaining the shape and mechanical integrity of tissue-engineering constructs [10]. However, PHA-based scaffolds undergo surface degradation after 2–3 months as opposed to hydrogel scaffolds that would crumble and break during this period. Hence, PHA-based scaffolds remain stable for longer periods of time and can be used for long-term studies.

MCL and SCL-PHAs have been studied previously and have exhibited excellent biocompatibility with different types of cell lines for various applications including tissue engineering and medical devices [70,71,88]. However, MCL and SCL PHA blends have not been explored previously for their suitability as 3D PHA scaffolds in cancer disease modelling. The scaffold was developed using P(3HB), which is a short-chain length PHA, known to be hard and brittle in nature and P(3HO-*co*-3HD), which is a medium-chain length PHA, known to be soft and elastomeric. Therefore, a ratio of 50:50 was chosen as a first example to study the feasibility of the PHA-based models. In the future, the ambition is to use a range of the ratios and hence obtain scaffolds with a varying range of mechanical properties. The MCL and SCL-PHAs used for the fabrication of 3D PHA scaffolds were produced by the fermentation of *P. mendocina* CH50 and *Bacillus subtilis* OK2, respectively, using glucose as the carbon source. It is well established that *Pseudomonas* sp. are capable of accumulating MCL-PHA copolymers when grown on structurally unrelated carbohydrates [89,90]. Similarly, *Bacillus* species are known to produce the P(3HB) homopolymer when grown on glucose as substrate [91]. GC-MS was used to identify the MCL-PHA as a P(3HO-*co*-3HD) copolymer and the SCL-PHA as a P(3HB) homopolymer.

The P(3HO-*co*-3HD)/P(3HB) blend was used to fabricate 3D PHA scaffolds using the salt-leaching technique [92]. SEM images presented a foam-like structure with well-defined pores formed due to the dissolution of sodium chloride particles (Figure 1e).The pore size in the fabricated 3D PHA scaffolds ranged from 30 to 300 μm, the pores of variable sizes were evenly distributed throughout the 3D PHA scaffold, the interconnected network of pores facilitated the infiltration of cells throughout the 3D PHA scaffold and the cells had enough space to be able to grow in colonies and form 3D tumour models similar to the in vivo models. The pores would act as channels to facilitate cellular interaction, nutrient and oxygen diffusion, as well as waste removal [93,94]. Previously, various breast cancer cell lines from different subtypes, such as MCF-7 (luminal A), BT474 (luminal B), SKBR3 (human epidermal growth factor receptor 2—HER2) and MDA-MB-231 (triple negative), have been cultured in 3D microenvironments. The 3D liver model of Alginate (1% and 0.5%): GelMA (gelatine methacrylate) fibres containing NIH-3T3 fibroblasts, HepG2s and HUVECs showed no change in cell viability [41]. UV light illumination always carries the risk of high cell death and DNA damage, which affects normal cellular function. The Alginate:GelMA mixture was optimised for normal cellular function at a 365 nm wavelength [41]. This was not a concern in our study, as the 3D PHA scaffolds were UV sterilised prior to cell seeding. In this study, breast cancer cell lines, such as MCF-7 and MDA-MB-231, and the colon cancer cell line, HCT116, were cultured on the PHA-based porous 3D PHA scaffolds, and growth was observed for 1 and 5 days. Cancer cell properties, such as morphology, proliferation pattern and tumorigenicity, were monitored until Day 5. Both the cell lines attached and proliferated on the 3D PHA scaffold over time. In comparison to tissue culture plates, MCF-7 and MDA MB -231 cells cultured on 3D PHA scaffolds exhibited delayed growth until Day 3 (Figure 4). Both the cell types continued to proliferate on the 3D PHA scaffolds until Day 7, while cells cultured on 2D tissue culture plastic (TCP) displayed no change in cell viability on Day 7 (Figure 4). This could be due to the higher surface area of 3D scaffolds as compared to the 2D cell culture, which has a major drawback of contact inhibition to sustain cell growth over long durations [10]. It is the 3D environment in which the cells grow and maintain a longer proliferation phase, and are hence growing under conditions that mimic in vivo conditions [95].

These observations were consistent with a similar study conducted by Florczyk et al., 2016, where three cell lines, TRAMP-C2 (prostate cancer), SK-Hep-1 (liver cancer) and MDA-MB-231 (breast cancer), were cultured on 2D tissue culture plastic and Chitosan-Alginate 3D scaffolds [96].

It was observed that in the 2D cell culture, cell lines proliferate rapidly and become confluent, whereas cells show delayed growth, forming tumour spheres on the 3D scaffolds [96]. In another study conducted by Chen et al., 2012, MCF-7 cells were cultured on 3D collagen scaffolds. MCF-7 cells proliferated on the 3D PHA scaffolds until the Day 13, whereas these cells did not proliferate beyond Day 7 on 2D tissue culture plates [63]. This was concurrent with results obtained in our study, pointing out that cells cultured on the 3D PHA scaffolds initially take time to attach and start proliferation, unlike 2D cell culture, where cells soon fail to proliferate due to contact inhibition.

SEM images revealed that MDA-MB-231 cells formed aggregates and exhibited rounded morphology in a 3D environment while displaying a spindle-shaped morphology in the 2D cell culture. Similar observations were made by Ivers et al., 2014, when MDA-MB-231 cells were cultured in 3D, using a reconstituted basement membrane matrix Geltrex® for 10 days. MDA-MB-231 form aggregates or spread in a dissociated manner showing elongated or round-shaped morphology, demonstrating the dynamic behaviour of the MDA-MB-231 in a 3D environment [97]. MCF-7 cells cultured on 3D PHA scaffolds displayed diverse morphologies in 3D (Figures 2a–d and 5a–d). They proliferated into sheets with some rounded cells. While HCT116 formed large colonies within the pockets of the 3D PHA scaffold by Day 5, this needs further analysis, as they might make hypoxic pockets within the 3D PHA scaffold. This was consistent with the observation made by Chen et al., 2012. Do Amaral et al., 2011, made an observation that MCF7 cells formed unusual spheroids when cultured in 3D for longer periods of time [98]. This could explain the absence of grape like cell clusters on Day 5.

Live cell imaging of cells cultured on 3D PHA scaffolds provides a better analysis of the growth pattern and viability of hard and soft cancer cell lines. On Day 1, the MCF7 cells were distributed evenly throughout the 3D PHA scaffold, and no clumps were observed (Figure 5a,b). The cells spread throughout the centre and until the bottom of the 3D PHA scaffolds. While HCT116 cells grew in pockets rather than being dispersed throughout the 3D PHA scaffolds (Figure 6a–c). This could be attributed to the fact that HCT116 cells also grow in colonies in 2D cultures. On Day 5, MCF7 cells showed an even distribution throughout the 3D PHA scaffold, and there was no significant cell death and they appeared to be denser than Day 1 (Figure 5c,d). However, HCT116 cells appeared to form bigger clusters on Day 5 as compared to Day 1 (Figure 7d–f), and colonies started appearing in the pockets within the 3D PHA scaffolds with no significant cell death because the cells had sufficient time to adhere and adapt to the 3D PHA scaffold's environment. Similarly, on Day 5, MCF7 cells formed an evenly dispersed layer, which covered most of the 3D PHA scaffold. HCT116 continued to form larger colonies within the pockets of the 3D PHA scaffold, which needs further analysis as they might make hypoxic pockets within the 3D PHA scaffold, as mentioned above. The porosity of the 3D PHA scaffolds facilitated the penetration of the cells to the bottom of the scaffolds and resulted in a consistent distribution of cells throughout the 3D PHA scaffold.

In chitosan alginate (CA) scaffolds, both hepatocellular carcinoma as well as human glioblastoma U-87 MG and U-118 MG cell lines showed increase in the expression levels of genes involved in EMT and cancer stem cells [99,100]. The mixed hydrogel of chitosan and hyaluronic acid (CH) used for human non-small cell lung cancer cells 3D spheroid formation showed increase in the expression level of EMT marker, stemness or drug resistance compared with those of cells in the 2D culture system [101]. 3D human glioblastoma cancer stem cells cultured in CH scaffolds also enhanced the expression of stem cell markers and drug resistance [102]. Gene expression profiles in hard cancer disease models were studied in a time-dependent manner. MCF7 and MDA-MB-231 cells cultured on 3D PHA scaffolds were analysed for mRNA expression levels of several EMT marker genes. *Wnt-11*, *E-cadherin*, *Vim* and *Snail* genes were monitored over a period of 14 days after seeding the cells in 3D PHA scaffolds. Our results demonstrated that MDA-MB-231 cells, when grown within 3D PHA scaffolds, express higher levels of *Wnt-11* and mesenchymal markers such as *Vim* and *Snail* mRNAs. Wnt signalling regulates a variety of cellular processes, including differentiation, cellular proliferation and stem cell pluripotency [74,103]. It has been reported that triple-negative breast cancer, which is an aggressive subtype of breast cancer, and expresses high levels of Wnt-11, which is accepted as a cancer stem cell (CSC)-like marker [60,61]. Epithelial cancer cells undergoing EMT adopt a cancer stem cell-like phenotype and are uniquely capable of seeding new tumours [104]. Moreover, it has been reported that TGFβ is linked to both EMT and Wnt-11. A recent study used TGF-β1 stimulation to investigate angiogenesis [105]. Future studies are required to determine the role of TGF-β1 in the PHA-based models.

E-cadherin is considered a pivotal marker in the EMT mechanism [106]; thus, we analysed the mRNA levels of *E-cadherin* in MCF7 and MDA-MB-231 cells as an "epithelial" marker at the molecular level. Both cell lines show increased *E cadherin* expression over a period of 14 days, which suggests gain of epithelial characteristics. This is inconsistent with the gain of EMT phenotypes as studies report *E-cadherin* to be linked to intercellular adhesion and to epithelial characteristics. However, most breast cancers are invasive ductal carcinoma and express *E-cadherin* in primary tumours and metastasis, which suggests that the PHA-based models resemble tumour expression profiles. We found increased *E-cadherin* mRNA levels in MCF-7 cells in a time-dependent manner, whereas MDA-MB-231 cells showed more 'mesenchymal' characteristics by increasing their *Snail* and *Vim* mRNA expressions. The complex genetic changes required to attain EMT-linked phenotypic changes are mediated by specific transcription factors including Snail (also known as Snail1). In 3D PHA scaffolds, the expression level of *E-cadherin* was significantly higher in MCF7 cells than MDA-MB-231 cells. Unlike MDA-MB-231 cells, MCF7 cell lines grown in 3D PHA scaffolds for 14 days show no significant change in *Vim* and *Snail*. This observation explains the significantly higher expression level of *E-cadherin* in MCF7 cells than MDA-MB-231 cells. The Snail acts as a transcriptional repressor of *E-cadherin* to regulate epithelial-mesenchymal transitions [107]. Snail 1 can be considered as one of the master EMT regulators and modulates cancer cell survival, cell cycle regulation, apoptosis evasion, cell adhesion, neuroendocrine differentiation and chemoresistance [106,108]. Snail also modulates the expression of a large number of genes directly or indirectly, associated with cancer invasion and metastasis to promote EMT in vitro [66,109]. Studies in tumour samples report that *Vim* is a downstream gene of Snail and is expressed by Snail to attain the EMT phenotype [107]. This explains why both breast cancer cell lines show a consistent change in their *Vim* and *Snail* mRNA levels, with increase in the transcription factor *Snail* gene expression and its downstream gene *vim* in MDA-MB-231, while in MCF7 cells, no change was observed in their expression. Studies of EMT markers in "basal-like" breast tumours reported that EMT markers (vimentin), as well as cadherin switching (reduced expression of *E-cadherin*), were significantly more frequent [110]. The epithelial components of breast carcinomas express *E- cadherin*, a proportion of them also show vimentin expression, while the mesenchymal components of breast carcinomas show vim expression [110].

These observations point out the resemblance in the EMT markers in breast cancer cell lines cultured in 3D PHA scaffolds. The EMT gene expression profiles of MCF7 and MDA-MB-231 cells cultured in 3D PHA scaffolds were similar to those seen in tumour samples, which is representative of better cancer models that can be further analysed for their tumour-like characteristics. The PHA-based cancer models result in better stemness characteristics and molecular marker resemblance to tumours. Our data confirmed that PHA-based 3D scaffolds allowed for breast cancer cells to grow in 3D, and the EMT/Wnt11 gene expressions increased significantly as shown by others in Matrigel-based 3D cultures. In summary, all the results are consistent with the fact that PHA-based 3D disease models are comparable to other biomaterials to study gene expression profiles as well as cellular behaviour as compared to 2D cell culture. In addition, in future, the 3D model developed in this work could be used to enhance the percentage of tumour development in vivo, especially with cells from patients, a truly bespoke model.

5. Conclusions

This is the first ever study investigating the suitability of 3D PHA scaffolds for the development of cancer models using breast cancer and colon cancer cell lines. We produced two types of PHAs, P(3HB) and P(3HO-*co*-3HD), using *Bacillus subtilis* and *Pseudomonas mendocina*, respectively, and with good yields. These were blended to form high-quality 3D scaffolds with controlled interconnected porosity. Three types of cancer cell lines, MCF7 cells, MDA-MB-231 (breast cancer cells) and HCT116 (colon cancer cells), were successfully grown within the PHA-based 3D scaffolds, exhibiting excellent proliferation and cellular morphology mimicking that of native cancer cells. These results confirmed that the PHA-based 3D scaffolds provided a suitable 3D environment for the cancer cells and would work very well to create functional 3D disease models to be used for in-depth understanding of the process of cancer development and for novel drug testing. Both Wnt-11 and EMT have been linked to the stem cell phenotype. These results could contribute to an understanding the cellular behaviour of cancer cells and help in finding better targeted therapies. Therefore, the role of increased EMT markers and Wnt-11 in cancer cells grown within the 3D disease models needs to be further explored. These models will be helpful in analysing the gene expression, cellular signalling pathways, angiogenesis and chemotherapy response more accurately than 2D and other currently available 3D models.

Author Contributions: The idea was conceived by U.P. and I.R. The experiments were conducted by A.T. and P.B. Polymer Production, GC-MC, NMR and scaffold construction was done by P.B., SEM analysis and Live Cell imaging was done by A.T. P.U.-O. did qRT-PCR analysis. A.T., P.B. and P.U.-O. interpreted the data. A.T. wrote the preliminary manuscript, which was further edited and revised by all under the supervision of U.P. and I.R. U.P. and I.R. acquired funding and provided mentoring. All authors have read and agreed to the published version of the manuscript.

Funding: This work is supported by UGC-UKIERI grant (UK-India Education and Research Initiative); Reference no. India-184-6/2017(IC) and UK-IND/CONT/G/20-21/01; awarded to U. Pati and I. Roy.

Institutional Review Board Statement: Not applicable.

Informed Consent Statement: Not applicable.

Data Availability Statement: All the data can be accessed in the main text.

Acknowledgments: The authors thank Prabhat Kumar, AIRF-JNU for Live cell imaging and SEM facility, University College London for NMR, SEM and Confocal imaging facility and UGC-UKIERI for funding.

Conflicts of Interest: The authors declare no conflict of interest.

References

1. Sung, H.; Ferlay, J.; Siegel, R.L.; Laversanne, M.; Soerjomataram, I.; Jemal, A.; Bray, F. Global Cancer Statistics 2020: GLOBOCAN Estimates of Incidence and Mortality Worldwide for 36 Cancers in 185 Countries. *CA Cancer J. Clin.* **2021**, *71*, 209–249. [CrossRef] [PubMed]
2. Liverani, C.; De Vita, A.; Minardi, S.; Kang, Y.; Mercatali, L.; Amadori, D.; Bongiovanni, A.; La Manna, F.; Ibrahim, T.; Tasciotti, E. A biomimetic 3D model of hypoxia-driven cancer progression. *Sci. Rep.* **2019**, *9*, 12263. [CrossRef] [PubMed]
3. Asghar, W.; El Assal, R.; Shafiee, H.; Pitteri, S.; Paulmurugan, R.; Demirci, U. Engineering cancer microenvironments for in vitro 3-D tumor models. *Mater. Today* **2015**, *18*, 539–553. [CrossRef]
4. Fletcher, D.A.; Mullins, R.D. Cell mechanics and the cytoskeleton. *Nature* **2010**, *463*, 485–492. [CrossRef] [PubMed]
5. Zhang, S. Beyond the Petri dish. *Nat. Biotechnol.* **2004**, *22*, 151–152. [CrossRef]
6. Benton, G.; George, J.; Kleinman, H.K.; Arnaoutova, I.P. Advancing science and technology via 3D culture on basement membrane matrix. *J. Cell. Physiol.* **2009**, *221*, 18–25. [CrossRef]
7. Yamada, K.M.; Cukierman, E. Modeling Tissue Morphogenesis and Cancer in 3D. *Cell* **2007**, *130*, 601–610. [CrossRef] [PubMed]
8. Lutolf, M.P.; Gilbert, P.M.; Blau, H.M. Designing materials to direct stem-cell fate. *Nature* **2009**, *462*, 433–441. [CrossRef]
9. Hutmacher, D.W.; Loessner, D.; Rizzi, S.; Kaplan, D.L.; Mooney, D.J.; Clements, J.A. Can tissue engineering concepts advance tumor biology research? *Trends Biotechnol.* **2010**, *28*, 125–133. [CrossRef]
10. Duval, K.; Grover, H.; Han, L.H.; Mou, Y.; Pegoraro, A.F.; Fredberg, J.; Chen, Z. Modeling physiological events in 2D vs. 3D cell culture. *Physiology* **2017**, *32*, 266–277. [CrossRef]

11. Kapałczyńska, M.; Kolenda, T.; Przybyła, W.; Zajączkowska, M.; Teresiak, A.; Filas, V.; Ibbs, M.; Bliźniak, R.; Łuczewski, Ł.; Lamperska, K. 2D and 3D cell cultures—A comparison of different types of cancer cell cultures. *Arch. Med. Sci.* **2018**, *14*, 910–919. [CrossRef]
12. Siolas, D.; Hannon, G.J. Patient-derived tumor xenografts: Transforming clinical samples into mouse models. *Cancer Res.* **2013**, *73*, 5315–5319. [CrossRef]
13. Wilding, J.L.; Bodmer, W.F. Cancer cell lines for drug discovery and development. *Cancer Res.* **2014**, *74*, 2377–2384. [CrossRef]
14. Han, J.; Chang, H.; Giricz, O.; Lee, G.Y.; Baehner, F.L.; Gray, J.W.; Bissell, M.J.; Kenny, P.A.; Parvin, B. Molecular predictors of 3D morphogenesis by breast cancer cell lines in 3D culture. *PLoS Comput. Biol.* **2010**, *6*, e1000684. [CrossRef]
15. Fischbach, C.; Chen, R.; Matsumoto, T.; Schmelzle, T.; Brugge, J.S.; Polverini, P.J.; Mooney, D.J. Engineering tumors with 3D scaffolds. *Nat. Methods* **2007**, *4*, 855–860. [CrossRef]
16. Fischbach, C.; Hyun, J.K.; Hsiong, S.X.; Evangelista, M.B.; Yuen, W.; Mooney, D.J. Cancer cell angiogenic capability is regulated by 3D culture and integrin engagement. *Proc. Natl. Acad. Sci. USA* **2009**, *106*, 399–404. [CrossRef]
17. Kenny, P.A.; Bissell, M.J. Tumor reversion: Correction of malignant behavior by microenvironmental cues. *Int. J. Cancer* **2003**, *107*, 688–695. [CrossRef]
18. Kenny, H.A.; Krausz, T.; Yamada, S.D.; Lengyel, E. Use of a novel 3D culture model to elucidate the role of mesothelial cells, fibroblasts and extra-cellular matrices on adhesion and invasion of ovarian cancer cells to the omentum. *Int. J. Cancer* **2007**, *121*, 1463–1472. [CrossRef] [PubMed]
19. Liang, Y.; Jeong, J.; DeVolder, R.J.; Cha, C.; Wang, F.; Tong, Y.W.; Kong, H. A cell-instructive hydrogel to regulate malignancy of 3D tumor spheroids with matrix rigidity. *Biomaterials* **2011**, *32*, 9308–9315. [CrossRef] [PubMed]
20. Loessner, D.; Stok, K.S.; Lutolf, M.P.; Hutmacher, D.W.; Clements, J.A.; Rizzi, S.C. Bioengineered 3D platform to explore cell-ECM interactions and drug resistance of epithelial ovarian cancer cells. *Biomaterials* **2010**, *31*, 8494–8506. [CrossRef]
21. Eke, I.; Cordes, N. Radiobiology goes 3D: How ECM and cell morphology impact on cell survival after irradiation. *Radiother. Oncol.* **2011**, *99*, 271–278. [CrossRef] [PubMed]
22. Weiss, M.S.; Bernabé, B.P.; Shikanov, A.; Bluver, D.A.; Mui, M.D.; Shin, S.; Broadbelt, L.J.; Shea, L.D. The impact of adhesion peptides within hydrogels on the phenotype and signaling of normal and cancerous mammary epithelial cells. *Biomaterials* **2012**, *33*, 3548–3559. [CrossRef] [PubMed]
23. Scanlon, S.E.; Glazer, P.M. Multifaceted control of DNA repair pathways by the hypoxic tumor microenvironment. *DNA Repair* **2015**, *32*, 180–189. [CrossRef] [PubMed]
24. Hammond, E.M.; Asselin, M.C.; Forster, D.; O'Connor, J.P.B.; Senra, J.M.; Williams, K.J. The Meaning, Measurement and Modification of Hypoxia in the Laboratory and the Clinic. *Clin. Oncol.* **2014**, *26*, 277–288. [CrossRef]
25. Wilson, W.R.; Hay, M.P. Targeting hypoxia in cancer therapy. *Nat. Rev. Cancer* **2011**, *11*, 393–410. [CrossRef]
26. Bristow, R.G. Hypoxia, DNA Repair, and Genetic Instability. *AACR Educ. Book* **2008**, *2008*, 287–291. [CrossRef]
27. Hanahan, D.; Weinberg, R.A. Hallmarks of cancer: The next generation. *Cell* **2011**, *144*, 646–674. [CrossRef]
28. Wang, M.; Zhao, J.; Zhang, L.; Wei, F.; Lian, Y.; Wu, Y.; Gong, Z.; Zhang, S.; Zhou, J.; Cao, K.; et al. Role of tumor microenvironment in tumorigenesis. *J. Cancer* **2017**, *8*, 761–773. [CrossRef]
29. Höckel, M.; Vaupel, P. Tumor hypoxia: Definitions and current clinical, biologic, and molecular aspects. *J. Natl. Cancer Inst.* **2001**, *93*, 266–276. [CrossRef]
30. Ricci, C.; Moroni, L.; Danti, S. Cancer tissue engineering new perspectives in understanding the biology of solid tumours a critical review. *OA Tissue Eng.* **2013**, *1*, 4. [CrossRef]
31. Hirt, C.; Papadimitropoulos, A.; Muraro, M.G.; Mele, V.; Panopoulos, E.; Cremonesi, E.; Ivanek, R.; Schultz-Thater, E.; Droeser, R.A.; Mengus, C.; et al. Bioreactor-engineered cancer tissue-like structures mimic phenotypes, gene expression profiles and drug resistance patterns observed "in vivo". *Biomaterials* **2015**, *62*, 138–146. [CrossRef]
32. Riedl, A.; Schlederer, M.; Pudelko, K.; Stadler, M.; Walter, S.; Unterleuthner, D.; Unger, C.; Kramer, N.; Hengstschläger, M.; Kenner, L.; et al. Comparison of cancer cells in 2D vs 3D culture reveals differences in AKT-mTOR-S6K signaling and drug responses. *J. Cell Sci.* **2017**, *130*, 203–218.
33. Cannon, T.M.; Shah, A.T.; Skala, M.C. Autofluorescence imaging captures heterogeneous drug response differences between 2D and 3D breast cancer cultures. *Biomed. Opt. Express* **2017**, *8*, 1911–1925. [CrossRef]
34. Hashemzadeh, A.H. Comparison between Three-Dimensional Spheroid and Two-Dimensional MonolayerA549 Lung Cancer and PC9 Normal Cell Lines under Treatment of Silver Nanoparticles. *J. Biotechnol.* **2019**, *10*, 573–580.
35. Saxton, R.A.; Sabatini, D.M. mTOR Signaling in Growth, Metabolism, and Disease. *Cell* **2017**, *168*, 960–976. [CrossRef]
36. Kenny, P.A.; Lee, G.Y.; Myers, C.A.; Neve, R.M.; Semeiks, J.R.; Spellman, P.T.; Lorenz, K.; Lee, E.H.; Barcellos-Hoff, M.H.; Petersen, O.W.; et al. The morphologies of breast cancer cell lines in three-dimensional assays correlate with their profiles of gene expression. *Mol. Oncol.* **2007**, *1*, 84–96. [CrossRef]
37. Rahmanzadeh, R.; Rai, P.; Celli, J.P.; Rizvi, I.; Baron-Lühr, B.; Gerdes, J.; Hasan, T. Ki-67 as a molecular target for therapy in an in vitro three-dimensional model for ovarian cancer. *Cancer Res.* **2010**, *70*, 9234–9242. [CrossRef]
38. Rizvi, I.; Celli, J.P.; Evans, C.L.; Abu-Yousif, A.O.; Muzikansky, A.; Pogue, B.W.; Finkelstein, D.; Hasan, T. Synergistic enhancement of carboplatin efficacy with photodynamic therapy in a three-dimensional model for micrometastatic ovarian cancer. *Cancer Res.* **2010**, *70*, 9319–9328. [CrossRef]

39. Lv, D.; Hu, Z.; Lu, L.; Lu, H.; Xu, X. Three-dimensional cell culture: A powerful tool in tumor research and drug discovery. *Oncol. Lett.* **2017**, *14*, 6999–7010. [CrossRef]
40. Nii, T.; Makino, K.; Tabata, Y. Three-dimensional culture system of cancer cells combined with biomaterials for drug screening. *Cancers* **2020**, *12*, 2754. [CrossRef]
41. Tamayol, A.; Najafabadi, A.H.; Aliakbarian, B.; Arab-Tehrany, E.; Akbari, M.; Annabi, N.; Juncker, D.; Khademhosseini, A. Hydrogel Templates for Rapid Manufacturing of Bioactive Fibers and 3D Constructs. *Adv. Healthc. Mater.* **2015**, *4*, 2146–2153. [CrossRef]
42. Liu, C.; Liu, Y.; Xu, X.X.; Wu, H.; Xie, H.G.; Chen, L.; Lu, T.; Yang, L.; Guo, X.; Sun, G.W.; et al. Potential effect of matrix stiffness on the enrichment of tumor initiating cells under three-dimensional culture conditions. *Exp. Cell Res.* **2015**, *330*, 123–134. [CrossRef] [PubMed]
43. DelNero, P.; Lane, M.; Verbridge, S.S.; Kwee, B.; Kermani, P.; Hempstead, B.; Stroock, A.; Fischbach, C. 3D culture broadly regulates tumor cell hypoxia response and angiogenesis via pro-inflammatory pathways. *Biomaterials* **2015**, *55*, 110–118. [CrossRef] [PubMed]
44. Hayashi, K.; Tabata, Y. Preparation of stem cell aggregates with gelatin microspheres to enhance biological functions. *Acta Biomater.* **2011**, *7*, 2797–2803. [CrossRef] [PubMed]
45. Nii, T.; Makino, K.; Tabata, Y. A Cancer Invasion Model Combined with Cancer-Associated Fibroblasts Aggregates Incorporating Gelatin Hydrogel Microspheres Containing a p53 Inhibitor. *Tissue Eng. Part C Methods* **2019**, *25*, 711–720. [CrossRef] [PubMed]
46. Dhiman, H.K.; Ray, A.R.; Panda, A.K. Characterization and evaluation of chitosan matrix for in vitro growth of MCF-7 breast cancer cell lines. *Biomaterials* **2004**, *25*, 5147–5154. [CrossRef] [PubMed]
47. Li, Z.; Ramay, H.R.; Hauch, K.D.; Xiao, D.; Zhang, M. Chitosan-alginate hybrid scaffolds for bone tissue engineering. *Biomaterials* **2005**, *26*, 3919–3928. [CrossRef] [PubMed]
48. Huang, G.S.; Dai, L.G.; Yen, B.L.; Hsu, S.H. Spheroid formation of mesenchymal stem cells on chitosan and chitosan-hyaluronan membranes. *Biomaterials* **2011**, *32*, 6929–6945. [CrossRef]
49. Stylianopoulos, T. The Solid Mechanics of Cancer and Strategies for Improved Therapy. *J. Biomech. Eng.* **2017**, *139*, 21004. [CrossRef]
50. Wei, S.C.; Fattet, L.; Tsai, J.H.; Guo, Y.; Pai, V.H.; Majeski, H.E.; Chen, A.C.; Sah, R.L.; Taylor, S.S.; Engler, A.J.; et al. Matrix stiffness drives epithelial-mesenchymal transition and tumour metastasis through a TWIST1-G3BP2 mechanotransduction pathway. *Nat. Cell Biol.* **2015**, *17*, 678–688. [CrossRef]
51. Samani, A.; Zubovits, J.; Plewes, D. Elastic moduli of normal and pathological human breast tissues: An inversion-technique-based investigation of 169 samples. *Phys. Med. Biol.* **2007**, *52*, 1565–1576. [CrossRef]
52. Kawano, S.; Kojima, M.; Higuchi, Y.; Sugimoto, M.; Ikeda, K.; Sakuyama, N.; Takahashi, S.; Hayashi, R.; Ochiai, A.; Saito, N. Assessment of elasticity of colorectal cancer tissue, clinical utility, pathological and phenotypical relevance. *Cancer Sci.* **2015**, *106*, 1232–1239. [CrossRef]
53. Muneer, F.; Rasul, I.; Azeem, F.; Siddique, M.H.; Zubair, M.; Nadeem, H. Microbial Polyhydroxyalkanoates (PHAs): Efficient Replacement of Synthetic Polymers. *J. Polym. Environ.* **2020**, *28*, 2301–2323. [CrossRef]
54. Shishatskaya, E.I.; Volova, T.G. A comparative investigation of biodegradable polyhydroxyalkanoate films as matrices for in vitro cell cultures. *J. Mater. Sci. Mater. Med.* **2004**, *15*, 915–923. [CrossRef]
55. Wu, Q.; Wang, Y.; Chen, G.Q. Medical application of microbial biopolyesters polyhydroxyalkanoates. *Artif. Cells Blood Substit. Biotechnol.* **2009**, *37*, 1–12. [CrossRef]
56. Misra, S.K.; Valappil, S.P.; Roy, I.; Boccaccini, A.R. Polyhydroxyalkanoate (PHA)/inorganic phase composites for tissue engineering applications. *Biomacromolecules* **2006**, *7*, 2249–2258. [CrossRef]
57. Basnett, P.; Marcello, E.; Lukasiewicz, B.; Nigmatullin, R.; Paxinou, A.; Ahmad, M.H.; Gurumayum, B.; Roy, I. Antimicrobial materials with lime oil and a poly(3-hydroxyalkanoate) produced via valorisation of sugar cane molasses. *J. Funct. Biomater.* **2020**, *11*, 24. [CrossRef]
58. Deng, Y.; Lin, X.S.; Zheng, Z.; Deng, J.G.; Chen, J.C.; Ma, H.; Chen, G.Q. Poly (hydroxybutyrate-co-hydroxyhexanoate) promoted production of extracellular matrix of articular cartilage chondrocytes in vitro. *Biomaterials* **2003**, *24*, 4273–4281. [CrossRef]
59. Wang, Y.; Bian, Y.Z.; Wu, Q.; Chen, G.Q. Evaluation of three-dimensional scaffolds prepared from poly (3-hydroxybutyrate-co-3-hydroxyhexanoate) for growth of allogeneic chondrocytes for cartilage repair in rabbits. *Biomaterials* **2008**, *29*, 2858–2868. [CrossRef]
60. Pohl, S.G.; Brook, N.; Agostino, M.; Arfuso, F.; Kumar, A.P.; Dharmarajan, A. Wnt signaling in triple-negative breast cancer. *Oncogenesis* **2017**, *6*, e310. [CrossRef]
61. Dwyer, M.A.; Joseph, J.D.; Wade, H.E.; Eaton, M.L.; Kunder, R.S.; Kazmin, D.; Chang, C.Y.; McDonnell, D.P. WNT11 expression is induced by estrogen-related receptor α and β-catenin and acts in an autocrine manner to increase cancer cell migration. *Cancer Res.* **2010**, *70*, 9298–9308. [CrossRef] [PubMed]
62. Murillo-Garzón, V.; Gorroño-Etxebarria, I.; Åkerfelt, M.; Puustinen, M.C.; Sistonen, L.; Nees, M.; Carton, J.; Waxman, J.; Kypta, R.M. Frizzled-8 integrates Wnt-11 and transforming growth factor-β signaling in prostate cancer. *Nat. Commun.* **2018**, *9*, 554–562. [CrossRef] [PubMed]
63. Chen, L.; Xiao, Z.; Meng, Y.; Zhao, Y.; Han, J.; Su, G.; Chen, B.; Dai, J. The enhancement of cancer stem cell properties of MCF-7 cells in 3D collagen scaffolds for modeling of cancer and anti-cancer drugs. *Biomaterials* **2012**, *33*, 1437–1444. [CrossRef] [PubMed]
64. Thiery, J.P.; Lim, C.T. Tumor dissemination: An EMT affair. *Cancer Cell* **2013**, *23*, 272–273. [CrossRef]

65. Mani, S.A.; Guo, W.; Liao, M.J.; Eaton, E.N.; Ayyanan, A.; Zhou, A.Y.; Brooks, M.; Reinhard, F.; Zhang, C.C.; Shipitsin, M.; et al. The Epithelial-Mesenchymal Transition Generates Cells with Properties of Stem Cells. *Cell* **2008**, *133*, 704–715. [CrossRef]
66. Moreno-Bueno, G.; Cubillo, E.; Sarrió, D.; Peinado, H.; Rodríguez-Pinilla, S.M.; Villa, S.; Bolós, V.; Jordá, M.; Fabra, A.; Portillo, F.; et al. Genetic profiling of epithelial cells expressing E-cadherin repressors reveals a distinct role for snail, Slug, and E47 Factors in epithelial- mesenchymal transition. *Cancer Res.* **2006**, *66*, 9543–9556. [CrossRef]
67. Wachsmannova, L.; Stevurkova, V.; Ciernikova, S. Changes in SNAI1 and VIM gene expression in Caco2 cells after co-cultivation with bacteria from colorectal cancer biopsies. *Neoplasma* **2019**, *66*, 271–275. [CrossRef]
68. Berx, G.; Cleton-Jansen, A.M.; Nollet, F.; De Leeuw, W.J.F.; Van De Vijver, M.J.; Cornelisse, C.; Van Roy, F. E-cadherin is a tumour/invasion suppressor gene mutated in human lobular breast cancers. *EMBO J.* **1995**, *14*, 6107–6115. [CrossRef]
69. Prat, A.; Perou, C.M. Deconstructing the molecular portraits of breast cancer. *Mol. Oncol.* **2011**, *5*, 5–23. [CrossRef]
70. Basnett, P.; Lukasiewicz, B.; Marcello, E.; Gura, H.K.; Knowles, J.C.; Roy, I. Production of a novel medium chain length poly (3-hydroxyalkanoate) using unprocessed biodiesel waste and its evaluation as a tissue engineering scaffold. *Microb. Biotechnol.* **2017**, *10*, 1384–1399. [CrossRef]
71. Basnett, P.; Marcello, E.; Lukasiewicz, B.; Panchal, B.; Nigmatullin, R.; Knowles, J.C.; Roy, I. Biosynthesis and characterization of a novel, biocompatible medium chain length polyhydroxyalkanoate by *Pseudomonas mendocina* CH50 using coconut oil as the carbon source. *J. Mater. Sci. Mater. Med.* **2018**, *29*, 179. [CrossRef]
72. Taylor, C.S.; Illangakoon, U.; Dawson, J.I.; Kanczler, J.M.; Behbehani, M.; Humphrey, E.; Majid, Q.; Lukasiewicz, B.; Nigmatullin, R.; Heseltine, P.; et al. Harnessing Polyhydroxyalkanoates and Pressurized Gyration for Hard and Soft Tissue Engineering. *ACS Appl. Mater. Interfaces* **2021**, *13*, 32624–32639.
73. Theodoridis, K.; Aggelidou, E.; Manthou, M.E.; Keklikoglou, K.; Tsimponis, A.; Demiri, E.; Bakopoulou, A.; Mihailidis, A.; Kritis, A. An effective device and method for enhanced cell growth in 3D scaffolds: Investigation of cell seeding and proliferation under static and dynamic conditions. *Mater. Sci. Eng. C* **2020**, *114*, 111060. [CrossRef]
74. Uysal-Onganer, P.; Kawano, Y.; Caro, M.; Walker, M.M.; Diez, S.; Darrington, R.S.; Waxman, J.; Kypta, R.M. Wnt-11 promotes neuroendocrine-like differentiation, survival and migration of prostate cancer cells. *Mol. Cancer* **2010**, *9*, 55. [CrossRef]
75. Nozato, M.; Kaneko, S.; Nakagawara, A.; Komuro, H. Epithelial-mesenchymal transition-related gene expression as a new prognostic marker for neuroblastoma. *Int. J. Oncol.* **2013**, *42*, 134–140. [CrossRef]
76. Gu, C.S.; Liu, L.Q.; Xu, C.; Zhao, Y.H.; Zhu, X.D.; Huang, S.Z. Reference gene selection for quantitative real-time RT-PCR normalization in Iris. lactea var. chinensis roots under cadmium, lead, and salt stress conditions. *Sci. World J.* **2014**, *2014*, 532713.
77. Arisan, E.D.; Rencuzogullari, O.; Freitas, I.L.; Radzali, S.; Keskin, B.; Kothari, A.; Warford, A.; Uysal-Onganer, P. Upregulated wnt-11 and mir-21 expression trigger epithelial mesenchymal transition in aggressive prostate cancer cells. *Biology* **2020**, *9*, 52. [CrossRef]
78. Xu, W.; Mezencev, R.; Kim, B.; Wang, L.; McDonald, J.; Sulchek, T. Cell Stiffness Is a Biomarker of the Metastatic Potential of Ovarian Cancer Cells. *PLoS ONE* **2012**, *7*, e46609. [CrossRef]
79. Kraning-Rush, C.M.; Califano, J.P.; Reinhart-King, C.A. Cellular traction stresses increase with increasing metastatic potential. *PLoS ONE* **2012**, *7*, e32572. [CrossRef]
80. Yousafzai, M.S.; Coceano, G.; Bonin, S.; Niemela, J.; Scoles, G.; Cojoc, D. Investigating the effect of cell substrate on cancer cell stiffness by optical tweezers. *J. Biomech.* **2017**, *60*, 266–269.
81. Jaganathan, H.; Gage, J.; Leonard, F.; Srinivasan, S.; Souza, G.R.; Dave, B.; Godin, B. Three-dimensional in vitro co-culture model of breast tumor using magnetic levitation. *Sci. Rep.* **2014**, *4*, 6468. [CrossRef]
82. Tryndyak, V.P.; Beland, F.A.; Pogribny, I.P. E-cadherin transcriptional down-regulation by epigenetic and microRNA-200 family alterations is related to mesenchymal and drug-resistant phenotypes in human breast cancer cells. *Int. J. Cancer* **2010**, *126*, 2575–2583. [CrossRef]
83. Narod, S.A.; Iqbal, J.; Jakubowska, A.; Huzarski, T.; Sun, P.; Cybulski, C.; Gronwald, J.; Byrski, T.; Lubinski, J. Are two-centimeter breast cancers large or small? *Curr. Oncol.* **2013**, *20*, 205–211. [CrossRef]
84. Iyer, J.G.; Storer, B.E.; Paulson, K.G.; Lemos, B.; Phillips, J.L.; Bichakjian, C.K.; Zeitouni, N.; Gershenwald, J.E.; Sondak, V.; Otley, C.C.; et al. Relationships between primary tumor size, number of involved nodes and survival among 8044 cases of Merkel cell carcinoma JG. *J. Am. Acad. Dermatol.* **2014**, *70*, 637–643. [CrossRef]
85. Cavo, M.; Fato, M.; Peñuela, L.; Beltrame, F.; Raiteri, R.; Scaglione, S. Microenvironment complexity and matrix stiffness regulate breast cancer cell activity in a 3D in vitro model. *Sci. Rep.* **2016**, *6*, 35367. [CrossRef]
86. Dwivedi, R.; Pandey, R.; Kumar, S.; Mehrotra, D. Poly hydroxyalkanoates (PHA): Role in bone scaffolds. *J. Oral Biol. Craniofacial Res.* **2020**, *10*, 389–392. [CrossRef]
87. Madl, C.M.; Katz, L.M.; Heilshorn, S.C. Tuning Bulk Hydrogel Degradation by Simultaneous Control of Proteolytic Cleavage Kinetics and Hydrogel Network Architecture. *ACS Macro Lett.* **2018**, *7*, 1302–1307. [CrossRef]
88. Constantinides, C.; Basnett, P.; Lukasiewicz, B.; Carnicer, R.; Swider, E.; Majid, Q.A.; Srinivas, M.; Carr, C.A.; Roy, I. In Vivo Tracking and $^{1}H/^{19}F$ Magnetic Resonance Imaging of Biodegradable Polyhydroxyalkanoate/Polycaprolactone Blend Scaffolds Seeded with Labeled Cardiac Stem Cells. *ACS Appl. Mater. Interfaces* **2018**, *10*, 25056–25068. [CrossRef]
89. Angelini, S.; National, I.; Cerruti, P.; National, I.; Santagata, G.; National, I.; Immirzi, B.; National, I. *Microbial Factories*; Springer: Berlin/Heidelberg, Germany, 2016.

90. Chen, Y.J.; Huang, Y.C.; Lee, C.Y. Production and characterization of medium-chain-length polyhydroxyalkanoates by *Pseudomonas mosselii* TO7. *J. Biosci. Bioeng.* **2014**, *118*, 145–152. [CrossRef]
91. Singh, M.; Patel, S.K.S.; Kalia, V.C. *Bacillus subtilis* as potential producer for polyhydroxyalkanoates. *Microb. Cell Factories* **2009**, *9*, 38. [CrossRef]
92. Sola, A.; Bertacchini, J.; D'Avella, D.; Anselmi, L.; Maraldi, T.; Marmiroli, S.; Messori, M. Development of solvent-casting particulate leaching (SCPL) polymer scaffolds as improved three-dimensional supports to mimic the bone marrow niche. *Mater. Sci. Eng. C* **2019**, *96*, 153–165. [CrossRef] [PubMed]
93. Choi, D.J.; Park, S.J.; Gu, B.K.; Kim, Y.J.; Chung, S.; Kim, C.H. Effect of the pore size in a 3D bioprinted gelatin scaffold on fibroblast proliferation. *J. Ind. Eng. Chem.* **2018**, *67*, 388–395. [CrossRef]
94. Loh, Q.L.; Choong, C. Three-dimensional scaffolds for tissue engineering applications: Role of porosity and pore size. *Tissue Eng. Part B Rev.* **2013**, *19*, 485–502. [CrossRef] [PubMed]
95. Gurski, L.A.; Petrelli, N.J.; Jia, X.; Farach-Carson, M.C. 3D Matrices for Anti-Cancer Drug Testing and Development. *Oncol. Issues* **2010**, *25*, 20–25. [CrossRef]
96. Florczyk, S.J.; Kievit, F.M.; Wang, K.; Erickson, A.E.; Ellenbogen, R.G.; Zhang, M. 3D porous chitosan-alginate scaffolds promote proliferation and enrichment of cancer stem-like cells. *J. Mater. Chem. B* **2016**, *4*, 6326–6334. [CrossRef] [PubMed]
97. Ivers, L.P.; Cummings, B.; Owolabi, F.; Welzel, K.; Klinger, R.; Saitoh, S.; O'Connor, D.; Fujita, Y.; Scholz, D.; Itasaki, N. Dynamic and influential interaction of cancer cells with normal epithelial cells in 3D culture. *Cancer Cell Int.* **2014**, *14*, 38. [CrossRef]
98. Do Amaral, J.B.; Rezende-Teixeira, P.; Freitas, V.M.; MacHado-Santelli, G.M. MCF-7 cells as a three-dimensional model for the study of human breast cancer. *Tissue Eng. Part C Methods* **2011**, *17*, 1097–1107. [CrossRef]
99. Kievit, F.M.; Florczyk, S.J.; Leung, M.C.; Veiseh, O.; Park, J.O.; Disis, M.L.; Zhang, M. Chitosan-alginate 3D scaffolds as a mimic of the glioma tumor microenvironment. *Biomaterials* **2010**, *31*, 5903–5910.
100. Pauff, S.M.; Miller, S.C. Proliferation and enrichment of CD133 + glioblastoma cancer stem cells on 3D chitosan-alginate scaffolds. *Bone* **2012**, *78*, 711–716.
101. Huang, Y.J.; Hsu, S.H. Acquisition of epithelial-mesenchymal transition and cancer stem-like phenotypes within chitosan-hyaluronan membrane-derived 3D tumor spheroids. *Biomaterials* **2014**, *35*, 10070–10079. [CrossRef]
102. Wang, K.; Kievit, F.M.; Erickson, A.E.; Silber, J.R.; Ellenbogen, R.G.; Zhang, M. Culture on 3D Chitosan-Hyaluronic Acid Scaffolds Enhances Stem Cell Marker Expression and Drug Resistance in Human Glioblastoma Cancer Stem Cells. *Adv. Healthc. Mater.* **2016**, *5*, 3173–3181. [CrossRef]
103. Arisan, E.D.; Rencuzogullari, O.; Cieza-borrella, C.; Miralles Arenas, F.; Dwek, M.; Lange, S.; Uysal-onganer, P. Mir-21 is required for the epithelial–mesenchymal transition in mda-mb-231 breast cancer cells. *Int. J. Mol. Sci.* **2021**, *22*, 1557. [CrossRef]
104. Scheel, C.; Weinberg, R.A. Cancer stem cells and epithelial-mesenchymal transition: Concepts and molecular links. *Semin. Cancer Biol.* **2012**, *22*, 396–403. [CrossRef]
105. Liu, H.; Wang, H.; Chen, D.; Gu, C.; Huang, J.; Mi, K. Endoplasmic reticulum stress inhibits 3D Matrigel-induced vasculogenic mimicry of breast cancer cells via TGF-β1/Smad2/3 and β-catenin signaling. *FEBS Open Bio* **2021**, *11*, 2607–2618. [CrossRef]
106. Peinado, H.; Olmeda, D.; Cano, A. Snail, ZEB and bHLH factors in tumour progression: An alliance against the epithelial phenotype? *Nat. Rev. Cancer* **2007**, *7*, 415–428. [CrossRef]
107. Cano, A.; Pérez-Moreno, M.A.; Rodrigo, I.; Locascio, A.; Blanco, M.J.; Del Barrio, M.G.; Portillo, F.; Nieto, M.A. The transcription factor Snail controls epithelial-mesenchymal transitions by repressing E-cadherin expression. *Nat. Cell Biol.* **2000**, *2*, 76–83. [CrossRef]
108. Fabregat, I.; Malfettone, A.; Soukupova, J. New insights into the crossroads between EMT and stemness in the context of cancer. *J. Clin. Med.* **2016**, *5*, 37. [CrossRef]
109. De Craene, B.; Gilbert, B.; Stove, C.; Bruyneel, E.; Van Roy, F.; Berx, G. The transcription factor snail induces tumor cell invasion through modulation of the epithelial cell differentiation program. *Cancer Res.* **2005**, *65*, 6237–6244. [CrossRef]
110. Sarrió, D.; Rodriguez-Pinilla, S.M.; Hardisson, D.; Cano, A.; Moreno-Bueno, G.; Palacios, J. Epithelial-mesenchymal transition in breast cancer relates to the basal-like phenotype. *Cancer Res.* **2008**, *68*, 989–997. [CrossRef]

cancers

MDPI

Article

Brain and Breast Cancer Cells with PTEN Loss of Function Reveal Enhanced Durotaxis and RHOB Dependent Amoeboid Migration Utilizing 3D Scaffolds and Aligned Microfiber Tracts

Annalena Wieland [1], Pamela L. Strissel [1,2,3], Hannah Schorle [1], Ezgi Bakirci [4], Dieter Janzen [5], Matthias W. Beckmann [1], Markus Eckstein [2], Paul D. Dalton [4,6] and Reiner Strick [1,*]

1 Laboratory for Molecular Medicine, Comprehensive Cancer Center Erlangen-EMN (CCC ER-EMN), Department of Gynecology and Obstetrics, University Hospital Erlangen, Friedrich-Alexander University Erlangen-Nuernberg, Universitaetsstrasse 21-23, 91054 Erlangen, Germany; Annalena.Wieland@uk-erlangen.de (A.W.); Pamela.Strissel@uk-erlangen.de (P.L.S.); hannah.schorle@fau.de (H.S.); fk-direktion@uk-erlangen.de (M.W.B.)

2 Institute of Pathology, University Hospital Erlangen, Friedrich-Alexander-Universität Erlangen-Nürnberg, Krankenhausstrasse 8-10, 91054 Erlangen, Germany; Markus.Eckstein@uk-erlangen.de

3 Adjunct Affiliation with Department of Radiation Oncology, University of Maryland School of Medicine, Baltimore, MD 21201, USA

4 Department of Functional Materials in Medicine and Dentistry, Bavarian Polymer Institute, University Hospital Wuerzburg, Pleicherwall 2, 97070 Wuerzburg, Germany; ezgi.bakirci@fmz.uni-wuerzburg.de (E.B.); daltonlab@gmail.com (P.D.D.)

5 Institute for Clinical Neurobiology, University Hospital Wuerzburg, Versbacherstrasse 5, 97078 Wuerzburg, Germany; Janzen_D@ukw.de

6 Knight Campus for Accelerating Scientific Impact, University of Oregon, 1505 Franklin Boulevard, Eugene, OR 97403-6231, USA

* Correspondence: Reiner.Strick@uk-erlangen.de; Tel.: +49-9131-8536671

Citation: Wieland, A.; Strissel, P.L.; Schorle, H.; Bakirci, E.; Janzen, D.; Beckmann, M.W.; Eckstein, M.; Dalton, P.D.; Strick, R. Brain and Breast Cancer Cells with PTEN Loss of Function Reveal Enhanced Durotaxis and RHOB Dependent Amoeboid Migration Utilizing 3D Scaffolds and Aligned Microfiber Tracts. *Cancers* **2021**, *13*, 5144. https://doi.org/10.3390/cancers13205144

Academic Editors: Serena Danti, Nicola Contessi Negrini and Alessandro Franchi

Received: 27 August 2021
Accepted: 7 October 2021
Published: 14 October 2021

Simple Summary: Glioblastoma multiforme (GBM) and metastatic triple-negative breast cancer (TNBC) with *PTEN* mutations are associated with brain tumor spreading and poor patient outcomes. GBM, and possibly TNBC, migrate on axons and blood vessels to disseminate in the brain; however, the mechanism is unresolved. There is a need for new therapeutic targets to blunt brain tumor spreading. Using 3D aligned printed microfibers mimicking brain structures proved that RHOB, in addition to ROCK and PTEN signaling, were essential for GBM and TNBC 3D cell migration. GBM and TNBC cell lines with PTEN loss of function and high RHOB expression exhibited amoeboid morphology with increased durotaxis, binding and migration speed on 3D microfibers, in contrast to the PTEN wildtype. Depending on the *PTEN* genotype, RHO-ROCK-PTEN inhibitors or PTEN rescue significantly regulated these properties. Regarding GBM and brain metastasizing TNBC, we conclude that RHOB inhibitors could play a novel role for improved therapy response and patient outcome.

Abstract: Background: Glioblastoma multiforme (GBM) and metastatic triple-negative breast cancer (TNBC) with *PTEN* mutations often lead to brain dissemination with poor patient outcome, thus new therapeutic targets are needed. To understand signaling, controlling the dynamics and mechanics of brain tumor cell migration, we implemented GBM and TNBC cell lines and designed 3D aligned microfibers and scaffolds mimicking brain structures. Methods: 3D microfibers and scaffolds were printed using melt electrowriting. GBM and TNBC cell lines with opposing PTEN genotypes were analyzed with RHO-ROCK-PTEN inhibitors and PTEN rescue using live-cell imaging. RNA-sequencing and qPCR of tumor cells in 3D with microfibers were performed, while scanning electron microscopy and confocal microscopy addressed cell morphology. Results: In contrast to the PTEN wildtype, GBM and TNBC cells with PTEN loss of function yielded enhanced durotaxis, topotaxis, adhesion, amoeboid migration on 3D microfibers and significant high RHOB expression. Functional studies concerning RHOB-ROCK-PTEN signaling confirmed the essential role for the above cellular processes. Conclusions: This study demonstrates a significant role of the *PTEN* genotype and RHOB expression for durotaxis, adhesion and migration dependent on 3D. GBM and TNBC cells with PTEN loss of function have an affinity for stiff brain structures promoting metastasis. 3D microfibers represent an important tool to model brain metastasizing tumor cells, where RHO-inhibitors could play an essential role for improved therapy.

Keywords: 3D tumor model; 3D microfiber; amoeboid cell migration; brain cancer; breast cancer; PTEN; RHO; ROCK; durotaxis; topotaxis

1. Introduction

Cell migration and invasion are hallmarks of development and cancer [1]. For the ability of cells to migrate, dynamic and spatially regulated changes of the cytoskeleton, cell adhesion and cell interactions with the extracellular matrix (ECM) must occur [2]. Depending upon the cell type, two main migratory phenotypes can be differentiated: (1) single cells, which are amoeboid or mesenchymal, and (2) multicellular, or so-called collective cell migration [3]. Cell-cell as well as cell-ECM interactions are crucial for a cell to instantly respond to the environment and further propagate the signals that control cell shape and motility [4]. Specific tissue types harbor different ratios of ECM components, such as collagen, laminin, fibronectin and hyaluronic acid [5]. Especially in the brain, higher amounts of laminin and hyaluronic acid are found, where they contribute to normal brain function and tumor stiffness regulating tumor cell motility [6,7]. Although laminin is not a component of axons, it is embedded in an ECM enriched with laminin. Collagen is not an ECM component in normal brain but is found in the basement membrane of blood vessels. Primary GBM and human GBM xenografted mice showed increased levels of collagen in tumors from tissue microarrays using second-harmonic generation microscopy [8]. These authors found a significant association of disorganized collagen filaments in GBM with a poorer patient survival. Several different laminin proteins compose an inner blood vessel basement membrane and an outer one with capillaries. Interestingly, within the specialized perivascular spaces containing blood vessels surrounded by cerebral spinal fluid, laminin encases pericytes on capillaries and also locates at the outer surface [9]. Functionally, brain tumor cells sense ECM structures as migratory cues, where they become polarized by reorganizing the actin cytoskeleton to facilitate both a protrusive leading and a contractile trailing edge [10]. Cellular protrusions with organized actin filaments include lamellipodia, filopodia, invadopodia and blebs, which are regulated by various cytoskeletal and signaling pathways [11].

Glioblastoma multiforme (GBM) arise from glial cells of the central nervous system and represent the most aggressive type of human brain tumors [12–14]. Due to the diffuse infiltration of GBM throughout the brain, the inability to achieve complete surgical tumor resection leads to poor patient prognosis with a mean survival of about 15 months [14]. Although circulating GBM cells have been detected within the blood in 20% of patients at primary diagnosis, extracranial metastases of GBM are rarely found, most likely due to short patient survival [12]. Therefore, it is crucial to obtain a better understanding of GBM migration at a cellular and molecular level.

It is known that GBM cells are attracted to white matter tracts composed of myelinated axons organized into bundles, but interestingly also on blood vessels within the perivascular spaces where GBM cells migrate and metastasize surrounding brain regions [13,15,16]. One explanation for GBM cell attraction could relate to the high mechanical rigidity of axons and blood vessels. This type of physical attraction towards a more rigid environment where cells migrate to and "home in" is defined as durotaxis, or mechanical substrate compliance [7,17]. It has been shown that integrin-based focal adhesions are stably or dynamically tracking/sensing the stiffness critical for durotaxis and migration speed [18]. Other environmental cues influencing cell migration are as follows [18]: chemotaxis, where cells migrate towards diffusible chemical gradients; haptotaxis representing cell movement towards chemical cues on a surface, such as ECM proteins and, lastly, contact guidance, also called topotaxis or ratchetaxis, where cells use ultrastructures as guiding tracks for migration. Topotaxis was described as a mechanism where a single cell membrane and the cytoskeleton conform to the topography of the cell adhesion substrate. Using melanoma cells, it was shown that this adaptation to the topography was dependent mainly on genetic and signaling differences involving the PI(3)K and ROCK-dependent pathways [18]. In addition, it was shown that nanostructures, which translate topographical signals to cells, influenced directional migration [19,20].

Distant metastasis of primary tumors are known, especially triple negative breast cancer (TNBC), which has a brain metastasis incidence rate of 46% [21,22]. TNBC is a highly malignant and fast progressing subtype of breast cancer that is diagnostically negative for the expression of the estrogen and progesterone receptors with no overexpression of the human epidermal growth factor 2 receptor. Research is ongoing about how metastasis of circulating TNBC cells to the brain occurs, but how these cells enter and disseminate throughout the brain and if they share similar cellular activities to GBM cells is still unanswered.

At the molecular level, a key player involved in regulating cellular migration and invasion is the Phosphatase and Tensin Homolog (PTEN) protein. PTEN is a known tumor-suppressor antagonizing the oncogenic phosphoinositide 3-kinase (PI3K) pathway, where it dephosphorylates the metabolite phosphatidylinositol-3,4,5-triphosphate (PIP3) to phosphatidylinositol-4,5-bisphosphate, inhibiting downstream signaling. The mechanism of PTEN in motile cells was shown due to its cellular localization and dynamics with PIP3. PIP3 is located at the leading edge of migrating cells, stimulating polymerization of actin. PTEN suppresses cell motility by dephosphorylating PIP3 resulting in lower actin polymerization. A cross-inhibitory role of PIP3 and PTEN was proposed where a rise of PIP3 levels led to the dissociation of PTEN from the membrane, and this cycling process resulted in PIP3 waves [21,22]. PTEN loss of function results in constitutively active PI3K signaling and induces proliferation, migration and cell survival [23,24]. PTEN mutations are one of the most common genetic alterations in GBM and are directly associated with malignant transformation and metastasis, but also noted in up to 40% of primary breast cancers associated with therapeutic resistance and a shorter patient overall survival [25–29]. Interestingly, PI3K activation or loss of PTEN was found in 77% of brain metastases or 25–71% of breast tumors metastasized to the brain, respectively [30–32].

Other essential factors for cell migration are the ras homolog family member (RHO) GTPases, which include 20 mammalian genes and play a more determining role regulating actin polymerization, depolymerization and activity of actin-associated myosins during migration [33]. RHO GTPAses cycle between an inactive (GDP-bound) and active (GTP-bound) form, regulated by the activity of Rho-specific guanine nucleotide exchange factors (exchange of GDP with GTP) and GTPase-activating proteins that catalyze GTP-hydrolysis. In addition, most RHO GTPAses, such as RHOA, also translocate between cytosol (inactive) and membrane (active) forms, due to modification by isoprenyl lipids [33]. PTEN is required for directional movement of cells via different cellular localizations and activation through RHO GTPase and Rho associated coiled-coil containing protein kinase (ROCK) signaling [34]. Besides the main RHO GTPase RHOA, RHOB plays an essential role for actin cytoskeleton and especially for membrane blebbing and amoeboid migration [35].

Biomedical materials developed for use in tissue engineering are essential for modeling 3D tumor growth, migration and invasion [36,37]. A relatively new 3D printing technology, melt electrowriting, has been used to fabricate complex multiscaled architectural biomaterial structures [38,39] from microfibers [40], also used for clinical purposes. Melt electrowriting processes polymers, which can be implemented within a biological context for 3D cell cultures recapitulating aspects of the in vivo environment, are novel tools for 3D tumor cell modeling [41–43].

In this investigation two different 3D melt electrowriting poly-(ε-caprolactone) (PCL) structural designs were used as topographical guides to mimic axons and blood vessels in the brain to assess if GBM and TNBC cells have similar cellular activities. RNA sequencing (RNA-seq) was used to shed light on the gene pathways regulating adhesion, migration and cell morphology. Our main findings demonstrate that GBM and TNBC types have enhanced durotaxis and topotaxis and increased migration influenced by the PCL melt electrowriting fibers. We show new findings that enhancement of the above cellular activities stems from mutated *PTEN* variants with loss of function and de-regulated RHOB signaling, which is dependent on a 3D environment. This study validates a significant role for the *PTEN* loss of function genotype for these cellular processes and explains why these tumor cells have a greater affinity for stiffer structures, such as axons and blood vessels promoting dissemination throughout the brain.

2. Materials and Methods

2.1. Cell Lines and Cell Culture

All cell lines used in this investigation were purchased from the American Type Culture Collection (ATCC, Rockville MD, USA) and tested regularly for mycoplasma infection according to the manufacturer's instructions (Minerva Biolabs, Berlin, Germany). GBM cell lines U87 MG (U87) (ATCC® HTB 14™) and LN18 (ATCC® CRL-2610™) and the triple negative breast cancer (TNBC) cell lines MDA-MB-231 (ATCC® HTB26™) and MDA-MB-468 (ATCC® HTB132™) were grown and maintained as described below. U87 cells were cultured in Minimum Essential Medium (MEM); LN18 and MDA-MB-231were cultured in Dulbecco's modified Eagle's medium (DMEM) and MDA-MB-468 in Nutrient Mixture F-12 (DMEM/F12). All media were supplemented with 10% fetal bovine serum (FBS), 2 mM L-Glutamine, 10 mM HEPES Buffer and 1 x nonessential amino acids (all media, FBS, and supplements from Gibco™ Thermo Fisher Scientific, Waltham, MA, USA). Cells were maintained in 75 cm tissue culture flasks in a humid atmosphere with 5% CO_2 at 37 °C. Cells were passaged at 70–80% confluency using 0.0625% Trypsin. The U87 cell line has a *PTEN* gene splice donor mutation of exon three resulting in an in-frame exon three deletion and PTEN loss of function, whereas LN18 is *PTEN* wild type (wt) [44]. The TNBC MDA-MB-468 cell line has a *PTEN* gene splice donor mutation of exon four resulting in an in-frame deletion of exon four and PTEN loss of function [44,45]. The MDA-MB-231 cell line is *PTEN* wt. For simplicity, both U87 and MDA-MB-468 cell lines are referred to as PTEN loss of function, whereas LN18 and MDA-MB-231 as PTEN wt.

2.2. Melt Electrowriting Printing and Glass Slide Treatment

A custom-built melt electrowriting printer was used to fabricate box-pore scaffolds as well as aligned and stacked microfiber tracts as previously described [38,46]. Briefly, melt electrowriting was performed using medical-grade PCL (PURASORB PC 12, Lot#1712002224, 05/2018, Corbion Inc, Amsterdam, Netherlands) at 21 ± 3 °C and a humidity of 40 ± 5%. Scaffolds were printed at 85 °C, 3 bar, 25 G nozzle, 4 mm collector distance and 6 kV voltage applied. A 48 mm × 96 mm rectangular mesh was direct-written and cut to 9 mm disks with an infrared laser. Aligned microfiber tracts were printed at 85 °C; 0.5 bar of air pressure; 22 G nozzle and 6.5 kV voltage applied across a 3.85 mm collector distance. The interfiber distance was set at 200 μm with 10 layers in G-code for both scaffolds. Scaffolds and aligned microfibers were printed onto a metal surface or glass coverslips, respectively, where the glass surface was previously coated with NCO-terminated, star-shaped poly(ethylene oxide-*stat*-propylene oxide) (sP(EO-*stat*-PO) (provided by DWI Leibnitz Institute for Interactive Materials, Aachen, Germany) to facilitate scaffold adherence but decrease surface protein adsorption. Prior to sP(EO-*stat*-PO) coating, coverslips were washed with acetone, water, and isopropanol and dried using an air pressure gun, treated with 100% oxygen plasma in a plasma generator (Pico low-pressure plasma system, Diener Electronic GmbH, Ebhausen, Germany) then incubated in a desiccator with 3-aminopropyltrimethoxysilane for surface activation. Coverslips were then homogenously rotationally spun and coated at 2500 rpm for 40 s with 10 mg mL^{-1} NCO-sP(EO-*stat*-PO) in 10% (v/v) tetrahydrofuran (Merck, Darmstadt, Germany)–MilliQ water. Coated coverslips were ready for melt electrowriting printing within 24 h.

2.3. Melt Electrowriting Scaffold Functionalization

3D box-pore scaffolds were cut into 9 mm discs and placed into a 15.6 mm culture dish (24-well plate) (Waltham, Massachusetts, MA, USA) and 3D-aligned microfiber tracts were placed into a 34.8 mm culture dish (6-well plate) (Corning, New York, NY, USA) then sterilized with UV-light for 30 min, incubated with 1 M NaOH for 10 min and finally washed five times with 1x PBS. For functionalization, scaffolds or aligned microfiber tracts were incubated with 10 µg mL^{-1} laminin-111 (BioLamina, Sundbyberg, Sweden) at 4 °C overnight.

2.4. U87 Cells Stably Express the Farnesylated-tdTomato Fluorescent Protein

A stable transfected U87 cell line was established expressing the farnesylated-tdTomato fluorescent protein in order to track the 3D movement of cells within Matrigel with or without scaffolds (for cell network formation assay see below). For cloning of the td-farnesyl-vector, the cDNA of the farnesylated-tdTomato fluorescent protein was reversed transcribed from the tdTomato-farnesyl-5 gene (Addgene, Watertown, MA, USA) implementing primers containing a Bam HI and NotI restriction enzyme sites in order to replace the AcGFP cDNA in the pLVX-AcGFP-N1 vector (Clontech, Mountain View, CA, USA). Following ligation into the pLVX-AcGFP-N1 vector, bacterial transformation, purified plasmid DNA was then expanded using the NucleoBond Xtra Maxi Kit (Macherey-Nagel, Düren, Germany) and verified for the correct tdTomato-farnesyl-5 gene sequence of the construct (Eurofins Genomics Germany GmbH, Ebersberg, Germany). The pLVX-tdTomato-farnesyl-5-N1 vector was then cotransfected with the packaging plasmid psPAX.2 (Addgene, Watertown, MA, USA) and the envelope vector VSV-G (Addgene, Watertown, MA, USA) into LentiX 293 T cells (Takara Bio Inc., Kusatsu, Shiga, Japan) using the Lipofectamine 2000 reagent (Thermo Fisher Scientific, Waltham, MA, USA). Forty-eight hours after transfection, the lentivirus-containing supernatant was harvested and briefly centrifuged before being concentrated with the LentiX concentrator (Takara Bio Inc., Kusatsu, Shiga, Japan). For subsequent reverse transduction of U87 cells, the concentrated virus (107 infectious units mL^{-1}) was dispensed at 100,000 cells per 9.6 cm^2. Successfully transduced U87 cells were further selected with 3 µm mL^{-1} Puromycin (Thermo Fisher, Waltham, MA, USA).

2.5. Cell Network Loop and Branch Formation Quantification

In order to perform the cell network formation assay, each cell suspension contained 50,000 U87 farnesylated-tdTomato cells and Matrigel (Corning, Corning, New York, NY, USA) at a final concentration of 4.5 mg mL^{-1} in a final volume of 150 µL, and then was pipetted onto a laminin-coated scaffold in a 15.6 mm culture dish. After 30 min incubation at 37 °C, 100 µL culture medium was added. Laminin-coated scaffolds embedded in Matrigel were imaged every 10 min at 568 nm for a total of 20 h using an inverted microscope (Olympus IX83, cellSens Software V1.16, Olympus Corporation, Tokyo, Japan) using a live cell imaging chamber at a temperature of 37 °C and at 5% CO_2. All cell culture experiments were maintained in this live cell imaging chamber during imaging. Six experiments were performed for U87 cells with scaffolds and eight experiments with U87 cells in Matrigel alone. The AVI-videos were converted into TIF-files using a Python script. The cell network formation assay was analyzed using the online Tool WimTube (Wimasis, Onimagin Technologies SCA, Cordoba, Spain). We previously implemented the WimTube program to analyze primary breast adipose stem cells, which form and resemble endothelial capillary like structures with Matrigel [43]. The WimTube-Tool software performs quantification of tube-like formations from microscopic images. It is an automated process that recognizes cellular networks, which form either closed tube-like or loop network structures (counted independently with numbers indicated in yellow), but also quantifies branch-like structures (marked in white), which are not closed in formation. In this present study, the following structures were quantified using WimTube for the cell network formation assay with U87 cells: (1) the total number of cell loop structures and (2) the total number of branch-like structures.

2.6. Cell Migration Assessment Following Treatment with PTEN, RHO and ROCK Inhibitors or Rescuing PTEN Function with Live Cell Imaging

Live cell imaging was performed using an inverted microscope and phase contrast (Olympus IX83, cellSens Software V1.16, Olympus Corporation, Tokyo, Japan). All cell culture experiments were maintained in this live cell imaging chamber during imaging. For each migration experiment, 20,000 cells of a respective cell line were initially seeded onto laminin-coated aligned microfiber tracts and, after a 15 min incubation at 37 °C, 100 µL of culture medium was added. Aligned microfibers were imaged every 10 min for up to 15 h in an enclosed chamber with constant equilibration at 37 °C and 5% CO2. The AVI-videos were converted into TIF-files using a Python script. To inhibit the PTEN wt protein, LN18 cells (300,000 per 9.6 cm^2) were treated with the PTEN inhibitor SF-1670 (Sellekchem, Houston, TX, USA) at a concentration of 3 µM starting 48 h prior to migration and live cell imaging (n = 7 experiments). To rescue U87 cells with a PTEN loss of function, we cloned the human *PTEN* wt gene under control of CMV in the overexpressing pRK5 vector. *PTEN* cDNA was reversed transcribed from human whole blood RNA using primers containing an EcoR1 and EcoR1/Sal1 sites. The PTEN PCR product (1230 bp) was ligated into the pRK5 Vector (pRK5-PTEN) and confirmed by DNA sequencing. U87 cells with a PTEN loss of function were transfected with 3 µg pRK5-PTEN vector using JetPEI® according to the manufacturer's instructions (VWR, Radnor, Pennsylvania, PA, USA). The U87 PTEN transfected cells were assayed for migration on laminin-coated aligned microfibers using live cell imaging after 48 h (n = 4 experiments for control; n = 3 experiments for transfection). To inhibit the RHO-associated, coiled-coil containing protein kinase (ROCK), we treated U87 cells (350,000 per 9.6 cm^2) with the ROCK inhibitor Y-27632 (Stemcell Technologies, Vancouver, Canada) at a concentration of 5 µM, starting 24 h prior to the migration assay using live cell imaging (n = 7 control and n = 9 ROCK inhibitor experiments). For RHO inhibition of respective cell lines (350,000 per 9.6 cm^2), 3 µg/mL of the RHO Inhibitor (C3 Trans based) (Biozol, Eching, Germany) was added to culture media containing 5% FCS, starting 6 h prior to the migration assay live-cell imaging. Cells were then incubated with laminin-coated aligned microfiber tracts and imaged every 10 min for a total of 20 h (U87 PTEN loss of function: n = 4 experiments, MDA-MB-468 PTEN loss of function: n = 3 experiments). AVI-videos were converted into TIF files using a Python script. The migration speed in µm/h was calculated using the manual tracking tool of ImageJ/Fiji.

2.7. Cell Viability

Cell viability after treating cell lines with the RHO inhibitor was assessed using a Live/Dead assay. In brief, cells were incubated for 30 min at 21 °C for 30 min with 2×10^{-6} M Calcein-AM (green = living cells; Thermo Fisher Scientific, Waltham, MA, USA) and 2×10^{-6} M ethidium homodimer I (red = dead cells, Sigma-Aldrich, St. Louis, MO, USA) in 1x PBS.

2.8. Immunocytochemistry and F-Actin Staining

U87 cells (50,000 cells grown in 4.5 mg mL^{-1} Matrigel (Corning, NY, USA) with scaffolds were stained for ß-Tubulin and microtubule-associated protein 2 (MAP2) to assess 3D cell growth and network formation at day 1, day 3 and day 6. All steps were performed at 21 °C. Cells were fixed for 15 min with 2% paraformaldehyde (PFA), washed with 1x PBS, and blocked/permeabilized for 3 min with 5% goat serum and 0.2% Triton X-100 in 1x PBS. Cells were incubated with the primary antibodies Anti-ß-Tubulin (1: 500, Abcam, Cambridge, UK) and Anti-MAP2 for 2 h (1: 500, Sigma Aldrich, St. Louis, MI, USA). Following washing, cells were incubated for 45 min with secondary Dylight-488 Alexa and Dylight-594 (both 1:200 dilutions; Sigma Aldrich, St. Louis, MI, USA). For F-actin staining, Actin Green 488 ready probes were used (Thermo Fisher Scientific, Waltham, MA, USA). To stain cell nuclei DRAQ5 (BioStatus, Shepshed, UK) or Hoechst 33342 (Thermo Fisher Scientific, Waltham, MA, USA) were used according to the manufacturer's instructions. The computer software Imaris was used for 3D reconstructions (Oxford Instruments,

Abingdon, UK). Cells were either incubated with Actin Red 555 Ready Probes Reagent (2 µg/mL) (Rhodamine phalloidin) or Actin Green 488 ready probes (both from Thermo Fisher Scientific, Waltham, MA, USA) specific for F-actin, and DRAQ5 (Biostatus Ltd., Shepshed, UK) for cell nuclei according to the manufacturer's instructions.

2.9. Confocal and Scanning Electron Microscopy

Cells were imaged using a confocal laser scanning microscope (Leica, SP5X, 20 x dip-in water-immersion objective; numerical aperture (NA) 1.0) with laser power set to 1.2 mW to avoid photobleaching. F-Actin Alexa 488 was detected with an Argon laser at 488 nm; F-Actin Red 555 (Tetramethylrhodamine (TRITC)) with an Argon laser at 540 nm, and DRAQ5 with a He-Ne laser at 633 nm. Hoechst 33342 was detected with a UV laser at 350 nm. Images of cells with scaffolds represent an overlay of 35–40 z-stack sections. Confocal images of cells on aligned microfibers represent an overlay ranging from 18–21 z-stack sections (each step 0.99 µm equaling 17.82–20.79 µm with a line average of 3).

For scanning electron microscopy, a Zeiss Crossbeam 340 scanning electron microscope (Carl Zeiss Microscopy GmbH, Oberkochen, Germany) was used. Aligned microfibers were washed with PBS and incubated in 6% glutaraldehyde (Sigma-Aldrich, Schnelldorf, Germany) for 15 min. After a washing step in ice-cold PBS, the samples were dehydrated by increasing concentrations of ethanol from 50 to 100%. The samples were incubated in hexamethyldisilazane (Sigma-Aldrich, Schnelldorf, Germany) for 15 min and dried overnight. All samples were sputter-coated with a 4 nm layer of platinum with a Leica EM ACE600 (Leica Microsystems, Wetzlar, Germany) before imaging.

2.10. RNA Extraction, qPCR and RNA-Sequencing

For RNA-extraction, 50,000 cells of representative cell lines were grown in Matrigel in the presence or absence of scaffolds compared with cells grown on 2D culture dishes alone or coated with Matrigel at 0.1 mg mL^{-1}. RNA was extracted at day 3 using TriFast® PeqGOLD (PEQLAB Biotechnologie, Erlangen, Germany) according to the manufacturer's instructions. All samples were incubated with DNase I to further fractionate the RNA and then precipitated in the presence of glycogen (both from Sigma-Aldrich, St. Louis, MI, USA). Human brain total RNA from a single healthy normal donor from freshly harvested tissue was purchased (Amsbio, catalog number HR-201). All RNAs were stored at −80° C. RNA was reversed-transcribed into cDNA using the High-Capacity cDNA Reverse Transcription Kit (Thermo Fisher Scientific, Waltham, MA, USA). For quantitative real time PCR, SYBR Select Master Mix (Life Technologies, 4472919) and MicroAMPFast 96-well Reaction-plates (Applied Biosystems, Forster City, CA, USA) were used with a StepOne™ Real-Time PCR System (Thermo Fisher Scientific, Waltham, MA, USA). All kits were used accordingly to the manufacturer's instructions. Primer Sequences: RPL13-TF (5′-CTGCTGAAGAACTGAAACTGGC-3′), RPL13-BR (5′-CTCTTCCTCAGTGATGACTGGA-3′), RHOA-TF (5′-TTCCCAAGAAACTGG-3′), RHOA-BR (5′-CATACACCTCTGGGA-3′), RHOB-TF (5′-CATCCAAGCCTACGA-3′), RHOB-BR (5′-CAGTTGATGCAGCCGTTCTG-3′), MAP2-TF (5′ AAGAGAATGGGATCAACGGA-GAG), MAP2-BR (5′ TTGTTCACCTTTCAGGACTGCT).

For RNA-sequencing (RNA-seq) 200 ng poly-A purified mRNA was isolated from U87 cells grown in 2D and in 3D with Matrigel and scaffolds from day 3 ($n = 3$ independent experiments were pooled). RNA-seq was performed by Eurofins Genomics Germany (Ebersberg, Germany). Paired-end read RNA-seq (2 × 150 bp) was executed via Illumina with 31,779,803 clean reads for U87 in 2D and 33,331,487 for U87 in 3D and scaffolds. Transcriptome analyses were performed by mapping and quantification of transcripts against a reference genome and a pairwise comparison of expression determining significant fold changes.

2.11. Western Blot

Cell lysates were loaded onto Mini-Protean TGX Precast protein gels (BioRad, Hercules, CA, USA) and run at 100 V for 1 h. Gels were semi-dry transferred onto PVDF membranes. The membrane was incubated in 1 x PBS + 0.1% Tween-20 buffer (1x PBST) with 3% milk powder (blocking buffer) for 1 h at room temperature. A PTEN antibody (Cell Signaling, Danvers, Massachusetts, MA, USA) and a GAPDH antibody (Santa Cruz Biotechnologies, Dallas, Texas, TX, USA) were diluted 1: 750 in blocking buffer and incubated overnight at 4° C. Washes were performed with PBST before the addition of goat-anti-rabbit-HRP (Cell Signaling, Danvers, MA, USA) (1:500) for 1 h at room temperature. For protein detection, the membrane was incubated for 3 min in HRP juice (PJK GmbH, Kleinbittersdorf, Germany) and visualized using the Amersham™ Imager 600 (GE Lifesciences, Little Chalfont, UK). Original Western blot images can be found in Figures S4 and S5.

2.12. Statistical Analysis and Data Presentation

Statistical analyses were performed using GraphPad Prism 8 for Windows (GraphPad Software Inc., San Diego, CA, USA) with nonparametric statistical tests (Mann-Whitney-Test). Breast and glioblastoma pan cancer atlas, and The Cancer Genome Atlas (TCGA) data sets were accessed via cbioportal [47]. Categorized mutational data for *PTEN* (PTEN mutation = pathogenic and likely pathogenic mutations; *PTEN* splice variants and wt) and *RHOB/RHOA* gene expression were downloaded and correlated with each other. For statistical testing nonparametric Mann-Whitney tests were used.

3. Results

3.1. Design and Fabrication of 3D Scaffolds and Aligned Microfibers for Cell Functional Studies

Previously we demonstrated that 8-chamber radial structures using melt electrowriting and filled with different concentrations of 3D matrices were important tools to determine the optimal matrix and concentration for growth of the U87 human cell line [48]. In a different study, aligned solution electrospun nanofiber sheets enhanced migration of U87 cells, xenografted in mice, away from the primary brain tumor site to an extracortical site [49]. Thus, tailoring melt electrowritten printed fiber structural designs is helpful to address biological functions of tumor cells. Here we designed melt electrowritten 3D box-pore scaffolds or aligned microfibers to resemble axons and blood vessel tracts (Figure 1). Scanning electron microscopy measurements for both scaffolds and aligned microfibers consisted of 10 layers of single aligned PCL microfibers with an average fiber diameter of 10 μm, an inter-fiber bundle distance of 200 μm and an overall height of approximately 100 μm (Figure 1B,C). Additionally, for scaffolds, ten layers of microfibers were deposited in each direction (0° and 90°) resulting in a 20-fiber overlap at intersections (Figure 1B).

3.2. U87 Cells Show an Enhanced Scaffold Directed Durotaxis, Adhesion and Network Formation

Since GBM, and possibly breast tumor cells, can disseminate throughout the brain via axons and blood vessels, we first tested if U87 cells have an affinity for the scaffolds. Growing U87 cells in 3D with Matrigel (4.5 mg mL^{-1}) and laminin-coated scaffolds resulted in an increased cellular affinity for the scaffold within 24 h and continued until 72 h (Figure S1). We observed at 24 to 72 h distinct amoeboid cells, migrating towards and associated with the scaffold (Figure S1A). By day six, U87 cells became more extended, where they formed distinct cellular networks between the scaffolds, but were also wrapped around aligned microfibers (Figures 2A and S1A). In order to directly test an affinity of GBM cells for scaffolds, we live-imaged and tracked U87-td-farnesyl expressing GBM cells, which emit a red cytosolic fluorescence, in a time kinetic manner up to 18 h in Matrigel with laminin-coated scaffolds or Matrigel alone (Figure 2B, Videos S1 and S2). In the presence of scaffolds, tumor cells actively migrated toward and adhered to the scaffold forming loop and branch-like cellular networks (Figure 2B,C). In contrast, although U87 cells migrated throughout the Matrigel alone, no loop cellular networks were noted. Additionally, cell attraction appeared to occur equally along aligned microfibers and at the corners of scaffold structures, thus showing no topographical preference for a particular scaffold region. The total number of U87 cellular loops and branches in the presence of the scaffold significantly peaked at 10 h and 15 h compared to Matrigel alone (Figure 2C).

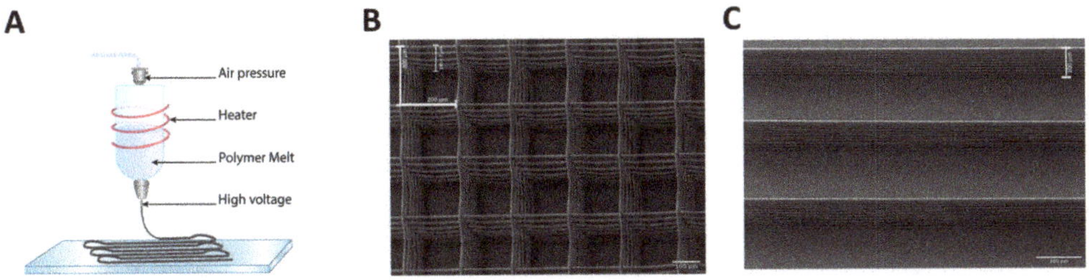

Figure 1. Fabrication and design of 3D scaffolds and aligned microfibers have similar properties. (**A**) Schematic representation of melt electrowriting. Medical grade PCL melt was extruded through an electrified nozzle into a jet that solidified upon reaching the collector. Repeated deposition enabled the fabrication of a 9 mm diameter scaffold or an 8 mm long array of aligned microfibers. (**B**) Scanning electron microscopy image of a scaffold consisting of 10 stacked aligned fibers along the length and 20 stacked overlapping aligned fibers at each corner, where each fiber has a diameter of 10 µm. The interfiber bundle distance is 200 µm (indicated as a white bar) with an overall box form scaffold structure with the total height of the fiber walls ~ 140 ± 7 µm as indicated (in white). (**C**) Scanning electron microscopy image of aligned microfibers. Each fiber bundle tract consists of 10 stacked aligned fibers where each PCL fiber is 10 µm in diameter with the total height of the fiber walls 106 ± 15 µm (indicated as a white bar). The interfiber bundle distance is 200 µm.

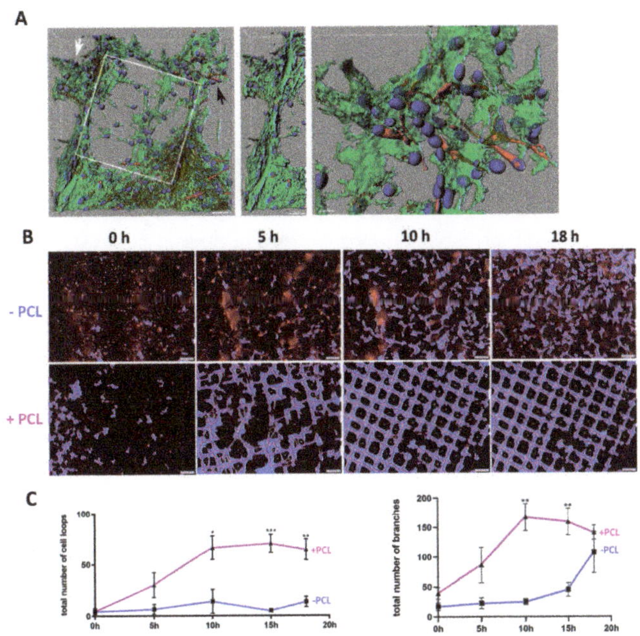

Figure 2. U87 durotaxis and cell network formation with scaffolds in 3D cultures. (**A**) Left image of a 3D reconstruction of an immunocytochemical staining of U87 cells at day 6 with 4.5 mg mL^{-1} Matrigel and a single box pore (200 µm) scaffold (white solid square) (white scale bar = 30 µm) (ß-tubulin, green; MAP2, red; Hoechst 33342, blue nuclei). The middle image shows the same region from the left image (white arrow) but without the scaffold to show the wrapping of U87 cells around

the aligned microfibers. The right image is a magnification of the left image (black arrow) to demonstrate U87 cell interactions at day 6 (right scale bar = 10 µm) (ß-tubulin, green; MAP2, red; Hoechst 33342, blue nuclei). (**B**) Cell network formation assay. U87-td-farnesyl expressing cells grown in 3D with 4.5 mg mL^{-1} Matrigel alone (minus (−)PCL above) or with a scaffold (+PCL bottom) and imaged over time up to 18 h. Images were analyzed for cell branches or loop-like structures (blue with red lines) using the WimTube computer program. White scale bar = 200 µm. (**C**) Graphs show cell network formation assay (+PCL) scaffolds top red line (n = 6 experiments) and (-PCL) scaffolds bottom blue line (n = 8 experiments) quantified as the total number of cell loops (left graph, Y-axis) and branches (right graph, Y-axis) formed over 18 h. Significance was found comparing +PCL to -PCL; cell loops at 10 h * p = 0.0113; 15 h *** p = 0.0007; 18 h ** p = 0.0023; cell branches at 10 h ** p = 0.002; 15 h ** p = 0.0079).

Previously it was shown that the stiffness of Matrigel (at 4.5 mg mL^{-1}) represents a soft matrix of <100 Pa (Young's modulus), but scaffolds exhibited a high stiffness of ~15 kPa [41,42]. A combination of hydrogel with scaffolds positively influences the overall handling by increasing stiffness synergistically [41,42]. In general, a higher matrix rigidity activates durotaxis-directed cell migration towards stiffer substrates [50,51]. Using the same conditions as above for Matrigel and scaffolds, we interpret our findings that a durotactic tumor cell response is occurring, where U87 cells migrate from a lower stiffness toward a substrate of a higher stiffness, where scaffolds having aligned microfibers represent a strong physical cue (Figures 1 and 2). However, our findings demonstrating the wrapping of U87 cells around melt electrowritten fibers also suggest that the fiber topography affects cell behavior (Figure 2A).

3.3. GBM and Breast Cancer Cell Migratory Behavior with 3D-Aligned Microfibers

Following our findings demonstrating a strong durotactic response of U87 cells for scaffolds, we next implemented aligned microfibers (Figure 1C) to analyze the cell migratory behavior of GBM and TNBC cell lines in detail using live cell imaging (Figures 3 and 4). We first tested U87 cell migration on aligned microfibers coated with laminin-111 (Figures 3 and S1B), as laminin-111 is the most prominent brain ECM subtype and also enriched on small blood vessels [9,52]. Similar to scaffolds, an enhanced durotactic response of U87 cells was also observed for aligned microfibers as well as active cell migration (Figure 3A, Video S3). Furthermore, scanning electron microscopy indicated that most U87 cells were morphologically round and organized onto single microfibers supporting a topographical cue for movement (Figure 3B).

In striking contrast to U87 cells, LN18 cells showed no active durotactic response for laminin-coated aligned microfibers, and no migration (Figure 4A, Video S4). However, over a 10 h period some cells could adhere to aligned microfibers but did not migrate. We hypothesized that these differences between U87 and LN18 cells could be due to their opposite *PTEN* genotypes (U87 PTEN loss of function with no PTEN protein expression; LN18 PTEN wt) (Figure S2A,B). Based on the known regulatory role of PTEN in migration, we performed live cell imaging experiments modulating PTEN function. Transfecting and rescuing U87 cells (PTEN loss of function) with a *PTEN* wt overexpressing vector (Figure S2B) significantly blunted migration speed from 32.4 µm/h to 29.5 µm/h (p = 0.0341) (Figure 4C. In contrast, treatment of LN18 cells (PTEN wt) with a PTEN inhibitor restored a durotactic response, adherence and migration on laminin-coated aligned microfibers (Figure 4A, Video S5). A total rescue of cell migration with a speed of 16.4 µm/h, compared to no migration without inhibitor treatment, was found (Figure 4D).

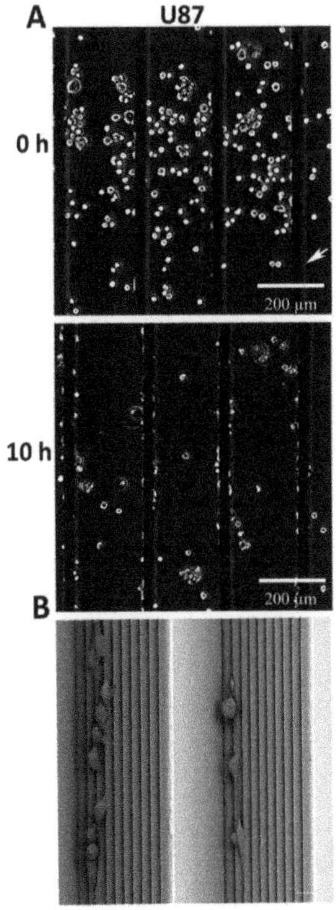

Figure 3. U87 durotaxis and cell migration with aligned microfibers. (**A**) Representative images of U87 cells on laminin-coated aligned microfibers at 0 h and 10 h; scale bar = 200 μm. (**B**) Representative scanning electron microscopy image of U87 cells on stacked aligned microfibers (total of 10 melt electrowritten printed fibers: scale bar = 20 μm).

We next tested if two TNBC cell lines with opposing *PTEN* genotypes behaved similar to GBM cells for durotaxis, cell adhesion and migration, using the same experimental conditions. Results showed that MDA-MB-468 cells (PTEN loss of function) and MDA-MB-231 cells (PTEN wt) were comparable to GBM cell lines, but with some differences (Figure 4B,E). For example, MDA-MB-468 cells (PTEN loss of function) like U87 cells demonstrated a strong durotactic response for aligned microfibers, cell adhesion and migration (Video S6). Although both PTEN wt, in contrast to LN18 and MDA-MB-231 cells, showed more collective cell adhesion, migration only occurred as single cells (Video S7). However, MDA-MB-231 migrated significantly slower at 17 μm/h on aligned microfibers than MDA-MB-468 (PTEN loss of function) at 26.5 μm/h ($p < 0.0001$) (Figure 4B,E).

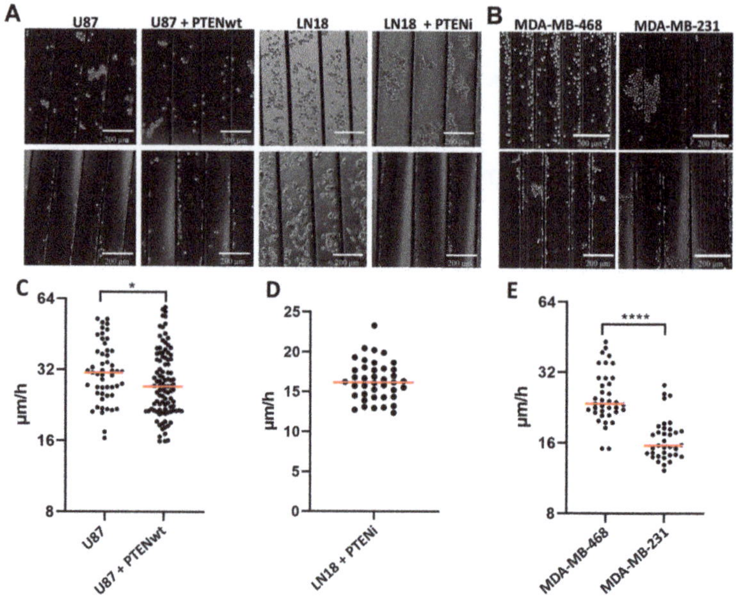

Figure 4. PTEN regulates GBM and TNBC cell durotaxis and migration. (**A**) Representative images from left to right of U87 PTEN loss of function cells (1st column, control) and U87 cells after *PTEN* wt transfection and rescue (2nd column) on laminin-coated aligned microfibers at 0 h (top) and 10 h (bottom) (scale bar = 200 µm). LN18 PTEN wt cells (3rd column, control), and LN18 cells (4th column) following PTEN inhibitor (PTENi) SF1670 treatment on laminin-coated aligned microfibers at 0 h (top) and 10 h (bottom). (**B**) Representative images of MDA-MB-468 PTEN loss of function cells and MDA-MB-231 PTEN wt cells on laminin-coated aligned microfibers at 0 h (top) and 10 h (bottom) (scale bar = 200 µm). Note the specific collective adherence at 0 h of MDA-MB-231 PTEN wt cells. (**C**) Aligned microfiber migration speed (µm/h) of U87 PTEN loss of function cells (control) (n = 4 experiments; n = 51 cells) and U87 cells after *PTEN* wt transfection and rescue (n = 3 experiments; n = 101 cells) * $p = 0.0341$. (**D**) Migration speed (µm/h) of LN18 PTEN wt cells with PTEN inhibitor (PTENi) SF1670; (n = 7 experiments; n = 40 cells). (**E**) Aligned microfiber migration speed (µm/h) of MDA-MB-468 PTEN loss of function cells (n = three experiments; n = 37 cells) and MDA-MB-231 PTEN wt cells (n = four experiments, n = 35 cells) **** $p < 0.0001$.

3.4. *RHOB as a Novel Regulator of Migration*

To further explain PTEN regulation of GBM and TNBC cell migration, we focused on the RHO-ROCK signaling pathway (Figure 5B). Next to RHOB and RHOC, RHOA has been described as the most prominent regulator of the migration signaling pathway [53]. To gain insight into the genes including RHO-ROCK signaling and other pathways, we performed RNA-seq from U87 cells grown in 3D with Matrigel and scaffolds compared to 2D controls. Results showed a highly significant 46.15-fold increase of RHOB expression in 3D versus 2D ($p = 0.0001$) (Figure 5A, Table S1). Differential gene expression of RHOB signaling members, as well as actin and myosin genes, support regulation of migration (Figure 5C). On the other hand, genes involved in other migration modes, such as neuronal migration via nucleokinesis (nuclear piston migration) and astrocyte migration (astrogliosis) were not increased, supporting the hypothesis that GBM cell migration is predominantly regulated by RHO/ROCK signaling (Figure 5C) [54].

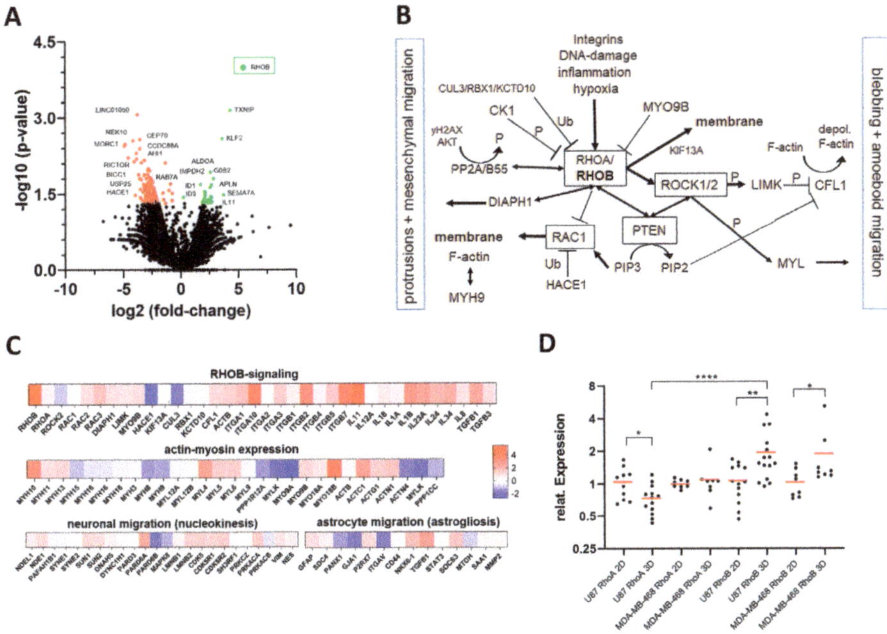

Figure 5. RHOB signaling and regulation of migration. (**A**) RNA-seq data (Volcano-plot) of U87 PTEN loss of function cells grown in 3D with Matrigel and scaffolds versus 2D (3 days). Red dots (left): significantly ($p < 0.05$) down regulated genes; green (right) dots significantly ($p < 0.05$) up-regulated genes. *RHOB* was 5.528-fold log2 up regulated in U87 3D cultures and scaffolds ($p = 0.0001$). Other significantly up-regulated genes include *G0S2* (6.4-fold, $p = 0.011$), induced in glioma with high invasion [55], *APLN* (4.55-fold, $p = 0.0307$) induces colon cancer cell migration [56], *ALDOA* (3.37-fold, $p = 0.0202$) promotes lamellipodia [57], *SEMA7A* (3.84-fold, $p = 0.05$) via ITGB1 [58] and *IMPDH2* (1.25-fold, $p = 0.036$) regulates colorectal cancer cell migration [59] (Supplemental Tables S1 and S2). (**B**) Overview of the RHOB signaling pathway. RAC1 promotes an elongated cell phenotype (mesenchymal) migration (left), which is inhibited by RHOB. RHO activates ROCK leading to blebbing and amoeboid migration (right) and is negatively regulated by PTEN dephosphorylating phosphatidylinositol-3,4,5-triphosphate (PIP3) to phosphatidylinositol-4,5-bisphosphate (PIP2). Ub (ubiquitinylated); P (phosphorylated). (**C**) RNA-seq gene expression differences (heatmaps) of U87 PTEN loss of function cells in 3D with Matrigel and scaffolds versus U87 in 2D (red = log2-fold up regulated and blue= log2-fold down regulated). Top heat map: RHOB signaling genes, e.g., up-regulation of Integrins (esp. *ITGA10* and *ITGB2*) and inflammation genes (e.g., *IL11*, *IL1B*) stimulating RHOB; middle heatmap: actin-myosin expression; bottom heatmap: neuronal migration (left) and astrocyte migration (right). (**D**) QPCR and relative gene expression (2-DDCt) of *RHOA* and *RHOB* in U87 and MDA-MB-468 PTEN loss of function cells in 2D vs. U87 and MDA-MB-468 cells in 3D with Matrigel and scaffolds (3 days). Significant values from left to right: U87 *RHOA* expression in 2D vs. 3D: * $p = 0.0316$; U87 *RHOB* expression in 2D vs. 3D: ** $p = 0.0031$, MDA-MB-468 *RHOB* expression in 2D vs. 3D: * $p = 0.0409$. U87 *RHOA* 3D vs. U87 *RHOB* 3D: **** $p < 0.0001$.

The significant role of RHOB was validated using quantitative real time PCR (qPCR) for both U87 and MDA-MB-468 cells using the same 3D and 2D culture conditions as for RNA-seq (Figure 5D). Furthermore, RHOB expression was significantly higher than RHOA for U87 cells in 3D. In contrast, both PTEN wt LN18 and MDA-MB-231 cells showed no significant increase of RHOB gene expression compared to 2D (Figure S2C). Importantly, U87 cells in 3D Matrigel without PCL also showed a significant induction of RHOB, but not RHOA expression, compared to U87 cells seeded on 2D or 2D Matrigel coating (Figure S2D). In contrast, LN18 showed no induction of RHOB expression in 3D Matrigel alone. Our results support that tumor cells with a PTEN loss of function induce high levels of RHOB expression and enhanced migration dependent on a 3D environment; however, durotaxis is only active in the presence of PCL (Figure 2B,C).

Based upon our RNA-Seq and qPCR results we initiated functional cell culture studies to further unravel the role of RHOB and pathway members in the migratory behavior of U87 and MDA-MB-468 cells (Figure 5B). Treatment of U87 cells with a ROCK inhibitor demonstrated a highly significant decrease of migration speed to 24.5 μm/h ($p < 0.0001$) on aligned microfibers compared to control, where cells migrated at an average speed of 31.5 μm/h (Figure 6A,B). Further in line with RHO/ROCK signaling, treating both PTEN loss of function U87 and MDA-MB-468 cells with a RHO inhibitor blunted durotaxis and adhesion, and completely inhibited cell migration, while both cell types remained viable (Figure 6C, Video S8). This result supports that blocking RHO kinase primarily inhibits cell migration, which is also necessary for the durotactic response.

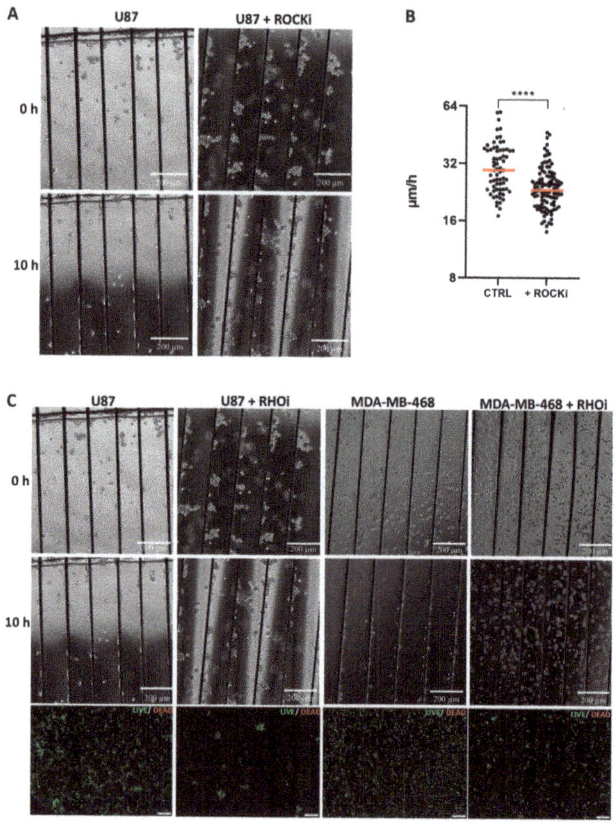

Figure 6. RHO-ROCK regulates tumor cell durotaxis and migration. (**A**) Representative images of U87 PTEN loss of function control cells (left) or cells treated with the ROCK inhibitor (ROCKi) Y-27632 at 0 h and 10 h on laminin-coated aligned microfibers. (**B**) Migration speed (μm/h) of U87 cells (control, CTRL) ($n = 7$ experiments; $n = 67$ cells) vs. U87 cells with the ROCK inhibitor (ROCKi)Y-27632 ($n = 9$ experiments, $n = 107$ cells); **** $p < 0.0002$. (**C**) Representative images (from left to right) of U87 control (1st column) or cells treated with RHO inhibitor (RHOi) (2nd column) and MDA-MB-468 PTEN wt cells control (3rd column) or treated with the RHO inhibitor (RHOi) (4th column) on aligned microfibers coated with laminin at 0 h and 10 h (U87: $n = 4$ experiments; MDA-MB-468: $n = 3$ experiments). Lower panels are examples of cell viability analyses of the respective tumor cells above after RHO inhibitor treatment. Green: live cells; red: dead cells.

Finally, we asked if there was a clinical correlation between primary GBM and breast cancers with *PTEN* wt or *PTEN* splice or other mutations and RHOB expression using

The Cancer Genome Atlas (TCGA). Although rare, primary GBM and breast tumors with *PTEN* splice mutations had significantly higher RHOB gene expression versus tumors with *PTEN* wt or other *PTEN* mutations (Figure 7A). Upon further analyses, in seven of ten of the primary GBM and breast cancer tumors with *PTEN* splice mutations and high RHOB expression a homozygote *PTEN* loss of function was found. In line with TCGA primary tumors, both U87 and MDA-MB-468 cells harbor *PTEN* splice variants leading to a homozygote PTEN loss of function (Figure S2A for U87). In contrast, analyzing RHOA expression in primary GBM and breast tumors, using the same TCGA cohorts, showed no significant RHOA expression for tumors with *PTEN* splice variants. However, a significant higher expression of RHOA (~10%) was found in tumors with *PTEN* mutations versus tumors with *PTEN* wt (p = 0.001). The above findings, along with our cell culture studies, support a prominent role for RHOB and *PTEN* splice variants with PTEN loss of function in regulating tumor cell migration of GBM and breast cancers.

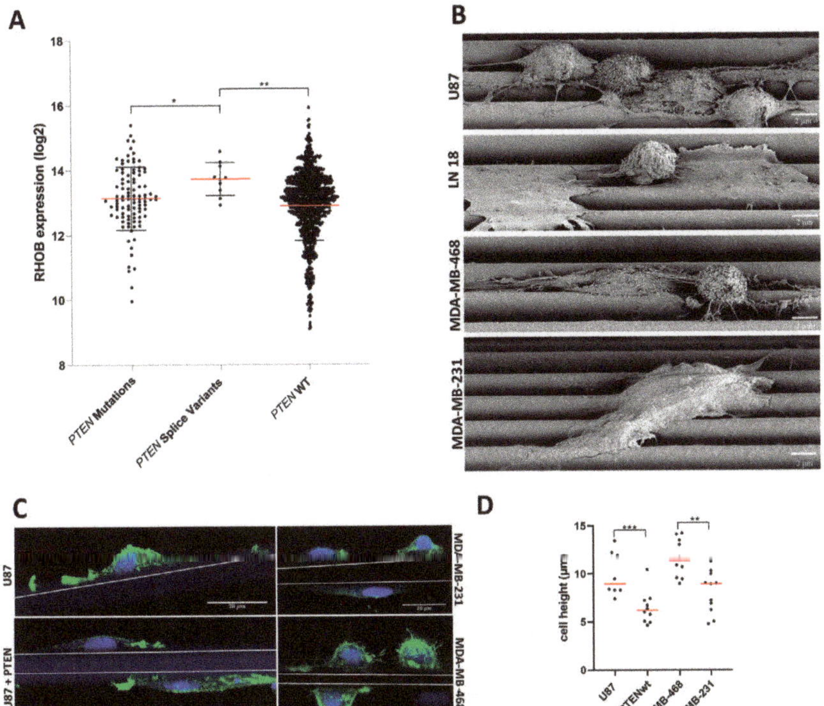

Figure 7. RHOB signaling in primary GBM and breast tumors and RHOB regulation of cell morphology. (**A**) Correlation of *PTEN* wt or *PTEN* splice or other mutations with *RHOB* expression from primary GBM and breast cancers (TCGA); * p = 0.031; ** p = 0.003; *PTEN* mutations: n = 98 tumors; including GBM (n = 46 tumors) and breast cancer (n = 52 tumors). *PTEN* splice variants: n = 10 tumors; including GBM (n = 4) and breast cancer (n = 6). *PTEN* wt: n = 1102 tumors including GBM (n = 97) and breast cancer (n = 1005 tumors). (**B**) Representative scanning electron microscopy images on laminin-coated aligned microfibers with U87 PTEN loss of function, LN18 PTEN wt, MDA-MB-468 PTEN loss of function and MDA-MB-231 PTEN wt cells at 24 h after seeding (scale bar = 2 µm). (**C**) Confocal z-stack images of F-actin Alexa 488 staining of U87 PTEN loss of function control cells (top left) vs. U87 cells after *PTEN* wt transfection and rescue (bottom left); and MDA-MB-231 PTEN wt (top) vs. MDA-MB-468 PTEN loss of function cells (bottom) on laminin-coated aligned microfibers at 24 h after seeding (green: F-actin; blue: DRAQ5 indicates nuclei) (scale bar = 20 µm). (**D**) Quantification of cell height from confocal images (µm) (Y-axis) of U87 PTEN loss of function (control) vs. U87 after *PTEN* wt transfection and rescue and MDA-MB-468 PTEN loss of function vs. MDA-MB-231 PTEN wt cells. *** p = 0.0010; ** p = 0.0016.

3.5. RHOB Signaling as a Regulator of Cell Morphology

A more in-depth view of RHOB signaling as regulated by PTEN supports a cellular decision of either an amoeboid or mesenchymal cell shape as well as movement (Figure 5B). Microscopic analyses comparing both GBM and TNBC cell lines with opposing *PTEN* genotypes on laminin-coated aligned microfibers noted significant differences in cell morphology. For example, scanning electron microscopy demonstrated that high RHOB-expressing U87 and MDA-MB-468 cells with *PTEN* loss of function were amoeboid and primarily located mainly on single fibers (Figure 7B). In contrast, both low RHOB-expressing and PTEN wt LN18 and MDA-MB-231 cells were flat and spread over 2–3 fibers. Measuring the cell heights confirmed that amoeboid U87 and MDA-MB-468 cells had average cell heights of 9.9 μm and 12.2 μm, respectively (Figure 7D). These values were significantly higher compared to MDA-MB-231 *PTEN* wt cells and U87 cells transfected with a *PTEN* wt overexpressing vector, which had average cell heights of 8.3 μm and 6.5 μm, respectively.

Using confocal laser microscopy, detecting F-actin confirmed cell morphologies as observed by scanning electron microscopy (Figure 7B,C). Congruent with an amoeboid morphology, U87 cells and MDA-MB-468 cells showed distinct F-actin protrusions with different sizes at the leading edge, but also small actin blebs on the surface of MDA-MB-468 cells (Figures 7C and 8C,G with white arrows). Prominent F-actin (Phalloidin positive) staining was also noted at the cell surface for both cell lines along the membrane with different intensities (Figures 7C and 8C,G with red arrows). The nucleus of the U87 cells was primarily located at the rear of the cell, opposite to the direction of movement (Figures 7C and 8C).

3.6. Skating Snail-Like Cell Migration and the Motor Clutch Model

To unravel a more detailed analysis of the migration mode, we focused on U87 and MDA-MB-468 tumor cells with PTEN loss of function and high RHOB expression. Using live cell imaging with a higher magnification, we compared cell morphology, plasticity modes, the total migratory distance and persistence behavior on aligned microfibers. Distinct amoeboid plasticity modes resembling "snail-like" crawling, followed by alternating fast movements, was common for both tumor cell types (Figures 7C, 8A–G and S3, Videos S9 and S10). These skating snail-like cellular forms resemble how speed skaters on ice initially propel their forces forward to accelerate and then skate in a gliding manner with higher speed.

For example, one mode showed that skating snail-like U87 cells were initially slow, then accelerated (between 9–15 min) and slowed down again (between 30–42 min) (Figures 8A,D,G top and S3B). Another skating snail-like U87 cell mode demonstrated an instant acceleration to a fast mode then slowed down and stopped (Figures 8E,G bottom and S3B). Lastly, a skating snail-like U87 cell mode showed a slow "crawling" and then a faster sliding with long cellular protrusions at the leading edge (Figures 8B,C,F and S3B). These changes in cellular plasticity modes dramatically point to an intrinsic characteristic of these tumor cells. However, when cells paused their migration on aligned microfibers, both U87 and MDA-MB-468 cells displayed a more elongated and flattened morphology (Figures 8A,B,D–F and S3A,B). Analyzing the cell surface length of moving cells on aligned microfibers demonstrated that cells during fast movements had the shortest length compared to slow and halted cells (Figure 8H). Next, we addressed the total migratory distance and persistent migration of U87 and MDA-MB-468 cells. Comparing the total time that cells migrated on aligned microfibers revealed that U87 cells significantly migrated for longer times (27.06 $\pm$ 1.65 min) compared to MDA-MB-468 cells (21.08 $\pm$ 1.09 min) (Figure 8I). Addressing the total time of reverse migration occurring in one single cellular movement, U87 cells demonstrated significantly fewer reverse movements (6.22 $\pm$ 0.83 min) compared to MDA-MB-468 cells (9.96 $\pm$ 1.53 min) (Figure 8J). The above findings support that U87 cells significantly migrated more persistently than MDA-MB-468.

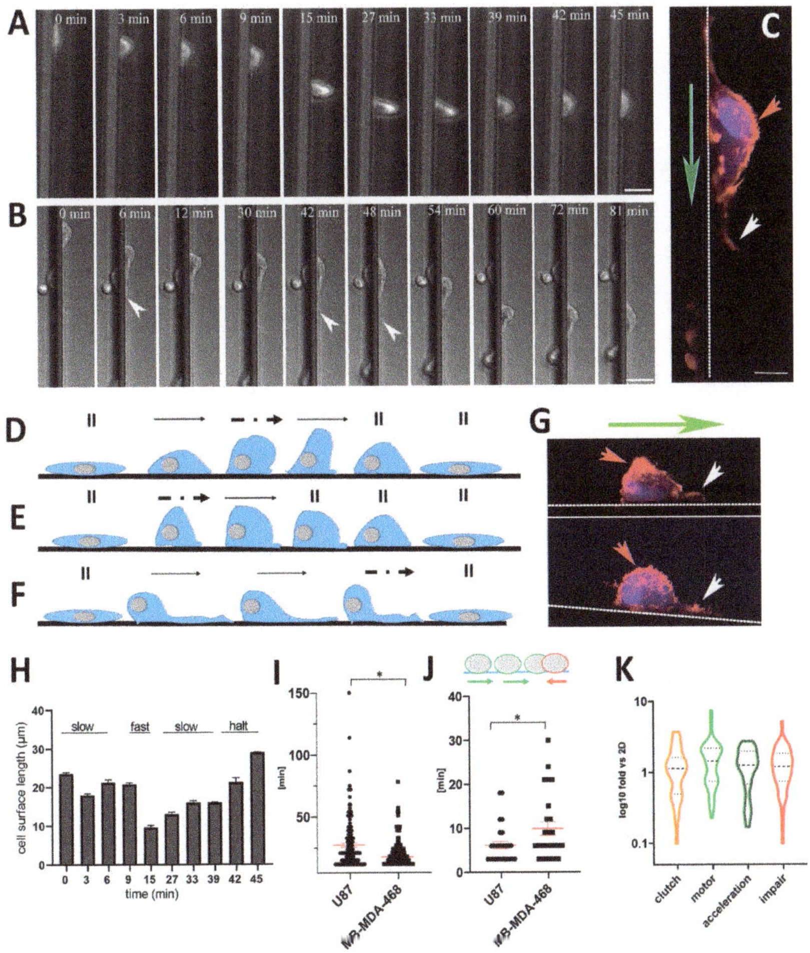

Figure 8. Skating snail-like cellular plasticity modes and persistence on aligned microfibers. (**A**,**B**) Consecutive images of U87 cell plasticity modes and migration over time; white arrows indicate leading edge cellular protrusions (scale bar = 25 µm). (**C**) Confocal image of z-stacks showing F-actin Red staining of an amoeboid U87 PTEN loss of function cell (red: F-actin, blue DRAQ5 indicates nuclei), red arrow indicates cell nucleus at cell rear, white arrow indicates leading edge cellular protrusion, green arrow indicates direction of movement (scale bar = 10 µm). (**D**–**F**) Schematic images of migration cell plasticity modes of "skating snail-like" movements for U87 cells. II = nonmoving resting cells; thin arrow = slow cell movement; bold dotted-thick arrow = fast cell movement (**G**) Confocal image of z-stacks of two U87 cells (top, bottom) (red: F-actin, blue DRAQ5 indicates nuclei), red arrow indicates strong F-actin stain around the cell circumference, white arrow indicates cellular protrusion, and green arrow indicates direction of movement. (**H**) Quantification of cell surface length on aligned microfibers (Y-axis) of different U87 migration modes (X-axis). Shortest bar graph = fastest mode of movement. (**I**) Quantification of total U87 and MDA-MB-468 cell migration time (min) for single cell movements along aligned microfibers, FigU87: 27.06 ± 1.65 min; MDA-MB-468: 21.08 ± 1.09 min; * $p = 0.0142$. (**J**) Quantification of reverse migration (min) of single movements of U87 and MDA-MB-468 after continual forward movements (green circle and arrow = forward movement and red circle and arrow = reverse movement), U87: 6.22 ± 0.83 min, MDA-MB-468: 9.96 ± 1.53 min; * $p = 0.0462$. A total of 258 single cell movements were analyzed for both 8I and 8J. (**K**) RNA-seq gene expression differences of U87 cells in 3D with Matrigel and scaffolds versus U87 in 2D. Expression of clutch genes ($n = 35$), motor genes ($n = 40$), acceleration genes ($n = 50$) and migration impairment genes ($n = 58$) [60–62].

The motor clutch model describes the physical and molecular mechanics of how cells move on substrates [63]. This complex process involves actin polymerization at the leading cell edge driving forward cell movement, which is coordinated with actin depolymerization proximally and centrally along with localized myosin motors that pull and exert force on this F-actin network, resulting in retrograde actin flow [64]. Clutch adaptor proteins such as cell receptor integrins and other clutch components, such as talin and vinculin, interact with F-actin, transmit forces to adhesion sites and regulate the cellular grip on the extracellular matrix controlling cell movement [64,65]. Previously it was shown that the molecular processes describing the motor-clutch model implemented many different genes [60–62] (n = genes) for clutch (n = 35), motor (n = 40), acceleration (n = 50) and migration impairment (n = 58). In line with the motor clutch model, we analyzed the relevant proteins mediating the motor clutch mechanism, as well as proposed proteins responsible for acceleration and impairment of migration according to our RNA-seq data [60–62]. Although some motor clutch specific genes were found more highly expressed in U87 grown in 3D with Matrigel and laminin-coated scaffolds and 2D cultures (Figure 5C), our expression analyses using RNA-seq of the above 183 genes showed no significant differences (Figure 8K).

4. Discussion

One of the first comparative histological analyses by Scherer in 1938 showed that GBM growth and dissemination throughout the brain depends on architectural organization [15]. Histologically, Scherer discovered that GBM cells were mainly associated with white matter axon bundles, and could even follow dendrites when invading the cortex. Importantly, secondary tumors to the brain were mainly detected in border zone regions, such as the cerebral vascular supply and gray and white matter junctions, further supporting that metastatic tumor cells travel along the arterial tree [66]. Brain structures in terms of stiffness are still under investigation. For example, local viscoelastic properties of neurons and glial cells using scanning force microscopy showed that the elastic storage modulus (E′) of astrocytes (glial cells) was between ~300 Pa (30 Hz) to ~520 Pa (200 Hz), whereas that of neurons ranged from ~650 Pa (30 Hz) to ~1,590 Pa (200 Hz) [67]. These values defined the cells of the central nervous system (CNS) as very soft compared to other cells, such as fibroblasts (E′ ~3 kPa at 200 Hz) [68]. Axon bundle tracts have widths ranging from 0.1–10 μm [69] and stiffnesses recorded of 4.6 ± 1.5 kPa (Young's modulus) [70]. Interestingly, using waveguide elastography with healthy individuals' *in vivo*, similar values for white matter tracts of 4–4.6 kPa (shear modulus) were found [71].

Evaluation of stiffness of arteries and veins is ongoing in the literature; however, stiffness analyses of brain vessels are still needed. For example, the Young's elastic modulus of human arteries has been measured as 679 kPa ± 304 kPa standard deviation with an adjusted arterial diameter of the left carotid artery of 410 μm ± 130 μm standard deviation (n = 6441 individuals) [72]. Diameters of smaller human blood vessels are ~8 μm for capillaries and 20–30 μm for venules and arterioles using magnetic resonance imaging with computed tomography [73]. Multiphoton laser-scanning microscopy of mice brains showed that the brain vessels ranged between ~1 and 20 μm in diameter [74]. Most importantly, using live imaging of GBM cells invading and growing in mice brains, it was shown that GBM cells migrated primarily along blood microvessels of the peritumoral area, especially when microvessels ran parallel to each other [74].

Isolated human normal brain and GBM have similar shear storage moduli at a low strain (elastic moduli of ~300 Pa) [75]. However, normal brain and GBM shear moduli increased to kPa range when uniaxially compressed ex vivo [75]. ECM protein components are increased in GBM, resulting in higher stiffness [76]. In contrast to the above, magnetic resonance elastography (MRE), which measures stiffnesses via mechanical shear waves in vivo, showed that GBM is softer than normal brain [77]. However, MRE results especially of GBM, have to be critically evaluated due to different cerebral blood flows, and lower stiffness measurements could reflect the overall heterogeneity including hemorrhaging, edema and necrosis [78]. Importantly, vascularization and interstitial pressure are very high in GBM (and other tumors), due to the fact that GBM shows the highest microvascular proliferation and microvessel density of all brain cancers [79]. What is clearly needed are measurements of localized solid tumor regions away from necrosis and edema to assess solid tumor stiffness. Thus, it is highly likely that GBM has essentially a higher stiffness then normal brain *in vivo*.

It is well known that breast cancer tumors are stiffer (E′ = 4.04 ± 0.9 kPa) than normal mammary tissue (E′ = 0.167 ± 0.031 kPa) [80] supporting tumor cell mediated ECM remodeling. On the other hand, single primary breast cancer cells are very soft (~300 Pa), similar to the CNS cells described above [81]. It is conceivable that tumor cells with different mechanical conditioning target different metastatic sites, as shown for breast cancer cells [82]. Since it is known that cells have a high affinity for stiffer substrates [17], it is plausible that tumor cells demonstrate a durotactic activity for stiffer white matter tracts and blood vessels in the brain. Support for this stems from our experiments, where both TNBC and GBM cells with PTEN loss of function demonstrated enhanced affinity for aligned microfiber tracts seeded in a nonstiff fluid environment. This scenario could especially represent circulating TNBC cells entering the blood/brain barrier and thus seeking stiffer substrates.

In the present study, we implemented scaffolds and aligned microfibers (each 10 µm) with measured compression stiffnesses of ~15 kPa including porous spaces [41] to simulate axon tracts and blood vessels within the brain. Although, the aligned microfibers used in this study are in line with the diameters of axon bundles and small blood vessels and ECM components, the high stiffness of microfibers may be relevant with small blood vessels but higher than axons measured in normal brain. However, axon stiffness in the GBM tumor environment with increased levels of ECM is unknown. Importantly, we present novel findings demonstrating that both GBM and TNBC cells with homozygote *PTEN* loss of function and high RHOB gene expression show similar 3D migration properties in contrast to GBM and TNBC cells with PTEN wt and low RHOB expression. Furthermore, using TCGA analyses, 70% of primary GBM and breast cancer tumors with *PTEN* splice mutations had homozygote *PTEN* loss of function with significantly high RHOB expression (Figure 7A). These primary tumors are similar to the cell lines used in this study, supporting the clinical relevance of our experimental findings regarding 3D migration. Thus, we conclude that the status of a specific tumor cell genotype together with expression of RHO/ROCK-signaling genes represent "molecular cues" for 3D structures regulating durotaxis, cell adhesion and migration.

Only GBM and TNBC cells with *PTEN* loss of function and high RHOB gene expression resulted in an enhanced single cell durotactic response, promoting cells to migrate towards stiffer scaffolds or aligned microfibers. Furthermore, following adherence these tumor cells significantly had a faster migration speed on aligned microfibers than PTEN wt and low RHOB expressing cells. Recently, two other GBM cell lines (U251 and GL15), with either a PTEN mutation or a deletion, also demonstrated durotaxis with a polydimethylsiloxane substrate [83]. It is noteworthy that MDA-MB-231 cells with PTEN wt and low RHOB expression showed a specific mode of "collective cell adherence" with aligned microfibers, where single cells could load onto and migrate at a slower speed. In contrast no durotaxis or migration occurred with LN18 PTEN wt. Interestingly, an essential role for *Pten* controlling collective cell migration was also found during mouse embryonic development [84]. It was demonstrated that TGFB induced migration via PTEN suppression, where TGFB1 switched breast cancer cells from a collective to a single cell migration mode via RHO/ROCK signaling [85]. It will be important to further study the collective nature of MDA-MB-231 binding to aligned microfibers.

Another study using solution electrospun nanofiber scaffolds with a higher stiffness, of up to 166 kPa, demonstrated increased migration of glioma stem cells depending on multibranched N-glycans metabolized by MGAT5 [86]. According to our RNA-seq data of U87 cells grown in 3D and scaffolds, with a ~10-fold lower stiffness to the above study, the MGAT5 gene as well as other MGAT genes were not overexpressed, implying that multibranched N-glycans were probably not responsible for migration. In contrast to glioma stem cells, U87 cells showed no overexpression of stem cell markers, such as SOX2, ZEB1, ATXN1, ALCAM, CD9, ITGA7, CD44 and CHI3L1.

Although much is known regarding PTEN regulating chemotaxis and migration [87], we prove a pivotal role for PTEN controlling durotaxis and migration speed using aligned microfibers. Manipulating PTEN with a PTEN inhibitor, or rescuing PTEN function, completely recovered durotaxis and migration of LN18 PTEN wt cells or blunted cell speed of U87, respectively. The 2D studies of Tamura et al. showed that NIH3T3 mouse embryonic fibroblasts and U87 cells overexpressing *PTEN* lowered 2D migration via reduced integrin-mediated cell spreading and focal adhesions [88]. The latter study also showed that *PTEN* inhibition enhanced cell migration using 2D scratch assays. Furthermore, our findings reveal an essential role of RHO or downstream ROCK kinase activity for durotaxis and migration. Corroborating our U87 RNA-seq, where RHOB was the highest significantly expressed gene (46.15-fold), which was confirmed with qPCR for U87 and MDA-MB-468, inhibition of RHO halted both durotaxis and migration without cell death. In addition, inhibiting ROCK also significantly blunted migration speed. Another ROCK inhibitor, Fasudil®, reduced invasion of GBM cells xenografted into mice [89]. In contrast, using MDA-MB-231 cells in confined environments, such as 3 μm channels or GBM cells on 2D surfaces, showed no inhibition of migration and speed with ROCK inhibitor Y-27632, supporting no Myosin II involvement using these conditions [90,91]. Taken together, our above findings, along with the literature, support an essential regulatory role for the RHOB/ROCK/PTEN signaling pathway controlling durotaxis and migration.

Concerning cell morphology, Gong et al. showed that RHOB overexpression, along with RHOB shuttling from endosomes to the plasma membrane, led to cell blebbing and increased amoeboid migration in 3D collagen [35]. In contrast, a morphological switch from an amoeboid to a more elongated lamellipodial shaped cell involved RHOA signaling [54]. We also detected a significant switch of PTEN rescued U87 from an amoeboid shape to an elongated flat phenotype with a lower migration speed. Furthermore, we discovered that U87 and MDA-MB-468 cells with homozygote *PTEN* loss of function and increased *RHOB* expression had distinct amoeboid cell shapes migrating through a 3D matrix towards scaffolds (Figure S1A), and also on aligned microfibers (Figures 7B,C and 8A–G). Further support for an amoeboid shape showed significant increased cell heights for U87 and MDA-MB-468 cells, compared to more elongated cells with PTEN wt and RHOB low expression. Normal mammary epithelial cells spread more on a stiff matrix (5000 Pa) compared to a soft matrix of 140 Pa, which was shown to be dependent on RHO, ROCK or Myosin [92].

Our confocal imaging of amoeboid shaped cells also showed enhancement of F-actin along the cell rim or cortex, which extended into different sized protrusions at the leading edge of the cell. The nucleus was mainly positioned in the cell rear. An enriched F-actin cortex has previously been noted in spherical-like shaped cells, presenting a single barrier along the cell circumference [93]. It is notable that RHOB recruits DIAPH1 to endosomes forming an F-actin coat [94]. It would be interesting to analyze if RHOB and DIAPH1 also play a role in F-actin membrane coating, due to the fact that RHOB shuttles between endosomes and plasma membrane [35]. Increased actin-myosin contraction also associates with highly motile round cancer cells along with decreased adhesion, thus supporting a basis for our observations of fast amoeboid cell movements on aligned microfibers with less cell surface area [95].

It is known that different cellular protrusions can be found associated with amoeboid-shaped cells, contributing to cell movement [96]. It was shown that single cellular focal adhesions autonomously sense the environment for stiffness and rigidity [17]. The latter study proved that the sensing of focal adhesions via the Paxillin/Vinculin/Focal adhesion kinase (PTK2) pathway was essential for durotaxis, but not chemotaxis. Our RNA-seq of U87 cells in 3D with scaffolds versus 2D showed an induction of the adhesion adaptor gene Paxillin (2.46-fold), but not of Vinculin and Focal adhesion kinase. Additionally, it was shown that soft matrices suppressed actin-myosin assembly and orientation. Specifically, myosin heavy chain IIA (MYH9) initiated formation of actin-myosin filaments and myosin heavy chain IIB (MYH10) bound and stabilized these filaments. In contrast to soft matrices, MYH9 did not polarize, but MYH10 was induced during durotaxis with stiff substrates [97], which we found to be 7.5-fold upregulated with our RNA-seq of U87 in 3D with scaffolds. Lastly, regarding the nucleus position of migrating cells, it was shown that the direction of cell migration appeared to be driven by positioning the nucleus towards the cell rear in multiple cell types via microtubules, with a possible extension of the leading edge [98]. Rearward orientation of the nucleus was also regulated by activation of RHO signaling pathways [99].

Our scanning electron microscopy findings showing alignment of single amoeboid cells on individual 10 μm fibers within a single 100 μm aligned microfiber tract, support a topographical cue for migration (Figures 3B and 7B). On the other hand, elongated cells appeared to topographically sense and stretch over several PCL fibers. Although the majority of topotaxis studies address cells sensing specific 3D nanostructures [20], our laminin-coated aligned microfibers represent smooth 10 μm round tube-like structures. Lastly, for both amoeboid U87 and MDA-MB-468 cells with high RHOB expression and PTEN loss of function, we recorded different plasticity modes, which we coined "skating snails". For example, U87 cell surface lengths along the aligned microfibers corresponded with different speeds. Fast migrating U87 cells showed a surface length of 10 μm, whereas slow and halted U87 cells were 20 μm or 30 μm in length, respectively (Figure 8H). In line with our findings, Maiuri et al. showed that cell speed and persistence are exponentially coupled via actin flow [100]. Although U87 and MDA-MB-468 showed similar plasticity modes, U87 cells were more persistent, "skating" for longer distances than MDA-MB-468 cells. Thus, we support the idea that these cellular plasticity modes reflect intrinsic genetic differences between GBM and TNBC cells.

Other publications have also implemented biomaterials or ECM fiber structures to mimic brain structures, such as axons, to investigate tumor cell migration in the brain [49,101–103]. All these publications tested fiber sizes, fiber widths, single, stacked or parallel fibers to resemble axon and blood vessel structures. One common finding among all studies was the striking different plasticity modes of cells migrating on various fiber constructs. Doyle et al. demonstrated that cells migrating on 1.5 μm single fibrillar lines (also called 1D) composed of fibronectin were identical to cells migrating in a 3D fibrillar matrix but were significantly slower on 2D fibronectin coated substrates [101]. The authors concluded that both 1D and 3D cellular migration were comparable based upon different properties of migrating cells. For example, cells had a uniaxial cell morphology with a single lamellipodia and a coordinated protrusion-retraction cycle, which resulted in a unidirectional rapid migration independent of ECM density, but dependent on myosin II contractility.

5. Conclusions

Our findings validate a significant role of specific *PTEN* genotypes along with RHOB signaling controlling cell morphology, durotaxis, adhesion, cellular plasticity and migration speed of tumor cells. For both GBM and TNBC types with homozygote *PTEN* loss of function and high RHOB expression, we propose that the cellular plasticity modes of amoeboid cells promote enhanced durotaxis, topotaxis, increased migratory speeds and longer traveling distances, which could lead to faster dissemination throughout the brain. The fact that PTEN regulates the cell polarity of the phosphatidylinositol-3,4,5-triphosphate/phosphatidylinositol-4,5-bisphosphate gradient supports the idea that an aberrant gradient due to PTEN loss of function could be responsible for the above deregulations. Presently clinical treatments have not yet achieved the goal of extending the survival of patients with GBM or breast cancer patients with brain metastasizing tumors. RHO GTPase inhibitors have been implemented in preclinical xenografted animal studies demonstrating anti-tumor effects [104]. Data from TCGA studies support that RHOA is mainly deleted among all major cancer types, thus being a possible tumor suppressor such as PTEN and would not be a proper candidate for inhibitors [105]. On the other hand, TCGA data supports RHOB as a proto-oncogene amplified in the majority of all major tumors, including GBM and invasive breast cancers [105]. Thus, RHOB could represent a target for therapeutic treatment. This opens the door for new therapeutic drugs regarding tumors overexpressing RHOB targeting cell migration. Our study demonstrates the importance of PTEN and RHOB not only for primary and metastasizing tumors, but also for circulating tumor cells upon entry into a stiffer environment, where RHOB inhibitors could play an essential role for improved therapy.

Supplementary Materials: The following are available online at https://www.mdpi.com/article/10.3390/cancers13205144/s1. Figure S1: Durotaxis, topotaxis and cell adhesion/network formation on 3D scaffolds, Figure S2: PTEN splice mutation analysis and RHOA/RHOB expression in GBM and TNBC cells, Figure S3: The skating snail-like cellular plasticity modes of migration, Figure S4: Uncropped Figure S2B (upper), Figure S5: Uncropped Figure S2B (down), Table S1: Ten most significant up-regulated genes of U87 in 3D vs. 2D using RNA-seq, Table S2: Ten most significant down-regulated genes of U87 in 3D vs. 2D using RNA-seq, Video S1: U87-td-farnesyl expressing cells in 3D Matrigel without scaffold, Video S2: U87-td-farnesyl in Matrigel with in 3D Matrigel with scaffold, Video S3: U87 on aligned 3D microfibers, Video S4: LN18 on aligned 3D microfibers, Video S5: LN18 in the presence of PTEN inhibitor on 3D aligned microfibers, Video S6: MDA-MB-468 on aligned 3D microfibers, Video S7: MDA-MB-231 on aligned 3D microfibers, Video S8: U87 in the presence of RHO inhibitor on 3D aligned microfiber, Video S9: U87 skating snail on 3D aligned microfibers, Video S10: MDA-MB-468 skating snail on 3D aligned microfibers.

Author Contributions: R.S. and P.L.S. conceived, designed and coordinated the study; A.W. and H.S. designed and performed the experiments; E.B. and P.D.D. designed and produced the biomaterials; D.J. performed 3D computer reconstructions from confocal images; A.W. and R.S. prepared the figures; M.E. performed the TCGA analysis; R.S. and P.L.S. wrote and A.W. helped to finalize the manuscript; A.W., H.S., E.B., P.D.D., D.J., M.E., M.W.B., R.S. and P.L.S. edited the manuscript. All authors have read and agreed to the published version of the manuscript.

Funding: This research project was funded by the Deutsche Forschungsgemeinschaft (DFG, German Research Foundation)—Project number 326998133—TRR 225 (subproject B01).

Institutional Review Board Statement: Not applicable, because no primary patient material was used in this study.

Informed Consent Statement: Not applicable, because no primary patient material was used in this study.

Data Availability Statement: The datasets used and/or analyzed during the current study are available from the corresponding author on reasonable request.

Acknowledgments: Technical assistance of Carmen Villmann, Natascha Schaefer, Lena Fischer and Adrian Weich is greatly appreciated. The present work was performed by A.W. in partial fulfillment of the requirements for obtaining the degree "Dr. rer. nat." at the Friedrich-Alexander-University Erlangen-Nuernberg (FAU), Germany.

Conflicts of Interest: The authors declare no conflict of interest. No funding from commercial sources was obtained for the present study. However, ME has financial relations to disclose: he received speaker's bureau honoraria from Janssen, Roche, Astellas, Diaceutics, MSD, AstraZeneca; grant support from AstraZeneca, Janssen, STRATIFYER, and Diaceutics; advisory honoraria, employment and personal fees from Diaceutics, AstraZeneca, GenomicHealth, and Janssen.

References

1. Hanahan, D.; Weinberg, R.A. Hallmarks of cancer: The next generation. *Cell* **2011**, *144*, 646–674. [CrossRef] [PubMed]
2. Ridley, A.J. Rho GTPase signalling in cell migration. *Curr. Opin. Cell Biol.* **2015**, *36*, 103–112. [CrossRef] [PubMed]
3. Friedl, P.; Sahai, E.; Weiss, S.; Yamada, K.M. New dimensions in cell migration. *Nat. Rev. Mol. Cell Biol.* **2012**, *13*, 743–747. [CrossRef] [PubMed]
4. Papusheva, E.; Heisenberg, C.P. Spatial organization of adhesion: Force-dependent regulation and function in tissue morphogenesis. *EMBO J.* **2010**, *29*, 2753–2768. [CrossRef]

5. Frantz, C.; Stewart, K.M.; Weaver, V.M. The extracellular matrix at a glance. *J. Cell Sci.* **2010**, *123*, 4195–4200. [CrossRef]
6. Pedron, S.; Becka, E.; Harley, B.A. Regulation of glioma cell phenotype in 3D matrices by hyaluronic acid. *Biomaterials* **2013**, *34*, 7408–7417. [CrossRef] [PubMed]
7. Ulrich, T.A.; de Juan Pardo, E.M.; Kumar, S. The mechanical rigidity of the extracellular matrix regulates the structure, motility, and proliferation of glioma cells. *Cancer Res.* **2009**, *69*, 4167–4174. [CrossRef] [PubMed]
8. Pointer, K.B.; Clark, P.A.; Schroeder, A.B.; Salamat, M.S.; Eliceiri, K.W.; Kuo, J.S. Association of collagen architecture with glioblastoma patient survival. *J. Neurosurg.* **2017**, *126*, 1812–1821. [CrossRef]
9. Yousif, L.F.; Di Russo, J.; Sorokin, L. Laminin isoforms in endothelial and perivascular basement membranes. *Cell Adhes. Migr.* **2013**, *7*, 101–110. [CrossRef]
10. Mair, D.B.; Ames, H.M.; Li, R. Mechanisms of invasion and motility of high-grade gliomas in the brain. *Mol. Biol. Cell* **2018**, *29*, 2509–2515. [CrossRef]
11. Vega, F.M.; Fruhwirth, G.; Ng, T.; Ridley, A.J. RhoA and RhoC have distinct roles in migration and invasion by acting through different targets. *J. Cell Biol.* **2011**, *193*, 655–665. [CrossRef]
12. Rossi, J.; Giaccherini, L.; Cavallieri, F.; Napoli, M.; Moratti, C.; Froio, E.; Serra, S.; Fraternali, A.; Ghadirpour, R.; Cozzi, S.; et al. Extracranial metastases in secondary glioblastoma multiforme: A case report. *BMC Neurol.* **2020**, *20*, 382. [CrossRef] [PubMed]
13. Hara, A.; Kanayama, T.; Noguchi, K.; Niwa, A.; Miyai, M.; Kawaguchi, M.; Ishida, K.; Hatano, Y.; Niwa, M.; Tomita, H. Treatment Strategies Based on Histological Targets against Invasive and Resistant Glioblastoma. *J. Oncol.* **2019**, *2019*, 2964783. [CrossRef] [PubMed]
14. Thakkar, J.P.; Dolecek, T.A.; Horbinski, C.; Ostrom, Q.T.; Lightner, D.D.; Barnholtz-Sloan, J.S.; Villano, J.L. Epidemiologic and molecular prognostic review of glioblastoma. *Cancer Epidemiol. Biomark. Prev.* **2014**, *23*, 1985–1996. [CrossRef]
15. Scherer, H.J. Structural development in gliomas. *Am. J. Cancer* **1938**, *34*, 333–351.
16. Farin, A.; Suzuki, S.O.; Weiker, M.; Goldman, J.E.; Bruce, J.N.; Canoll, P. Transplanted glioma cells migrate and proliferate on host brain vasculature: A dynamic analysis. *Glia* **2006**, *53*, 799–808. [CrossRef] [PubMed]
17. Plotnikov, S.V.; Pasapera, A.M.; Sabass, B.; Waterman, C.M. Force fluctuations within focal adhesions mediate ECM-rigidity sensing to guide directed cell migration. *Cell* **2012**, *151*, 1513–1527. [CrossRef] [PubMed]
18. SenGupta, S.; Parent, C.A.; Bear, J.E. The principles of directed cell migration. *Nat. Rev. Mol. Cell Biol.* **2021**, *22*, 529–547. [CrossRef]
19. Caballero, D.; Comelles, J.; Piel, M.; Voituriez, R.; Riveline, D. Ratchetaxis: Long-Range Directed Cell Migration by Local Cues. *Trends Cell Biol.* **2015**, *25*, 815–827. [CrossRef] [PubMed]
20. Park, J.; Kim, D.H.; Kim, H.N.; Wang, C.J.; Kwak, M.K.; Hur, E.; Suh, K.Y.; An, S.S.; Levchenko, A. Directed migration of cancer cells guided by the graded texture of the underlying matrix. *Nat. Mater.* **2016**, *15*, 792–801. [CrossRef] [PubMed]
21. Lv, Y.; Ma, X.; Du, Y.; Feng, J. Understanding Patterns of Brain Metastasis in Triple-Negative Breast Cancer and Exploring Potential Thhuaerapeutic Targets. *OncoTargets Ther.* **2021**, *14*, 589–607. [CrossRef]
22. Brosnan, E.M.; Anders, C.K. Understanding patterns of brain metastasis in breast cancer and designing rational therapeutic strategies. *Ann. Transl. Med.* **2018**, *6*, 163. [CrossRef] [PubMed]
23. Leslie, N.R.; Yang, X.; Downes, C.P.; Weijer, C.J. PtdIns(3,4,5)P(3)-dependent and -independent roles for PTEN in the control of cell migration. *Curr. Biol.* **2007**, *17*, 115–125. [CrossRef] [PubMed]
24. Gerisch, G.; Schroth-Diez, B.; Muller-Taubenberger, A.; Ecke, M. PIP3 waves and PTEN dynamics in the emergence of cell polarity. *Biophys. J.* **2012**, *103*, 1170–1178. [CrossRef]
25. Ohgaki, H.; Dessen, P.; Jourde, B.; Horstmann, S.; Nishikawa, T.; Di Patre, P.L.; Burkhard, C.; Schuler, D.; Probst-Hensch, N.M.; Maiorka, P.C.; et al. Genetic pathways to glioblastoma: A population-based study. *Cancer Res.* **2004**, *64*, 6892–6899. [CrossRef]
26. Furukawa, K.; Kumon, Y.; Harada, H.; Kohno, S.; Nagato, S.; Teraoka, M.; Fujiwara, S.; Nakagawa, K.; Hamada, K.; Ohnishi, T. PTEN gene transfer suppresses the invasive potential of human malignant gliomas by regulating cell invasion-related molecules. *Int. J. Oncol.* **2006**, *29*, 73–81. [CrossRef]
27. Witzel, I.; Oliveira-Ferrer, L.; Pantel, K.; Muller, V.; Wikman, H. Breast cancer brain metastases: Biology and new clinical perspectives. *Breast Cancer Res.* **2016**, *18*, 8. [CrossRef] [PubMed]
28. Csolle, M.P.; Ooms, L.M.; Papa, A.; Mitchell, C.A. PTEN and Other PtdIns(3,4,5)P3 Lipid Phosphatases in Breast Cancer. *Int. J. Mol. Sci.* **2020**, *21*, 9189. [CrossRef]
29. Milella, M.; Falcone, I.; Conciatori, F.; Cesta Incani, U.; Del Curatolo, A.; Inzerilli, N.; Nuzzo, C.M.; Vaccaro, V.; Vari, S.; Cognetti, F.; et al. PTEN: Multiple Functions in Human Malignant Tumors. *Front. Oncol.* **2015**, *5*, 24. [CrossRef]
30. Hohensee, I.; Chuang, H.N.; Grottke, A.; Werner, S.; Schulte, A.; Horn, S.; Lamszus, K.; Bartkowiak, K.; Witzel, I.; Westphal, M.; et al. PTEN mediates the cross talk between breast and glial cells in brain metastases leading to rapid disease progression. *Oncotarget* **2017**, *8*, 6155–6168. [CrossRef] [PubMed]
31. Wikman, H.; Lamszus, K.; Detels, N.; Uslar, L.; Wrage, M.; Benner, C.; Hohensee, I.; Ylstra, B.; Eylmann, K.; Zapatka, M.; et al. Relevance of PTEN loss in brain metastasis formation in breast cancer patients. *Breast Cancer Res.* **2012**, *14*, R49. [CrossRef] [PubMed]
32. Adamo, B.; Deal, A.M.; Burrows, E.; Geradts, J.; Hamilton, E.; Blackwell, K.L.; Livasy, C.; Fritchie, K.; Prat, A.; Harrell, J.C.; et al. Phosphatidylinositol 3-kinase pathway activation in breast cancer brain metastases. *Breast Cancer Res.* **2011**, *13*, R125. [CrossRef] [PubMed]

33. Lawson, C.D.; Ridley, A.J. Rho GTPase signaling complexes in cell migration and invasion. *J. Cell Biol.* **2018**, *217*, 447–457. [CrossRef] [PubMed]
34. Li, Z.; Dong, X.; Wang, Z.; Liu, W.; Deng, N.; Ding, Y.; Tang, L.; Hla, T.; Zeng, R.; Li, L.; et al. Regulation of PTEN by Rho small GTPases. *Nat. Cell Biol.* **2005**, *7*, 399–404. [CrossRef]
35. Gong, X.; Didan, Y.; Lock, J.G.; Stromblad, S. KIF13A-regulated RhoB plasma membrane localization governs membrane blebbing and blebby amoeboid cell migration. *EMBO J.* **2018**, *37*, e98994. [CrossRef] [PubMed]
36. Loessner, D.; Rockstroh, A.; Shokoohmand, A.; Holzapfel, B.M.; Wagner, F.; Baldwin, J.; Boxberg, M.; Schmalfeldt, B.; Lengyel, E.; Clements, J.A.; et al. A 3D tumor microenvironment regulates cell proliferation, peritoneal growth and expression patterns. *Biomaterials* **2019**, *190-191*, 63–75. [CrossRef] [PubMed]
37. Hutmacher, D.W.; Loessner, D.; Rizzi, S.; Kaplan, D.L.; Mooney, D.J.; Clements, J.A. Can tissue engineering concepts advance tumor biology research? *Trends Biotechnol.* **2010**, *28*, 125–133. [CrossRef]
38. Kim, J.; Bakirci, E.; O'Neill, K.L.; Hrynevich, A.; Dalton, P.D. Fiber Bridging during Melt Electrowriting of Poly(epsilon-Caprolactone) and the Influence of Fiber Diameter and Wall Height. *Macromol. Mater. Eng.* **2021**, *306*. [CrossRef]
39. Pham, Q.P.; Sharma, U.; Mikos, A.G. Electrospinning of polymeric nanofibers for tissue engineering applications: A review. *Tissue Eng.* **2006**, *12*, 1197–1211. [CrossRef] [PubMed]
40. Robinson, T.M.; Hutmacher, D.W.; Dalton, P.D. The Next Frontier in Melt Electrospinning: Taming the Jet. *Adv. Funct. Mater.* **2019**, *29*, 1904664. [CrossRef]
41. Visser, J.; Melchels, F.P.; Jeon, J.E.; van Bussel, E.M.; Kimpton, L.S.; Byrne, H.M.; Dhert, W.J.; Dalton, P.D.; Hutmacher, D.W.; Malda, J. Reinforcement of hydrogels using three-dimensionally printed microfibres. *Nat. Commun.* **2015**, *6*, 6933. [CrossRef] [PubMed]
42. Janzen, D.; Bakirci, E.; Wieland, A.; Martin, C.; Dalton, P.D.; Villmann, C. Cortical Neurons form a Functional Neuronal Network in a 3D Printed Reinforced Matrix. *Adv. Healthc. Mater.* **2020**, *9*, e1901630. [CrossRef] [PubMed]
43. Weigand, A.; Boos, A.M.; Tasbihi, K.; Beier, J.P.; Dalton, P.D.; Schrauder, M.; Horch, R.E.; Beckmann, M.W.; Strissel, P.L.; Strick, R. Selective isolation and characterization of primary cells from normal breast and tumors reveal plasticity of adipose derived stem cells. *Breast Cancer Res.* **2016**, *18*, 32. [CrossRef]
44. Furnari, F.B.; Lin, H.; Huang, H.S.; Cavenee, W.K. Growth suppression of glioma cells by PTEN requires a functional phosphatase catalytic domain. *Proc. Natl. Acad. Sci. USA* **1997**, *94*, 12479–12484. [CrossRef]
45. Hollestelle, A.; Elstrodt, F.; Nagel, J.H.; Kallemeijn, W.W.; Schutte, M. Phosphatidylinositol-3-OH kinase or RAS pathway mutations in human breast cancer cell lines. *Mol. Cancer Res.* **2007**, *5*, 195–201. [CrossRef] [PubMed]
46. Grosshaus, C.; Bakirci, E.; Berthel, M.; Hrynevich, A.; Kade, J.C.; Hochleitner, G.; Groll, J.; Dalton, P.D. Melt Electrospinning of Nanofibers from Medical-Grade Poly(epsilon-Caprolactone) with a Modified Nozzle. *Small* **2020**, *16*, e2003471. [CrossRef] [PubMed]
47. Cerami, E.; Gao, J.; Dogrusoz, U.; Gross, B.E.; Sumer, S.O.; Aksoy, B.A.; Jacobsen, A.; Byrne, C.J.; Heuer, M.L.; Larsson, E.; et al. The cBio cancer genomics portal: An open platform for exploring multidimensional cancer genomics data. *Cancer Discov.* **2012**, *2*, 401–404. [CrossRef] [PubMed]
48. Bakirci, E.; Schaefer, N.; Dahri, O.; Hrynevich, A.; Strissel, P.; Strick, R.; Dalton, P.D.; Villmann, C. Melt Electrowritten In Vitro Radial Device to Study Cell Growth and Migration. *Adv. Biosyst.* **2020**, *4*, e2000077. [CrossRef]
49. Jain, A.; Betancur, M.; Patel, G.D.; Valmikinathan, C.M.; Mukhatyar, V.J.; Vakharia, A.; Pai, S.B.; Brahma, B.; MacDonald, T.J.; Bellamkonda, R.V. Guiding intracortical brain tumour cells to an extracortical cytotoxic hydrogel using aligned polymeric nanofibres. *Nat. Mater.* **2014**, *13*, 308–316. [CrossRef] [PubMed]
50. Lo, C.M.; Wang, H.B.; Dembo, M.; Wang, Y.L. Cell movement is guided by the rigidity of the substrate. *Biophys. J.* **2000**, *79*, 144–152. [CrossRef]
51. DuChez, B.J.; Doyle, A.D.; Dimitriadis, E.K.; Yamada, K.M. Durotaxis by Human Cancer Cells. *Biophys. J.* **2019**, *116*, 670–683. [CrossRef]
52. Lau, L.W.; Cua, R.; Keough, M.B.; Haylock-Jacobs, S.; Yong, V.W. Pathophysiology of the brain extracellular matrix: A new target for remyelination. *Nat. Rev. Neurosci.* **2013**, *14*, 722–729. [CrossRef] [PubMed]
53. Gadea, G.; de Toledo, M.; Anguille, C.; Roux, P. Loss of p53 promotes RhoA-ROCK-dependent cell migration and invasion in 3D matrices. *J. Cell Biol.* **2007**, *178*, 23–30. [CrossRef] [PubMed]
54. Yamada, K.M.; Sixt, M. Mechanisms of 3D cell migration. *Nat. Rev. Mol. Cell Biol.* **2019**, *20*, 738–752. [CrossRef] [PubMed]
55. Fukunaga, T.; Fujita, Y.; Kishima, H.; Yamashita, T. Methylation dependent down-regulation of G0S2 leads to suppression of invasion and improved prognosis of IDH1-mutant glioma. *PLoS ONE* **2018**, *13*, e0206552. [CrossRef] [PubMed]
56. Podgorska, M.; Pietraszek-Gremplewicz, K.; Nowak, D. Apelin Effects Migration and Invasion Abilities of Colon Cancer Cells. *Cells* **2018**, *7*, 113. [CrossRef]
57. Tochio, T.; Tanaka, H.; Nakata, S.; Hosoya, H. Fructose-1,6-bisphosphate aldolase A is involved in HaCaT cell migration by inducing lamellipodia formation. *J. Dermatol. Sci.* **2010**, *58*, 123–129. [CrossRef] [PubMed]
58. Messina, A.; Ferraris, N.; Wray, S.; Cagnoni, G.; Donohue, D.E.; Casoni, F.; Kramer, P.R.; Derijck, A.A.; Adolfs, Y.; Fasolo, A.; et al. Dysregulation of Semaphorin7A/beta1-integrin signaling leads to defective GnRH-1 cell migration, abnormal gonadal development and altered fertility. *Hum. Mol. Genet.* **2011**, *20*, 4759–4774. [CrossRef]

59. Duan, S.; Huang, W.; Liu, X.; Liu, X.; Chen, N.; Xu, Q.; Hu, Y.; Song, W.; Zhou, J. IMPDH2 promotes colorectal cancer progression through activation of the PI3K/AKT/mTOR and PI3K/AKT/FOXO1 signaling pathways. *J. Exp. Clin. Cancer Res.* **2018**, *37*, 304. [CrossRef]
60. Bangasser, B.L.; Odde, D.J. Master equation-based analysis of a motor-clutch model for cell traction force. *Cell. Mol. Bioeng.* **2013**, *6*, 449–459. [CrossRef]
61. Prahl, L.S.; Bangasser, P.F.; Stopfer, L.E.; Hemmat, M.; White, F.M.; Rosenfeld, S.S.; Odde, D.J. Microtubule-Based Control of Motor-Clutch System Mechanics in Glioma Cell Migration. *Cell. Rep.* **2018**, *25*, 2591–2604.e2598. [CrossRef]
62. Simpson, K.J.; Selfors, L.M.; Bui, J.; Reynolds, A.; Leake, D.; Khvorova, A.; Brugge, J.S. Identification of genes that regulate epithelial cell migration using an siRNA screening approach. *Nat. Cell Biol.* **2008**, *10*, 1027–1038. [CrossRef] [PubMed]
63. Mitchison, T.; Kirschner, M. Cytoskeletal dynamics and nerve growth. *Neuron* **1988**, *1*, 761–772. [CrossRef]
64. Giannone, G.; Dubin-Thaler, B.J.; Rossier, O.; Cai, Y.; Chaga, O.; Jiang, G.; Beaver, W.; Dobereiner, H.G.; Freund, Y.; Borisy, G.; et al. Lamellipodial actin mechanically links myosin activity with adhesion-site formation. *Cell* **2007**, *128*, 561–575. [CrossRef] [PubMed]
65. Riveline, D.; Zamir, E.; Balaban, N.Q.; Schwarz, U.S.; Ishizaki, T.; Narumiya, S.; Kam, Z.; Geiger, B.; Bershadsky, A.D. Focal contacts as mechanosensors: Externally applied local mechanical force induces growth of focal contacts by an mDia1-dependent and ROCK-independent mechanism. *J. Cell Biol.* **2001**, *153*, 1175–1186. [CrossRef] [PubMed]
66. Hwang, T.L.; Close, T.P.; Grego, J.M.; Brannon, W.L.; Gonzales, F. Predilection of brain metastasis in gray and white matter junction and vascular border zones. *Cancer* **1996**, *77*, 1551–1555. [CrossRef]
67. Lu, Y.B.; Franze, K.; Seifert, G.; Steinhauser, C.; Kirchhoff, F.; Wolburg, H.; Guck, J.; Janmey, P.; Wei, E.Q.; Kas, J.; et al. Viscoelastic properties of individual glial cells and neurons in the CNS. *Proc. Natl. Acad. Sci. USA* **2006**, *103*, 17759–17764. [CrossRef] [PubMed]
68. Mahaffy, R.E.; Shih, C.K.; MacKintosh, F.C.; Kas, J. Scanning probe-based frequency-dependent microrheology of polymer gels and biological cells. *Phys. Rev. Lett.* **2000**, *85*, 880–883. [CrossRef]
69. Costa, A.R.; Pinto-Costa, R.; Sousa, S.C.; Sousa, M.M. The Regulation of Axon Diameter: From Axonal Circumferential Contractility to Activity-Dependent Axon Swelling. *Front. Mol. Neurosci.* **2018**, *11*, 319. [CrossRef]
70. Zhang, Y.; Abiraman, K.; Li, H.; Pierce, D.M.; Tzingounis, A.V.; Lykotrafitis, G. Modeling of the axon membrane skeleton structure and implications for its mechanical properties. *PLoS Comput. Biol.* **2017**, *13*, e1005407. [CrossRef]
71. Romano, A.; Scheel, M.; Hirsch, S.; Braun, J.; Sack, I. In vivo waveguide elastography of white matter tracts in the human brain. *Magn. Reson. Med.* **2012**, *68*, 1410–1422. [CrossRef] [PubMed]
72. Liao, D.; Arnett, D.K.; Tyroler, H.A.; Riley, W.A.; Chambless, L.E.; Szklo, M.; Heiss, G. Arterial stiffness and the development of hypertension. The ARIC study. *Hypertension* **1999**, *34*, 201–206. [CrossRef] [PubMed]
73. Mueller, B.; Lang, S.; Dominietto, M.; Rudin, M.; Schulz, G.; Deyhle, H.; Germann, M.; Pfeiffer, F.; David, C.; Weitkamp, T. High-resolution tomographic imaging of microvessels. *Proc. SPIE* **2008**, *7078*, 70780B. [CrossRef]
74. Winkler, F.; Kienast, Y.; Fuhrmann, M.; Von Baumgarten, L.; Burgold, S.; Mitteregger, G.; Kretzschmar, H.; Herms, J. Imaging glioma cell invasion in vivo reveals mechanisms of dissemination and peritumoral angiogenesis. *Glia* **2009**, *57*, 1306–1315. [CrossRef] [PubMed]
75. Pogoda, K.; Chin, L.; Georges, P.C.; Byfield, F.J.; Bucki, R.; Kim, R.; Weaver, M.; Wells, R.G.; Marcinkiewicz, C.; Janmey, P.A. Compression stiffening of brain and its effect on mechanosensing by glioma cells. *New J. Phys.* **2014**, *16*, 075002. [CrossRef] [PubMed]
76. Mohiuddin, E.; Wakimoto, H. Extracellular matrix in glioblastoma: Opportunities for emerging therapeutic approaches. *Am. J. Cancer Res.* **2021**, *11*, 3742–3754.
77. Pepin, K.M.; McGee, K.P.; Arani, A.; Lake, D.S.; Glaser, K.J.; Manduca, A.; Parney, I.F.; Ehman, R.L.; Huston, J., 3rd. MR Elastography Analysis of Glioma Stiffness and IDH1-Mutation Status. *AJNR Am. J. Neuroradiol.* **2018**, *39*, 31–36. [CrossRef] [PubMed]
78. Hanif, F.; Muzaffar, K.; Perveen, K.; Malhi, S.M.; Simjee, S.U. Glioblastoma Multiforme: A Review of its Epidemiology and Pathogenesis through Clinical Presentation and Treatment. *Asian Pac. J. Cancer Prev.* **2017**, *18*, 3–9. [CrossRef]
79. Lebelt, A.; Dzieciol, J.; Guzinska-Ustymowicz, K.; Lemancewicz, D.; Zimnoch, L.; Czykier, E. Angiogenesis in gliomas. *Folia Histochem. Cytobiol.* **2008**, *46*, 69–72. [CrossRef] [PubMed]
80. Paszek, M.J.; Zahir, N.; Johnson, K.R.; Lakins, J.N.; Rozenberg, G.I.; Gefen, A.; Reinhart-King, C.A.; Margulies, S.S.; Dembo, M.; Boettiger, D.; et al. Tensional homeostasis and the maligant phenotype. *Cancer Cell* **2005**, *8*, 241–254. [CrossRef] [PubMed]
81. Lautscham, L.A.; Kammerer, C.; Lange, J.R.; Kolb, T.; Mark, C.; Schilling, A.; Strissel, P.L.; Strick, R.; Gluth, C.; Rowat, A.C.; et al. Migration in Confined 3D Environments Is Determined by a Combination of Adhesiveness, Nuclear Volume, Contractility, and Cell Stiffness. *Biophys. J.* **2015**, *109*, 900–913. [CrossRef] [PubMed]
82. Watson, A.W.; Grant, A.D.; Parker, S.S.; Hill, S.; Whalen, M.B.; Chakrabarti, J.; Harman, M.W.; Roman, M.R.; Forte, B.L.; Gowan, C.C.; et al. Breast tumor stiffness instructs bone metastasis via maintenance of mechanical conditioning. *Cell Rep.* **2021**, *35*, 109293. [CrossRef] [PubMed]
83. Palama, I.E.; D'Amone, S.; Ratano, P.; Donatelli, A.; Liscio, A.; Antonacci, G.; Testini, M.; Di Angelantonio, S.; Ragozzino, D.; Cortese, B. Mechanical Durotactic Environment Enhances Specific Glioblastoma Cell Responses. *Cancers* **2019**, *11*, 643. [CrossRef]

84. Bloomekatz, J.; Grego-Bessa, J.; Migeotte, I.; Anderson, K.V. Pten regulates collective cell migration during specification of the anterior-posterior axis of the mouse embryo. *Dev. Biol.* **2012**, *364*, 192–201. [CrossRef] [PubMed]
85. Giampieri, S.; Manning, C.; Hooper, S.; Jones, L.; Hill, C.S.; Sahai, E. Localized and reversible TGFbeta signalling switches breast cancer cells from cohesive to single cell motility. *Nat. Cell Biol.* **2009**, *11*, 1287–1296. [CrossRef] [PubMed]
86. Marhuenda, E.; Fabre, C.; Zhang, C.; Martin-Fernandez, M.; Iskratsch, T.; Saleh, A.; Bauchet, L.; Cambedouzou, J.; Hugnot, J.P.; Duffau, H.; et al. Glioma stem cells invasive phenotype at optimal stiffness is driven by MGAT5 dependent mechanosensing. *J. Exp. Clin. Cancer Res.* **2021**, *40*, 139. [CrossRef] [PubMed]
87. Funamoto, S.; Meili, R.; Lee, S.; Parry, L.; Firtel, R.A. Spatial and temporal regulation of 3-phosphoinositides by PI 3-kinase and PTEN mediates chemotaxis. *Cell* **2002**, *109*, 611–623. [CrossRef]
88. Tamura, M.; Gu, J.; Matsumoto, K.; Aota, S.; Parsons, R.; Yamada, K.M. Inhibition of cell migration, spreading, and focal adhesions by tumor suppressor PTEN. *Science* **1998**, *280*, 1614–1617. [CrossRef] [PubMed]
89. Deng, L.; Li, G.; Li, R.; Liu, Q.; He, Q.; Zhang, J. Rho-kinase inhibitor, fasudil, suppresses glioblastoma cell line progression in vitro and in vivo. *Cancer Biol. Ther.* **2010**, *9*, 875–884. [CrossRef] [PubMed]
90. Balzer, E.M.; Tong, Z.; Paul, C.D.; Hung, W.C.; Stroka, K.M.; Boggs, A.E.; Martin, S.S.; Konstantopoulos, K. Physical confinement alters tumor cell adhesion and migration phenotypes. *FASEB J.* **2012**, *26*, 4045–4056. [CrossRef] [PubMed]
91. Smith, C.L.; Kilic, O.; Schiapparelli, P.; Guerrero-Cazares, H.; Kim, D.H.; Sedora-Roman, N.I.; Gupta, S.; O'Donnell, T.; Chaichana, K.L.; Rodriguez, F.J.; et al. Migration Phenotype of Brain-Cancer Cells Predicts Patient Outcomes. *Cell Rep.* **2016**, *15*, 2616–2624. [CrossRef] [PubMed]
92. Butcher, D.T.; Alliston, T.; Weaver, V.M. A tense situation: Forcing tumour progression. *Nat. Rev. Cancer* **2009**, *9*, 108–122. [CrossRef] [PubMed]
93. Charras, G.T.; Hu, C.K.; Coughlin, M.; Mitchison, T.J. Reassembly of contractile actin cortex in cell blebs. *J. Cell Biol.* **2006**, *175*, 477–490. [CrossRef]
94. Fernandez-Borja, M.; Janssen, L.; Verwoerd, D.; Hordijk, P.; Neefjes, J. RhoB regulates endosome transport by promoting actin assembly on endosomal membranes through Dia1. *J. Cell Sci.* **2005**, *118*, 2661–2670. [CrossRef] [PubMed]
95. Deakin, N.O.; Turner, C.E. Distinct roles for paxillin and Hic-5 in regulating breast cancer cell morphology, invasion, and metastasis. *Mol. Biol. Cell* **2011**, *22*, 327–341. [CrossRef]
96. Petrie, R.J.; Harlin, H.M.; Korsak, L.I.; Yamada, K.M. Activating the nuclear piston mechanism of 3D migration in tumor cells. *J. Cell Biol.* **2017**, *216*, 93–100. [CrossRef] [PubMed]
97. Raab, M.; Swift, J.; Dingal, P.C.; Shah, P.; Shin, J.W.; Discher, D.E. Crawling from soft to stiff matrix polarizes the cytoskeleton and phosphoregulates myosin-II heavy chain. *J. Cell Biol.* **2012**, *199*, 669–683. [CrossRef] [PubMed]
98. Gundersen, G.G.; Worman, H.J. Nuclear positioning. *Cell* **2013**, *152*, 1376–1389. [CrossRef] [PubMed]
99. Maninova, M.; Klimova, Z.; Parsons, J.T.; Weber, M.J.; Iwanicki, M.P.; Vomastek, T. The reorientation of cell nucleus promotes the establishment of front-rear polarity in migrating fibroblasts. *J. Mol. Biol.* **2013**, *425*, 2039–2055. [CrossRef] [PubMed]
100. Maiuri, P.; Rupprecht, J.F.; Wieser, S.; Ruprecht, V.; Benichou, O.; Carpi, N.; Coppey, M.; De Beco, S.; Gov, N.; Heisenberg, C.P.; et al. Actin flows mediate a universal coupling between cell speed and cell persistence. *Cell* **2015**, *161*, 374–386. [CrossRef] [PubMed]
101. Doyle, A.D.; Wang, F.W.; Matsumoto, K.; Yamada, K.M. One-dimensional topography underlies three-dimensional fibrillar cell migration. *J. Cell Biol.* **2009**, *184*, 481–490. [CrossRef] [PubMed]
102. Sharma, P.; Sheets, K.; Elankumaran, S.; Nain, A.S. The mechanistic influence of aligned nanofibers on cell shape, migration and blebbing dynamics of glioma cells. *Integr. Biol.* **2013**, *5*, 1036–1044. [CrossRef]
103. Estabridis, H.M.; Jana, A.; Nain, A.; Odde, D.J. Cell Migration in 1D and 2D Nanofiber Microenvironments. *Ann. Biomed. Eng.* **2018**, *46*, 392–403. [CrossRef] [PubMed]
104. Clayton, N.S.; Ridley, A.J. Targeting Rho GTPase Signaling Networks in Cancer. *Front. Cell Dev. Biol.* **2020**, *8*, 222. [CrossRef] [PubMed]
105. Svensmark, J.H.; Brakebusch, C. Rho GTPases in cancer: Friend or foe? *Oncogene* **2019**, *38*, 7447–7456. [CrossRef] [PubMed]

cancers

Review

3D Printing and Bioprinting to Model Bone Cancer: The Role of Materials and Nanoscale Cues in Directing Cell Behavior

Tiziana Fischetti [1], Gemma Di Pompo [2], Nicola Baldini [1,2,3,*], Sofia Avnet [1,†] and Gabriela Graziani [3,*,†]

1 Department of Biomedical and Neuromotor Sciences, Alma Mater Studiorum-Università di Bologna, 40138 Bologna, Italy; tiziana.fischetti2@unibo.it (T.F.); sofia.avnet3@unibo.it (S.A.)
2 Biomedical Science and Technologies Lab, IRCCS Istituto Ortopedico Rizzoli, 40136 Bologna, Italy; gemma.dipompo@ior.it
3 Laboratory of NanoBiotechnology, IRCCS Istituto Ortopedico Rizzoli, 40136 Bologna, Italy
* Correspondence: Nicola.baldini@ior.it (N.B.); gabriela.graziani@ior.it (G.G.)
† Equal contribution.

Simple Summary: Three-dimensional bioprinting is a promising tool for the study of cancer development and progression in bone, as it permits modeling the complexity of the microenvironment and cell-to-cell interactions. To this aim, an ideal model should combine a proper structure design, biomaterials selection, and the cellular counterpart. In this review, 3D-bioprinted bone systems obtained by different bioinks, and strategies, are discussed, aimed at mimicking the bone cancer microenvironment. The main challenges and unmet needs to reach perfect biomimicry are highlighted.

Abstract: Bone cancer, both primary and metastatic, is characterized by a low survival rate. Currently, available models lack in mimicking the complexity of bone, of cancer, and of their microenvironment, leading to poor predictivity. Three-dimensional technologies can help address this need, by developing predictive models that can recapitulate the conditions for cancer development and progression. Among the existing tools to obtain suitable 3D models of bone cancer, 3D printing and bioprinting appear very promising, as they enable combining cells, biomolecules, and biomaterials into organized and complex structures that can reproduce the main characteristic of bone. The challenge is to recapitulate a bone-like microenvironment for analysis of stromal–cancer cell interactions and biological mechanics leading to tumor progression. In this review, existing approaches to obtain in vitro 3D-printed and -bioprinted bone models are discussed, with a focus on the role of biomaterials selection in determining the behavior of the models and its degree of customization. To obtain a reliable 3D bone model, the evaluation of different polymeric matrices and the inclusion of ceramic fillers is of paramount importance, as they help reproduce the behavior of both normal and cancer cells in the bone microenvironment. Open challenges and future perspectives are discussed to solve existing shortcomings and to pave the way for potential development strategies.

Keywords: 3D printing; 3D bioprinting; bone cancer; calcium phosphates; bone model; orthopedics

Citation: Fischetti, T.; Di Pompo, G.; Baldini, N.; Avnet, S.; Graziani, G. 3D Printing and Bioprinting to Model Bone Cancer: The Role of Materials and Nanoscale Cues in Directing Cell Behavior. *Cancers* **2021**, *13*, 4065. https://doi.org/10.3390/cancers13164065

Academic Editors: Nicola Contessi Negrini, Alessandro Franchi and Serena Danti

Received: 12 July 2021
Accepted: 6 August 2021
Published: 12 August 2021

1. Introduction

Bone cancer can arise as primary (sarcomas) or metastatic lesions. Bone sarcomas, including osteosarcoma and Ewing's sarcoma, are highly aggressive tumors, mainly affecting pediatric patients and young adults [1]. Although the advent of chemotherapy has considerably prolonged life expectancy, bone sarcomas are still associated with a 5-year survival rate of approximately 50–60% due to their resistant or recurrent nature, thus representing a leading cause of cancer-related death in young people [2]. Bone is also the third most common metastatic site in patients affected by breast, prostate, lung, and renal carcinoma. Bone metastatic progression leads to 90% of death from cancer [3,4] and is associated with a significant decrease in the 5-year survival rates [5–9] and severe morbidities, including pain, fracture, hypercalcemia, and spinal cord compression [7,10,11].

At this stage, the disease is usually considered incurable, and treatment is only palliative, consisting of pain-relieving medication, radiation therapy, and surgery [12–14].

To date, the availability of tissue samples of bone cancers has been limited by the difficulty of reaching the bone site. Furthermore, adjuvant therapy is often administered prior to surgery. Because the therapy has some cytotoxicity, it may alter the integrity of DNA, RNA, and proteins, or interfere with the metabolic and the proliferation activities of cells of the tumor microenvironment, all prior to tissue sampling. As a consequence, this may alter the native characteristics and behavior of cells, thus affecting the relative molecular and morpho-histological analyses [15]. Consequently, the study of the biological mechanisms underlying bone tumors and the development of successful strategies for their treatment and prevention are very difficult. In this context, preclinical modeling of bone microenvironment appears to be a crucial and promising challenge. For decades, evaluation of cancer cell proliferation, migration, invasion, and drug response has relied on two-dimensional (2D) in vitro cell culture systems. However, such models fail to mimic the spatial, biochemical, and mechanical complexity of the native three-dimensional (3D) tumor microenvironment, that is, tissue architecture, severely limiting their interpretation in the study of primary and secondary bone cancer [2]. In vivo models overcome these drawbacks by mimicking the physiological context of tumor growth and progression, thus being more predictive of drug response compared to 2D cultures. Nevertheless, studies on animals are limited by ethical concerns, species-specific differences, and high costs. In addition, non-spontaneous cancer models, such as syngeneic, xenografts, or orthotropic models, also fail to recapitulate the paracrine circuits by which the bone niche modulates bone cancer progression and response to treatments. This is because they often develop too rapidly, which impedes the establishment of the natural interactions between cancer cells and stromal cells that occur in vivo [16–18].

Three-dimensional in vitro models can help to bridge the gap between preclinical in vitro and in vivo models, as they are highly reproducible, affordable, support the use of human cells, and can recapitulate the key features of the bone tumor niche, such as 3D cell–cell and cell–extracellular matrix (ECM) interactions, therefore facilitating mechanistic and drug response studies [19,20]. To this aim, to date, various 3D techniques have been developed, including multicellular spheroids, microfluidic chips, cell patterning techniques, and 3D printing [21]. Thanks to the combination of these advanced technologies with different types of biomaterials, versatile approaches can be obtained to develop 3D cellular constructs that can recapitulate the tumor microenvironment complexity.

To study bone tumors, the model shall mimic as closely as possible the composition and properties of the native bone tissue, merging biological and materials science-related requirements. Bone, however, is a complex tissue, composed of a mineral and an organic phase, and by cells, all arranged in a highly hierarchical structure [22–24]. Therefore, the model shall also possess a certain degree of complexity and fulfill several requirements. Among these, (i) excellent biocompatibility, (ii) suitable surface properties, (iii) adequate mechanical properties, (iv) a porous structure that can allow cell colonization and vascularization, and (v) tailored degradability are those identified as mandatory in the literature [22,25,26].

In this scenario, 3D bioprinting offers new perspectives, as it allows easily producing porous structures having finely tunable architecture, mechanical properties, and composition [27,28]. In these models, surface characteristics (morphology and roughness) can be modulated from the macro to the submicrometric scale by tailoring the model shape and porosity. By loading the models with nanoscale materials (nanocoatings or nanoparticles), features at the nanoscale can also be obtained while creating a hybrid organic/inorganic composition. The use of inorganic micro- or nano-fillers permits increasing printing fidelity, and further modulates degradability, stability, mechanical properties, and interactions with host cells [29–33]. However, in 3D printing, cells can be seeded onto the scaffolds but cannot be incorporated in the fibers, hindering the study of cell–cell and cell–ECM interactions [34,35]. To address these limitations, 3D bioprinting can be used to create complex in vitro cancer and bone models that can replicate different aspects of the 3D tumor microenvironment [36–38] and may be useful for understanding tumor heterogeneity and identify those mechanisms responsible for tumor progression and resistance to therapy [36,39]. More in detail, 3D bioprinting permits: (i) printing multiple cell types, including cancer and normal cells associated with the tumor microenvironment [40]; (ii) enabling the formation of vessel-like structures that are crucial to study the metastatic process and to assess anti-cancer drug delivery and responses [41]; (iii) modulating the composition of the exogenous ECM for what regards both the inorganic matrix and the various growth factors or signaling molecules [36,38]. This is important as it permits simulating the loose or dense connective tissues surrounding the cells in the tumor microenvironment, thus providing a reliable re-establishment of the existing crosstalk between cancer cells and neighboring matrices [21,38]. Consequently, both 3D printing and bioprinting, combined with nanoscale materials, appear promising to simulate specific properties of the bone tissue and study cancer.

In summary, biofabrication technologies, combined with specifically engineered materials, enable the printing of biomimetic 3D structures with detailed morphological features, from the millimeter to nanometer range [42], and fine control over the spatial positioning of cells during the bioprinting process [43–46]. For this reason, 3D printing and bioprinting have opened new routes, overcoming the limitations of overly simplified traditional 2D cultures in mimicking heterogeneity and complexity of both native [40] and tumor tissues [36,38].

In this review, we will focus on new trends in the development and manufacturing of 3D osteomimetic scaffolds obtained through additive manufacturing techniques for the study of osteosarcoma and bone metastases. The state of the art and advances regarding novel organic/composite bioinks, inorganic fillers, and new strategies for biomimetic scaffold development are systematically reviewed, and the main challenges, opportunities, and future perspectives are highlighted.

2. The Bone Microenvironment: Key Features for 3D Modeling of Bone Cancer

Bone sarcomas and bone metastases share the same environment and niche. Once established in bone, they interact with normal resident cells and with physical stimuli, including mechanical stress and hypoxia [47]. When cancer cells colonize the bone compartment, they start to proliferate, invade, and disrupt the normal bone matrix and acquire an osteomimetic phenotype, thereby interacting with bone cells [48–52].

2.1. Interaction with Normal Cells

The cellular environment of the bone tumor niche is comprised of complex and dynamic interactions between tumor and normal resident cells, including osteoblasts, osteoclasts, endothelial, immune, and hematopoietic cells, all of which are implicated in the pathogenesis of bone cancers [48]. Invading cancer cells strongly affect the activity of osteoblasts and osteoclasts (bone-forming cells and bone-resorbing cells, respectively), thereby disrupting the physiological balance of bone remodeling. Therefore, abnormal bone tissue formation and/or dysregulated bone resorption may occur, resulting in osteoblastic or osteolytic lesions [53] (Figure 1).

In osteoblastic lesions (i.e., osteosarcoma and prostate cancer metastases), tumor-derived growth factors (i.e., insulin growth factors (IGF)-1 and -2, transforming growth factor-beta (TGF-β), bone morphogenetic proteins (BMPs), platelet-derived growth factor (PDGF), endothelin-1 (ET-1), and fibroblasts growth factors (FGFs)) stimulate the differentiation and bone-forming activity of osteoblasts. In turn, osteoblasts produce growth factors that further stimulate tumor growth, such as interleukin-6 (IL-6), monocyte chemoattractant protein 1 (MCP-1), or vascular endothelial growth factor (VEGF) [54].

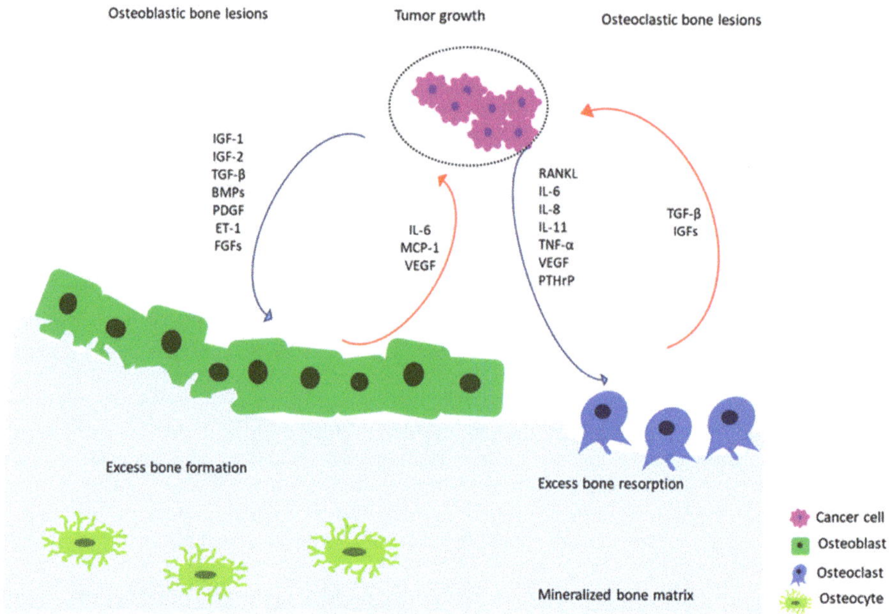

Figure 1. Schematic depiction of vicious cycle of bone metastasis involving the complex mutual interactions between tumor cells and bone cells in osteoblastic and osteoclastic bone lesions. Tumor cells secrete pro-osteoblastic (i.e., IGF-1 and -2, TGF-β, BMPs, PDGF, ET-1, and FGFs) or pro-osteoclastic (i.e., RANKL, IL-6, IL-8, IL-11, TNF-α, VEGF, and PTHrP) mediators (blue arrows) that induce bone formation or bone resorption, respectively. In turn, in osteoblastic lesions, osteoblasts produce pro-tumor growth factors (i.e., IL-6, MCP-1, and VEGF) that further stimulate the growth of cancer cells (red arrows). In osteolytic lesions, osteoclast-mediated bone resorption induced by cancer cells triggers the release of pro-tumor growth factors (i.e., IGFs and TGF-β) from the bone matrix, thus fueling the vicious cycle of cancer growth (red arrows).

In cancer-induced osteolytic bone disease, such as breast cancer metastases, cancer cells secrete a variety of cytokines and growth factors, including receptor activator of nuclear factor kappa-B ligand (RANKL), IL-6, IL-8, interleukin 11 (IL-11), tumor-necrosis factor-alpha (TNF-α), vascular endothelial growth factor (VEGF), and parathyroid hormone-related protein (PTHrP) [32], which directly or indirectly stimulate osteoclasts to resorb bone. The process of bone resorption, in turn, causes the release of additional growth factors from the matrix, such as IGFs and TGF-β, that can also favor cancer progression [48]. Among the paracrine pro-osteoclastogenic factors produced by cancer cells, RANKL, PTHrP, IL-11, and VEGF have particular relevance [55]. Furthermore, in the acidic tumor microenvironment, tumor-activated osteoblasts or mesenchymal stromal cells (MSCs) can secrete inflammatory cytokines, such as IL-6 and IL-8, that, in turn, further boost bone disruption and tumor progression [56].

However, only taking into consideration the interactions of osteoblasts and osteoclasts with tumor cells is an oversimplification. In fact, as for lung [57,58], brain [59–61], colon [62,63], and breast cancer [41,59], tumor-induced vasculature is a critical factor for the survival and proliferation of cancer cells in bone. During the uncontrolled growth of the tumor, oxygen and nutrient deprivation strongly stimulate the pro-angiogenic activity of tumors cells, inducing the secretion of angiogenic growth factors and cytokines, such VEGF and IL-8, into the surrounding extracellular microenvironment. This dysregulated signaling pathway activates the adjacent endothelial cells and perivascular cells, causing the recruitment of new blood vessels, which further support the growth of the tumor [64,65].

In summary, modeling the complex interactions between resident bone cells and tumor cells by considering all the different cellular players is crucial to recapitulate the molecular and cellular mechanisms of bone cancers in vitro.

2.2. Interaction with the Bone Matrix

Upon spreading to the skeleton, cancer cells not only interact with bone cells but also with the ECM. The latter is a dynamic structure including both organic and inorganic components that contribute to the functioning of the musculoskeletal system [66]. Although it is mainly composed of type I collagen (≈90%) [67], bone ECM also comprises several non-collagenous proteins, including fibronectin and lysyl oxidase (LOX). The inorganic phase of bone (≈75–80 wt.%), instead, is constituted by biogenic apatite (BA) nanocrystals (also called biological hydroxyapatite, or bone apatite), which allow for mineral exchange.

Biophysical properties of bone ECM are crucial in determining cell phenotype and behavior, both in normal and cancer cells. In fact, (i) the cell–matrix interactions in bone can affect cell migration, proliferation, survival, and remodeling [68], and (ii) ECM cues can promote tumor growth and decrease the response to therapeutics [69]. In addition, (iii) ECM stiffness can modulate the stemness and the expression of epithelial–mesenchymal transition (EMT) markers in osteosarcoma cells [70]. Finally, (iv) several studies have shown that hydroxyapatite (HA, $Ca_{10}(PO_4)_6(OH)_2$) can affect the behavior of normal and cancer cells [29,71], thereby validating the importance of including the ceramic counterpart in 3D in vitro tumor models [72]. In conclusion, reproducing a composition as similar as possible to that of native tissue is crucial to investigate how cues provided by the bone matrix can modulate cancer cell phenotype, growth, and chemoresistance.

3. Additive Manufacturing for Printing Bone-like Tissues

In bone oncology, the development of 3D models by additive manufacturing is still at its early stages, and several key issues are yet to be investigated. On the contrary, in orthopedics, biomimetic 3D constructs have been largely applied for the regeneration and repair of native bone tissue [44,73–78]. The knowledge acquired in this field can thus be advantageously translated to create models and solutions for the study of cancer cell development and progression in bone.

3.1. Printing Versus Bioprinting

Additive manufacturing is a very promising and versatile technique that allows the development of 3D constructs through a layer-by-layer process in which various biomaterials can be combined and possibly mixed with different cell types and/or growth factors [79,80]. Additive manufacturing is referred to as 3D bioprinting or printing, respectively, depending on whether cells are included in the printing process. Embedding the cells into the ink has some advantages and drawbacks that depend on the tissue to model and its specific characteristic. Models manufactured by 3D printing require the cell to be seeded on the surface of 3D constructs, so the technique is also known as "indirect bioprinting". These models permit high freedom in the choice of the materials to be printed, so they better mimic mechanical and structural properties of bone, as well as its degradation profile [80]. The so-obtained models may have long-term stability and can be inserted into bioreactors. However, seeding cells onto the 3D constructs does not allow for homogeneous cell dispersion and scaffold colonization [81,82], thus partially allowing for the simulation of cell–cell and cell–ECM interactions.

Instead, 3D bioprinting permits the creation of a defined distribution of cells and/or biomolecules inside the ink and hence across the fibers of the whole scaffold [26,45,83], which is necessary to mimic the biological complexity of cancer [44,84]. In addition, the use of natural hydrogels, having high water content, guarantees high biocompatibility and the possibility to tune the chemical and physical characteristics of the ink (including viscosity, crosslinking, and concentration, all determining shear stress) by selecting the appropriate polymeric matrix. As a consequence, they permit creating a microenvironment compatible with the medium-term survival of the cells embedded in the ink [85]. However, different from 3D-printed scaffolds, 3D-bioprinted constructs show limited mechanical properties and lower stability, so the models are not suitable for applications that require mimicking mechanical stress or are for use in bioreactors. In bone 3D printing, the most used synthetic polymers are polycaprolactone (PCL) [86–90], polyethylene glycol diacrylate (PEGDA) [31,78], and polylactic acid (PLA) [41], because of their mechanical strength, structural properties, and biocompatibility, both in vitro and in vivo. These polymers, largely used for 3D printing, have high rigidity and slow degradation rate [28,87,91]. However, they need to be processed in aggressive conditions (dilution in acid/toxic solvents and/or at high temperature) and cannot incorporate cell media [28,87]. Hence, cell loading of the polymeric fibers is impeded, which makes them unsuitable for bioprinting.

An optimal ink for 3D bioprinting shall fulfill the needs of high printing fidelity, shape maintenance, cell viability, and function [79]. These outcomes are affected by several parameters, both depending on the ink, such as: (i) chemical composition, including polymer concentration and molecular weight [85,92]; (ii) viscosity (hydrogels with shear-thinning characteristics are desired [92,93]), and cell density (suitable cell concentration in the order of 10^6 cells/mL, corresponding to approximately 5% of the total bioink volume [44]). The chemical composition of the ink is the main parameter regulating cells response, as described in Section 3.3.2. However, the ink's physical characteristics (i.e., polymer concentration, viscosity, and crosslinking mechanism) also have importance in determining cells viability in the short, medium, and long term [79,85,94,95].

More in detail: (i) high polymer concentrations permit obtaining dense polymer chains, resulting in increased mechanical properties and stiffness. However, increased density causes a lower diffusion rate of the nutrients, which reduces cell viability and proliferation [79,96]. (ii) High viscosity increases printability and shape fidelity but also shear stress, which negatively impacts cells viability [79,95,97]. Similarly, (iii) a high degree of crosslinking increases mechanical properties but decreases cell viability [79,85,95]. Consequently, the bioink choice is not trivial, as it depends on multiple and opposing parameters (Figure 2). For an extensive overview, see [68,69].

Based on these considerations, in 3D bone bioprinting, natural polymers are preferred, such as alginate [73,98–101], gelatin [98,100,102], and gelatin methacrylate (GelMA) [103–106], silk fibroin [107,108], chitosan [75,109,110], hyaluronic acid [76,111], fibrin [86,112], and collagen [31,109,113]. Considering biomimicry, collagen and its denatured counterpart, gelatin, are the most promising, although alginate is often preferred due to the easiness of printing. The selection of the hydrogel is of paramount importance, as it determines printability (shape fidelity and printing conditions), cell survival, and cell–cell interactions. Furthermore, for the development of bone tumor models, it must be considered that 70–75% of bone is composed of a mineralized phase [24,67,114,115], BA, which is a multi-substituted nanocrystalline HA. This phase has a strong influence on tumor and bone cell behavior (viability, morphology, differentiation, etc.) and promotes cancer cells' proliferation and release of IL-8 [29,71]. For this reason, a mineralized model can be more suitable than a polymeric one for the study of bone cancer. As already reported, there are two ways to include a mineralized fraction in 3D-printed and bioprinted scaffold: directly by adding micro/nanoparticles in the osteomimetic ink [30,31,116,117] or by grafting on the surface as a coating [117,118], the latter being more diffused in 3D printing technology [119].

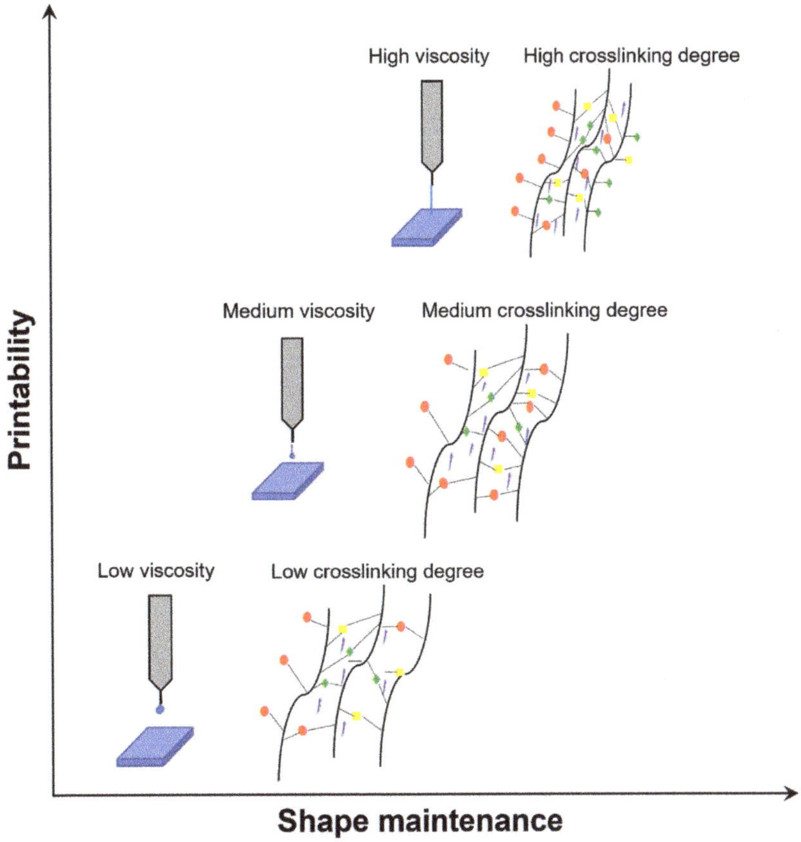

Figure 2. Trend of printability and shape maintenance depending on bioink viscosity (related to ink concentration) and crosslinking degree. Bioink type and the reported parameters need evaluation for each 3D bioprinting experiment. Generally, low/intermediate values of these parameters are preferable to guarantee cell viability.

Although a wide and increasing number of studies investigate 3D printing of biomimetic inks for applications in bone tissue regeneration, to date, only a few focus on 3D printing for bone tumor modeling, and an even lower number takes into account bioprinted constructs for tumor modeling (Figure 3). Moreover, not all the performed studies consider the inclusion of a biomimetic or non-biomimetic ceramic phase. Detailed examples and results will be reported in the next paragraphs. However, the increasing trend of research studies on these topics clearly shows their relevance.

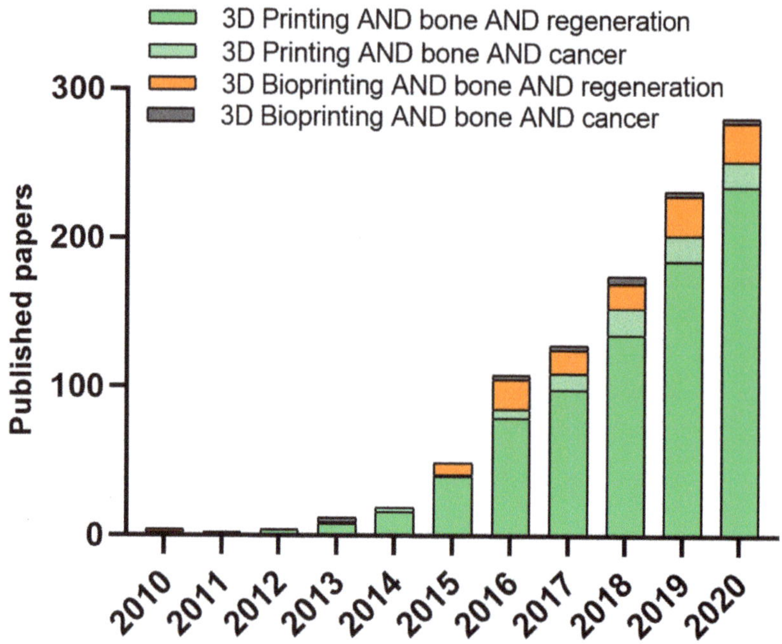

Figure 3. Comparison of uses of 3D printing and 3D bioprinting approaches over time (based on Web of Science, type of document was article, keywords for "3D printing" AND "bone" AND "regeneration", "3D printing" AND "bone" AND "cancer", "3D bioprinting" AND "bone" AND "regeneration" and "3D bioprinting" AND "bone" AND "cancer").

3.2. *Bioactive and Bioinert Bioceramic Fillers in Bioprinting*

Bioceramic can be used to functionalize both natural and synthetic polymers and tune mechanical properties, viscosity, and/or stability of the ink, as well as its architecture and biological properties [74,120–122]. In particular, the addition of bioceramic fillers, independently of their composition, provides increased mechanical properties [121,123,124], partially overcoming the intrinsic limitations of the hydrogels. At the same time, they affect the rheological characteristics of the inks and permit higher shape fidelity and stability over time [31,125]. To these aims, both bioinert and bioactive compounds can be selected.

However, the use of bioactive compounds (calcium phosphates CaPs and/or bioactive glasses—BG) can provide additional benefits: (i) the release of ions (such as Ca, P, Mg, Na, etc.) that are present in bone, simulating the tissue environment and interacting with healthy and tumor cells; (ii) topographical cues both at the micro- and the nano-scale, that can support cells adhesion to the models surface and directly influence their behavior (for instance, Nano-HA particles/coatings has been reported to direct early differentiation of MSCs [126–128]). Increased adhesion, in turn, facilitates seeding, cell spreading, and proper colonization of all parts of the model.

Among the most investigated bioceramic (for a detailed description of different types of bioceramic, see [129,130]), hydroxyapatite [31,75,99] is the most used because of its similarity to the inorganic phase of bone. However, although bone apatite is somehow similar to pure HA, they differ in terms of composition, ion doping, stoichiometry, crystallinity degree, crystal size/morphology, and, consequently, solubility and ions release into the biological medium [131]. Indeed, BA is characterized by low crystallinity and a high solubility and is ion-doped. In BA, carbonate ions substitute for hydroxide and phosphate ions, changing into the formula $Ca_{10-2x/3}(PO_4)_{6-x}(CO_3)_x(OH)_{2-x/3}$. Besides carbonates, BA contains a significant amount of foreign ions, such as magnesium, fluorine, strontium, silicate, zinc, and manganese, all having specific and significant biological roles [72]. Thus, in comparison to pure or stoichiometric HA, ion-substituted or BA better mimics normal bone tissue [72]. As a demonstration, a large number of studies on bone tissue regeneration have confirmed that BA improves adhesion and proliferation of osteoblast-like cells and osteogenic differentiation of osteoblast precursors, as demonstrated by increased alkaline phosphatase (ALP) and mineralization activity, both in vitro and in vivo [31,116,124,132–141].

Due to its putative pro-tumorigenic effect [142], HA has also been used to model bone cancers, in particular, osteolytic bone metastases from breast carcinoma [106,116]. Typically, secondary tumor formation in bone is considered a function of bone resorption because the degradation of bone mineral matrix releases bioactive ions and soluble growth factors that, in turn, are critical for the proliferation of both normal and cancer cells [143]. However, insoluble cues inherent to the inorganic component of the bone mineral matrix may also regulate metastatic growth by promoting adhesion, proliferation, and colonization of tumor cells, as highlighted in [116] when comparing a mineralized and a non-mineralized scaffold for breast cancer modeling. In the same study, the presence of HA in the mineralized scaffolds also promoted the release of IL-8 from breast cancer cells that, in turn, exerted pro-tumorigenic and pro-osteolytic effects [71], thereby supporting the vicious cycle of tumor growth and bone resorption. Hence, incorporation of a bone-like mineralized component into engineered cancer models may allow the study of the molecular mechanisms behind HA-induced metastasis in bone or, possibly, the study of HA-promoted drug resistance of cancer cells in bone. In another study on the same type of cancer [29], the effects of HA particles were studied by varying their size, crystallinity, and synthesis route, and assessing their effects on protein adsorption, cancer cells adhesion, growth, and IL-8 secretion. Protein adsorption, cell adhesion, and proliferation increased with decreasing HA crystallinity and crystal size. In contrast, IL-8 secretion reached the highest level in scaffolds with highly crystalline HA [29]. Data obtained by this study are very interesting, and it would be worthy to investigate the same behavior by using other cancer types that are prone to grow or metastasize to bone. However, although a large body of literature is already available on the use of functionalized nanoparticles and their biological role [31,117,124,144–146], research on printing of ion-doped CaPs is only in its early infancy, and further development is expected in the next years. Finally, of note, all the studied particles were used at one same scale (e.g., nanoparticles as nanoscale cues to modulate cells colonization in the scaffolds, but also to boost early differentiation and influence morphology), whereas the study of the effects of multi-scale particles is still unexplored.

3.3. Bioprinting: Cells and Bioinks

3.3.1. Normal Cells Used in 3D Bioprinting for Mimicking the Bone Microenvironment

The most used cells for 3D bone bioprinting in orthopedics are murine or human MSCs from either bone marrow or adipose tissue [87,110,113,147–151], murine calvarial MC3T3-E1 pre-osteoblast cells [46,68,152], and human fetal osteoblasts [105,147,153]. Printing of osteocytes has also been recently proposed [154]. These cells, when cultured in osteoinductive media or in media added with growth factors (i.e., bone morphogenic protein-2 BMP-2 [76,132,155–158], FGF-2 [112,159], VEGF [74,147]) and/or other additives (i.e., Ca^{++}) [102,107], express osteogenic markers (i.e., ALP, osteonectin (ON), osteopontin (OPN), osteocalcin (OCN)) and markers of late osteocyte phenotype (i.e., podoplanin (PDPN) and sclerostin (SOST)), and are able to mineralize.

3.3.2. Biomimetic Inks

An ideal biomimetic ink for cancer modeling should mimic the structural, physico-chemical, and biological properties of ECM. Indeed, to study the mechanisms underlying tumorigenesis and cancer progression, biomimetic inks should mimic the substrate on which the tumor develops and grows. On the other hand, they should reproduce ECM–cells interactions, thereby allowing the study of the mechanisms occurring in osteolytic or osteoblastic lesions, both in primary and secondary bone tumors.

Silk fibroin and chitosan [75,107,110,120,158,160,161], collagen [113,133,145,150,154] and hyaluronic acid [76,78,116,148,151], chemically modified (i.e., methacrylation reaction) or used in blends [98,105,116,161] are among the most widely natural hydrogels employed to induce bone formation. These polymers have biological and chemical features resembling the organic ECM components of bone native tissue. To best mimic bone ECM, the addition of inorganic counterparts as bioceramic (i.e., bioactive glass [30,31,87,100,156], β-tricalcium phosphate (TCP) [89,130,162], HAs [31,73,75,99,122,149], and nanoclays in powders or micron (nanoparticles) [74,78]) has been fully investigated. Three-dimensional bioprinting is promising in the obtainment of 3D cell-laden constructs based on osteomimetic inks combining polymeric matrix and ceramic fillers. To date, positive results have already been obtained regarding either the process, such as printability and extrudability, and the properties resulting from the obtained constructs, such as mechanical strength and stability maintenance. For instance, increased osteogenic ability and mineralization were obtained by using Laponite nanosilicates [74,78], Poly–Ca^{2+}complex (i.e., PolyP mixed with $CaCl_2$) [30,98] and β-TCP particles [89,130,162], even in absence of an osteogenic medium. These data offer important insights for the development of biomimetic models, as the same technologies can be adapted to reproduce "synthetic bone" with mineralized fractions. To this regard, the studies carried out for bone regeneration also stress the importance of an accurate selection of the following material characteristics: (i) composition (for instance, differences in biological behavior were assessed between scaffolds doped with different BG, with higher mineralization for borate instead of silicate glasses [30,71,152]); (ii) morphology (particle shape, dimensions, and surface features affect the overall behavior of the scaffold); and (iii) mechanical properties.

Most of the studies focus on HA, as it more closely resembles the composition of bone. HA positively influences mechanical and biological properties of the constructs, including the extent of mineralization and collagen production exerted by host cells [73,75,99,156], besides their osteogenic differentiation [19]. In these terms, HA shows more promising results compared to BG nanoparticles [31]. It has also been observed that results in osteoinduction and mineralization may be affected by the hydrogel combination with HA and by the specific characteristics of HA particles (for instance, carbonated HA nanoparticles show increased solubility and hence bioactivity, compared to the stoichiometric counterpart [75]; see Section 3.2).

3.3.3. Vascularized 3D-Bioprinted Bone-like Constructs

Along with the addition of inorganic fillers, a bioprinted model of bone cancer should incorporate vasculature, as it is essential to mimic both normal and cancer cell behavior. In normal musculoskeletal development and regenerative processes, blood vessels have different functions: (i) providing an efficient transport network for molecules and hematopoietic cells, (ii) nourishing niches for hematopoietic stem cells that reside within the bone marrow, and (iii) supporting bone formation and homeostasis [163]. On the other hand, during cancer progression, tumor-induced vascularization fosters tumor growth and dissemination by providing oxygen and nutrients and by supporting the intravasation and extravasation of cancer cells [164]. Tumor angiogenesis is initiated by environmental stresses, such as hypoxia and acidosis, leading to a disequilibrium in the pro-/anti-angiogenic balance and consequently to the increased expression of pro-angiogenic factors, including hypoxia-induced factor (HIF) and VEGF. Although the formation of a tumor vascular network starts from the existing healthy blood vessels, its expansion may be aided by additional processes, such as vasculogenesis and vascular mimicry [165].

Therefore, the development of 3D-bioprinted bone-like constructs incorporating vasculature is essential to recapitulate and study the multistep process of cancer development in bone, both for primary tumors and metastases. To reproduce the osteogenic and vasculogenic niches of bone in vascularized bone constructs, different approaches have been reported [86,104,118,152,165–168]. These include different combinations of composite materials (i.e., natural hydrogels blended with rigid polymers and bioceramic fillers) and cell types (i.e., MSCs and human umbilical vein endothelial cell HUVEC). Hence, a functional vascularized bone model should possess: (i) high mechanical stability and durability, (ii) specific biological cues, and (iii) osteoconductive properties, which can be obtained by the combined use of rigid/synthetic polymers, natural hydrogels, and bioceramic fillers, respectively [169]. The fulfillment of these requirements has shown promising results in the expression of osteogenic and angiogenic markers (i.e., Angiopoietin-1 (Ang-1), FGF-2 and VEGF). Furthermore, it has been reported that co-culturing HUVECs and MSCs boosted cell proliferation and vascular network development [166]. Finally, dynamic perfusion culture through a bioreactor system [168] can be beneficial for both bone and vascular regions, as the combination of liquid flux and mechanical cues (e.g., shear stress) enhance osteogenic differentiation, mineralization, and VEGF expression. The here-reported strategies, though lacking in reproducing the complexity in the combination of the vascular and bone region, are promising for the development of the vascular network in 3D bone construction.

4. 3D-Bioprinted Models of Bone Cancers

To date, only a few 3D-bioprinted in vitro cancer models have been proposed, and an even lower number has been published on bone sarcomas and metastases (Table 1). Among these, the majority exploits indirect 3D bioprinting, where cells (either tumor cells, bone cells, or co-cultures) are not embedded in the scaffold fibers but seeded onto its surface. These studies focus on: (i) the effects of scaffolds geometry and composition on cancer cells proliferation, (ii) cancer cell chemoresistance compared to 2D cultures, and (iii) the effects of the direct and indirect interplay between stromal and cancer cells.

Regarding bone sarcomas, a few studies have shown the effect of bioceramic fillers on proliferation and mineralization of 3D-bioprinted Saos-2 osteosarcoma cell lines (Section 3.3.2) [30,98]. Notably, although these studies consider SaOS-2 as osteoblast-like cells for bone tissue engineering applications, the obtained results can be directly translated to models for osteosarcoma growth in bone.

Among the different types of carcinomas that metastasize to bone, breast cancer is the most frequent and, hence, the most studied in the field of 3D bioprinting. In particular, breast cancer cells are often co-cultured with stromal cells of the bone microenvironment, such as the MSCs and the osteoblasts [19,78,147,170–174], since they support the key events in breast carcinoma metastasization and progression, including migration and drug resistance [175,176].

More in detail, Holmes et al. [173] used fused deposition modeling-based 3D bioprinting for studying bone colonization by breast cancer cells. Three dimensional bone scaffolds were obtained by PLA, then modified through carboxyl nanocrystalline HA coatings. Square and hexagon patterns (250 and 150 μm size) were chosen because they mimic the random orientations of ECM in bone. Among the chosen patterns, small hexagonal pores were the ones that allowed the highest proliferation of breast cancer cells. This study confirmed that the nanosurface texturization provided by HA offers a biomimetic and tunable bone model that can effectively simulate bone invasion and colonization by metastatic carcinoma cells [173].

Table 1. Summary of reports on 3D bone bioprinting for the development of 3D bone tumor in vitro models.

3D Printing Technology	Materials	Type of Cells	Results	Ref.
Extrusion bioprinter	Alginate, gelatin Overlay with agarose layer and PolyP-Ca^{++} complex (100 μM)	SaOS-2 (5×10^5 cells/mL)	- PolyPCa^{2+} enhanced structure stability - PolyPCa^{2+} metabolic degradation by cells - PolyPCa^{2+} modulator of gene expression in SaOS-2	[98]
Extrusion bioprinter	Alginate, gelatin Addition of PolyP, silica, or biosilica + BG nanoparticles (55 nm)	SaOS-2 (5×10^5 cells/mL)	- Formation of mineral nodules composed of Ca-phosphate, Ca-carbonate	[30]
Fused deposition modeling	PLA, HA coating (wet deposition)	MDA-MB-231, MSCs	- Young's moduli between 30 and 50 MPa, suitable for biomimetic mechanical cues - Effective adhesion of breast cancer cells on HA-coated scaffolds	[173]
Stereolithography bioprinter	Polyethylene glycol (PEG), PEG-DA nHA 10 wt% (wet deposition) Grain size: width = 25 nm, length = 50–100 nm	MDA-MB-231 (5×10^5 cells/scaffold, MSCs (1.5×10^5 cells/scaffold)	- 3D-printed scaffold retains native characteristics of in vivo tumor - Homogenous dispersion of HA nanoparticles in the scaffold - Larger number of spheroids and enhanced migration when HA was added to the scaffolds	[146]
Stereolithography bioprinter	PEG, PEG-DA nHA 10 wt% (wet deposition) Grain size: width = 25 nm, length = 50–100 nm	MDA-MB-231 (5×10^5 cells/scaffold), Human fetal osteoblasts (hFOBs) (5×10^5 cells/scaffold)	- Homogeneous dispersion of HA within the matrix - nHA-PEG suitable microenvironment for cell attachment and proliferation - Multicellular spheroids similar to natural tumor structure	[170]
Stereolithography bioprinter	GelMA (different concentrations), nHA 10 wt% (wet deposition) Grain size: width = 25 nm, length = 50–100 nm	MSCs or osteoblasts (1×10^6 cells/mL) MDA-MB-231 (1×10^6 cells/mL)	- Uniform porosity and good dispersion of nHA within the scaffolds - GelMA + nHA suitable for studying MSCs/breast cancer and osteoblasts/breast cancer cells in vitro	[147]
Stereolithography bioprinter	GelMA, PEGDA (different concentrations) nHA (different concentrations) (wet deposition) Grain size: width = 25 nm, length = 50–100 nm	MDA-MB-231 Endothelial cells hFOBs (1×10^4 cells/mL)	- Multi-interaction of tri-culture (cancer–vessel–tissue) - Mechanical properties lower than physiological range but suitable for bone cells growth - Vascular environment important for directional migration of cancer cells	[41]

Conversely, Zhu et al. used stereolithography-based 3D printing to create 3D bone models with 500 μm and 250 μm square and hexagonal pores and co-culture human MSCs and MDA-MB-231 breast cancer cells on the scaffold. In this study, the authors demonstrated that pattern geometry greatly influences cell proliferation. Small square patterns produced the strongest mitogenic effect. In this study, PEG and PEGDA resins were functionalized by HA nanoparticles and printed. MDA-MB-231 cells cultured on the 3D scaffolds were able to migrate and form distinct and spheroidal 3D structures (Figure 4a(i)), which was not observed in 2D culture. The obtained spheroidal morphology was emphasized when MDA-MB-231 were co-cultured with MSCs, thus showing the effect of the tumor-associated mesenchymal stroma in regulating cancer cell behavior (Figure 4a(ii)). Furthermore, the addition of HA nanoparticles promoted cell–matrix interactions and the formation of MDA-MB-231 larger spheroids compared to the bare 3D matrix. Finally, MDA-MB-231 cultured on the 3D scaffold showed a phenotype more resistant to the anti-cancer drug 5-fluorouracil compared to 2D matrices, possibly due to a reduced drug penetration in the 3D in vitro tumor microenvironment [146]. These findings further corroborate the existence of differences in the drug sensitivity of cancer cells when cultured in 3D instead of 2D models. Indeed, it has been widely demonstrated that 3D models recapitulate cell–matrix interactions and enhanced ECM synthesis, thereby mimicking the in vivo tumor microenvironment. In turn, ECM deposition reduces the penetration of drugs into the tumor mass [145], while 2D monolayered cell cultures are directly exposed to drug treatment.

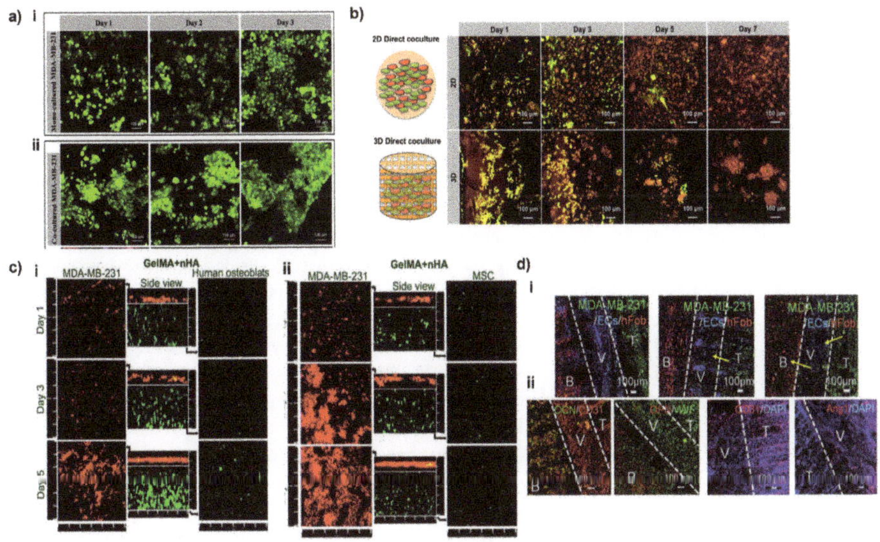

Figure 4. Effect of tumor-healthy cell interactions in co-culture systems and in 3D vs. 2D models. (**a**) Morphology of breast cancer cells cultured alone or with MSCs. (i) Confocal images of MDA-MB-231 alone, and (ii) in co-culture with MSCs; green fluorescence represents Cell Tracker Green™ stained breast cancer cells. Reproduced with the permission of © 2015 Elsevier Inc. All rights reserved [146]. (**b**) Enhanced spheroid formation by direct co-culture of hFOB and MDA-MB-231 cells on the 3D matrix in comparison to monolayer culture. hFOB and MDA-MB-231 were pre-stained with cell tracker green and orange, respectively. Reproduced with the permission of © IOP Publishing. All rights reserved [170]. (**c**) Confocal micrographs of osteoblasts/breast cancer cells (i) and MSCSs/breast cancer cells (ii) co-cultured in the 3D-bioprinted matrix after 1, 3, and 5 days. The middle columns represent the cross-sectional views. Osteoblasts and breast cancer cells were stained by Cell Tracker Green CMFDA dye (green) and Orange CMTMR dye (red), respectively. Reproduced with the permission of © 2016, American Chemical Society [147]. (**d**) Development of MDA-MB-231 cells metastasis and colonization toward bone over 14 d of the culture period. Cell tracker imaging was conducted to monitor the BrCa invasive process, including breast cancer growth, transendothelial migration, and colonization. The yellow arrows indicate the migration of invasive breast cancer cells. (i) Immunofluorescent images of hFOB and MDA-MB-231 cells in a vascular environment with DAPI staining after 14 d of culture. CD31 and vWF staining were used to identify both EC and breast cancer cells. (ii) Osteogenesis of hFOB was characterized by OCN and OPN staining. Combining CD31 and Ang1 was used to distinguish the breast cancer cells and endothelial cells. B: bone tissue, V: vessel, T: tumor tissue. Reproduced with the permission of Wiley-VCH GmbH [41].

Further confirmation of the possibility to modulate drug response by a 3D biomimetic environment was shown by Han et al., who demonstrated the ability of a 3D-printed biomimetic model of the bone niche to host metastatic breast cancer cells isolated from patient-derived xenografts (PDX). These models showed a drug response to cisplatin similar to the in vivo model, thus supporting the use of 3D printing for drug testing. This possibility was also confirmed in other types of cancers, such as cervical [177], brain [60,178], lung [179], and bladder [179,180]. The quick rising of novel bioprinted models of several types of cancer for drug screening and personalized medicine approaches, as well as the increasing trend of publications on bone models for oncology (Figure 3), clearly indicates that numerous studies will be published in the upcoming years for bone tumors as well.

In another study, PEG/PEGDA + nHA scaffolds were investigated to assess the interactions between hFOBs and MDA-MB-231 cells. The authors used a stereolithography-based 3D bioprinter to create 3D bone models with square pore patterns and a transwell culture system to evaluate the crosstalk between MDA-MB-231 and osteoblasts. In this system, the two cell populations were physically separated but able to exchange medium and secreted cytokines. This study aimed to recreate the microenvironment of bone metastases and to study the effect of bone-invading breast cancer cells on osteoblast activity, with a specific focus on their effect on cell proliferation, on the synthesis of proteins necessary for bone repair, and on the secretion of inflammatory cytokines, which may stimulate osteoclasts activity and are relevant to breast cancer progression in bone [15,181]. The co-culturing induced a significant effect on cells proliferation, increasing proliferation of MDA-MB-231 cells and decreasing that of hFOBs, respectively. Furthermore, co-culturing MDA-MB-231 and hFOB cells led to an increase in the secretion of IL-8, both by MDA-MB-231 and by hFOBs, up to three-fold higher for hFOBs, when in the presence of MDA-MB-231. Comparing 2D and 3D direct co-culture models, differences were also observed in cancer cell growth. While in 2D models, cancer cells grew in monolayer regardless of hFOBs presence (data acquired at 7 days), in 3D-printed matrices, they arranged in spherical aggregates, forming spheroid-like structures ($\approx$100 µm diameter) even at early culture time points (Figure 4b) [170]. Data reported enlightened the importance of both compositional and morphological cues, alongside the possibility to tune cells response by patterning and adding bio-and nano-bioceramic. These results pave the way for a systematic study of these aspects in combination with 3D printing and bioprinting.

The interactions between cancer and stromal cells were also studied by Zhou et al., who used a 3D stereolithography-based bioprinting technique to fabricate a 3D biomimetic bone matrix able to recreate a bone-like microenvironment. In this study, for the first time, MSCs and osteoblasts were embedded in matrices composed of GelMA and nanocrystalline HA, later seeded with breast cells. The model's aim was to develop a 3D bioprinting bone tumor model. As for Zhu et al. [170], the addition of cancer cells in the model reduced the proliferation of both osteoblasts (Figure 4c(i)) and MSCs (Figure 4c(ii)), while the macromolecules that these two cell types secrete promote cancer cells growth [147]. Additionally, the secretion of VEGF, a crucial regulator of angiogenesis, was overexpressed in tumor cells in co-cultures with MSCs and osteoblasts, whereas, in the same culture, ALP activity, a marker of osteogenesis, was decreased for both MSCs and osteoblasts.

Finally, a further example of bioprinting for modeling and studying breast cancer and the metastasization process to the bone is the study of Cui et al. [41], who used a 3D stereolithography technology for the creation of a model with three distinct regions: (i) a compartment enriched with breast cancer cells, (ii) a vessel with endothelial cells, and (iii) a zone mimicking micro-vascularized bone. The model was developed to study the metastatic process of carcinoma cells intravasating through the endothelial barrier and then extravasating to the bone region. To create the distinct regions, GelMA and PEGDA were used in different concentrations for cancer and bone matrices, and nHA was added to the latter to simulate the inorganic phase of bone. GelMA was also employed to print the vessel interposed between the bone and the cancer matrix. To foster cell seeding on the printed matrices (hFOBs/endothelial cells and MDA-MB-231 cell on bone and cancer matrices, respectively), 3D-printed matrices were mounted on GelMA, and then the vessel was printed in the central part. It was observed that, over a 7-day culture period, MDA-MB-231 cells migrated to the bone matrix. Migration was even accelerated by the presence of endothelial cells in the central vessel, showing the crucial role of these cells in cancer progression. When MDA-MB-231 cells colonized the bone matrix, hFOBs showed a decreased proliferation, while MDA-MB-231 were strongly stimulated to a mitogenic phenotype possibly by the cytokines secreted by osteoblasts, as observed in [170]. Moreover, the growth of endothelial cells in co-culture was slower than in monoculture, suggesting that the factors secreted by MDA-MB-231 inhibited endothelial cell proliferation (Figure 4d(i)). Furthermore, the upregulation of CD31 angiogenic marker and the downregulation of OPN and OCN (Figure 4d(ii)) confirmed the pro-angiogenic activity and the osteogenic inhibition of cancer cells [41].

Overall, the reported 3D models show accurate and reproducible results in terms of mimicking cancer and stromal cell behavior in the bone metastasis microenvironment, thus representing valuable research tools for bone cancer research.

5. Conclusions and Future Perspectives

Bone cancer (sarcoma and metastases) are associated with high mortality and complications rate, so more predictive models are needed to study the progression of the disease and the efficacy of therapy.

Higher predictivity requires a better mimicry of the characteristics of the bone and the tumor microenvironment, and new biomaterials-assisted strategies can help overcome this unmet challenge. Among the new strategies, 3D-printed and bioprinted models can offer new perspectives for mimicking the composition, architecture, and physicomechanical characteristics of bone. In addition, the recent development of multi-material bioprinting systems appears promising to allow the simultaneous deposit of the desired cell types (i.e., stromal cells, immune cells, cancer-associated fibroblasts, and microvascular cells [21,36,44]) and the recapitulation of the cancer microenvironment at different stages of cancer progression. Moreover, preliminary results regarding the inclusion of a vascularization compartment indicate that this approach could significantly improve the biological and physiological relevance of 3D in vitro cancer models.

However, the development of 3D models is still at its early stages, with very few results being published. Although the different strategies and results obtained in bone tissue engineering are a useful benchmark for a deeper understanding and further development of in vitro bone models, many challenges are yet to be addressed:

Increased cell model complexity. The cellular environment of the bone tumor niche is comprised of complex and dynamic interactions between tumor and normal resident cells, but the available studies only consider the interactions of osteoblasts and osteoclasts with tumor cells, which is an oversimplification.

Better mimicry of ECM complexity in terms of composition, stiffness, and complexity of the organic and mineral phases. To this aim, the inclusion of HA in the models appears important, as it dictates normal and tumor cells behavior. At the same time, biomimicry could be further increased, by incorporating ion-substituted or BA in the models, to recapitulate the crystallinity, solubility, and ion-availability of the bone environment.

Increased use of ceramic to merge nanoscale morphological cues and biomimetic composition. This approach allows the model ions naturally present in bone to have an important biological role. Ceramic permits increasing the stability and mechanical properties of the models and guarantees higher adhesion of cells to the scaffold's surface. Thanks to the tunable (and not yet exploited) properties of CaPs, several parameters of the model can be regulated, including printability, viscosity, and shear stress during printing (depending on particles shape and dimensions, which can be determined by selecting the specific CaP and its ion-doping), and CaPs stability/solubility (depending on CaP type and ion-doping, crystallinity, and specific surface).

For 3D printing, a more detailed study on the strategies for the incorporation of ceramics in the model, for instance, by nanostructured coatings and/or more controlled distribution of the nanoparticles. Coatings obtained by wet synthesis might lack homogeneity and adhesion to the substrate. On the other side, nanoparticles might unevenly distribute and generally remain in the bulk of the fiber, with a few available on the surface for interaction with the surrounding microenvironment.

A combination of 3D printing and bioprinting strategies merges the advantages of the two techniques.

More detailed studies show pores' shape/size and surface patterning. Available studies show that patterning can improve the characteristics of the model, but a systematic investigation is needed to determine which are the best architectural parameters to be selected.

Vascularization. This aspect is highly neglected, but it influences several parameters that are of paramount importance in tumor progression and drug response, including hypoxia, nutrients and oxygen diffusion, and shear stress.

Author Contributions: All authors listed have made a substantial, direct, and intellectual contribution to the work and approved it for publication. All authors have read and agreed to the published version of the manuscript.

Funding: This work was supported by the Eranet-lac (PER-2012-ELAC2015/T07-0713 to N.B.), the Italian Association for Cancer Research (AIRC IG n. 21403 to N.B.), the Ministry of Health (project Starting Grant SG-2018-12367059, BANDO RICERCA FINALIZZATA 2018 to G.G.). Financial support for Scientific Research 5xMille to N.B.

Conflicts of Interest: The authors declare no conflict of interest.

Abbreviations

2D	Two-dimensional
3D	Three-dimensional
Ang-1	Angiopoietin-1
ALP	Alkaline phosphatase
BA	Biogenic apatite
BG	Bioactive glass
BMP-2	Bone morphogenic protein-2
CaPs	Calcium phosphates
ET-1	Endothelin-1
EMT	Epithelial–mesenchymal transition
ECM	Extracellular matrix
FGF	Fibroblast growth factor
GelMA	Gelatin methacrylate
HA	Hydroxyapatite
hFOB	Human fetal osteoblasts
HIF	Hypoxia-induced factor
HUVEC	Human umbilical vascular endothelial cells
IGF	Insulin growth factor
IL-6,8,11	Interleukin-6,8,11
LOX	Lysyl oxidase
MSCs	Mesenchymal stromal cells
nHA	Hydroxyapatite nanoparticles
OCN	Osteocalcin
ON	Osteonectin
OPN	Osteopontin
PCL	Polycaprolactone
PDGF	Platelet-derived growth factor
PDPN	Podoplanin

PEG	Polyethylene glycol
PEGDA	Polyethylene glycol diacrylate
PTHrP	Parathyroid hormone-related protein
PLA	Polylactic acid
RANKL	Receptor activator of nuclear factor-kappa B ligand
SOST	Sclerostin
TCP	Tricalcium phosphate
TGF-β	Transforming growth factor-beta
VEGF	Vascular endothelial growth factor

References

1. Grimer, R.J.; Hogendoorn, P.C.W.; Vanel, D. Tumours of Bone: Introduction. In *WHO Classification of Tumours of Soft Tissue and Bone*; IARC: Lyon, France, 2013; pp. 244–247.
2. Cortini, M.; Baldini, N.; Avnet, S. New Advances in the Study of Bone Tumors: A Lesson From the 3D Environment. *Front. Physiol.* **2019**, *10*, 814. [CrossRef]
3. Ma, X.; Liu, J.; Zhu, W.; Tang, M.; Lawrence, N.; Yu, C.; Gou, M.; Chen, S. 3D bioprinting of functional tissue models for personalized drug screening and in vitro disease modeling. *Adv. Drug Deliv. Rev.* **2018**, *132*, 235–251. [CrossRef]
4. Díaz, E.C.G.; Sinha, S.; Avedian, R.S.; Yang, F. Tissue-engineered 3D models for elucidating primary and metastatic bone cancer progression. *Acta Biomater.* **2019**, *99*, 18–32. [CrossRef]
5. Yuasa, T.; Urakami, S.; Yamamoto, S.; Yonese, J.; Saito, K.; Takahashi, S.; Hatake, K.; Fukui, I. Treatment outcome and prognostic factors in renal cell cancer patients with bone metastasis. *Clin. Exp. Metastasis* **2011**, *28*, 405–411. [CrossRef]
6. Santini, D.; Procopio, G.; Porta, C.; Ibrahim, T.; Barni, S.; Mazzara, C.; Fontana, A.; Berruti, A.; Berardi, R.; Vincenzi, B.; et al. Natural History of Malignant Bone Disease in Renal Cancer: Final Results of an Italian Bone Metastasis Survey. *PLoS ONE* **2013**, *8*, e83026. [CrossRef]
7. Coleman, R.E. Clinical Features of Metastatic Bone Disease and Risk of Skeletal Morbidity. *Clin. Cancer Res.* **2006**, *12*, 6243s–6249s. [CrossRef]
8. Santoni, M.; Conti, A.; Procopio, G.; Porta, C.; Ibrahim, T.; Barni, S.; Guida, F.M.; Fontana, A.; Berruti, A.; Berardi, R.; et al. Bone metastases in patients with metastatic renal cell carcinoma: Are they always associated with poor prognosis? *J. Exp. Clin. Cancer Res.* **2015**, *34*, 1–9. [CrossRef]
9. Drzymalski, D.M.; Oh, W.; Werner, L.; Regan, M.M.; Kantoff, P.; Tuli, S. Predictors of survival in patients with prostate cancer and spinal metastasis. *J. Neurosurg. Spine* **2010**, *13*, 789–794. [CrossRef] [PubMed]
10. Kimura, T. Multidisciplinary Approach for Bone Metastasis: A Review. *Cancers* **2018**, *10*, 156. [CrossRef] [PubMed]
11. Katakami, N.; Kunikane, H.; Takeda, K.; Takayama, K.; Sawa, T.; Saito, H.; Harada, M.; Yokota, S.; Ando, K.; Saito, Y.; et al. Prospective Study on the Incidence of Bone Metastasis (BM) and Skeletal-Related Events (SREs) in Patients (pts) with Stage IIIB and IV Lung Cancer—CSP-HOR 13. *J. Thorac. Oncol.* **2014**, *9*, 231–238. [CrossRef] [PubMed]
12. Lote, K.; Walløe, A.; Bjersand, A. Bone Metastasis Prognosis, Diagnosis and Treatment. *Acta Radiol. Oncol.* **1986**, *25*, 227–232. [CrossRef]
13. Saarto, T.; Janes, R.; Tenhunen, M.; Kouri, M. Palliative radiotherapy in the treatment of skeletal metastases. *Eur. J. Pain* **2002**, *6*, 323–330. [CrossRef]
14. Coleman, R. Metastatic bone disease: Clinical features, pathophysiology and treatment strategies. *Cancer Treat. Rev.* **2001**, *27*, 165–176. [CrossRef]
15. Krishnan, V.; Vogler, E.A.; Sosnoski, D.M.; Mastro, A. In Vitro Mimics of Bone Remodeling and the Vicious Cycle of Cancer in Bone. *J. Cell. Physiol.* **2013**, *229*, 453–462. [CrossRef]
16. Tan, P.H.; Aung, K.; Toh, S.; Goh, J.C.; Nathan, S. Three-dimensional porous silk tumor constructs in the approximation of in vivo osteosarcoma physiology. *Biomaterials* **2011**, *32*, 6131–6137. [CrossRef]
17. Yamada, K.; Cukierman, E. Modeling Tissue Morphogenesis and Cancer in 3D. *Cell* **2007**, *130*, 601–610. [CrossRef] [PubMed]
18. da Rocha, E.L.; Porto, L.; Rambo, C. Nanotechnology meets 3D in vitro models: Tissue engineered tumors and cancer therapies. *Mater. Sci. Eng. C* **2014**, *34*, 270–279. [CrossRef] [PubMed]
19. Avnet, S.; Lemma, S.; Cortini, M.; Di Pompo, G.; Perut, F.; Baldini, N. Pre-clinical Models for Studying the Interaction Between Mesenchymal Stromal Cells and Cancer Cells and the Induction of Stemness. *Front. Oncol.* **2019**, *9*, 305. [CrossRef]
20. Xu, X.; Farach-Carson, M.; Jia, X. Three-dimensional in vitro tumor models for cancer research and drug evaluation. *Biotechnol. Adv.* **2014**, *32*, 1256–1268. [CrossRef]
21. Song, H.-H.G.; Park, K.M.; Gerecht, S. Hydrogels to model 3D in vitro microenvironment of tumor vascularization. *Adv. Drug Deliv. Rev.* **2014**, *79–80*, 19–29. [CrossRef]
22. Scheinpflug, J.; Pfeiffenberger, M.; Damerau, A.; Schwarz, F.; Textor, M.; Lang, A.; Schulze, F. Journey into Bone Models: A Review. *Genes* **2018**, *9*, 247. [CrossRef]
23. Rho, J.-Y.; Kuhn-Spearing, L.; Zioupos, P. Mechanical properties and the hierarchical structure of bone. *Med. Eng. Phys.* **1998**, *20*, 92–102. [CrossRef]

24. Reznikov, N.; Shahar, R.; Weiner, S. Bone hierarchical structure in three dimensions. *Acta Biomater.* **2014**, *10*, 3815–3826. [CrossRef] [PubMed]
25. Arrigoni, C.; Gilardi, M.; Bersini, S.; Candrian, C.; Moretti, M. Bioprinting and Organ-on-Chip Applications Towards Personalized Medicine for Bone Diseases. *Stem Cell Rev. Rep.* **2017**, *13*, 407–417. [CrossRef] [PubMed]
26. Mandrycky, C.; Wang, Z.; Kim, K.; Kim, D.-H. 3D bioprinting for engineering complex tissues. *Biotechnol. Adv.* **2016**, *34*, 422–434. [CrossRef]
27. Haleem, A.; Javaid, M.; Khan, R.H.; Suman, R. 3D printing applications in bone tissue engineering. *J. Clin. Orthop. Trauma* **2019**, *11*, S118–S124. [CrossRef]
28. Zhang, L.; Yang, G.; Johnson, B.N.; Jia, X. Three-dimensional (3D) printed scaffold and material selection for bone repair. *Acta Biomater.* **2018**, *84*, 16–33. [CrossRef]
29. Pathi, S.P.; Lin, D.D.; Dorvee, J.R.; Estroff, L.A.; Fischbach, C. Hydroxyapatite nanoparticle-containing scaffolds for the study of breast cancer bone metastasis. *Biomaterials* **2011**, *32*, 5112–5122. [CrossRef]
30. Wang, X.; Tolba, E.; Schröder, H.C.; Neufurth, M.; Feng, Q.; Diehl-Seifert, B.; Müller, W.E.G. Effect of Bioglass on Growth and Biomineralization of SaOS-2 Cells in Hydrogel after 3D Cell Bioprinting. *PLoS ONE* **2014**, *9*, e112497. [CrossRef]
31. Gao, G.; Schilling, A.F.; Yonezawa, T.; Wang, J.; Dai, G.; Cui, X. Bioactive nanoparticles stimulate bone tissue formation in bioprinted three-dimensional scaffold and human mesenchymal stem cells. *Biotechnol. J.* **2014**, *9*, 1304–1311. [CrossRef]
32. Kim, W.; Kim, G. Collagen/bioceramic-based composite bioink to fabricate a porous 3D hASCs-laden structure for bone tissue regeneration. *Biofabrication* **2019**, *12*, 015007. [CrossRef]
33. Ashammakhi, N.; Hasan, A.; Kaarela, O.; Byambaa, B.; Sheikhi, A.; Gaharwar, A.K.; Khademhosseini, A. Advancing Frontiers in Bone Bioprinting. *Adv. Health Mater.* **2019**, *8*, e1801048. [CrossRef] [PubMed]
34. Pati, F.; Gantelius, J.; Svahn, H.A. 3D Bioprinting of Tissue/Organ Models. *Angew. Chem. Int. Ed.* **2016**, *55*, 4650–4665. [CrossRef]
35. Zhu, W.; Ma, X.; Gou, M.; Mei, D.; Zhang, K.; Chen, S. 3D printing of functional biomaterials for tissue engineering. *Curr. Opin. Biotechnol.* **2016**, *40*, 103–112. [CrossRef]
36. Zhang, Y.S.; Duchamp, M.; Oklu, R.; Ellisen, L.W.; Langer, R.; Khademhosseini, A. Bioprinting the Cancer Microenvironment. *ACS Biomater. Sci. Eng.* **2016**, *2*, 1710–1721. [CrossRef] [PubMed]
37. Samavedi, S.; Joy, N. 3D printing for the development of in vitro cancer models. *Curr. Opin. Biomed. Eng.* **2017**, *2*, 35–42. [CrossRef]
38. Albritton, J.L.; Miller, J.S. 3D bioprinting: Improving in vitro models of metastasis with heterogeneous tumor microenvironments. *Dis. Model. Mech.* **2017**, *10*, 3–14. [CrossRef]
39. Wang, Y.; Shi, W.; Kuss, M.A.; Mirza, S.; Qi, D.; Krasnoslobodtsev, A.; Zeng, J.; Band, H.; Band, V.; Duan, B. 3D Bioprinting of Breast Cancer Models for Drug Resistance Study. *ACS Biomater. Sci. Eng.* **2018**, *4*, 4401–4411. [CrossRef]
40. Ashammakhi, N.; Ahadian, S.; Xu, C.; Montazerian, H.; Ko, H.; Nasiri, R.; Barros, N.; Khademhosseini, A. Bioinks and bioprinting technologies to make heterogeneous and biomimetic tissue constructs. *Mater. Today Bio* **2019**, *1*, 100008. [CrossRef] [PubMed]
41. Cui, H.; Esworthy, T.; Zhou, X.; Hann, S.Y.; Glazer, R.I.; Li, R.; Zhang, L.G. Engineering a Novel 3D Printed Vascularized Tissue Model for Investigating Breast Cancer Metastasis to Bone. *Adv. Health Mater.* **2019**, *9*, e1900924. [CrossRef]
42. Do, A.-V.; Khorsand, B.; Geary, S.M.; Salem, A.K. 3D Printing of Scaffolds for Tissue Regeneration Applications. *Adv. Health Mater.* **2015**, *4*, 1742–1762. [CrossRef]
43. Abdulghani, S.; Morouço, P.G.; Abdulghani, S. Biofabrication for osteochondral tissue regeneration: Bioink printability requirements. *J. Mater. Sci. Mater. Med.* **2019**, *30*, 20. [CrossRef]
44. Moroni, L.; Burdick, J.A.; Highley, C.; Lee, S.J.; Morimoto, Y.; Takeuchi, S.; Yoo, J.J. Biofabrication strategies for 3D in vitro models and regenerative medicine. *Nat. Rev. Mater.* **2018**, *3*, 21–37. [CrossRef]
45. Melchels, F.; Domingos, M.; Klein, T.; Malda, J.; Bartolo, P.; Hutmacher, D.W. Additive manufacturing of tissues and organs. *Prog. Polym. Sci.* **2012**, *37*, 1079–1104. [CrossRef]
46. Li, J.; Chen, M.; Fan, X.; Zhou, H. Recent advances in bioprinting techniques: Approaches, applications and future prospects. *J. Transl. Med.* **2016**, *14*, 1–15. [CrossRef]
47. Kwakwa, K.A.; Vanderburgh, J.P.; Guelcher, S.A.; Sterling, J.A. Engineering 3D Models of Tumors and Bone to Understand Tumor-Induced Bone Disease and Improve Treatments. *Curr. Osteoporos. Rep.* **2017**, *15*, 247–254. [CrossRef] [PubMed]
48. Weilbaecher, K.N.; Guise, T.A.; McCauley, L.K. Cancer to bone: A fatal attraction. *Nat. Rev. Cancer* **2011**, *11*, 411–425. [CrossRef]
49. Kolb, A.D.; Shupp, A.B.; Mukhopadhyay, D.; Marini, F.C.; Bussard, K.M. Osteoblasts are "educated" by crosstalk with metastatic breast cancer cells in the bone tumor microenvironment. *Breast Cancer Res.* **2019**, *21*, 31. [CrossRef] [PubMed]
50. Patel, L.R.; Camacho, D.F.; Shiozawa, Y.; Pienta, K.J.; Taichman, R.S. Mechanisms of cancer cell metastasis to the bone: A multistep process. *Futur. Oncol.* **2011**, *7*, 1285–1297. [CrossRef] [PubMed]
51. Shupp, A.B.; Kolb, A.D.; Mukhopadhyay, D.; Bussard, K.M. Cancer Metastases to Bone: Concepts, Mechanisms, and Interactions with Bone Osteoblasts. *Cancers* **2018**, *10*, 182. [CrossRef]
52. Karnoub, A.E.; Dash, A.B.; Vo, A.P.; Sullivan, A.; Brooks, M.W.; Bell, G.W.; Richardson, A.L.; Polyak, K.; Tubo, R.; Weinberg, R.A. Mesenchymal stem cells within tumour stroma promote breast cancer metastasis. *Nat. Cell Biol.* **2007**, *449*, 557–563. [CrossRef]
53. Mundy, G.R. Metastasis to bone: Causes, consequences and therapeutic opportunities. *Nat. Rev. Cancer* **2002**, *2*, 584–593. [CrossRef]
54. Theresa, A.; Guise, G.R.M. Cancer and bone. *Endocr. Rev.* **1998**, *19*, 18–54.

55. Kingsley, L.A.; Fournier, P.; Chirgwin, J.M.; Guise, T.A. Molecular Biology of Bone Metastasis. *Mol. Cancer Ther.* **2007**, *6*, 2609–2617. [CrossRef]
56. Avnet, S.; Di Pompo, G.; Chano, T.; Errani, C.; Ibrahim-Hashim, A.; Gillies, R.J.; Donati, D.M.; Baldini, N. Cancer-associated mesenchymal stroma fosters the stemness of osteosarcoma cells in response to intratumoral acidosis via NF-κB activation. *Int. J. Cancer* **2016**, *140*, 1331–1345. [CrossRef]
57. Heist, R.S.; Duda, D.G.; Sahani, D.V.; Ancukiewicz, M.; Fidias, P.; Sequist, L.V.; Temel, J.S.; Shaw, A.T.; Pennell, N.; Neal, J.W.; et al. Improved tumor vascularization after anti-VEGF therapy with carboplatin and nab-paclitaxel associates with survival in lung cancer. *Proc. Natl. Acad. Sci. USA* **2015**, *112*, 1547–1552. [CrossRef]
58. Szabo, V.; Bugyik, E.; Dezso, K.; Ecker, N.; Nagy, P.; Timar, J.; Tovari, J.; Laszlo, V.; Bridgeman, V.L.; Wan, E.; et al. Mechanism of tumour vascularization in experimental lung metastases. *J. Pathol.* **2014**, *235*, 384–396. [CrossRef]
59. Téglási, V.; Csűry, D.T.; Dezső, K.; Bugyik, E.; Szabó, V.; Szallasi, Z.; Paku, S.; Reiniger, L. Origin and Distribution of Connective Tissue and Pericytes Impacting Vascularization in Brain Metastases With Different Growth Patterns. *J. Neuropathol. Exp. Neurol.* **2019**, *78*, 326–339. [CrossRef]
60. Dai, X.; Ma, C.; Lan, Q.; Xu, T. 3D bioprinted glioma stem cells for brain tumor model and applications of drug susceptibility. *Biofabrication* **2016**, *8*, 045005. [CrossRef]
61. Vitiani, L.R.; Pallini, R.; Biffoni, M.; Todaro, M.; Invernici, G.; Cenci, T.; Maira, G.; Parati, E.A.; Stassi, G.; Larocca, L.M.; et al. Tumour vascularization via endothelial differentiation of glioblastoma stem-like cells. *Nat. Cell Biol.* **2010**, *468*, 824–828. [CrossRef]
62. Stoeltzing, O.; Liu, W.; Reinmuth, N.; Parikh, A.; Ahmad, S.A.; Jung, Y.D.; Fan, F.; Ellis, L.M. New Approaches to the Treatment of Hepatic Malignancies Angiogenesis and Antiangiogenic Therapy of Colon Cancer Liver Metastasis. *Ann. Surg. Oncol.* **2003**, *10*, 722–733. [CrossRef]
63. Abou-Elkacem, L.; Arns, S.; Brix, G.; Gremse, F.; Zopf, D.; Kiessling, F.; Lederle, W. Regorafenib Inhibits Growth, Angiogenesis, and Metastasis in a Highly Aggressive, Orthotopic Colon Cancer Model. *Mol. Cancer Ther.* **2013**, *12*, 1322–1331. [CrossRef]
64. Bussard, K.; Gay, C.V.; Mastro, A.M. The bone microenvironment in metastasis; what is special about bone? *Cancer Metastasis Rev.* **2007**, *27*, 41–55. [CrossRef] [PubMed]
65. Hillen, F.; Griffioen, A.W. Tumour vascularization: Sprouting angiogenesis and beyond. *Cancer Metastasis Rev.* **2007**, *26*, 489–502. [CrossRef]
66. Alford, A.I.; Kozloff, K.M.; Hankenson, K.D. Extracellular matrix networks in bone remodeling. *Int. J. Biochem. Cell Biol.* **2015**, *65*, 20–31. [CrossRef] [PubMed]
67. Lin, X.; Patil, S.; Gao, Y.-G.; Qian, A. The Bone Extracellular Matrix in Bone Formation and Regeneration. *Front. Pharmacol.* **2020**, *11*, 757. [CrossRef]
68. Lu, P.; Weaver, V.M.; Werb, Z. The extracellular matrix: A dynamic niche in cancer progression. *J. Cell Biol.* **2012**, *196*, 395–406. [CrossRef] [PubMed]
69. Henke, E.; Nandigama, R.; Ergün, S. Extracellular Matrix in the Tumor Microenvironment and Its Impact on Cancer Therapy. *Front. Mol. Biosci.* **2020**, *6*, 160. [CrossRef]
70. Jabbari, E.; Sarvestani, S.K.; Daneshian, L.; Moeinzadeh, S. Optimum 3D Matrix Stiffness for Maintenance of Cancer Stem Cells Is Dependent on Tissue Origin of Cancer Cells. *PLoS ONE* **2015**, *10*, e0132377. [CrossRef]
71. Pathi, S.P.; Kowalczewski, C.; Tadipatri, R.; Fischbach, C. A Novel 3-D Mineralized Tumor Model to Study Breast Cancer Bone Metastasis. *PLoS ONE* **2010**, *5*, e8849. [CrossRef]
72. Graziani, G.; Boi, M.; Bianchi, M. A Review on Ionic Substitutions in Hydroxyapatite Thin Films: Towards Complete Biomimetism. *Coatings* **2018**, *8*, 269. [CrossRef]
73. Bendtsen, S.T.; Quinnell, S.P.; Wei, M. Development of a novel alginate-polyvinyl alcohol-hydroxyapatite hydrogel for 3D bioprinting bone tissue engineered scaffolds. *J. Biomed. Mater. Res. Part A* **2017**, *105*, 1457–1468. [CrossRef]
74. Cidonio, G.; Alcala-Orozco, C.R.; Lim, K.S.; Glinka, M.; Mutreja, I.; Kim, Y.-H.; Dawson, J.I.; Woodfield, T.B.F.; Oreffo, R.O.C. Osteogenic and angiogenic tissue formation in high fidelity nanocomposite Laponite-gelatin bioinks. *Biofabrication* **2019**, *11*, 035027. [CrossRef]
75. Demirtaş, T.T.; Irmak, G.; Gümüşderelioğlu, M. A bioprintable form of chitosan hydrogel for bone tissue engineering. *Biofabrication* **2017**, *9*, 035003. [CrossRef]
76. Poldervaart, M.T.; Goversen, B.; de Ruijter, M.; Abbadessa, A.; Melchels, F.; Öner, F.C.; Dhert, W.; Vermonden, T.; Alblas, J. 3D bioprinting of methacrylated hyaluronic acid (MeHA) hydrogel with intrinsic osteogenicity. *PLoS ONE* **2017**, *12*, e0177628. [CrossRef]
77. Xing, F.; Xiang, Z.; Rommens, P.M.; Ritz, U. 3D Bioprinting for Vascularized Tissue-Engineered Bone Fabrication. *Materials* **2020**, *13*, 2278. [CrossRef]
78. Zhai, X.; Ruan, C.; Ma, Y.; Cheng, D.; Wu, M.; Liu, W.; Zhao, X.; Pan, H.; Lu, W.W. 3D-Bioprinted Osteoblast-Laden Nanocomposite Hydrogel Constructs with Induced Microenvironments Promote Cell Viability, Differentiation, and Osteogenesis both In Vitro and In Vivo. *Adv. Sci.* **2017**, *5*, 1700550. [CrossRef] [PubMed]
79. Malda, J.; Visser, J.; Melchels, F.P.; Jüngst, T.; Hennink, W.E.; Dhert, W.; Groll, J.; Hutmacher, D.W. 25th Anniversary Article: Engineering Hydrogels for Biofabrication. *Adv. Mater.* **2013**, *25*, 5011–5028. [CrossRef] [PubMed]
80. Murphy, S.V.; Atala, A. 3D bioprinting of tissues and organs. *Nat. Biotechnol.* **2014**, *32*, 773–785. [CrossRef]

81. Melchels, F.; Tonnarelli, B.; Olivares, A.L.; Martin, I.; Lacroix, D.; Feijen, J.; Wendt, D.J.; Grijpma, D.W. The influence of the scaffold design on the distribution of adhering cells after perfusion cell seeding. *Biomaterials* **2011**, *32*, 2878–2884. [CrossRef] [PubMed]
82. Cámara-Torres, M.; Sinha, R.; Mota, C.; Moroni, L. Improving cell distribution on 3D additive manufactured scaffolds through engineered seeding media density and viscosity. *Acta Biomater.* **2020**, *101*, 183–195. [CrossRef]
83. Ji, S.; Guvendiren, M. Recent Advances in Bioink Design for 3D Bioprinting of Tissues and Organs. *Front. Bioeng. Biotechnol.* **2017**, *5*, 23. [CrossRef]
84. Angeloni, V.; Contessi, N.; De Marco, C.; Bertoldi, S.; Tanzi, M.C.; Daidone, M.G.; Farè, S. Polyurethane foam scaffold as in vitro model for breast cancer bone metastasis. *Acta Biomater.* **2017**, *63*, 306–316. [CrossRef]
85. Hölzl, K.; Lin, S.; Tytgat, L.; Van Vlierberghe, S.; Gu, L.; Ovsianikov, A. Bioink properties before, during and after 3D bioprinting. *Biofabrication* **2016**, *8*, 032002. [CrossRef]
86. Piard, C.; Baker, H.; Kamalitdinov, T.; Fisher, J. Bioprinted osteon-like scaffolds enhance in vivo neovascularization. *Biofabrication* **2019**, *11*, 025013. [CrossRef]
87. Murphy, C.; Kolan, K.; Li, W.; Semon, J.; Day, D.; Leu, M. 3D bioprinting of stem cells and polymer/bioactive glass composite scaffolds for tissue engineering. *Int. J. Bioprinting* **2017**, *3*, 54–64. [CrossRef]
88. Daly, A.; Cunniffe, G.M.; Sathy, B.N.; Jeon, O.; Alsberg, E.; Kelly, D.J. 3D Bioprinting of Developmentally Inspired Templates for Whole Bone Organ Engineering. *Adv. Health Mater.* **2016**, *5*, 2353–2362. [CrossRef]
89. Park, S.A.; Lee, H.-J.; Kim, K.-S.; Lee, S.J.; Lee, J.-T.; Kim, S.-Y.; Chang, N.-H.; Park, S.-Y. In Vivo Evaluation of 3D-Printed Polycaprolactone Scaffold Implantation Combined with β-TCP Powder for Alveolar Bone Augmentation in a Beagle Defect Model. *Materials* **2018**, *11*, 238. [CrossRef]
90. Zhang, J.; Zhao, S.; Zhu, M.; Zhu, Y.; Zhang, Y.; Liu, Z.; Zhang, C. 3D-printed magnetic Fe3O4/MBG/PCL composite scaffolds with multifunctionality of bone regeneration, local anticancer drug delivery and hyperthermia. *J. Mater. Chem. B* **2014**, *2*, 7583–7595. [CrossRef]
91. Bose, S.; Vahabzadeh, S.; Bandyopadhyay, A. Bone tissue engineering using 3D printing. *Mater. Today* **2013**, *16*, 496–504. [CrossRef]
92. Paxton, N.; Smolan, W.; Böck, T.; Melchels, F.; Groll, J.; Jungst, T. Proposal to assess printability of bioinks for extrusion-based bioprinting and evaluation of rheological properties governing bioprintability. *Biofabrication* **2017**, *9*, 044107. [CrossRef] [PubMed]
93. Derakhshanfar, S.; Mbeleck, R.; Xu, K.; Zhang, X.; Zhong, W.; Xing, M. 3D bioprinting for biomedical devices and tissue engineering: A review of recent trends and advances. *Bioact. Mater.* **2018**, *3*, 144–156. [CrossRef] [PubMed]
94. Park, J.; Lee, S.J.; Chung, S.; Lee, J.H.; Kim, W.D.; Lee, J.Y.; Park, S.A. Cell-laden 3D bioprinting hydrogel matrix depending on different compositions for soft tissue engineering: Characterization and evaluation. *Mater. Sci. Eng. C* **2017**, *71*, 678–684. [CrossRef] [PubMed]
95. Theus, A.S.; Ning, L.; Hwang, B.; Gil, C.; Chen, S.; Wombwell, A.; Mehta, R.; Serpooshan, V. Bioprintability: Physiomechanical and Biological Requirements of Materials for 3D Bioprinting Processes. *Polymers* **2020**, *12*, 2262. [CrossRef]
96. Zhang, J.; Wehrle, E.; Vetsch, J.R.; Paul, G.R.; Rubert, M.; Müller, R.; Mueller, R. Alginate dependent changes of physical properties in 3D bioprinted cell-laden porous scaffolds affect cell viability and cell morphology. *Biomed. Mater.* **2019**, *14*, 065009. [CrossRef]
97. Ouyang, L.; Yao, R.; Zhao, Y.; Sun, W. Effect of bioink properties on printability and cell viability for 3D bioplotting of embryonic stem cells. *Biofabrication* **2016**, *8*, 035020. [CrossRef]
98. Neufurth, M.; Wang, X.; Schröder, H.C.; Feng, Q.; Diehl-Seifert, B.; Ziebart, T.; Steffen, R.; Wang, S.; Müller, W.E. Engineering a morphogenetically active hydrogel for bioprinting of bioartificial tissue derived from human osteoblast-like SaOS-2 cells. *Biomaterials* **2014**, *35*, 8810–8819. [CrossRef]
99. Bendtsen, S.T.; Wei, M. In vitro evaluation of 3D bioprinted tri-polymer network scaffolds for bone tissue regeneration. *J. Biomed. Mater. Res. Part A* **2017**, *105*, 3262–3272. [CrossRef]
100. Ojansivu, M.; Rashad, A.; Ahlinder, A.E.; Massera, J.; Mishra, A.; Syverud, K.; Finne-Wistrand, A.; Miettinen, S.; Mustafa, K. Wood-based nanocellulose and bioactive glass modified gelatin–alginate bioinks for 3D bioprinting of bone cells. *Biofabrication* **2019**, *11*, 035010. [CrossRef]
101. Chung, J.H.Y.; Naficy, S.; Yue, Z.; Kapsa, R.; Quigley, A.; Moulton, S.; Wallace, G. Bio-ink properties and printability for extrusion printing living cells. *Biomater. Sci.* **2013**, *1*, 763–773. [CrossRef]
102. Alcala-Orozco, C.R.; Mutreja, I.; Cui, X.; Kumar, D.; Hooper, G.J.; Lim, K.S.; Woodfield, T.B. Design and characterisation of multi-functional strontium-gelatin nanocomposite bioinks with improved print fidelity and osteogenic capacity. *Bioprinting* **2019**, *18*, e00073. [CrossRef]
103. Anada, T.; Pan, C.-C.; Stahl, A.M.; Mori, S.; Fukuda, J.; Suzuki, O.; Yang, Y. Vascularized Bone-Mimetic Hydrogel Constructs by 3D Bioprinting to Promote Osteogenesis and Angiogenesis. *Int. J. Mol. Sci.* **2019**, *20*, 1096. [CrossRef]
104. Byambaa, B.; Annabi, N.; Yue, K.; de Santiago, G.T.; Alvarez, M.M.; Jia, W.; Kazemzadeh-Narbat, M.; Shin, S.R.; Tamayol, A.; Khademhosseini, A. Bioprinted Osteogenic and Vasculogenic Patterns for Engineering 3D Bone Tissue. *Adv. Health Mater.* **2017**, *6*, 15. [CrossRef] [PubMed]
105. McBeth, C.; Lauer, J.; Ottersbach, M.; Campbell, J.; Sharon, A.; Sauer-Budge, A.F. 3D bioprinting of GelMA scaffolds triggers mineral deposition by primary human osteoblasts. *Biofabrication* **2017**, *9*, 015009. [CrossRef]
106. Yin, J.; Yan, M.; Wang, Y.; Fu, J.; Suo, H. 3D Bioprinting of Low-Concentration Cell-Laden Gelatin Methacrylate (GelMA) Bioinks with a Two-Step Cross-linking Strategy. *ACS Appl. Mater. Interfaces* **2018**, *10*, 6849–6857. [CrossRef]

107. Sharma, A.; Desando, G.; Petretta, M.; Chawla, S.; Bartolotti, I.; Manferdini, C.; Paolella, F.; Gabusi, E.; Trucco, D.; Ghosh, S.; et al. Investigating the Role of Sustained Calcium Release in Silk-Gelatin-Based Three-Dimensional Bioprinted Constructs for Enhancing the Osteogenic Differentiation of Human Bone Marrow Derived Mesenchymal Stromal Cells. *ACS Biomater. Sci. Eng.* **2019**, *5*, 1518–1533. [CrossRef] [PubMed]
108. Das, S.; Pati, F.; Choi, Y.-J.; Rijal, G.; Shim, J.-H.; Kim, S.W.; Ray, A.R.; Cho, D.-W.; Ghosh, S. Bioprintable, cell-laden silk fibroin–gelatin hydrogel supporting multilineage differentiation of stem cells for fabrication of three-dimensional tissue constructs. *Acta Biomater.* **2015**, *11*, 233–246. [CrossRef]
109. Moreira, C.; Carvalho, S.M.; Mansur, H.; Pereira, M.M. Thermogelling chitosan–collagen–bioactive glass nanoparticle hybrids as potential injectable systems for tissue engineering. *Mater. Sci. Eng. C* **2016**, *58*, 1207–1216. [CrossRef] [PubMed]
110. Machado, C.B.; Ventura, J.; Lemos, A.; Ferreira, J.; Leite, M.F.; Goes, A.M. 3D chitosan–gelatin–chondroitin porous scaffold improves osteogenic differentiation of mesenchymal stem cells. *Biomed. Mater.* **2007**, *2*, 124–131. [CrossRef]
111. Noh, I.; Kim, N.; Tran, H.N.; Lee, J.; Lee, C. 3D printable hyaluronic acid-based hydrogel for its potential application as a bioink in tissue engineering. *Biomater. Res.* **2019**, *23*, 1–9. [CrossRef]
112. Ker, D.F.E.; Nain, A.S.; Weiss, L.E.; Wang, J.; Suhan, J.; Amon, C.H.; Campbell, P.G. Bioprinting of growth factors onto aligned sub-micron fibrous scaffolds for simultaneous control of cell differentiation and alignment. *Biomaterials* **2011**, *32*, 8097–8107. [CrossRef] [PubMed]
113. Campos, D.F.D.; Blaeser, A.; Buellesbach, K.; Sen, K.S.; Xun, W.; Tillmann, W.; Fischer, H. Bioprinting Organotypic Hydrogels with Improved Mesenchymal Stem Cell Remodeling and Mineralization Properties for Bone Tissue Engineering. *Adv. Health Mater.* **2016**, *5*, 1336–1345. [CrossRef]
114. Melke, J.; Midha, S.; Ghosh, S.; Ito, K.; Hofmann, S. Silk fibroin as biomaterial for bone tissue engineering. *Acta Biomater.* **2016**, *31*, 1–16. [CrossRef]
115. Nguyen, T.B.L.; Lee, B.-T. A Combination of Biphasic Calcium Phosphate Scaffold with Hyaluronic Acid-Gelatin Hydrogel as a New Tool for Bone Regeneration. *Tissue Eng. Part A* **2014**, *20*, 1993–2004. [CrossRef]
116. Wenz, A.; Borchers, K.; Tovar, G.E.M.; Kluger, P.J. Bone matrix production in hydroxyapatite-modified hydrogels suitable for bone bioprinting. *Biofabrication* **2017**, *9*, 044103. [CrossRef]
117. Fedorovich, N.E.; Leeuwenburgh, S.C.; Van Der Helm, Y.J.M.; Alblas, J.; Dhert, W. The osteoinductive potential of printable, cell-laden hydrogel-ceramic composites. *J. Biomed. Mater. Res. Part A* **2012**, *100*, 2412–2420. [CrossRef]
118. Tang, D.; Tare, R.; Yang, L.-Y.; Williams, D.F.; Ou, K.-L.; Oreffo, R.O. Biofabrication of bone tissue: Approaches, challenges and translation for bone regeneration. *Biomaterials* **2016**, *83*, 363–382. [CrossRef]
119. Zare, R.N.; Doustkhah, E.; Assadi, M.H.N. Three-dimensional bone printing using hydroxyapatite-PLA composite. *Mater. Today Proc.* **2019**, *42*, 1531–1533. [CrossRef]
120. Farokhi, M.; Mottaghitalab, F.; Samani, S.; Shokrgozar, M.A.; Kundu, S.C.; Reis, R.L.; Fatahi, Y.; Kaplan, D.L. Silk fibroin/hydroxyapatite composites for bone tissue engineering. *Biotechnol. Adv.* **2017**, *36*, 68–91. [CrossRef]
121. Causa, F.; Netti, P.; Ambrosio, L.; Ciapetti, G.; Baldini, N.; Pagani, S.; Martini, D.; Giunti, A. Poly-ε-caprolactone/hydroxyapatite composites for bone regeneration: In vitro characterization and human osteoblast response. *J. Biomed. Mater. Res. Part A* **2006**, *76*, 151–162. [CrossRef]
122. Park, S.A.; Lee, S.H.; Kim, W.D. Fabrication of porous polycaprolactone/hydroxyapatite (PCL/HA) blend scaffolds using a 3D plotting system for bone tissue engineering. *Bioprocess Biosyst. Eng.* **2010**, *34*, 505–513. [CrossRef]
123. Mi, H.-Y.; Jing, X.; Salick, M.R.; Cordie, T.M.; Peng, X.-F.; Turng, L.-S. Morphology, mechanical properties, and mineralization of rigid thermoplastic polyurethane/hydroxyapatite scaffolds for bone tissue applications: Effects of fabrication approaches and hydroxyapatite size. *J. Mater. Sci.* **2013**, *49*, 2324–2337. [CrossRef]
124. Jamshidi, P.; Chouhan, G.; Williams, R.L.; Cox, S.; Grover, L.M.; Birdi, G. Modification of gellan gum with nanocrystalline hydroxyapatite facilitates cell expansion and spontaneous osteogenesis. *Biotechnol. Bioeng.* **2016**, *113*, 1568–1576. [CrossRef]
125. Comeau, P.; Willett, T. Triethyleneglycol dimethacrylate addition improves the 3D-printability and construct properties of a GelMA-nHA composite system towards tissue engineering applications. *Mater. Sci. Eng. C* **2020**, *112*, 110937. [CrossRef]
126. Kim, K.; Dean, D.; Lu, A.; Mikos, A.G.; Fisher, J.P. Early osteogenic signal expression of rat bone marrow stromal cells is influenced by both hydroxyapatite nanoparticle content and initial cell seeding density in biodegradable nanocomposite scaffolds. *Acta Biomater.* **2011**, *7*, 1249–1264. [CrossRef]
127. Sartori, M.; Graziani, G.; Sassoni, E.; Pagani, S.; Boi, M.; Maltarello, M.C.; Baldini, N.; Fini, M. Nanostructure and biomimetics orchestrate mesenchymal stromal cell differentiation: An in vitro bioactivity study on new coatings for orthopedic applications. *Mater. Sci. Eng. C* **2021**, *123*, 112031. [CrossRef] [PubMed]
128. de Peppo, G.M.; Agheli, H.; Karlsson, C.; Ekström, K.; Brisby, H.; Lenneras, M.; Gustafsson, S.; Sjovall, P.; Johansson, A.; Olsson, E.; et al. Osteogenic response of human mesenchymal stem cells to well-defined nanoscale topography in vitro. *Int. J. Nanomed.* **2014**, *9*, 2499–2515. [CrossRef]
129. Ahlfeld, T.; Doberenz, F.; Kilian, D.; Vater, C.; Korn, P.; Lauer, G.; Lode, A.; Gelinsky, M. Bioprinting of mineralized constructs utilizing multichannel plotting of a self-setting calcium phosphate cement and a cell-laden bioink. *Biofabrication* **2018**, *10*, 045002. [CrossRef]

130. Ahlfeld, T.; Cubo, N.; Cometta, S.; Guduric, V.; Vater, C.; Bernhardt, A.; Akkineni, A.R.; Lode, A.; Gelinsky, M. A Novel Plasma-Based Bioink Stimulates Cell Proliferation and Differentiation in Bioprinted, Mineralized Constructs. *ACS Appl. Mater. Interfaces* **2020**, *12*, 12557–12572. [CrossRef]
131. Graziani, G.; Bianchi, M.; Sassoni, E.; Russo, A.; Marcacci, M. Ion-substituted calcium phosphate coatings deposited by plasma-assisted techniques: A review. *Mater. Sci. Eng. C* **2017**, *74*, 219–229. [CrossRef]
132. Jang, H.J.; Lee, E.C.; Kwon, G.J.; Seo, Y.K. The effect of coated nano-hydroxyapatite concentration on scaffolds for osteogenesis. *J. Biomater. Appl.* **2019**, *34*, 827–839. [CrossRef]
133. Li, Q.; Lei, X.; Wang, X.; Cai, Z.; Lyu, P.; Zhang, G. Hydroxyapatite/Collagen Three-Dimensional Printed Scaffolds and Their Osteogenic Effects on Human Bone Marrow-Derived Mesenchymal Stem Cells. *Tissue Eng. Part A* **2019**, *25*, 1261–1271. [CrossRef]
134. Vines, J.B.; Lim, D.-J.; Anderson, J.M.; Jun, H.-W. Hydroxyapatite nanoparticle reinforced peptide amphiphile nanomatrix enhances the osteogenic differentiation of mesenchymal stem cells by compositional ratios. *Acta Biomater.* **2012**, *8*, 4053–4063. [CrossRef]
135. Xie, J.; Baumann, M.J.; McCabe, L.R. Osteoblasts respond to hydroxyapatite surfaces with immediate changes in gene expression. *J. Biomed. Mater. Res.* **2004**, *71*, 108–117. [CrossRef]
136. Calabrese, G.; Giuffrida, R.; Fabbi, C.; Figallo, E.; Furno, D.L.; Gulino, R.; Colarossi, C.; Fullone, F.; Giuffrida, R.; Parenti, R.; et al. Collagen-Hydroxyapatite Scaffolds Induce Human Adipose Derived Stem Cells Osteogenic Differentiation In Vitro. *PLoS ONE* **2016**, *11*, e0151181. [CrossRef] [PubMed]
137. Ren, X.; Tuo, Q.; Tian, K.; Huang, G.; Li, J.; Xu, T.; Lv, X.; Wu, J.; Chen, Z.; Weng, J.; et al. Enhancement of osteogenesis using a novel porous hydroxyapatite scaffold in vivo and vitro. *Ceram. Int.* **2018**, *44*, 21656–21665. [CrossRef]
138. Zhao, C.; Wang, X.; Gao, L.; Jing, L.; Zhou, Q.; Chang, J. The role of the micro-pattern and nano-topography of hydroxyapatite bioceramics on stimulating osteogenic differentiation of mesenchymal stem cells. *Acta Biomater.* **2018**, *73*, 509–521. [CrossRef]
139. Lyu, L.-X.; Zhang, X.-F.; Deegan, A.J.; Liang, G.-F.; Yang, H.-N.; Hu, S.-Q.; Yan, X.L.; Huang, N.-P.; Xu, T.; Lü, L.-X. Comparing hydroxyapatite with osteogenic medium for the osteogenic differentiation of mesenchymal stem cells on PHBV nanofibrous scaffolds. *J. Biomater. Sci. Polym. Ed.* **2019**, *30*, 150–161. [CrossRef]
140. Fayyazbakhsh, F.; Solati-Hashjin, M.; Keshtkar, A.; Shokrgozar, M.A.; Dehghan, M.M.; Larijani, B. Novel layered double hydroxides-hydroxyapatite/gelatin bone tissue engineering scaffolds: Fabrication, characterization, and in vivo study. *Mater. Sci. Eng. C* **2017**, *76*, 701–714. [CrossRef]
141. Kim, H.-W.; Kim, H.-E.; Salih, V. Stimulation of osteoblast responses to biomimetic nanocomposites of gelatin–hydroxyapatite for tissue engineering scaffolds. *Biomaterials* **2005**, *26*, 5221–5230. [CrossRef]
142. Morgan, M.; Cooke, M.M.; Christopherson, P.A.; Westfall, P.R.; McCarthy, G. Calcium hydroxyapatite promotes mitogenesis and matrix metalloproteinase expression in human breast cancer cell lines. *Mol. Carcinog.* **2001**, *32*, 111–117. [CrossRef]
143. Kozlow, W.; Guise, T.A. Breast Cancer Metastasis to Bone: Mechanisms of Osteolysis and Implications for Therapy. *J. Mammary Gland. Biol. Neoplasia* **2005**, *10*, 169–180. [CrossRef]
144. Chen, S.; Shi, Y.; Zhang, X.; Ma, J. Biomimetic synthesis of Mg-substituted hydroxyapatite nanocomposites and three-dimensional printing of composite scaffolds for bone regeneration. *J. Biomed. Mater. Res. Part A* **2019**, *107*, 2512–2521. [CrossRef]
145. Luo, Y.; Chen, S.; Shi, Y.; Ma, J. 3D printing of strontium-doped hydroxyapatite based composite scaffolds for repairing critical-sized rabbit calvarial defects. *Biomed. Mater.* **2018**, *13*, 065004. [CrossRef] [PubMed]
146. Zhu, W.; Holmes, B.; Glazer, R.I.; Zhang, L.G. 3D printed nanocomposite matrix for the study of breast cancer bone metastasis. *Nanomed. Nanotechnol. Biol. Med.* **2016**, *12*, 69–79. [CrossRef] [PubMed]
147. Zhou, X.; Zhu, W.; Nowicki, M.; Miao, S.; Cui, H.; Holmes, B.; Glazer, R.I.; Zhang, L.G. 3D Bioprinting a Cell-Laden Bone Matrix for Breast Cancer Metastasis Study. *ACS Appl. Mater. Interfaces* **2016**, *8*, 30017–30026. [CrossRef]
148. Kim, J.; Kim, I.S.; Cho, T.H.; Lee, K.B.; Hwang, S.J.; Tae, G.; Noh, I.; Lee, S.H.; Park, Y.; Sun, K. Bone regeneration using hyaluronic acid-based hydrogel with bone morphogenic protein-2 and human mesenchymal stem cells. *Biomaterials* **2007**, *28*, 1830–1837. [CrossRef]
149. Keriquel, V.; Oliveira, H.; Rémy, M.; Ziane, S.; Delmond, S.; Rousseau, B.; Rey, S.; Catros, S.; Amédée, J.; Guillemot, F.; et al. In situ printing of mesenchymal stromal cells, by laser-assisted bioprinting, for in vivo bone regeneration applications. *Sci. Rep.* **2017**, *7*, 1–10. [CrossRef]
150. Holmes, B.; Zhu, W.; Li, J.; Lee, J.D.; Zhang, L.G. Development of Novel Three-Dimensional Printed Scaffolds for Osteochondral Regeneration. *Tissue Eng. Part A* **2015**, *21*, 403–415. [CrossRef]
151. Costantini, M.; Idaszek, J.; Szöke, K.; Jaroszewicz, J.; Dentini, M.; Barbetta, A.; Brinchmann, J.E.; Swieszkowski, W. 3D bioprinting of BM-MSCs-loaded ECM biomimetic hydrogels for in vitro neocartilage formation. *Biofabrication* **2016**, *8*, 035002. [CrossRef]
152. Ibrahim, A. 3D bioprinting bone. In *3D Bioprinting for Reconstructive Surgery*; Thomas, D.J., Jessop, Z.M., Whitaker, I.S., Eds.; Elsevier: Amsterdam, The Netherlands, 2018; pp. 245–275.
153. Adepu, S.; Dhiman, N.; Laha, A.; Sharma, C.; Ramakrishna, S.; Khandelwal, M. Three-dimensional bioprinting for bone tissue regeneration. *Curr. Opin. Biomed. Eng.* **2017**, *2*, 22–28. [CrossRef]
154. Yang, Y.; Wang, M.; Yang, S.; Lin, Y.; Zhou, Q.; Li, H.; Tang, T. Bioprinting of an osteocyte network for biomimetic mineralization. *Biofabrication* **2020**, *12*, 045013. [CrossRef]
155. Lin, H.; Tang, Y.; Lozito, T.P.; Oyster, N.; Kang, R.B.; Fritch, M.R.; Wang, B.; Tuan, R.S. Projection Stereolithographic Fabrication of BMP-2 Gene-activated Matrix for Bone Tissue Engineering. *Sci. Rep.* **2017**, *7*, 11327. [CrossRef] [PubMed]

156. Kolan, K.C.R.; Huang, Y.-W.; Semon, J.A.; Leu, M.C. 3D-printed Biomimetic Bioactive Glass Scaffolds for Bone Regeneration in Rat Calvarial Defects. *Int. J. Bioprinting* **2020**, *6*, 274. [CrossRef]
157. Kumar, P.S.; Hashimi, S.M.; Saifzadeh, S.; Ivanovski, S.; Vaquette, C. Additively manufactured biphasic construct loaded with BMP-2 for vertical bone regeneration: A pilot study in rabbit. *Mater. Sci. Eng. C* **2018**, *92*, 554–564. [CrossRef]
158. Karageorgiou, V.; Tomkins, M.; Fajardo, R.; Meinel, L.; Snyder, B.; Wade, K.; Chen, J.; Vunjak-Novakovic, G.; Kaplan, D.L. Porous silk fibroin 3-D scaffolds for delivery of bone morphogenetic protein-2in vitro and in vivo. *J. Biomed. Mater. Res. Part A* **2006**, *78*, 324–334. [CrossRef]
159. Jo, Y.-Y.; Kim, S.-G.; Kwon, K.-J.; Kweon, H.; Chae, W.-S.; Yang, W.-G.; Lee, E.-Y.; Seok, H. Silk Fibroin-Alginate-Hydroxyapatite Composite Particles in Bone Tissue Engineering Applications In Vivo. *Int. J. Mol. Sci.* **2017**, *18*, 858. [CrossRef]
160. Shi, W.; Sun, M.; Hu, X.; Ren, B.; Cheng, J.; Li, C.; Duan, X.; Fu, X.; Zhang, J.; Chen, H.; et al. Structurally and Functionally Optimized Silk-Fibroin-Gelatin Scaffold Using 3D Printing to Repair Cartilage Injury In Vitro and In Vivo. *Adv. Mater.* **2017**, *29*. [CrossRef]
161. Huang, J.; Fu, H.; Wang, Z.; Meng, Q.; Liu, S.; Wang, H.; Zheng, X.; Dai, J.; Zhang, Z. BMSCs-laden gelatin/sodium alginate/carboxymethyl chitosan hydrogel for 3D bioprinting. *RSC Adv.* **2016**, *6*, 108423–108430. [CrossRef]
162. Serra, I.; Fradique, R.; Vallejo, M.; Correia, T.; Miguel, S.A.P.; Correia, I. Production and characterization of chitosan/gelatin/β-TCP scaffolds for improved bone tissue regeneration. *Mater. Sci. Eng. C* **2015**, *55*, 592–604. [CrossRef] [PubMed]
163. Sivaraj, K.K.; Adams, R.H. Blood vessel formation and function in bone. *Development* **2016**, *143*, 2706–2715. [CrossRef]
164. Hanahan, D.; Weinberg, R.A. Hallmarks of Cancer: The Next Generation. *Cell* **2011**, *144*, 646–674. [CrossRef]
165. Carmeliet, P.; Jain, R.K. Molecular mechanisms and clinical applications of angiogenesis. *Nat. Cell Biol.* **2011**, *473*, 298–307. [CrossRef]
166. Liu, B.; Li, J.; Lei, X.; Cheng, P.; Song, Y.; Gao, Y.; Hu, J.; Wang, C.; Zhang, S.; Li, D.; et al. 3D-bioprinted functional and biomimetic hydrogel scaffolds incorporated with nanosilicates to promote bone healing in rat calvarial defect model. *Mater. Sci. Eng. C* **2020**, *112*, 110905. [CrossRef]
167. Chen, Y.-W.; Shen, Y.-F.; Ho, C.-C.; Yu, J.; Wu, Y.-H.A.; Wang, K.; Shih, C.-T.; Shie, M.-Y. Osteogenic and angiogenic potentials of the cell-laden hydrogel/mussel-inspired calcium silicate complex hierarchical porous scaffold fabricated by 3D bioprinting. *Mater. Sci. Eng. C* **2018**, *91*, 679–687. [CrossRef] [PubMed]
168. Cui, H.; Zhu, W.; Nowicki, M.; Zhou, X.; Khademhosseini, A.; Zhang, L.G. Hierarchical Fabrication of Engineered Vascularized Bone Biphasic Constructs via Dual 3D Bioprinting: Integrating Regional Bioactive Factors into Architectural Design. *Adv. Health Mater.* **2016**, *5*, 2174–2181. [CrossRef]
169. Shahabipour, F.; Ashammakhi, N.; Oskuee, R.K.; Bonakdar, S.; Hoffman, T.; Shokrgozar, M.A.; Khademhosseini, A. Key components of engineering vascularized 3-dimensional bioprinted bone constructs. *Transl. Res.* **2019**, *216*, 57–76. [CrossRef] [PubMed]
170. Zhu, W.; Castro, N.; Cui, H.; Zhou, X.; Boualam, B.; McGrane, R.; Glazer, R.I.; Zhang, L.G. A 3D printed nano bone matrix for characterization of breast cancer cell and osteoblast interactions. *Nanotechnology* **2016**, *27*, 315103. [CrossRef]
171. Alemany-Ribes, M.; Semino, C.E. Bioengineering 3D environments for cancer models. *Adv. Drug Deliv. Rev.* **2014**, *79–80*, 40–49. [CrossRef] [PubMed]
172. Vanderburgh, J.P.; Guelcher, S.A.; Sterling, J.A. 3D bone models to study the complex physical and cellular interactions between tumor and the bone microenvironment. *J. Cell. Biochem.* **2018**, *119*, 5053–5059. [CrossRef]
173. Holmes, B.; Zhu, W.; Zhang, L.G. Development of a Novel 3D Bioprinted In Vitro Nano Bone Model for Breast Cancer Bone Metastasis Study. *MRS Proc. Library Arch.* **2014**, *1724*, 1–6. [CrossRef]
174. Qiao, H.; Tang, T. Engineering 3D approaches to model the dynamic microenvironments of cancer bone metastasis. *Bone Res.* **2018**, *6*, 1–12. [CrossRef]
175. Graham, N.; Qian, B.-Z. Mesenchymal Stromal Cells: Emerging Roles in Bone Metastasis. *Int. J. Mol. Sci.* **2018**, *19*, 1121. [CrossRef]
176. Hughes, R.; Chen, X.; Cowley, N.; Ottewell, P.; Hawkins, R.; Hunter, K.; Hobbs, J.; Brown, N.; Holen, I. Osteoblast-Derived Paracrine and Juxtacrine Signals Protect Disseminated Breast Cancer Cells from Stress. *Cancers* **2021**, *13*, 1366. [CrossRef] [PubMed]
177. Zhao, Y.; Yao, R.; Ouyang, L.; Ding, H.; Zhang, T.; Zhang, K.; Cheng, S.; Sun, W. Three-dimensional printing of Hela cells for cervical tumor model in vitro. *Biofabrication* **2014**, *6*, 035001. [CrossRef]
178. Smits, I.P.; Blaschuk, O.W.; Willerth, S.M. Novel N-cadherin antagonist causes glioblastoma cell death in a 3D bioprinted co-culture model. *Biochem. Biophys. Res. Commun.* **2020**, *529*, 162–168. [CrossRef] [PubMed]
179. Gebeyehu, A.; Surapaneni, S.K.; Huang, J.; Mondal, A.; Wang, V.Z.; Haruna, N.F.; Bagde, A.; Arthur, P.; Kutlehria, S.; Patel, N.; et al. Polysaccharide hydrogel based 3D printed tumor models for chemotherapeutic drug screening. *Sci. Rep.* **2021**, *11*, 372. [CrossRef]
180. Miranda, M.A.; Marcato, P.D.; Mondal, A.; Chowdhury, N.; Gebeyehu, A.; Surapaneni, S.K.; Bentley, M.V.L.B.; Amaral, R.; Pan, C.-X.; Singh, M. Cytotoxic and chemosensitizing effects of glycoalkaloidic extract on 2D and 3D models using RT4 and patient derived xenografts bladder cancer cells. *Mater. Sci. Eng. C* **2020**, *119*, 111460. [CrossRef]
181. Wang, M.; Xia, F.; Wei, Y.; Wei, X. Molecular mechanisms and clinical management of cancer bone metastasis. *Bone Res.* **2020**, *8*, 1–20. [CrossRef]

cancers

MDPI

Review

Printing the Pathway Forward in Bone Metastatic Cancer Research: Applications of 3D Engineered Models and Bioprinted Scaffolds to Recapitulate the Bone–Tumor Niche

Anne M. Hughes [1], Alexus D. Kolb [2], Alison B. Shupp [2], Kristy M. Shine [3,*] and Karen M. Bussard [2,*]

[1] Department of Biomedical Engineering, Worcester Polytechnic Institute, Worcester, MA 01609, USA; amhughes@wpi.edu

[2] Department of Cancer Biology, Thomas Jefferson University, Philadelphia, PA 19107, USA; alexus.kolb@students.jefferson.edu (A.D.K.); alison.shupp@students.jefferson.edu (A.B.S.)

[3] Health Design Lab, Jefferson Bioprinting Lab, Department of Emergency Medicine, Thomas Jefferson University, Philadelphia, PA 19107, USA

* Correspondence: kristy.shine@jefferson.edu (K.M.S.); karen.bussard@jefferson.edu (K.M.B.)

Simple Summary: Breast cancer commonly migrates to the skeleton. Once patients have breast cancer in their bones, their quality of life is poor. Research has shown that the environment and structure of bone helps support metastatic breast cancer cell growth. Current models used in laboratories to study breast cancer that has migrated to bone do not include all the components of what happens in the human body. This is because technology in the past was limiting. Now, however, a new technology exists, called three-dimensional (3D) bioprinting, that "prints" living cells in geometries like the ones within the human body, to replicate a more realistic environment. In cancer research, 3D bioprinting allows researchers to more closely replicate events that occur when cancer cells migrate from a primary location to a secondary site, as well as model events in the environment of the secondary site. 3D bioprinting may also be used instead of animal models in some cases, and may speed up new drug discovery for cancer treatment.

Abstract: Breast cancer commonly metastasizes to bone, resulting in osteolytic lesions and poor patient quality of life. The bone extracellular matrix (ECM) plays a critical role in cancer cell metastasis by means of the physical and biochemical cues it provides to support cellular crosstalk. Current two-dimensional in-vitro models lack the spatial and biochemical complexities of the native ECM and do not fully recapitulate crosstalk that occurs between the tumor and endogenous stromal cells. Engineered models such as bone-on-a-chip, extramedullary bone, and bioreactors are presently used to model cellular crosstalk and bone–tumor cell interactions, but fall short of providing a bone-biomimetic microenvironment. Three-dimensional bioprinting allows for the deposition of biocompatible materials and living cells in complex architectures, as well as provides a means to better replicate biological tissue niches in-vitro. In cancer research specifically, 3D constructs have been instrumental in seminal work modeling cancer cell dissemination to bone and bone–tumor cell crosstalk in the skeleton. Furthermore, the use of biocompatible materials, such as hydroxyapatite, allows for printing of bone-like microenvironments with the ability to be implanted and studied in in-vivo animal models. Moreover, the use of bioprinted models could drive the development of novel cancer therapies and drug delivery vehicles.

Keywords: bioprinting; breast cancer; metastasis; bone; tissue engineering; 3D modeling; tumor microenvironment; extracellular matrix

Citation: Hughes, A.M.; Kolb, A.D.; Shupp, A.B.; Shine, K.M.; Bussard, K.M. Printing the Pathway Forward in Bone Metastatic Cancer Research: Applications of 3D Engineered Models and Bioprinted Scaffolds to Recapitulate the Bone–Tumor Niche. *Cancers* **2021**, *13*, 507. https://doi.org/10.3390/cancers13030507

Academic Editors: Serena Danti, Nicola Contessi Negrini and Alessandro Franchi

Received: 9 December 2020
Accepted: 25 January 2021
Published: 29 January 2021

Publisher's Note: MDPI stays neutral with regard to jurisdictional claims in published maps and institutional affiliations.

1. Introduction

Bone is a unique organ that provides, among other things, structural support and protection for the body. The bone, including the extracellular matrix (ECM), is continuously remodeled whereby components of the ECM are modified, secreted, or degraded [1].

The skeleton is composed of a variety of cells, including osteoblasts, osteocytes, osteoclasts, fibroblasts, endothelial cells, and mesenchymal stromal cells [2,3]. These cells are surrounded by organic and inorganic components such as type I collagen and apatite nanocrystals [2–4]. These organic and inorganic components constitute the bone matrix which is capable of providing biochemical and physiological cues that regulate bone modeling and mechanotransduction [3].

The skeleton is an attractive site for cancer cell dissemination and colonization due to its unique molecular and structural properties [5]. Although prognosis is excellent in patients with early non-invasive disease, 5-year mortality rates with treatment exceed 80% once cancer has invaded the bone [6]. Normal bone processes, such as bone remodeling, produce a plethora of growth factors and cytokines that are strong chemoattractants for metastatic cancer cells [7–11]. Once in the bone, metastatic cancer cells hijack the endogenous stromal cells in the bone microenvironment, including osteoblasts and osteoclasts, to remodel the bone matrix, leading to a disorganized basement membrane and disrupted cellular crosstalk, which ultimately promote tumor progression [12–14]. Thus, understanding the interactions between the ECM and metastatic cancer cells is crucial in order to both regulate and prevent metastatic cancer cell growth in bone. However, the study of cancer cell dissemination, seeding, colonization, and survival in bone is complicated by the vast structural and physiological complexities of the organ. Most model systems currently available only account for select aspects of the niche; thus, a combination of techniques is frequently utilized to wholly recapitulate events that occur during bone–tumor crosstalk.

Two-dimensional (2D) in-vitro cell culture environments have arisen as important platforms for the study of cancer cell proliferation, migration, and invasion as well as cancer cell response to therapeutic treatment. Although simple and low cost, the planar geometry and non-physiologic composition (i.e., tissue culture plastic) of 2D cultures can induce variable cellular morphologies and aberrant gene expression [15–17] that impact the behavior and cellular crosstalk of endogenous bone and tumor cells [15,18]. Moreover, mechanical cues translated by the microenvironment are absent [3]. Importantly, Bissell and colleagues discovered that normal breast epithelial cells required interactions with the basement membrane, as present in three-dimensional (3D) cultures, to display and maintain their normal phenotype. Otherwise, the normal breast epithelial cells behaved similarly to tumor cells when grown in 2D cultures [19]. Thus, such models inadequately represent the spatial, biochemical, and mechanical complexities of the native 3D tumor microenvironment, severely limiting their interpretation in the study of metastatic disease to the bone.

Recognizing these shortcomings, investigators are now developing 3D bone microenvironments such as bioreactors [20–26], scaffolds [27,28], extramedullary bone [29,30], and bone-on-a-chip [31,32] models for metastatic cancer research. Such novel systems allow for more relevant material composition and spatial relationships among cells and, in some cases, provide similar mechanical forces. However, these systems face individual challenges including reproducibility, scale, and ability to manipulate matrix biochemical composition and structural integrity, and can be difficult to seed with cells for culture [33]. Thus, a gold standard model for the study of cancer metastasis to bone remains elusive.

3D bioprinting, an evolving technology, may help to facilitate such research. A novel biofabrication technique that yields reproducible, biomimetic 3D environments composed of cells and native biomaterials, 3D printing offers distinct advantages over prior 2D and 3D cell culture models. ECM composition and matrix geometry and stiffness, as well as cell density and position, are all controllable parameters in bioprinted 3D scaffolds, supporting cellular proliferation, migration, phenotype development, and matrix and biochemical factor production [16]. 3D constructs can achieve biophysical characteristics similar to the native tissue microenvironments, allowing for the study of cancer progression and tumorigenesis [34]. Thus, 3D bioprinting may provide a more robust solution, yielding biomimetic microenvironments for the study of cell–cell and cell–matrix interactions at the forefront of cancer research.

In this review, we focus on cancer metastases to bone and provide an overview of current 2D and 3D in-vitro culture model systems, as well as 3D in-vivo and ex-vivo model systems to study cancer metastases to bone. We review the applications and limitations of current technologies, which include the inability of current models to wholly represent the spatial, biochemical, and mechanical complexities of the native 3D tumor microenvironment, causing limited interpretation of results generated in these systems. We next discuss an emerging technology called 3D bioprinting. 3D bioprinted bone-mimetic microenvironments offer distinct advantages over current 2D and 3D technologies, including reproducibility and the ability to incorporate native biophysical properties of the ECM, including cellular composition, stiffness, and geometry. As a result of these and other advantages, 3D bioprinting has the potential to elevate into the next generation current models in cancer research that mimic the bone–tumor niche.

2. Gross Anatomy of Bone and Bone Physiology

Bone is a metabolically active and dynamic organ that supports many body functions, including regulation of calcium blood levels, providing structural support, and protecting internal organs. To accomplish these and other functions, bone is composed of three main cell types: osteoblasts, osteoclasts, and osteocytes. Osteoblasts, the cells that synthesize new osteoid matrix, account for 4–6% of total cells in bone. Osteoblasts are derived from mesenchymal stromal cells present in the bone marrow [35]. After stimulation with bone morphogenetic proteins and growth factors, mesenchymal stromal cells proliferate to form pre-osteoblasts, which then differentiate into mature osteoblasts capable of matrix synthesis and deposition [36]. Newly formed bone is composed of type I collagen and non-collagenous proteins (proteoglycans and other extracellular matrix proteins) (~22%), hydroxyapatite (~70%), and water (~8% by weight) [35,37,38].

Osteoclasts, on the other hand, are the cells responsible for bone resorption. Osteoclasts account for 1–4% of total cells in the bone, and are derived from bone monocytes [39,40]. Bone marrow monocytes (i.e., precursor osteoclasts) express Receptor Activator for Nuclear factor Kappa beta (NF-κB)(RANK) on their cell surface, which binds to Receptor Activator for Nuclear factor Kappa beta Ligand (RANK-L) on osteoblasts to elicit osteoclastogenesis [41]. Several single-nucleated osteoclasts then fuse together to form one mature, multi-nucleated bone-resorbing osteoclast [41]. Upon subsequent activation, osteoclasts carry out a complex degradation process to secrete matrix degradation enzymes and resorb bone. This process exposes bones' organic matrix via degradation by cathepsin K and additional cysteine proteinases, lysozymal enzymes, phosphatases, and matrix metalloproteinases at the osteoclast–bone interface [41–46]. As bone is resorbed, growth factors stored in the bone matrix, including transforming growth factor-beta (TGF-B), insulin growth factor (IGF), and bone morphogenetic proteins (BMPs), are released into the microenvironment [47].

Osteocytes help to regulate the activities of osteoblasts and osteoclasts via mechanotransduction, and represent ~90–95% of cells in the bone [48,49]. Osteocytes are formed when osteoblasts become terminally differentiated and embedded in the bone matrix. Osteocytes reside in small cavities, called lacunae, which transmit signals to bone cells, including osteoblasts and osteoclasts, via tiny channels called canaliculi [49]. These signals regulate nutrient exchange in bone, as well as bone remodeling and sensation of mechanical stimuli [3,50–53]. As a result of their ability to sense mechanical load in the bone, osteocytes have earned the nickname the "mechanosensors of bone" [54–56].

3. Bone Remodeling during Disease

Bone resorption and deposition are held in a tightly regulated, homeostatic balance in which there is no net bone loss or gain. However, this balance is upset in pathological conditions, including infection (osteomyelitis), osteoarthritis, and bone metastatic cancers [57–60]. In each of these conditions, and especially in osteolytic disease related to cancer, osteoclasts are overstimulated to degrade bone. Osteoblasts do not deposit new bone; thus resulting in net bone loss [61]. In brief, osteomyelitis is a severe bone infection, caused by the bacterium *Staphylococcus aureus*, which can be fatal [62,63]. Osteomyelitis typically occurs from chronic inflammation as a result of osseointegrated implants (e.g., dental implants or femoral implants (i.e., artificial hip)) [64–66]. Osteomyelitis involves the internalization of *S. aureus* by osteoblasts, permitting *S. aureus* to evade immune detection and cause sustained inflammation [67]. Upon internalization of *S. aureus*, osteoblasts increase their production of inflammatory cytokines, including interleukin-6 (IL-6), monocyte chemoattractant protein-1 (MCP-1), regulated on activation normal T cell expressed and secreted (RANTES), and macrophage inflammatory protein-1 alpha (MIP-1 alpha), as well as factors that stimulate osteoclastogenesis, including granulocyte-colony stimulating factor (G-CSF), RANK-L, and interleukin-8 (IL-8), among others [68–74]. Sustained infection leads to a reduction in osteoblast proliferation and eventual osteoblast death. Subsequently, bone is resorbed at an increased rate by osteoclasts, leading to sustained bone loss [74]. Osteoarthritis is another bone disease characterized by chronic inflammation that impacts osteoblast function. In osteoarthritis, osteoblasts overexpress inflammatory cytokines, including IL-8, IL-6, vascular endothelial growth factor (VEGF), and prostaglandin E2 (PGE2), among others [59,60,75]. This results in a reduction of osteoblast bone deposition, and an imbalance in bone remodeling that results in sustained bone resorption. [76,77]. Interestingly, many of the same cytokines and inflammatory factors overexpressed by osteoblasts during osteomyelitis and osteoarthritis have also been known to be overexpressed during bone invasion by metastatic cancer cells, including breast, prostate, lung, and multiple myeloma [7,26,78–82].

4. Bone Is a Favored Site of Cancer Metastasis

Bone is a preferential site of cancer metastasis [8,9]. In an attempt to explain directional tropism of disseminated cancer cells for specific organs in the body, Stephen Paget made the following statement in 1889: "When a plant goes to seed, its seeds are carried in all directions; but they can only live and grow if they fall on congenial soil [83]". As suggested by Paget, disseminated tumor cells are the "seeds" and the bone niche, rich in growth factors, neovascularization factors, and cytokines, is the "congenial soil" necessary for cancer cell growth [84]. Paget's "seed and soil" hypothesis also explains the preferential metastasis of breast, prostate, lung, and multiple myeloma disseminated cancer cells to the bone. Furthermore, mounting evidence has implicated the cells of the bone responsible for remodeling, the osteoblasts and osteoclasts, as key players in bone metastatic cancer cell progression, including cancer cell homing to and seeding in bone, dormancy, cancer cell re-activation, and contribution to macrometastatic lesion growth [85,86].

Once shed from a primary tumor, disseminated tumor cells enter the circulation, ultimately traveling to large blood vessels found in the bone called vascular sinusoids. Blood flow within the sinusoids is sluggish, allowing for normal movement of immune, hematopoietic, and lymphoid cells into and out of the bone. This slow blood flow also enables cancer cells to easily invade bone simply due to the normal physiology [8,87]. The vascular sinusoids are the main entry point for disseminated tumor cells into the bones, with the majority of metastases occurring at the ends of long bones, including the femur, where there is a high rate of metabolic activity due to bone turnover by the osteoblasts and osteoclasts.

4.1. Breast Cancer Cells Preferentially Metastasize to Bone

Approximately 20-30% of breast cancer patients will develop metastatic lesions, with ~15% of those being bone metastases [88,89]. Approximately 50% of these metastases will involve bone as a primary secondary tissue, with nearly 80% being secondary or recurring metastatic sites [82,90–92]. Once breast cancer invades bone, the relative 5-year survival rate falls to less than 10% [90]. Lesions in the bone that form as a result of bone metastatic breast cancer are predominantly osteolytic in nature, however, some lesions can also be mixed lytic and blastic [10,87,91–93]. During the formation of osteolytic lesions, osteoclasts are constitutively activated to resorb bone, and osteoblasts do not deposit new bone, resulting in sustained bone degradation. This phenomenon has been described as the "vicious cycle" of breast cancer metastases to bone [9,25,94]. Occurring during advanced tumor progression, metastatic breast cancer cells produce parathyroid hormone-related protein (PTHrP), which stimulates osteoblasts to produce increased amounts of RANK-L. Osteoblasts that overexpress RANK-L bind to the receptor RANK on osteoclast precursors, stimulating osteoclastogenesis and increased bone resorption. Growth factors stored in bone are released, including TGF-beta, which are used by cancer cells to produce additional PTHrP [9,94], ultimately leading to sustained bone resorption and osteolytic lesion formation. Bone pain, fractures, hypercalcemia, and spinal cord compression subsequently occur [9,95,96]. Clinically, patients with osteolytic lesions are treated with drugs such as bisphosphonates that block the activity of osteoclasts, thus reducing bone degradation [97–100]. However, these drugs are not curative for the lesions already present; currently, no therapeutics are available that have the sole purpose to directly stimulate new bone deposition by osteoblasts.

4.2. Multiple Myeloma Metastases to the Bone

Multiple myeloma is another cancer that preferentially metastasizes to bone [101,102]. Approximately 70% of patients presenting with multiple myeloma have bone metastases upon diagnosis. Over time, over 90% of patients with multiple myeloma will develop bone metastases during the course of their disease [103]. The vast majority of these metastases will be osteolytic in nature, with bone being resorbed at a rate greater than it is deposited. Multiple studies by Roodman and colleagues have determined this is due to metastatic multiple myeloma cell suppression of osteoblast differentiation via a variety of factors [104–110]. At present, two main treatment strategies are being evaluated either in pre-clinical or clinical trials for the treatment of osteolytic metastases due to bone metastatic multiple myeloma. First, in pre-clinical models, treatment with an antibody to sclerostin, a Wnt/beta-catenin antagonist, reduced osteolytic lesion formation by 60% compared to wild-type mice. In addition, in tumor-bearing mice with a genetic deletion of sclerostin, there was a 50% increase in trabecular bone volume as well as an increase in the number of osteoblasts compared to tumor-bearing, wild-type mice [106]. Furthermore, as part of a phase I/II clinical trial, an inhibitor of DKK1, which is also involved in the Wnt signaling pathway, was found to increase osteoblast differentiation and calcium deposition in co-cultures of osteoblasts plus multiple myeloma cells in-vitro. That same DKK inhibitor increased both trabecular bone volume and Wnt/beta-catenin signaling (crucial for osteoblast differentiation) in tumor-bearing mice when compared to placebo [111]. Taken together, these results suggest that the inhibition of sclerostin and/or inhibitors of DKK1 may be promising tools to promote osteoblast activity, leading to increased subsequent bone deposition in multiple myeloma bone disease.

4.3. Prostate Cancer Colonization of the Skeleton

Prostate cancer has a predilection for bone metastases. Greater than 80% of patients with late-stage prostate cancer will develop bone metastases [9]. As opposed to other cancers with strong tropism for the bone, the vast majority of metastatic lesions that occur as a result of prostate cancer are osteoblastic in nature in which there is an excess of bone deposition [112]. This is due to prostate cancer cells preferentially homing to the ends of long bones where osteoblasts predominantly reside, leading to increases in both osteoblast and prostate cancer cell proliferation [80,113]. Interestingly, newly formed bone is disorganized, typically with low density, and is weak in strength [114,115]. This is due to a malalignment of osteoblasts along a collagen matrix, leading to the production of bone with a spongy structure as opposed to a hard and compact lamellar structure [116,117]. There are some reports, however, of bone metastatic prostate cancer lesions being mixed or osteolytic [118,119].

Similar to the "vicious cycle" of breast cancer metastases to the bone, bone metastatic prostate cancer cells are involved in a cycle that promotes excessive bone deposition by osteoblasts and continued proliferation of metastatic prostate cancer cells. Prostate cancer cells in the bone produce growth factors, including TGF-beta and IGF-1, as well as bone morphogenetic proteins that drive osteoblasts to deposit new bone [81,95,120,121]. In turn, bone resorption by osteoclasts is suppressed, leading to sustained net bone deposition [94,122]. In addition, activated osteoblasts overexpress inflammatory cytokines, including IL-6, IL-8, MCP-1, and VEGF, that fuel metastatic prostate cancer cell growth in bone [123]. A number of clinical trials are underway to treat osteoblastic lesions associated with bone metastatic prostate cancer. In a phase I/II clinical trial, atrasentan, an antagonist to the endothelin receptor A, reduced bone pain, slowed metastatic prostate tumor growth, and resulted in a significant decrease in median time to disease progression when compared to placebo [124,125]. In addition to atrasentan, Radium-223, an alpha particle emitter that preferentially accumulates at sites rich in osteoblast activity, has proven to be a promising therapeutic for patients with bone metastatic prostate cancer. In pre-clinical models, tumor-bearing mice exposed to Radium-223 had reduced bone growth, and preserved bone volume and architecture [126].

4.4. Lung Cancer Metastasizes to the Skeleton

Lung cancer is another cancer that has a tropism for the bone, although less so than breast cancer, prostate cancer, and multiple myeloma; lung cancer metastasizes to the bone approximately 34.3% of the time [127]. While exact mechanisms for preferential metastasis of lung cancer to bone are unclear, recent data implicate growth factors including TGF-beta [128], cytokines and chemokines such as stromal-derived factor-1 (SDF-1) and CXCR4 [129], and matrix metalloproteinases [129] that help drive lung cancer metastases to bone. Of the many types of lung cancer, small cell lung cancer and non-small cell lung cancer present mainly with osteolytic bone metastases in the spine and ribs [130–132]. Additional data suggest that lesions associated with bone metastatic lung cancer can also be mixed osteolytic and osteoblastic [130–132]. Clinically, markers of bone turnover, including osteopontin, collagen type I, and bone sialoprotein, can be used as biomarkers for the evaluation of bone metastatic lung cancer progression in human patients [133–140]. Osteopontin in particular has been associated with an aggressive lung cancer phenotype whereby increased osteopontin expression promotes lung cancer cell migration and invasion in bone via increased epithelial-to-mesenchymal transition [141,142].

5. Models to Investigate Tumor-Bone Cell Interactions

5.1. Model Systems for Early Cancer Cell Dissemination to Bone

Upon dissemination to bone, a cancer cell will either proliferate to form a metastatic lesion, or the cancer cell may enter a proliferatively quiescent or dormant state [143–145]. Sometimes also named "metastatic latency" [85], cellular dormancy can be defined as either proliferative arrest or a state of balanced proliferation and apoptosis where there is no net growth. In many situations, this is an adaptive response to microenvironmental stress at a secondary site whereby disseminated tumor cells must respond to unknown signals from the niche in order to survive [143]. Over time, dormant cancer cells become reawakened, leading to cancer cell proliferation and the formation of macrometastases in bone. At present, there is no model available that fully recapitulates all of the steps of early human cancer cell dissemination to bone. While a number of pre-clinical models are available that recapitulate specific steps of established disease, efforts to specifically study earlier events in bone metastatic cancer progression, especially early dissemination to bone, initial cancer cell seeding of bone, and events associated with metastatic latency, have been hampered by a lack of mouse and ex-vivo model systems due to substantial technical limitations [146]. Most importantly, there are no immune-competent mouse models capable of recapitulating all steps of the metastatic cascade available to investigate the role of the immune system during disease progression. As a result, knowledge of the early events of disease progression is severely limited.

Presently, only a handful of studies have attempted to investigate early steps of cancer metastasis to bone. Of those, osteoblasts in the endosteal bone niche have been identified as key mediators of cancer cell seeding. Scimeca et al. assessed human tissue samples of benign breast lesions, breast-infiltrating carcinomas, and bone metastatic breast cancer lesions [147]. The authors found a large number of breast cancer cells that underwent epithelial-to-mesenchymal transition (EMT) in the infiltrating carcinomas, which corresponded to a large number of breast–osteoblast-like cells that were positive for RANKL expression as well as the vitamin D receptor. These breast–osteoblast-like cells were also discovered in matched bone metastases [147]. These data suggest that osteoblasts and osteoblast-like cells may be used as an early predictor of bone metastatic breast cancer. Bodenstine et al. attempted to describe mechanisms that regulate osteoblast populations during the early stages of bone metastatic breast cancer [148]. An intratibial injection model was used to introduce osteoblasts plus bone-tropic breast cancer cells into the tibia to assess tumor growth. In mice injected with cancer cells alone (control), tumors that formed were smaller in size and remained contained in the bone cavity. By comparison, breast cancer cells co-injected with osteoblasts formed large palpable tumors nearly double the size of those formed with cancer cells injected alone, that migrated into the extra-osseous space [148]. These data suggest intracellular crosstalk between osteoblasts and bone metastatic cancer cells is important in cancer cell progression. However, while the mouse model used may not be fully representative of early-stage events, it may still help shed light on crosstalk that occurs between osteoblasts and metastatic breast cancer cells in bone.

Our laboratory additionally utilized a mouse model of intratibial injection to study interactions between osteoblasts and bone-tropic breast cancer cells in early disease progression. We identified a subset of osteoblasts present in the bone–tumor niche that regulate breast cancer progression in bone by producing inhibitory factors that reduce breast cancer cell growth [85]. We called this subset "educated" osteoblasts (EOs). Remarkably, tumors formed in mouse bones in the presence of EOs were smaller and grew at a slower rate than tumors composed of naive osteoblasts. We identified candidate proteins that distinguish "educated" osteoblasts (EOs) from naïve osteoblasts both in-vivo and ex-vivo. Using these markers, we interrogated human patient samples of bone metastatic breast cancer and identified EOs in patients with estrogen receptor positive (ER+) breast cancer [85,93]. Moreover, we demonstrated that exposure to EO conditioned medium reduced breast cancer cell proliferation and led to a reduction in the number of cells in the S phase of the cell cycle of both triple negative (TN) and ER+ breast cancer cells in-vitro. Furthermore, direct co-culture with EOs increased TN and ER+ breast cancer expression of p21 compared to cultures with naïve osteoblasts [85]. Similar to our findings in bone metastatic breast cancer, Lawson et al. found that osteoblasts and osteoclasts in the bone were capable of promoting (osteoblasts) or reawakening (osteoclasts) multiple myeloma cells from proliferative quiescence using an intravenous injection syngeneic mouse model as well as in-vitro analyses [86]. The authors used intravital imaging to track single multiple myeloma cells as they entered the bone niche and found that when fluorescently labeled multiple myeloma cells directly engaged with osteoblasts, myeloma cell growth was suppressed [86]. The authors recapitulated this in-vitro, whereby exposure to the conditioned medium of or co-culture with murine osteoblasts increased the number of dormant multiple myeloma cells. Conversely, when exposed to osteoclast-precursor conditioned media, multiple myeloma cells exhibited increased proliferation and growth [86]. Interestingly, daily injection of soluble RANK-L into tumor-bearing mice resulted in a reduction of dormant multiple myeloma cells and subsequent increase in osteoclastogenesis and bone resorption. In two additional studies, osteoblast expression of TGF-beta and Gas6 [149], as well as BMP7 [150], maintained bone metastatic prostate cancer cells in a dormant state, whereas reduction of those factors promoted prostate cancer cell growth. Thus, these data suggest engagement with osteoblasts either via direct or indirect means is key in the regulation of bone metastatic cancer cell progression.

5.2. Models to Investigate Late-Stage Metastatic Cancer Cell Invasion, Colonization, and Survival in Bone

Currently, there are no models available, in-vitro, ex-vivo, in-vivo, or otherwise, that fully recapitulate all the steps of the human bone metastatic cascade. This is mainly due to extensive technical hurdles and limitations, limited available human cell lines, limited immune-competent models, and sustaining long-term cell growth of primary human stromal cells [151–154]. As a result, multiple experimental models are used to investigate specific aspects of the metastatic cascade, with each mainly designed to investigate a specific stage of metastasis. Often, a combination of model systems is used to answer a given experimental question. For example, we and others have used complimentary approaches to investigate breast cancer metastasis in bone, including, but not limited to, multiple human cell lines [155,156], extended culture bioreactors [156], poly-ether-urethane foam scaffolds [157], biocompatible printed scaffolds [33], and novel engineered organotypic models [154].

5.2.1. In-Vivo Models of Cancer Metastasis to Bone

In-vivo mouse models are a common pre-clinical assay to study a variety of steps of the bone metastatic cascade. Syngeneic mouse models refer to murine cancer cells injected into mice (i.e., same genetic background as the host), whereas xenograft models describe a host that is genetically distinct from the cancer cells inoculated (i.e., human cancer cells inoculated into mice). Each model has its own advantages and disadvantages: syngeneic models permit the analysis of an immune-competent system, as well as allow investigation between inoculated cells and endogenous stromal cells within the microenvironment [158]. However, there are proteins and genes that are not analogous between humans and mice (e.g., IL-8 (human) and MIP-2 (murine); GRO-alpha (human) and KC (murine)). Therefore, mechanisms that may be identified via a syngeneic mouse model may not directly translate to humans [158]. Xenograft models permit the study of human cancer cells in an immune-compromised host such that mechanisms specific to human cancer cell progression may be studied. In many cases, human and mouse cells are capable of uninhibited crosstalk [158,159]. However, xenograft models do not permit the study of the immune system with cancer progression.

Syngeneic and xenograft models may further be broken down based on the route of injection, specifically via the orthotopic, intracardiac, or intratibial route [160]. In an orthotopic model, cancer cells or fragments are implanted into the same anatomic location from which the cancer originated; i.e., breast cancer cells inoculated into the mammary gland or prostate cancer cells inoculated into the prostate of a mouse. While orthotopic models are very beneficial in studying primary cancer progression, many times, a primary tumor may out-pace the growth of any metastases that may arise. In these cases, primary tumors may be removed upon reaching a certain size, subsequently allowing metastases in secondary sites to progress. Depending on the growth of the inoculated cancer cells, progression to metastasis in orthotopic models may be prolonged, requiring multiple months of study [161]. Experimental metastasis models permit the investigation of steps involving the trafficking of cancer cells to secondary sites, as well as investigation of late-stage metastases. Intracardiac injections recapitulate cancer cell trafficking and homing to secondary organs via the vasculature due to the cancer cells being directly injected into the left ventricle of the heart, which bypasses pulmonary circulation [160]. Several laboratories have developed bone-tropic cell lines utilizing intracardiac mouse models, which more closely mimic bone metastases in humans [162,163]. In the majority of cases, intracardiac injections are carried out in a xenograft model due to fluorescent labeling of inoculated cancer cells for tracking and retrieval for ex-vivo analysis. Finally, intratibial injections are a model of established disease whereby cancer cells are directly injected into the trabecular bone of the tibia of mice. Tumors rapidly form, permitting the study of the bone microenvironment during late-stage disease [160].

5.2.2. Three-Dimensional Tissue-Engineered Models to Study Cancer–Bone Interactions during Disease Progression

Current models to investigate complex cancer cell–bone interactions include tissue-engineered bone constructs (TEBCs), bone-on-a-chip (BC) bone tissue models, bioreactors for continuous culture, three-dimensional printed bone matrix scaffolds, hydrogels, and extramedullary bone models [164–166]. While these models replicate physiological systems better than standard two-dimensional cell culture due to their biophysical properties, all of these models fall short of being able to fully recapitulate the complexity and cellular crosstalk that occurs in the bone–tumor microenvironment. As a result, many investigators use a combination of 2D cell culture, 3D tissue-engineered models, and in-vivo models to investigate specific steps of the metastatic cascade.

Hydrogel Tissue Constructs

Hydrogels can be used in both two- and three-dimensional formats and are available as either natural (i.e., collagen) or synthetic (i.e., polyacrylamide) materials [167–170]. Of the types, three-dimensional hydrogels best mimic native bone, recapitulating tissue stiffness and bone's elastic modulus, which most closely represent the mechanotransductive properties of the skeleton [164]. Furthermore, three-dimensional hydrogels permit the embedding of cells to be studied, which facilitate the study of cellular migration and invasion through a matrix. This also enables seeded cells to resemble their normal morphology in-vivo, thus allowing for better representation of physiologic events that occur during cell–cell and cell–matrix contact. While beneficial, the increased complexities involved in the use of cell-containing three-dimensional hydrogels (e.g., viability-maintaining fabrication techniques, crosslinking, and nutrient diffusion) may increase resource needs relative to more simple two-dimensional models (e.g., gel-coated polystyrene) as well as create technically challenging analysis via full thickness penetration of the hydrogel for imaging [167,170]. Because of these limitations, two-dimensional hydrogels are beneficial for easier cell observation and analysis, as well as the ease of manipulation of microenvironmental conditions. Two-dimensional hydrogels are also well established. A major disadvantage of two-dimensional hydrogels, though, is the inability to permit three-dimensional tissue structure, including flattened cell morphology, forced cell polarity (i.e., cells seeded on top of a hydrogel as opposed to within), and a high matrix stiffness due to the conformation of the hydrogel being flat [164].

Bone-Like Scaffolds

Bone-like scaffolds are one 3D model system frequently used to model the bone microenvironment [107,117]. These models are capable of recapitulating the rigid extracellular matrix of bone. Polymers such as poly-(ϵ-caprolactone) (PCL) [114,115], polylactic acid (PLA), or PCL/PLA blends [116] have been used to print 3D scaffold structures onto which cells of interest are seeded. PCL, in particular, is an FDA-approved biocompatible and bioresorbable scaffold that has been used in the past as a three-dimensional scaffold for craniofacial bone grafts, but is now being used in other applications [171,172]. Our laboratory in particular has used PCL-based scaffolds as model systems in which GFP-labeled triple negative breast cancer cells were cultured [33]. We showed that the cancer cells were capable of infiltrating the depth of the scaffold while proliferating, and were also capable of establishing vascular networks through the formation of blood vessels [33]. Other studies have shown that PCL blended with PLA mimics bone architecture well, whereby osteosarcoma cells were capable of seeding and proliferating in scaffolds generated with these materials [165,173,174]. Thus, biomimetic synthetic polymers are a useful tool that effectively models both the biomechanical properties and architecture of the bone niche. Furthermore, these models allow for the manipulation of the niche via exchange of seeded cells of interest.

Bioreactor-Based Engineered Tissue

Bioreactors are another useful tool to specifically examine bone–tumor interactions under long-term culture conditions [118–120]. While these models are mainly limited to studying cellular behavior in-vitro, they do offer the flexibility of long-term cell growth/crosstalk (i.e., 6+ months). Vogler, Mastro, and colleagues designed a 3D bioreactor capable of generating crosstalk upon long-term culture (up to 120 days) of both murine and human osteoblast cell lines. Importantly, these cells displayed characteristics of normal osteoblasts throughout their growth [118]. Furthermore, that same model system was used to mimic breast cancer colonization in bone whereby a tri-culture of osteoblasts, osteoclasts, and human breast cancer cells recapitulated events observed during the "vicious cycle of breast cancer metastasis", including increased osteoclast activity, along with a decrease in osteoblast response, decreased matrix thickness, increased matrix resorption, and increased cancer cell proliferation [119]. Both human and murine cells were capable of growth in the bioreactor.

Humanized Biomaterial Implants

Finally, other groups have developed novel humanized biomaterial implants to serve as new platforms for the study of the bone niche. Andreeff and colleagues developed an extramedullary bone (EMB) model using mesenchymal stromal cells and endothelial colony-forming cells to recapitulate the hematopoietic and bone marrow microenvironments [30]. Importantly, this was the first development of a genetically controlled human bone marrow microenvironment capable of engrafting into NOD/SCID/IL-2r gamma null (NSG) immunocompromised mice. Unlike prior models to study the human bone marrow microenvironment, the EMB closely mimics human disease in-vivo, provides a robust hematopoietic environment, and can serve as a representative assay for modeling cancer metastases to bone [30]. Importantly, this model effectively recapitulated leukemia cell seeding and proliferation in a human bone marrow microenvironment, as well as permitted cancer cell metastasis to the ends of long bones; a preferential site of cancer cell invasion [30]. Several years later, Andreeff and colleagues refined this mouse model to include a humanized bone-chip implant whereby freshly isolated human bone fragments were collected from patients undergoing hip replacement surgery and mixed with Matrigel™ [29]. Four weeks post-implantation into the flanks for NSG mice, the human bone implants developed their own vasculature. To investigate leukemia engraftment and the impacts of leukemia on osteogenic differentiation, Molm13 leukemia cells expressing GFP and luciferase were injected via the tail vein and strongly engrafted into the human bone implant as early as 10 days post-injection [29]. Similarly, the Lee group utilized a bioengineering approach to create genetically engineered scaffold microenvironments using human bone stromal cells [27,28]. The human bone marrow stromal cells were genetically engineered to stably express human cytokines (including TNF-alpha) or human growth factors (including VEGF) important for bone remodeling and were embedded into 3D porous hydrogel scaffolds. This permitted the generation of a human soluble factor-enriched microenvironment. The seeded scaffolds were then implanted into immunocompromised mice and stromal cell engraftment and proliferation were investigated [27].

In an effort to merge principles from the EMB model and genetically engineered bone scaffold model, our laboratory seeded PCL scaffolds (Figure 1A) coated with Matrigel™ with GFP-labeled MDA-MB-231 human metastatic breast cancer cells admixed at a 1:1 ratio with either naïve MC3T3-E1 osteoblasts (OB) engineered to express tdTomato or educated osteoblast (EO) cells engineered to express tdTomato [175]. The seeded scaffolds were then cultured for 12 days during which we monitored the engraftment of both the MDA-MB-231GFP breast cancer cells and either MC3T3-E1-tdTomato osteoblasts or tdTomato-EO cells over time using fluorescence imaging. As seen in Figure 1, MDA-MB-231-GFP breast cancer cells and either MC3T3-E1-tdTomato osteoblasts (Figure 1B–E) or tdTomato-EO (Figure 1F–I) cells can clearly be seen proliferating, spreading out, and positioning themselves adjacent to each other within the scaffold over time [175].

Scaffolds were implanted into the flanks of NSG mice 13 days after cell seeding. Mice were monitored via an In-Vivo Imaging System (IVIS) non-invasive intravital imaging for 2 months, where cancer cell proliferation in the scaffold can be seen (Figure 2A,G) [175]. Mice were humanely euthanized, and scaffolds and femurs (site of preferential metastasis) harvested for analysis. Ex-vivo scaffold analysis revealed well-formed tumors composed of either MC3T3-E1-tdTomato osteoblasts (Figure 2B) or tdTomato-EO cells (Figure 2H). Interestingly, even though scaffolds were originally seeded with a 1:1 admix of GFP-cancer cells plus either MC3T3-E1-Tdtomato osteoblasts or tdTomato-EO cells, less proliferation of MC3T3-E1-tdTomato osteoblasts was apparent, as evidenced by less expression of tdTomato: ~38% of the tumor (and ~62% of the tumor composed of GFP-breast cancer cells; Figure 2B), when compared to tdTomato-EO cell engraftment of the tumor: ~55% of the tumor (and ~45% of the tumor composed of GFP-breast cancer cells; Figure 2H) [175]. These results corroborate our prior findings that EO cells suppress metastatic breast cancer cell growth [85]. Both tumors exhibited blood vessel formation (white arrows, Figure 2B,H). Surprisingly however, we were unable to recapitulate breast cancer cell metastasis to the bone, whereby no femurs showed evidence of MDA-MB-231GFP metastatic breast cancer cells by fluorescence microscopy (Figure 2C–F,I–L) [175]. This phenomenon had been observed as part of the EMB model as per Andreeff and colleagues [30].

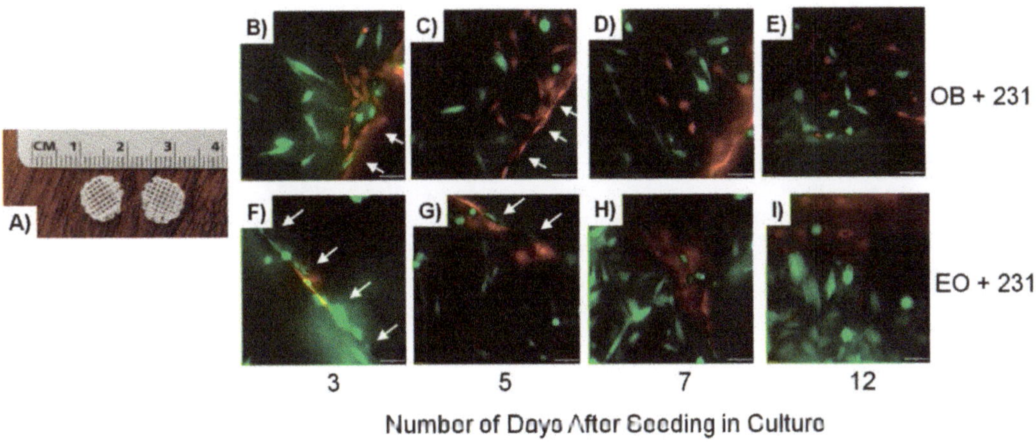

Figure 1. Co-cultures of osteoblasts and breast cancer cells attach and proliferate over time on polycaprolactone printed scaffolds. (**A**) Printed polycaprolactone scaffolds were coated with Matrigel™ then seeded with a 1:1 admix of GFP-MDA-MB-231 human breast cancer cells plus either (**B–E**) tdTomato MC3T3-E1 murine osteoblasts (OB) or (**F–I**) tdTomato murine educated osteoblast (EO) cells. Seeded scaffolds were imaged for fluorescence over the course of 12 days for cell seeding, growth, and spread on the scaffold. White arrows indicate edges of the criss-cross printed pattern of the scaffold. Scale bar on fluorescent images = 50 µm.

Whereas these model systems recapitulate select steps of the metastatic cascade or closely model aspects of human disease, none sufficiently combine both a humanized/human bone microenvironment with appropriate stromal cells, architecture, and mechanical forces along with a system that demonstrates all steps of a human metastatic cascade. Thus, three-dimensional bioprinted model systems address this gap in the research.

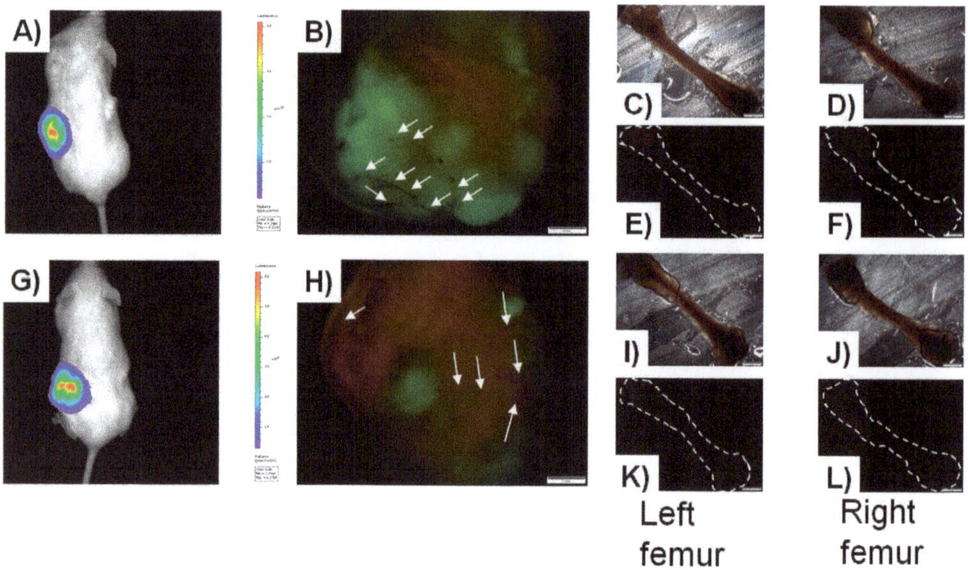

Figure 2. Printed polycaprolactone scaffolds seeded with osteoblasts plus breast cancer cells form tumors in-vivo. IVIS image of a NOD/SCID/IL-2r gamma null (NSG) mouse 2 months post-implant with a printed polycaprolactone scaffold coated with Matrigel™ then seeded with a 1:1 admix of GFP-MDA-MB-231 human breast cancer cells plus either (**A**) tdTomato MC3T3-E1 murine osteoblasts or (**G**) tdTomato murine EO cells. (**B,H**) Fluorescent microscope images of tumors from A and G ex-vivo; (**B**) 1:1 admix of GFP-MDA-MB-231 human breast cancer cells plus tdTomato MC3T3-E1 murine osteoblasts; (**H**) 1:1 admix of GFP-MDA-MB-231 human breast cancer cells plus tdTomato murine EO cells. White arrows indicate blood vessel formation in the tumors. Light microscopy images of (**C,I**) left and (**D,J**) right ex-vivo femurs of (**C,D**) NSG mouse implanted with printed polycaprolactone scaffold seeded with 1:1 admix of GFP-MDA-MB-231 human breast cancer cells plus tdTomato MC3T3-E1 murine osteoblasts, or (**I,J**) 1:1 admix of GFP-MDA-MB-231 human breast cancer cells plus tdTomato murine EO cells. Fluorescent microscopy images of (**E,K**) left and (**F,L**) right ex-vivo femurs of (**E,F**) NSG mouse implanted with printed polycaprolactone scaffold seeded with 1:1 admix of GFP-MDA-MB-231 human breast cancer cells plus tdTomato MC3T3-E1 murine osteoblasts, or (**K,L**) 1:1 admix of GFP-MDA-MB-231 human breast cancer cells plus tdTomato murine EO cells. Scale bar = 2 mm.

6. Bioprinting

6.1. Basis of Bioprinting

Bioprinting provides a novel additive manufacturing technique to fabricate prescribed architectures formed through layer-by-layer construction of biological materials and living cells [176]. Cells can be impregnated into the resultant extracellular matrices post-printing using traditional cell seeding techniques or incorporated directly into the printing process to produce precursor tissue constructs. Such constructs can then be directly implanted in-vivo for regenerative medicine applications or cultured in-vitro to produce mature, biomimetic tissues, holding promise for breast cancer applications [177].

Although bioprinting was initially established in the form of cytoscribing, a modified inkjet printing technique allowing the deposition of biologic materials for cell adhesion and growth into planar geometries that could be individually stacked and adhered [178], the field quickly evolved to produce in-situ three-dimensional structures. Along with developing inkjet techniques, the rise of rapid fabrication processes such as fused deposition modelling (FDM), stereolithography (SLA), and selective laser sintering (SLS) has allowed for more complex three-dimensional shapes [179]. Most recently, the incorporation of living cells into printable biological materials (i.e., bioinks) and modifications in the printing process to maintain cell viability have led to advanced constructs with patterned cellular and structural elements with bioactivity [179]. Thus, current bioprinting technologies allow for an expansive array of promising applications including providing research environments for the study of stem and cancer cell behavior [16,180–182], platforms for drug screening, delivery, and production [183,184], and engineered tissues/organs for transplantation [185].

The advantages of bioprinted constructs over traditional two-dimensional cell culture are numerous. The complex, three-dimensional architectures allow for micron-scale precision of the structural environment housing cells, thereby directing their native functions in a manner difficult to achieve with two-dimensional cell culture [186]. Macroscopic geometries can be customized in size and shape identical to native tissues and organs based on patient specific CT and/or MRI data. The biochemical properties of the substrate materials and potential for conjugation with other biomolecules not only support tissue-specific cell function, but can control vascularization, nerve integration, and maturation of engineered tissue [186]. In addition, the recent availability of desktop bioprinting systems with easily accessible hardware and software systems presents a practical option for a fast, low-cost 3D bioactive construct production [187], further supporting the development of personalized tissues for research and regenerative medicine [188].

6.2. Bioprinting Techniques

3D bioprinting offers the opportunity to manufacture an array of engineered tissues with varying physical and biochemical properties to suit individual applications. Methods for printing include inkjet, extrusion-based laser-induced printing, and projection stereolithography (Figure 3). Each method has unique benefits and limitations and the most appropriate printing technique is dependent on the specific operational requirements of the engineered tissue produced [189].

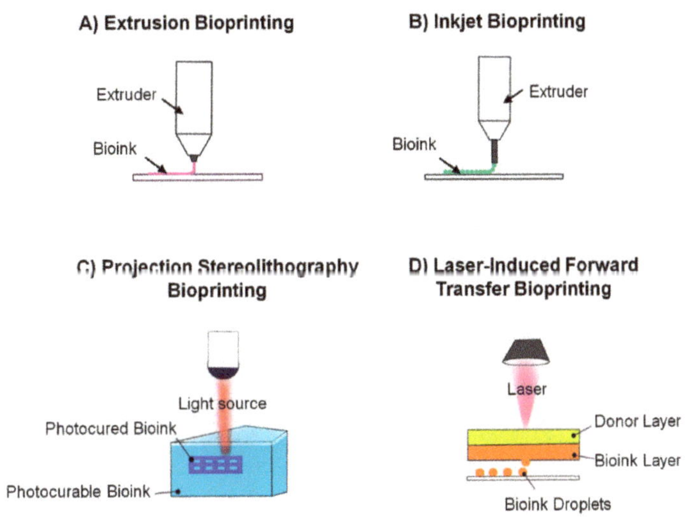

Figure 3. 3D bioprinting techniques. (**A**) Extrusion bioprinting: nozzle-based continuous layer-by-layer deposition of bioink using pneumatic pressure-, screw-, or pledged-based mechanisms. (**B**) Inkjet bioprinting: nozzle-based layer-by-layer deposition of bioink droplets using thermal heat or mechanical pulses to acoustic waves. (**C**) Projection stereolithography bioprinting: layer-by-layer photoinduced curing of bioink using UV light. (**D**) Laser-induced forward transfer bioprinting: laser-based depositing bioink droplets through energy transfer from the absorption donor layer to the bioink layer.

Extrusion-based manufacturing (Figure 3A) is the most common method of bioprinting, whereby the substrate cell +/− biomaterials are deposited through a thin nozzle in a continuous flowing fashion onto the build plate through the application of force on the bioink reservoir. Various methods for force generation exist, resulting in a range of extrusion-based bioprinter designs, including pneumatic-based (air pressure), piston-driven (vertical force), and screw-driven (rotational force) dispensing techniques [190]. Scaffold constructs can be printed from a homogeneous cell-laden biomaterial ink directly, with dual extrusion heads allowing for independent biomaterial deposition followed by targeted cell placement within the structure, or as a cell-free scaffold with manual cell seeding post-printing [190]. Thus, there is an expanded range of materials that can be used as substrates (e.g., those requiring washing prior to cell seeding to enhance viability) and more precise control over cell distribution with the resultant matrices. The resolution of micro-extrusion bioprinting is in general from 100–500 μm, though it can be as small as 50 μm with the incorporation of microfluidics in the system [179,191]. The extrusion-based method allows for fabrication of structures with bioinks of higher viscosities and cell densities (e.g., greater than 1×10^6 cells/mL [190]), however, as a nozzle-based system it is also associated with shear stress, potentially impacting cell viability [179].

Inkjet bioprinting (Figure 3B), the first method developed, involves dispensing bioink through a thin nozzle in 1–100 μL volumes in a precise three-dimensional pattern [192]. Also known as the drop-on-demand method, the droplets are discharged by either an internal vapor bubble generated by an external thermal heating element or an acoustic wave generated by mechanical pulses of a piezoelectric element allowing for a resolution of 50 μm [177,179]. Printing using this technique is fast (up to 10,000 droplets/second) and relatively low cost vs. other methods, but input substrate options and cell densities may be limited secondary to nozzle clogging with more viscous bioinks [177,179]. Moreover, the heat, acoustic pressure, and high shear rates generated can compromise cell viability, although limiting exposure time may improve cell survival [179,187].

Projection stereolithography (SLA; Figure 3C) is another layer-by-layer fabrication method for bioprinting which involves solidifying liquid biocompatible resins by photopolymerization under irradiation [192,193]. This method is therefore limited to specialized, photocurable bioinks [177]. Although SLA produces mechanically robust constructs, uncured bioink can seep into open spaces, making hollow structures (e.g., vessels) particularly challenging to execute [193].

Laser-induced printing (Figure 3D) does not involve a nozzle, circumventing some of the limitations of the previously described methods. Instead, a glass slide coated with an ink solution and a laser absorption layer consisting of a metal or metal oxide is ablated by a laser at distinct location (40–100 μm resolution), generating a pressure that ejects ink onto the substrate [189]. Also known as laser-induced forward transfer, it can print viscous bioinks at very high resolution, although at very high costs and temperatures that may risk heat-induced cell death [189].

In an effort to maintain the structural integrity of some complex 3D structures, support material may be necessary to improve the printability of certain bioinks and attain more accurate print results [194]. A suspension-based 3D bioprinting technique, known as freeform reversible embedding of suspended hydrogels (FRESH), allows for the printing of soft hydrogels in a support bath of a thermoreversible gel that can be subsequently melted away, as depicted in Figure 4 [195]. The extrusion and embedding of the bioinks into the gel in essence diminish gravitational effects, allowing for print geometries otherwise unattainable with certain soft biomaterials, as they would ordinarily collapse in air [196]. FRESH-printed structures require some post-print processing, including washing away the excess gelatin post-melting [196]. This noteworthy technique has led to improved print resolution and reliability, and the potential to engineer versatile unsupported print architectures and larger advanced tissue scaffolds [195].

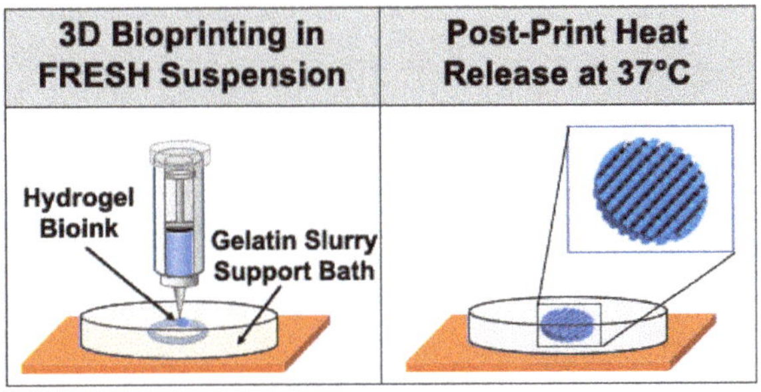

Figure 4. Freeform reversible embedding of suspended hydrogels (FRESH) bioprinting technique. Hydrogel bioink deposited via extrusion bioprinting into a thermoreversible gelatin slurry support bath with a Bingham plastic rheology allows for structural suspension of bioprinted tissue scaffolds to mitigate the force of gravity. An overnight post-print heat release in a 37 °C incubator melts the gelatin, leaving the 3D structure behind.

6.3. Bioinks

Bioinks are substrates used to produce 3D printed constructs containing cells and/or extracellular matrix components. The composition of the bioink plays a critical role in maintaining cell viability during the printing process and providing the structural integrity necessary to support function post-production [188]. For example, hydrogel-based cell-laden bioinks can protect cells from potentially harmful desiccation and shear forces during the printing [197]. In order to achieve the ideal balance between cell survival and the desired construct structural parameters, careful consideration of not only cytocompatibility, but biocompatibility, bioactivity, and local geometry must be considered to support cell differentiation and growth [189]. Moreover, considerations of printability (e.g., viscosity) along with final overall construct mechanical strength and degradation characteristics (e.g., crosslinking) also require attention when determining the resultant construct physical properties [189].

The three most common bioink options include cell-laden hydrogels, ECM-based inks, and cell suspensions [189]. Cell-laden hydrogels closely mimic/recapitulate the cellular microenvironment by incorporating bioactive compounds and growth factors and permitting the diffusion of nutrients [198]. Natural hydrogels, such as collagen, fibrin, gelatin, and hyaluronic acid are inherently bioactive and simulate the native structure of the ECM, whereas synthetic hydrogels, such as pluronic (poloxamer) and polyethylene glycol (PEG), are mechanically tunable and can support the delivery of added bioactive cues [189,199]. While such hydrogels allow for highly reproducible structures, their limited mechanical strength can be an issue, often augmented by FRESH techniques during and crosslinking after printing the polymers [199].

Single-component hydrogels prepared as bioinks may not adequately provide the composition and function of native ECM in 3D models, as they lack the complex environment that allows for cell engraftment, migration, signaling, and function [200,201]. Decellularized ECM (dECM) offers organ-specific biochemical cues from native ECM to improve cell proliferation and survival [200]. The use of dECM in bioinks can be enhanced by the addition of chemical and biological crosslinking agents to strengthen the scaffold mechanically and improve bioactivity [200]. Maintaining the viscoelastic materials of the tissue is vital to generating a sufficient model of the native tissue. Collagen is a common bioink and major ECM protein as it is easily crosslinked using methods, such as temperature and pH, however, its gelation time and unstable mechanical properties are barriers for use in 3D bioprinting [202]. dECM hydrogels are a promising method for constructing functional tissues and organs with multicellular compositions.

The incorporation of specific bioactive additives allows for manipulation of the mechanical and physiological properties of printed constructs [189]. Hydroxyapatite (HA) and bone morphogenetic protein 2 (BMP-2), two common bioink additives, are known for promoting osteoinductivity and osteoconductivity, and can also confer structural and mechanical properties to bioprinted bone constructs [179]. BMP-2 has been shown to induce mesenchymal stem cells toward osteogenic differentiation [203], and when loaded into hydrogel-based 3D printed constructs, demonstrated sustained release and improved osteogenic differentiation and bone formation effects in-vitro and in-vivo [204].

HA has been shown to regulate the behavior of cells, specifically osteoblast adhesion on nanocrystalline hydroxyapatite (nHA) nanoparticles [205], and to improve normal bone formation, through the stimulation of cell proliferation and osteogenic differentiation [206]. The incorporation of nHA with gelatin methacrylate (GelMA) into bioprinted bone matrices in a study by Xuan Zhou and colleagues showed osteoblast and mesenchymal stem cell (MSC) proliferation and improved overall compressive stress of the resultant matrix constructs [207]. Hyperelastic bone (HB), a recently commercialized extrudable bioink made up of hydroxyapatite and PCL or poly(lactic-co-glycolic acid) offers a promising material for bioprinted bone applications [208]. In addition to its local effect on cell function, printed HB constructs can be successfully stretched, cut, and sutured to a soft tissue (such as tendon), making it a promising candidate for applications in tissue replacement surgeries [208]. Synthetic scaffolds constructed from this porous substrate have been used to study bone regeneration in-vivo [209]. Fluffy–poly(lactic-co-glycolic acid) and hyperelastic bone scaffolds were compared against a negative control and positive control of autologous calvarial bone for the treatment of calvarial defects in rats. The hyperelastic bone was shown to be effective for bone regeneration and for inducing bone formation in-vivo in the defects. Hyperelastic bone has also been implanted in non-human primates to model biocompatibility and fusion rates, and successfully demonstrated vascularization and no significant immune response [210,211].

6.4. *Applications of Bioprinting in Cancer*

6.4.1. 3D Printed In-Vitro Models

Three-dimensional bioprinting presents an opportunity to develop functioning tissue beyond just structural scaffolds by incorporating cells into the printable bioinks [179]. Current applications of bioprinting include three-dimensional extracellular matrix components, organ structures, regenerative tissue grafts, and disease models [198]. FRESH 3D bioprinting can help to optimize the fabricated microenvironments, by controlling print resolution, improving cell viability, and supporting delicate architectures necessary for certain in-vitro cancer research models. For example, Lewicki et al. used the technique to optimize human neuroblastoma cell-laden hydrogels of low viscosity [212]. Applications of dECM-based bioinks hold particular promise in studying ECM components of the tumor microenvironment (TME) and tumor–ECM interplay [213]. Quickly polymerizable dECM bioinks allow for tunable stiffness in applications such as 3D bioprinted kidney cancer constructs in which 3D bioprinted dECM microtumor models were achieved with refined control over both the multicellular populations and dECM bioink deposition [213]. Notably, bioprinted matrices constructed from various substrates with or without FRESH methods can provide viable in-vitro models for understanding cell behaviors and interactions in cancer. For example, extrusion-based printing of bioinks composed of HeLa cells and gelatin/alginate/fibrinogen hydrogels allowed for the development of a 3D cervical tumor model [214]. Compared with cells in planar 2D cultures, HeLa cells in the 3D environment showed more behavioral similarity to native cancer cells, including higher cell proliferation, matrix metalloproteinase (MMP) protein expression, and chemoresistance [215].

Increasing numbers of bioprinted models are being developed as higher fidelity systems to study key aspects of cancer development and progression in a variety of tissues/organs. Tang et al. used bioprinting techniques to develop a complex in-vitro 3D glioblastoma model to allow for exploration of the role of immune components within the tissue microenvironment [216]. The glioblastoma stem cells, resident CNS cells, and macrophage precursors better resembled the invasive cell types of patient tumor tissue in the bioprinted models as opposed to sphere cultures [216]. Hakobyan et al. used laser-assisted bioprinting to generate 3D pancreatic cell spheroid arrays to replicate the initial stages of pancreatic ductal adenocarcinoma (PDAC) as a high-throughput, reproducible model for the study of pancreatic cancer progression and potential therapeutic approaches [217]. Novel model systems are emerging using 3D bioprinting techniques to study other neurological tumors, as well as liver, breast, and skin cancers, that allow for the in-vitro study of susceptibility and resistance to chemotherapeutics for predictable human response [218].

6.4.2. Bioprinted Models for Breast Cancer Metastasis

Bioprinting offers a biomimetic in-vitro three-dimensional cancer culture system capable of modeling cell colonization, tumor growth, and response to therapies. The accuracy of the architecture, vascularization, and composition of the tumor microenvironments demonstrated by bioprinted models reinforces the progress in this domain beyond conventional tissue-engineered constructs. However, cell viability presents a great challenge in the bioprinting process, as the high-throughput processing can directly impact cell proliferation. Recent advances in the use of bioprinting to model cancer microenvironments include tumor fabrication [51,52], allowing for successful modeling of cancer cell interactions.

One of the most crucial elements of an in-vitro cancer model is the scaffold, which emulates the extracellular matrix of the cancer microenvironment. Bioprinted matrices allow for the fabrication of those scaffolds, permitting continued study of the metastasis of breast cancer in bone. In-vitro models provide limitless opportunity for the study of cancer cell interactions and therapeutic advances. Current applications of bioprinting in the study of breast cancer metastasis in bone include bioprinted bone matrices to investigate cancer cell interactions, cell migration, and drug resistance.

Cell-laden bone matrix scaffolds to study the interaction of breast cancer cells with osteoblasts or mesenchymal stromal cells were developed by Zhou and colleagues [37]. The bone matrices were printed with a stereolithography bioprinter, using a bioink consisting of GelMA hydrogel and nanohydroxyapatite (nHA) (incorporated to simulate native bone tissue), and a printable bioink with a cell suspension consisting of osteoblasts or mesenchymal stromal cells. Breast cancer cells were seeded on the surface of the stromal cell-laden matrices and co-cultured with osteoblasts or mesenchymal stromal cells for 5 days. In the co-culture environment, breast cancer cells showed enhanced proliferation but drastically inhibited the osteoblast/mesenchymal stromal cell growth. These data suggest that nHA facilitates interactions between the bone stromal cells and breast cancer cells [37].

Exploring the crosstalk between different cell types, Zhu et al. conducted a study observing the interaction between human fetal osteoblasts (hFOBs) and metastatic breast cancer cells on a 3D bioprinted artificial bone matrix, as depicted in Figure 5 [219]. The matrix also contained the calcium phosphate nHA as it is associated with metastatic breast cancer progression to bone [219]. Human breast cancer cells co-cultured with hFOB cells on the matrix impacted the morphology and proliferation rate of both cell types, in addition to enhanced IL-8 secretion, a pro-inflammatory chemokine that contributes to angiogenesis and tumorigenesis. Notably, the presence of metastatic breast cancer cells induced heightened osteoblastic IL-8 expression. Furthermore, the addition of the nHA to the bone matrix increased the overall proliferation rate of breast cancer cells in a concentration-dependent manner [219]. Beyond examining cellular morphology, proliferation, and cytokine expression, indirect and 3D bioprinted constructs may also be used to assay alterations in cellular crosstalk, including osteoblast differentiation and cellular protein expression, as well as cellular engraftment into the 3D bone matrix (Figure 5). Overall, these data suggest that the composition of the 3D bioprinted matrix is crucial to fully recapitulating cellular in-vivo behavior of metastatic cancer cells in the bone tumor microenvironment.

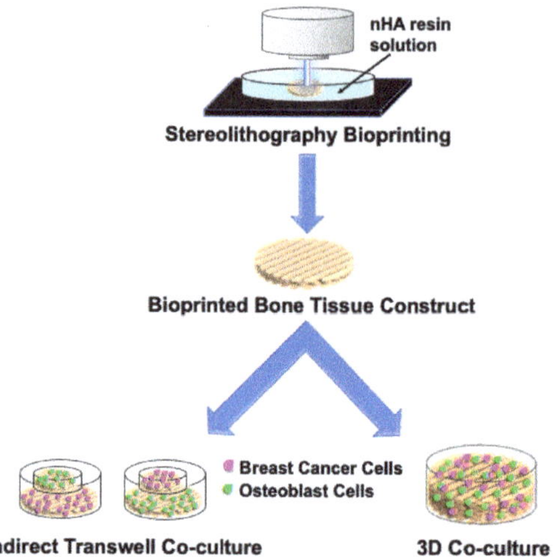

Figure 5. 3D bioprinted bone tissue constructs to study crosstalk between osteoblasts and breast cancer cells. Indirect transwell co-culture and 3D co-culture of breast cancer cells and osteoblasts on an artificial nanohydroxyapatite (nHA) bone matrix fabricated by stereolithography-based 3D printing. These model systems permit the study of cellular crosstalk between osteoblasts and breast cancer cells as it relates to cellular morphology, proliferation, osteoblast differentiation, expression of soluble factors, alterations in cellular protein expression, and cellular engraftment into the bone matrix.

7. Conclusions and Future Exploration

Although 3D bioprinting allows for the precise control of scaffold architecture, these techniques present many challenges in ensuring ideal cell function. Moving forward, physical and biophysical properties of 3D bioprinted scaffolds, including scaffold–matrix stiffness, "tunability" of the scaffold composition, which may include the addition of growth factors and metabolic products native to an endogenous extracellular matrix to facilitate cell growth, porosity of the scaffold to ensure successful cellular attachment and proliferation, alterations in scaffold size, and alterations in scaffold geometry (e.g., honeycomb vs. grid, etc.), as well as other mechanical properties, should be assessed to optimize these factors for successful cellular engraftment and growth [220–225]. Furthermore, optimization of a stable scaffold capable of in-vivo implantation with resistance to degradation over time may be critical to the development of long-term viable cellular systems and matrix vascularization.

Interestingly, it has been found that certain scaffold geometries are superior to others with respect to cellular engraftment and viability over time. In particular, Foresti et al. developed a well-defined scaffold with complex architecture that was vascularized and capable of long-term cellular viability [226]. The authors additionally investigated the biological responses of cells seeded on several different scaffolds of complex geometries and determined certain geometries were superior to others for cellular engraftment and proliferation as well as the ability of the scaffold to retain its shape—in some cases for up to 23 months while in culture [226]. Wang et al. additionally investigated how modulating the parameters of a gelatin/alginate hydrogel scaffold, including using various geometries, pore volumes, volume porosity, and surface areas, affected the cell viability, distribution, morphology, proliferation, and expression of cell-specific markers of C3A liver cells [223]. The authors determined that the geometry of PO250+ (fine checkerboard), with a pore size of 250 μm, was optimal towards eliciting strong cell proliferation and viability, as well as causing the expression of the liver-specific mRNA CYP3A4 and protein albumin [223]. With specific reference to the cellular behavior of cancer cells, Hanumantharao et al. carried out an interesting study examining how different topographical features and mechanical properties of PCL scaffolds influenced human ER+ and triple negative breast cancer cells [227]. The authors found that human MCF-7 ER+ breast cancer cells proliferated well on a variety of different scaffold topographies, where the MCF-7 cells had a higher rate of proliferation on scaffolds with low Young's modulus and stiffness. On the other hand, human MDA-MB-231 triple negative breast cancer cells preferred scaffolds that had a high matrix stiffness [227]. These results suggest that cancer cells respond to changes in scaffold stiffness, mechanical properties, and topography, which should be considered when developing a 3D bioprinted scaffold to model the tumor microenvironment.

Importantly, scaffold "tunability" may be extremely useful when investigating tumor–stromal cell crosstalk and the biophysical properties of the extracellular matrix. Recently, 3D bioprinted scaffolds have been developed that are capable of the controlled release of growth factors that affect angiogenesis and osteogenesis in the bone microenvironment. Specifically, Freeman et al. developed 3D bioprinted constructs to deliver VEGF and BMP-2 with distinct spatiotemporal release profiles to enhance the regeneration of large bone defects [228]. Importantly, the properties of the 3D bioprinted constructs were "tunable" in that (1) the release of VEGF and BMP-2 could be slowed or accelerated, and (2) the distribution of release of VEGF and BMP-2 could be localized or eluted in a spatial gradient [228]. In a similar fashion, Sun et al. developed a 3D bioprinted scaffold loaded with connective tissue growth factor and transforming growth factor beta 3, then seeded with mesenchymal stromal cells and implanted into mice to facilitate the regeneration of an intervertebral disc [229]. Importantly, the authors showed that the reconstructed intervertebral disc exhibited properties similar to a native disc bone, with corresponding histological and immunological phenotypes [229]. Given that 3D bioprinted scaffolds may be loaded with specific growth factors at different concentrations, these models represent a unique and novel way to study interactions between the tumor and stromal cells in different environmental conditions that recapitulate different stages of disease.

Furthermore, 3D bioprinted scaffolds may also be valuable for their ability to both elute drugs into the niche in a sustained manner over time and manipulate the expression of specific factors in the microenvironment. For example, Wu et al. developed a novel biocompatible scaffold for endothelial cell repair in cardiovascular disease [224]. The team loaded dimethyloxalylglycine (DMOG) into biocompatible ink with a final concentration of 30% (w/w). Over time, in culture with aortic endothelial cells, the scaffold exhibited sustained release of DMOG into the media, which induced the expression of HIF-1 alpha by the aortic endothelial cells. HIF-1 alpha then elicited the transcriptional activation of VEGF, a HIF-1 alpha target gene, suggesting that implantable 3D bioprinted scaffolds loaded with drugs can aid in the repair of endothelial cells in cardiovascular disease [224]. Using the same principle and taking it a step further, the same idea can be applied to manipulating the tumor microenvironment to elicit the expression of certain tumor suppressor molecules, promote an antitumor immune response, or deliver localized therapeutic treatment. The ability to tailor and customize both the environmental response as well as drug loaded into the printed scaffold would be of great benefit to more efficiently target specific cells in the tumor niche over time.

The use of bioprinted constructs introduces a significant high-throughput, low-cost advance in future cancer research that provides a novel perspective for cancer cell growth determinants and potential therapeutics. The incorporation of growth factors and biophysical properties of the ECM could allow researchers to better understand the 3D breast cancer microenvironment through these native-like tumor platforms.

As research progresses towards optimization of the matrix geometry, and more physiologically accurate ECM, researchers come closer to achieving in-vivo conditions for evaluating breast cancer cell behavior and tumor growth. Achieving a matrix that exhibits native characteristics of tumor–stromal cell crosstalk in-vivo will lead to opportunities to study interactions between breast cancer cells and bone stromal cells, metastatic progression, and response to drug therapy. Drug screening and therapeutic cancer drug response testing on 3D bioprinted bone matrices could also reduce the time to screen candidate therapeutics when compared to in-vivo systems and reduce the number of animal models used in testing, thus providing a viable alternative to traditional animal models.

Author Contributions: A.M.H. conceived and wrote the manuscript; A.D.K. and A.B.S. analyzed data and wrote the manuscript; K.M.S. conceived, wrote, and edited the manuscript; K.M.B. conceived, wrote, and edited the manuscript, provided funding for the project, and organized manuscript assembly. All authors have read and agreed to the published version of the manuscript.

Funding: This work was supported by NIH, NCI K99/R00 Pathway to Independence Grant R00CA178177 (K.M.B.) and Pennsylvania State Department of Health SAP 4100072566 (K.M.B.).

Data Availability Statement: The data presented in this study are available in Figures 1 and 2.

Acknowledgments: The authors wish to thank Dayna DiPiero for her assistance with figure preparation and the Jefferson Innovation team, specifically Robert Pugliese, for resource support.

Conflicts of Interest: The authors declare no conflict of interest.

References

1. Lu, P.; Takai, K.; Weaver, V.M.; Werb, Z. Extracellular matrix degradation and remodeling in development and disease. *Cold Spring Harb. Perspect. Biol.* **2011**, *3*, a005058. [CrossRef] [PubMed]
2. Miller, R.T. Mechanical properties of basement membrane in health and disease. *Matrix Biol.* **2017**, *57–58*, 366–373. [CrossRef] [PubMed]
3. Alford, A.I.; Kozloff, K.M.; Hankenson, K.D. Extracellular matrix networks in bone remodeling. *Int. J. Biochem. Cell Biol.* **2015**, *65*, 20–31. [CrossRef] [PubMed]
4. Gentili, C.; Cancedda, R. Cartilage and bone extracellular matrix. *Curr. Pharm. Des.* **2009**, *15*, 1334–1348. [CrossRef] [PubMed]
5. Bone Metastasis: Symptoms and Diagnosis. Metastatic Breast Cancer Symptoms and Diagnosis. 2018. Available online: https://www.breastcancer.org/symptoms/types/recur_metast/metastic/bone (accessed on 11 December 2020).
6. Miller, K.D.; Siegel, R.L.; Khan, R.; Jemal, A. Cancer statistics. *Cancer Rehabil.* **2018**, *70*, 7–30. [CrossRef]
7. Bussard, K.M.; Venzon, D.J.; Mastro, A.M. Osteoblasts are a major source of inflammatory cytokines in the tumor microenvironment of bone metastatic breast cancer. *J. Cell. Biochem.* **2010**, *111*, 1138–1148. [CrossRef]
8. Mastro, A.M.; Gay, C.V.; Welch, D. The skeleton as a unique environment for breast cancer cells. *Clin. Exp. Metastasis* **2003**, *20*, 275–284. [CrossRef]
9. Mundy, G.R. Metastasis to bone: Causes, consequences and therapeutic opportunities. *Nat. Rev. Cancer* **2002**, *2*, 584–593. [CrossRef]
10. Phadke, P.A.; Mercer, R.R.; Harms, J.F.; Jia, Y.; Frost, A.R.; Jewell, J.L.; Bussard, K.M.; Nelson, S.; Moore, C.; Kappes, J.C.; et al. Kinetics of metastatic breast cancer cell trafficking in bone. *Clin. Cancer Res.* **2006**, *12*, 1431–1440. [CrossRef]
11. Sun, Y.-X.; Schneider, A.; Jung, Y.; Wang, J.; Dai, J.; Wang, J.; Cook, K.; Osman, N.I.; Koh-Paige, A.J.; Shim, H.; et al. Skeletal localization and neutralization of the SDF-1(CXCL12)/CXCR4 axis blocks prostate cancer metastasis and growth in osseous sites in vivo. *J. Bone Miner. Res.* **2004**, *20*, 318–329. [CrossRef]
12. Lu, P.; Weaver, V.M.; Werb, Z. The extracellular matrix: A dynamic niche in cancer progression. *J. Cell Biol.* **2012**, *196*, 395–406. [CrossRef] [PubMed]
13. Walker, C.; Mojares, E.; Hernández, A.D.R. Role of extracellular matrix in development and cancer progression. *Int. J. Mol. Sci.* **2018**, *19*, 3028. [CrossRef] [PubMed]
14. Eble, J.A.; Niland, S. The extracellular matrix in tumor progression and metastasis. *Clin. Exp. Metastasis* **2019**, *36*, 171–198. [CrossRef]
15. Leonard, F.; Godin, B. 3D in vitro model for breast cancer research using magnetic levitation and bioprinting method. *Toxic. Assess.* **2016**, *1406*, 239–251. [CrossRef]
16. Reid, J.A.; Mollica, P.A.; Bruno, R.D.; Sachs, P.C. Consistent and reproducible cultures of large-scale 3D mammary epithelial structures using an accessible bioprinting platform. *Breast Cancer Res.* **2018**, *20*, 122. [CrossRef] [PubMed]
17. Wergedal, J.E.; Mohan, S.; Lundy, M.; Baylink, D.J. Skeletal growth factor and other growth factors known to be present in bone matrix stimulate proliferation and protein synthesis in human bone cells. *J. Bone Miner. Res.* **1990**, *5*, 179–186. [CrossRef] [PubMed]
18. Morgan, M.P.; Cooke, M.M.; Christopherson, P.A.; Westfall, P.R.; McCarthy, G.M. Calcium hydroxyapatite promotes mitogenesis and matrix metalloproteinase expression in human breast cancer cell lines. *Mol. Carcinog.* **2001**, *32*, 111–117. [CrossRef]
19. Petersen, O.W.; Rønnov-Jessen, L.; Howlett, A.R.; Bissell, M.J. Interaction with basement membrane serves to rapidly distinguish growth and differentiation pattern of normal and malignant human breast epithelial cells. *Proc. Natl. Acad. Sci. USA* **1992**, *89*, 9064. [CrossRef]
20. Dhurjati, R.; Krishnan, V.; Shuman, L.A.; Mastro, A.M.; Vogler, E.A. Metastatic breast cancer cells colonize and degrade three-dimensional osteoblastic tissue in vitro. *Clin. Exp. Metastasis* **2008**, *25*, 741–752. [CrossRef]
21. Dhurjati, R.; Liu, X.; Gay, C.V.; Mastro, A.M.; Vogler, E.A. Extended-Term culture of bone cells in a compartmentalized bioreactor. *Tissue Eng.* **2006**, *12*, 3045–3054. [CrossRef]

22. Krishnan, V.; Dhurjati, R.; Vogler, E.A.; Mastro, A.M. Osteogenesis in vitro: From pre-osteoblasts to osteocytes: A contribution from the Osteobiology Research Group, The Pennsylvania State University. *In Vitro Cell. Dev. Biol. Anim.* **2010**, *46*, 28–35. [CrossRef] [PubMed]
23. Krishnan, V.; Shuman, L.A.; Sosnoski, D.M.; Dhurjati, R.; Vogler, E.A.; Mastro, A.M. Dynamic interaction between breast cancer cells and osteoblastic tissue: Comparison of Two- and Three-dimensional cultures. *J. Cell. Physiol.* **2011**, *226*, 2150–2158. [CrossRef] [PubMed]
24. Krishnan, V.; Vogler, E.A.; Mastro, A.M. Three-Dimensional in vitro model to study osteobiology and osteopathology. *J. Cell. Biochem.* **2015**, *116*, 2715–2723. [CrossRef] [PubMed]
25. Krishnan, V.; Vogler, E.A.; Sosnoski, D.M.; Mastro, A.M. In vitro mimics of bone remodeling and the vicious cycle of cancer in bone. *J. Cell. Physiol.* **2013**, *229*, 453–462. [CrossRef]
26. Sosnoski, D.M.; Krishnan, V.; Kraemer, W.J.; Dunn-Lewis, C.; Mastro, A.M. Changes in cytokines of the bone microenvironment during breast cancer metastasis. *Int. J. Breast Cancer* **2012**, *2012*. [CrossRef]
27. Lee, J.; Heckl, D.; Parekkadan, B. Multiple genetically engineered humanized microenvironments in a single mouse. *Biomater. Res.* **2016**, *20*, 19. [CrossRef]
28. Lee, J.; Li, M.; Milwid, J.; Dunham, J.; Vinegoni, C.; Gorbatov, R.; Iwamoto, Y.; Wang, F.; Shen, K.; Hatfield, K.; et al. Implantable microenvironments to attract hematopoietic stem/cancer cells. *Proc. Natl. Acad. Sci. USA* **2012**, *109*, 19638–19643. [CrossRef]
29. Battula, V.L.; Le, P.M.; Sun, J.C.; Nguyen, K.; Yuan, B.; Zhou, X.; Sonnylal, S.; McQueen, T.; Ruvolo, V.; Michel, K.A.; et al. AML-induced osteogenic differentiation in mesenchymal stromal cells supports leukemia growth. *JCI Insight* **2017**, *2*, 90036. [CrossRef]
30. Chen, Y.; Jacamo, R.; Shi, Y.-X.; Wang, R.-Y.; Battula, V.L.; Konoplev, S.; Strunk, D.; Hofmann, N.A.; Reinisch, A.; Konopleva, M.; et al. Human extramedullary bone marrow in mice: A novel in vivo model of genetically controlled hematopoietic microenvironment. *Blood* **2012**, *119*, 4971–4980. [CrossRef]
31. Marturano-Kruik, A.; Nava, M.M.; Yeager, K.; Chramiec, A.; Hao, L.; Robinson, S.; Guo, E.; Raimondi, M.T.; Vunjak-Novakovic, G. Human bone perivascular niche-on-a-chip for studying metastatic colonization. *Proc. Natl. Acad. Sci. USA* **2018**, *115*, 1256–1261. [CrossRef]
32. Hao, S.; Ha, L.; Cheng, G.; Wan, Y.; Xia, Y.; Sosnoski, D.M.; Mastro, A.M.; Zheng, S.-Y. A spontaneous 3D bone-on-a-chip for bone metastasis study of breast cancer cells. *Small* **2018**, *14*, e1702787. [CrossRef] [PubMed]
33. Shupp, A.B.; Kolb, A.D.; Bussard, K.M. *Novel Techniques to Study the Bone-Tumor Microenvironment, in Tumor Microenvironment: Advances in Experimental Medicine and Biology*; Birbrair, A., Ed.; Springer: New York, NY, USA, 2020.
34. Belgodere, J.A.; King, C.T.; Bursavich, J.B.; Burow, M.E.; Martin, E.C.; Jung, J.P. Engineering breast cancer microenvironments and 3D bioprinting. *Front. Bioeng. Biotechnol.* **2018**, *6*, 66. [CrossRef] [PubMed]
35. Marks, S.C., Jr.; Odgren, P.R. *Structure and Development of the Skeleton, in Principles of Bone Biology*; Bilezikian, J.P., Raisz, L.G., Rodan, G.A., Eds.; Academic Press: San Diego, CA, USA, 2002; pp. 3–16.
36. Minguell, J.J.; Erices, A.; Conget, P. Mesenchymal stem cells. *Exp. Biol. Med.* **2001**, *226*, 507–520. [CrossRef] [PubMed]
37. Augat, P.; Schorlemmer, S. The role of cortical bone and its microstructure in bone strength. *Age Ageing* **2006**, *35*, ii27–ii31. [CrossRef] [PubMed]
38. Florencio-Silva, R.; Sasso, G.R.d.S.; Sasso-Cerri, E.; Simões, M.J.; Cerri, P.S. Biology of bone tissue: Structure, function, and factors that influence bone cells. *BioMed Res. Int.* **2015**, *2015*, 421746. [CrossRef] [PubMed]
39. Capulli, M.; Paone, R.; Rucci, N. Osteoblast and osteocyte: Games without frontiers. *Arch. Biochem. Biophys.* **2014**, *561*, 3–12. [CrossRef] [PubMed]
40. Mullender, M.M.; Van Der Meer, D.; Huiskes, H.R.; Lips, P. Osteocyte density changes in aging and osteoporosis. *Bone* **1996**, *18*, 109–113. [CrossRef]
41. Takahashi, N.; Udagawa, N.; Takami, M.; Suda, T. Cells of bone: Osteoclast generation. In *Principles of Bone Biology*; Bilezikian, J.P., Raisz, L.G., Rodan, G.A., Eds.; Academic Press: San Diego, CA, USA, 2002; pp. 109–126.
42. Kanis, J.A.; McCloskey, E.V. Bone turnover and biochemical markers in malignancy. *Cancer* **1997**, *80*, 1538–1545. [CrossRef]
43. Wilson, S.R.; Peters, C.; Saftig, P.; Brömme, D.; Cathepsin, K. Activity-Dependent regulation of osteoclast actin ring formation and bone resorption. *J. Biol. Chem.* **2009**, *284*, 2584–2592. [CrossRef]
44. Blair, H.C. How the osteoclast degrades bone. *Bioessays* **1998**, *20*, 837–846. [CrossRef]
45. Teitelbaum, S.L. Therapeutic implications of suppressing osteoclast formation versus function. *Rheumatology* **2016**, *55*, ii61–ii63. [CrossRef] [PubMed]
46. Stenbeck, G. Formation and function of the ruffled border in osteoclasts. *Semin. Cell Dev. Biol.* **2002**, *13*, 285–292. [CrossRef] [PubMed]
47. Hauschka, P.V.; Mavrakos, A.E.; Iafrati, M.D.; Doleman, S.E.; Klagsbrun, M. Growth factors in bone matrix. Isolation of multiple types by affinity chromatography on heparin-Sepharose. *J. Biol. Chem.* **1986**, *261*, 12665–12674. [CrossRef]
48. Raggatt, L.J.; Partridge, N.C. Cellular and molecular mechanisms of bone remodeling. *J. Biol. Chem.* **2010**, *285*, 25103–25108. [CrossRef] [PubMed]
49. Dallas, S.L.; Prideaux, M.; Bonewald, L.F. The osteocyte: An endocrine cell ... and more. *Endocr. Rev.* **2013**, *34*, 658–690. [CrossRef]
50. Sottnik, J.L.; Dai, J.; Zhang, H.; Campbell, B.; Keller, E.T. Tumor-Induced pressure in the bone microenvironment causes osteocytes to promote the growth of prostate cancer bone metastases. *Cancer Res.* **2015**, *75*, 2151–2158. [CrossRef] [PubMed]

51. Mullen, C.; Haugh, M.; Schaffler, M.; Majeska, R.; McNamara, L.M. Osteocyte differentiation is regulated by extracellular matrix stiffness and intercellular separation. *J. Mech. Behav. Biomed. Mater.* **2013**, *28*, 183–194. [CrossRef]
52. Robling, A.G.; Turner, C.H. Mechanical signaling for bone modeling and remodeling. *Crit. Rev. Eukaryot. Gene Expr.* **2009**, *19*, 319–338. [CrossRef]
53. Alberts, B.; Johnson, A.; Lewis, J.; Morgan, D.; Raff, M.; Roberts, K.; Walter, P. *Molecular Biology of the Cell*, 6th ed.; Garland Science: New York, NY, USA, 2015.
54. Wolff, J. *Das Gesetz der Transformation der Knochen*; Hirschwald: Berlin, Germany, 1892.
55. Klein-Nulend, J.; Bacabac, R.G.; Bakker, A.D. Mechanical loading and how it affects bone cells: The role of the osteocyte cytoskeleton in maintaining our skeleton. *Eur. Cells Mater.* **2012**, *24*, 278–291. [CrossRef]
56. Klein-Nulend, J.; Van Der Plas, A.; Semeins, C.M.; Ajubi, N.E.; Erangos, J.A.; Nijweide, P.J.; Burger, E. Sensitivity of osteocytes to biomechanical stress in vitro. *FASEB J.* **1995**, *9*, 441–445. [CrossRef]
57. Mastro, A.M.; Gay, C.V.; Welch, D.; Donahue, H.J.; Jewell, J.; Mercer, R.; DiGirolamo, D.; Chislock, E.M.; Guttridge, K. Breast cancer cells induce osteoblast apoptosis: A possible contributor to bone degradation. *J. Cell. Biochem.* **2004**, *91*, 265–276. [CrossRef] [PubMed]
58. Josse, J.; Velard, F.; Gangloff, S. Staphylococcus aureus vs. osteoblast: Relationship and consequences in osteomyelitis. *Front. Cell. Infect. Microbiol.* **2015**, *5*, 85. [CrossRef] [PubMed]
59. Sanchez, C.; DeBerg, M.A.; Bellahcène, A.; Castronovo, V.; Msika, P.; Delcour, J.P.; Crielaard, J.M.; Henrotin, Y. Phenotypic characterization of osteoblasts from the sclerotic zones of osteoarthritic subchondral bone. *Arthritis Rheum.* **2008**, *58*, 442–455. [CrossRef] [PubMed]
60. Sanchez, C.; Mazzucchelli, G.; Lambert, C.; Comblain, F.; Depauw, E.; Henrotin, Y. Comparison of secretome from osteoblasts derived from sclerotic versus non-sclerotic subchondral bone in OA: A pilot study. *PLoS ONE* **2018**, *13*, e0194591. [CrossRef]
61. Yoneda, T. Mechanisms of preferential metastasis of breast cancer to bone—(Review). *Int. J. Oncol.* **1996**, *9*, 103–109. [CrossRef]
62. Lentino, J.R. Prosthetic joint infections: Bane of orthopedists, challenge for infectious disease specialists. *Clin. Infect. Dis.* **2003**, *36*, 1157–1161. [CrossRef]
63. Berendt, T.; Byren, I. Bone and joint infection. *Clin. Med.* **2004**, *4*, 510–518. [CrossRef]
64. Tillander, J.; Hagberg, K.; Berlin, Ö.; Hagberg, L.; Brånemark, R. Osteomyelitis risk in patients with transfemoral amputations treated with osseointegration prostheses. *Clin. Orthop. Relat. Res.* **2017**, *475*, 3100–3108. [CrossRef]
65. Kellesarian, S.V.; Javed, F.; Romanos, G.E. Osteomyelitis arising around osseointegrated dental implants: A systematic review. *Implant Dent.* **2018**, *27*, 226–235. [CrossRef]
66. Semel, G.; Wolff, A.; Shilo, D.; Akrish, S.; Emodi, O.; Rachmiel, A. Mandibular osteomyelitis associated with dental implants. A case series. *Eur. J. Oral Implantol.* **2016**, *9*, 435–442.
67. Bosse, M.J.; Gruber, H.E.; Ramp, W.K. Internalization of bacteria by osteoblasts in a patient with recurrent, long-term osteomyelitis: A case report. *JBJS Case Connect.* **2005**, *87*, 1343–1347. [CrossRef] [PubMed]
68. Dapunt, U.; Maurer, S.; Giese, T.; Gaida, M.M.; Hänsch, G.M. The macrophage inflammatory proteins MIP1α(CCL3) and MIP2α(CXCL2) in implant-associated osteomyelitis: Linking inflammation to bone degradation. *Mediat. Inflamm.* **2014**, *2014*, 1–10. [CrossRef] [PubMed]
69. Bost, K.L.; Bento, J.L.; Ellington, J.K.; Marriott, I.; Hudson, M.C. Induction of colony-stimulating factor expression following staphylococcus or salmonellainteraction with mouse or human osteoblasts. *Infect. Immun.* **2000**, *68*, 4834–4837. [CrossRef] [PubMed]
70. Wright, K.M.; Friedland, J.S. Regulation of chemokine gene expression and secretion in Staphylococcus aureus-infected osteoblasts. *Microbes Infect.* **2004**, *6*, 844–852. [CrossRef] [PubMed]
71. Ning, R.; Zhang, X.; Guo, X.; Li, Q. Staphylococcus aureus regulates secretion of interleukin-6 and monocyte chemoattractant protein-1 through activation of nuclear factor kappaB signaling pathway in human osteoblasts. *Braz. J. Infect. Dis.* **2011**, *15*, 189–194.
72. Gasper, N.A.; Petty, C.C.; Schrum, L.W.; Marriott, I.; Bost, K.L. Bacterium-Induced CXCL10 secretion by osteoblasts can be mediated in part through toll-like receptor 4. *Infect. Immun.* **2002**, *70*, 4075–4082. [CrossRef] [PubMed]
73. Somayaji, S.N.; Ritchie, S.; Sahraei, M.; Marriott, I.; Hudson, M.C. Staphylococcus aureus induces expression of receptor activator of NF-kappaB ligand and prostaglandin E2 in infected murine osteoblasts. *Infect. Immun.* **2008**, *76*, 5120–5126. [CrossRef]
74. Widaa, A.; Claro, T.; Foster, T.J.; O'Brien, F.J.; Kerrigan, S.W. Staphylococcus aureus protein a plays a critical role in mediating bone destruction and bone loss in osteomyelitis. *PLoS ONE* **2012**, *7*, e40586. [CrossRef]
75. Sanchez, C.; Pesesse, L.; Gabay, O.; Delcour, J.-P.; Msika, P.; Baudouin, C.; Henrotin, Y. Regulation of subchondral bone osteoblast metabolism by cyclic compression. *Arthritis Rheum.* **2012**, *64*, 1193–1203. [CrossRef]
76. Bianco, D.; Todorov, A.; Cengic, T.; Pagenstert, G.; Schären, S.; Netzer, C.; Hügle, T.; Geurts, J. Alterations of subchondral bone progenitor cells in human knee and hip osteoarthritis lead to a bone sclerosis phenotype. *Int. J. Mol. Sci.* **2018**, *19*, 475. [CrossRef]
77. Martineau, X.; Abed, É.; Martel-Pelletier, J.; Pelletier, J.-P.; Lajeunesse, D. Alteration of Wnt5a expression and of the non-canonical Wnt/PCP and Wnt/PKC-Ca2+ pathways in human osteoarthritis osteoblasts. *PLoS ONE* **2017**, *12*, e0180711. [CrossRef] [PubMed]
78. Kinder, M.; Chislock, E.M.; Bussard, K.M.; Shuman, L.; Mastro, A.M. Metastatic breast cancer induces an osteoblast inflammatory response. *Exp. Cell Res.* **2008**, *314*, 173–183. [CrossRef] [PubMed]

79. Bussard, K.M.; Gay, C.V.; Mastro, A.M. The bone microenvironment in metastasis; what is special about bone? *Cancer Metastasis Rev.* **2007**, *27*, 41–55. [CrossRef] [PubMed]
80. Logothetis, C.; Morris, M.J.; Den, R.; Coleman, R.E. Current perspectives on bone metastases in castrate-resistant prostate cancer. *Cancer Metastasis Rev.* **2018**, *37*, 189–196. [CrossRef] [PubMed]
81. Logothetis, C.J.; Lin, S.-H. Osteoblasts in prostate cancer metastasis to bone. *Nat. Rev. Cancer* **2005**, *5*, 21–28. [CrossRef]
82. Roodman, G.D. Mechanisms of bone metastasis. *N. Engl. J. Med.* **2004**, *350*, 1655–1664. [CrossRef]
83. Paget, S. The distribution of secondary growths in cancer of the breast. *Lancet* **1889**, *133*, 571–573. [CrossRef]
84. Paget, S. The distribution of secondary growths in cancer of the breast. *Cancer Metastasis Rev.* **1989**, *8*, 98–101. [CrossRef]
85. Kolb, A.D.; Shupp, A.B.; Mukhopadhyay, D.; Marini, F.C.; Bussard, K.M. Osteoblasts are "educated" by crosstalk with metastatic breast cancer cells in the bone tumor microenvironment. *Breast Cancer Res.* **2019**, *21*, 31. [CrossRef]
86. Lawson, M.A.; McDonald, M.M.; Kovacic, N.; Khoo, W.H.; Terry, R.L.; Down, J.M.; Kaplan, W.; Paton-Hough, J.; Fellows, C.; Pettitt, J.A.; et al. Osteoclasts control reactivation of dormant myeloma cells by remodelling the endosteal niche. *Nat. Commun.* **2015**, *6*, 8983. [CrossRef]
87. Mundy, G.R. *Bone Remodeling and Its Disorders*; Martin Dunitz Ltd.: London, UK, 1999.
88. Kennecke, H.; Yerushalmi, R.; Woods, R.; Cheang, M.C.U.; Voduc, D.; Speers, C.H.; Nielsen, T.O.; Gelmon, K. Metastatic behavior of breast cancer subtypes. *J. Clin. Oncol.* **2010**, *28*, 3271–3277. [CrossRef] [PubMed]
89. Liede, A.; Jerzak, K.J.; Hernandez, R.K.; Wade, S.W.; Sun, P.; Narod, S.A. The incidence of bone metastasis after early-stage breast cancer in Canada. *Breast Cancer Res. Treat.* **2016**, *156*, 587–595. [CrossRef] [PubMed]
90. Siegel, R.L.; Miller, K.D.M.; Jemal, A. Cancer statistics, 2018. *CA Cancer J. Clin.* **2018**, *68*, 7–30. [CrossRef] [PubMed]
91. Lipton, A.; Uzzo, R.; Amato, R.J.; Ellis, G.K.; Hakimian, B.; Roodman, G.D.; Smith, M.R. The science and practice of bone health in oncology: Managing bone loss and metastasis in patients with solid tumors. *J. Natl. Compr. Cancer Netw.* **2009**, *7* (Suppl. S7), S-1. [CrossRef] [PubMed]
92. Manders, K.; Van De Poll-Franse, L.V.; Creemers, G.-J.; Vreugdenhil, G.; Van Der Sangen, M.J.C.; Nieuwenhuijzen, G.A.P.; Roumen, R.M.H.; Voogd, A.C. Clinical management of women with metastatic breast cancer: A descriptive study according to age group. *BMC Cancer* **2006**, *6*, 179. [CrossRef]
93. Coleman, R.E. Skeletal complications of malignancy. *Cancer* **1997**, *80*, 1588–1594. [CrossRef]
94. Guise, T.A.; Mundy, G.R. Cancer and bone. *Endocr. Rev.* **1998**, *19*, 18–54.
95. Guise, T.A.; Mohammad, K.S.; Clines, G.; Stebbins, E.G.; Wong, D.H.; Higgins, L.S.; Vessella, R.; Corey, E.; Padalecki, S.; Suva, L.; et al. Basic mechanisms responsible for osteolytic and osteoblastic bone metastases: Fig. 1. *Clin. Cancer Res.* **2006**, *12*, 6213s–6216s. [CrossRef]
96. Guise, T.A. Molecular mechanisms of osteolytic bone metastases. *Cancer* **2000**, *88*, 2892–2898. [CrossRef]
97. Marathe, D.D.; Marathe, A.; Mager, D.E. Integrated model for denosumab and ibandronate pharmacodynamics in postmenopausal women. *Biopharm. Drug Dispos.* **2011**, *32*, 471–481. [CrossRef]
98. Anagnostis, P.; Vakalopoulou, S.; Christoulas, D.; Paschou, S.A.; Papatheodorou, A.; Garipidou, V.; Kokkoris, P.; Terpos, E. The role of sclerostin/dickkopf-1 and receptor activator of nuclear factor kB ligand/osteoprotegerin signalling pathways in the development of osteoporosis in patients with haemophilia A and B: A cross-sectional study. *Haemophilia* **2017**, *24*, 316–322. [CrossRef] [PubMed]
99. Geng, C.-J.; Liang, Q.; Zhong, J.-H.; Zhu, M.; Meng, F.-Y.; Wu, N.; Liang, R.; Yuan, B.-Y. Ibandronate to treat skeletal-related events and bone pain in metastatic bone disease or multiple myeloma: A meta-analysis of randomised clinical trials. *BMJ Open* **2015**, *5*, e007258. [CrossRef] [PubMed]
100. Tu, K.N.; Lie, J.D.; Wan, C.K.V.; Cameron, M.; Austel, A.G.; Nguyen, J.K.; Van, K.; Hyun, D. Osteoporosis: A review of treatment options. *Pharm. Ther.* **2018**, *43*, 92–104.
101. Greenberg, A.J.; Rajkumar, S.V.; Therneau, T.M.; Singh, P.; Dispenzieri, A.; Kumar, S.K. Relationship between initial clinical presentation and the molecular cytogenetic classification of myeloma. *Leukemia* **2014**, *28*, 398–403. [CrossRef] [PubMed]
102. Marino, S.; Roodman, G.D. Multiple myeloma and bone: The fatal interaction. *Cold Spring Harb. Perspect. Med.* **2018**, *8*, a031286. [CrossRef]
103. Bataille, R.; Chappard, D.; Marcelli, C.; Dessauw, P.; Sany, J.; Baldet, P.; Alexandre, C. Mechanisms of bone destruction in multiple myeloma: The importance of an unbalanced process in determining the severity of lytic bone disease. *J. Clin. Oncol.* **1989**, *7*, 1909–1914. [CrossRef]
104. Ehrlich, L.A.; Chung, H.Y.; Ghobrial, I.; Choi, S.J.; Morandi, F.; Colla, S.; Rizzoli, V.; Roodman, G.D.; Giuliani, N. IL-3 is a potential inhibitor of osteoblast differentiation in multiple myeloma. *Blood* **2005**, *106*, 1407–1414. [CrossRef]
105. Silbermann, R.; Bolzoni, M.; Storti, P.; Guasco, D.; Bonomini, S.; Zhou, D.; Wu, J.; Anderson, J.L.; Windle, J.J.; Aversa, F.; et al. Bone marrow monocyte-/macrophage-derived activin A mediates the osteoclastogenic effect of IL-3 in multiple myeloma. *Leukemia* **2014**, *28*, 951–954. [CrossRef]
106. Delgado-Calle, J.; Anderson, J.; Cregor, M.D.; Condon, K.W.; Kuhstoss, S.A.; Plotkin, L.I.; Bellido, T.; Roodman, G.D. Genetic deletion of Sost or pharmacological inhibition of sclerostin prevent multiple myeloma-induced bone disease without affecting tumor growth. *Leukemia* **2017**, *31*, 2686–2694. [CrossRef]

107. Waning, D.L.; Mohammad, S.K.; Reiken, S.; Xie, W.; Andersson, D.C.; John, S.; Chiechi, A.; Wright, L.E.; Umanskaya, A.; Niewolna, M.; et al. Excess TGF-beta mediates muscle weakness associated with bone metastases in mice. *Nat. Med.* **2015**, *21*, 1262–1271. [CrossRef]
108. Nyman, J.S.; Merkel, A.R.; Uppuganti, S.; Nayak, B.; Rowland, B.; Makowski, A.J.; Oyajobi, B.O.; Sterling, J.A. Combined treatment with a transforming growth factor beta inhibitor (1D11) and bortezomib improves bone architecture in a mouse model of myeloma-induced bone disease. *Bone* **2016**, *91*, 81–91. [CrossRef] [PubMed]
109. D'Souza, S.; Del Prete, D.; Jin, S.; Sun, Q.; Huston, A.J.; Kostov, F.E.; Sammut, B.; Hong, C.-S.; Anderson, J.L.; Patrene, K.D.; et al. Gfi1 expressed in bone marrow stromal cells is a novel osteoblast suppressor in patients with multiple myeloma bone disease. *Blood* **2011**, *118*, 6871–6880. [CrossRef] [PubMed]
110. Pozzi, S.; Fulciniti, M.; Yan, H.; Vallet, S.; Eda, H.; Patel, K.; Santo, L.; Cirstea, D.; Hideshima, T.; Schirtzinge, L.; et al. In vivo and in vitro effects of a novel anti-Dkk1 neutralizing antibody in multiple myeloma. *Bone* **2013**, *53*, 487–496. [CrossRef] [PubMed]
111. Fulciniti, M.; Tassone, P.; Hideshima, T.; Vallet, S.; Nanjappa, P.; Ettenberg, S.A.; Shen, Z.; Patel, N.; Tai, Y.-T.; Chauhan, D.; et al. Anti-DKK1 mAb (BHQ880) as a potential therapeutic agent for multiple myeloma. *Blood* **2009**, *114*, 371–379. [CrossRef] [PubMed]
112. Morris, M.J.; Scher, H.I. Clinical approaches to osseous metastases in prostate cancer. *Oncologist* **2003**, *8*, 161–173. [CrossRef] [PubMed]
113. Wang, N.; Docherty, F.E.; Brown, H.K.; Reeves, K.J.; Fowles, A.C.M.; Ottewell, P.D.; Dear, T.N.; Holen, I.; Croucher, P.I.; Eaton, C.L. Prostate cancer cells preferentially home to osteoblast-rich areas in the early stages of bone metastasis: Evidence from in vivo models. *J. Bone Miner. Res.* **2014**, *29*, 2688–2696. [CrossRef]
114. Roudier, M.P.; Morrissey, C.; True, L.D.; Higano, C.S.; Vessella, R.L.; Ott, S.M. Histopathological assessment of prostate cancer bone osteoblastic metastases. *J. Urol.* **2008**, *180*, 1154–1160. [CrossRef]
115. Eastham, J.A. Bone health in men receiving androgen deprivation therapy for prostate cancer. *J. Urol.* **2007**, *177*, 17–24. [CrossRef]
116. Sekita, A.; Matsugaki, A.; Nakano, T. Disruption of collagen/apatite alignment impairs bone mechanical function in osteoblastic metastasis induced by prostate cancer. *Bone* **2017**, *97*, 83–93. [CrossRef]
117. Matsugaki, A.; Aramoto, G.; Ninomiya, T.; Sawada, H.; Hata, S.; Nakano, T. Abnormal arrangement of a collagen/apatite extracellular matrix orthogonal to osteoblast alignment is constructed by a nanoscale periodic surface structure. *Biomaterials* **2015**, *37*, 134–143. [CrossRef]
118. Charhon, S.A.; Chapuy, M.C.; Delvin, E.E.; Valentin-Opran, A.; Edouard, C.M.; Meunier, P.J. Histomorphometric analysis of sclerotic bone metastases from prostatic carcinoma with special reference to osteomalacia. *Cancer* **1983**, *51*, 918–924. [CrossRef]
119. Clarke, N.W.; McClure, J.; George, N.J.R. Morphometric evidence for bone resorption and replacement in prostate cancer. *BJU Int.* **1991**, *68*, 74–80. [CrossRef]
120. Wan, X.; Corn, P.G.; Yang, J.; Palanisamy, N.; Starbuck, M.W.; Efstathiou, E.; Li-Ning-Tapia, E.M.; Zurita, A.J.; Aparicio, A.; Ravoori, M.K.; et al. Prostate cancer cell–stromal cell crosstalk via FGFR1 mediates antitumor activity of dovitinib in bone metastases. *Sci. Transl. Med.* **2014**, *6*, 252ra122. [CrossRef]
121. Fizazi, K.; Yang, J.; Peleg, S.; Sikes, C.R.; Kreimann, E.L.; Daliani, D.; Olive, M.; Raymond, K.A.; Janus, T.J.; Logothetis, C.; et al. Prostate cancer cells-osteoblast interaction shifts expression of growth/survival-related genes in prostate cancer and reduces expression of osteoprotegerin in osteoblasts. *Clin. Cancer Res.* **2003**, *9*, 2587–2597. [PubMed]
122. Oberneder, R.; Riesenberg, R.; Kriegmair, M.; Bitzer, U.; Klammert, R.; Schneede, P.; Hofstetter, A.; Pantel, K. Immunocytochemical detection and phenotypic characterization of micrometastatic tumour cells in bone marrow of patients with prostate cancer. *Urol. Res.* **1994**, *22*, 3–8. [CrossRef]
123. Ottewell, P.D. The role of osteoblasts in bone metastasis. *J. Bone Oncol.* **2016**, *5*, 124–127. [CrossRef] [PubMed]
124. Carducci, M.A.; Nelson, J.B.; Bowling, M.K.; Rogers, T.; Eisenberger, M.A.; Sinibaldi, V.; Donehower, R.; Leahy, T.L.; Carr, R.A.; Isaacson, J.D.; et al. Atrasentan, an endothelin-receptor antagonist for refractory adenocarcinomas: Safety and pharmacokinetics. *J. Clin. Oncol.* **2002**, *20*, 2171–2180. [CrossRef]
125. Carducci, M.A.; Padley, R.J.; Breul, J.; Vogelzang, N.J.; Zonnenberg, B.A.; Daliani, D.D.; Schulman, C.C.; Nabulsi, A.A.; Humerickhouse, R.A.; Weinberg, M.A.; et al. Effect of endothelin-a receptor blockade with atrasentan on tumor progression in men with hormone-refractory prostate cancer: A randomized, phase ii, placebo-controlled trial. *J. Clin. Oncol.* **2003**, *21*, 679–689. [CrossRef]
126. Suominen, M.I.; Fagerlund, K.M.; Rissanen, J.P.; Konkol, Y.M.; Morko, J.P.; Peng, Z.; Alhoniemi, E.J.; Laine, S.K.; Corey, E.; Mumberg, D.; et al. Radium-223 inhibits osseous prostate cancer growth by dual targeting of cancer cells and bone microenvironment in mouse models. *Clin. Cancer Res.* **2017**, *23*, 4335–4346. [CrossRef]
127. Popper, H.H. Progression and metastasis of lung cancer. *Cancer Metastasis Rev.* **2016**, *35*, 75–91. [CrossRef]
128. Baker, J.; Falconer, A.M.D.; Wilkinson, D.J.; Europe-Finner, G.N.; Litherland, G.J.; Rowan, A.D. Protein kinase D3 modulates MMP1 and MMP13 expression in human chondrocytes. *PLoS ONE* **2018**, *13*, e0195864. [CrossRef] [PubMed]
129. Tang, C.-H.; Tan, T.-W.; Fu, W.-M.; Yang, R.-S. Involvement of matrix metalloproteinase-9 in stromal cell-derived factor-1/CXCR4 pathway of lung cancer metastasis. *Carcinogenesis* **2007**, *29*, 35–43. [CrossRef] [PubMed]
130. Sugiura, H.; Yamada, K.; Sugiura, T.; Hida, T.; Mitsudomi, T. Predictors of surivival in patients with bone metastasis of lung cancer. *Clin. Ortho. Relat. Res.* **2008**, *466*, 729–736. [CrossRef] [PubMed]
131. Coleman, R. Metastatic bone disease: Clinical features, pathophysiology and treatment strategies. *Cancer Treat. Rev.* **2001**, *27*, 165–176. [CrossRef]

132. Hill, C.A. Bronchioloalveolar carcinoma: A review. *Radiology* **1984**, *150*, 15–20. [CrossRef]
133. Gordon, J.A.R.; Tye, C.E.; Sampaio, A.V.; Underhill, T.M.; Hunter, G.K.; Goldberg, H.A. Bone sialoprotein expression enhances osteoblast differentiation and matrix mineralization in vitro. *Bone* **2007**, *41*, 462–473. [CrossRef]
134. Kang, E.J.; Lee, S.Y.; Kim, H.J.; Min, K.H.; Hur, G.-Y.; Shim, J.J.; Kang, K.H.; Oh, S.C.; Seo, J.H.; Lee, S.; et al. Prognostic factors and skeletal-related events in patients with small cell lung cancer with bone metastases at the time of diagnosis. *Oncologist* **2016**, *90*, 103–111. [CrossRef]
135. Lang, J.; Zhao, Q.; He, Y.; Yu, X. Bone turnover markers and novel biomarkers in lung cancer bone metastases. *Biomarkers* **2018**, *23*, 518–526. [CrossRef]
136. Papotti, M.; Kalebic, T.; Volante, M.; Chiusa, L.; Bacillo, E.; Cappia, S.; Lausi, P.O.; Novello, S.; Borasio, P.; Scagliotti, G.V. Bone sialoprotein is predictive of bone metastases in resectable non–small-cell lung cancer: A retrospective case-control study. *J. Clin. Oncol.* **2006**, *24*, 4818–4824. [CrossRef]
137. Brown, J.E.; Cook, R.J.; Major, P.; Lipton, A.; Saad, F.; Smith, M.R.; Lee, K.-A.; Zheng, M.; Hei, Y.-J.; Coleman, R.E. Bone turnover markers as predictors of skeletal complications in prostate cancer, lung cancer, and other solid tumors. *J. Natl. Cancer Inst.* **2005**, *97*, 59–69. [CrossRef]
138. Costa, L.; Demers, L.M.; Gouveia-Oliveira, A.; Schaller, J.; Costa, E.B.; De Moura, M.C.; Lipton, A. Prospective evaluation of the peptide-bound collagen type I cross-links N-telopeptide and C-telopeptide in predicting bone metastases status. *J. Clin. Oncol.* **2002**, *20*, 850–856. [PubMed]
139. Liu, B.; Zhao, Y.; Yuan, J.; Zeng, L.; Sun, R.; Meng, X.; Yang, S. Elevated N-telopeptide as a potential diagnostic marker for bone metastasis in lung cancer: A meta-analysis. *PLoS ONE* **2017**, *12*, e0187860. [CrossRef]
140. Zhang, Y.; Yan, S.; Su, Y.; Chen, H.; Wang, S.; Sun, B.; Zhang, L. Serum cross-linked N-telopeptide of type I collagen as a biomarker of bone metastases for patients with lung cancer: A meta-analysis. *Int. J. Clin. Exp. Med.* **2018**, *11*, 12864–12869.
141. Hu, Z.; Lin, D.; Yuan, J.; Xiao, T.; Zhang, H.; Sun, W.; Han, N.; Ma, Y.; Di, X.; Gao, M.; et al. Overexpression of osteopontin is associated with more aggressive phenotypes in human non-small cell lung cancer. *Clin. Cancer Res.* **2005**, *11*, 4646–4652. [CrossRef] [PubMed]
142. Kothari, A.N.; Arffa, M.; Chang, V.; Blackwell, R.H.; Syn, W.-K.; Zhang, J.; Mi, Z.; Kuo, P.C. Osteopontin—A master regulator of epithelial-mesenchymal transition. *J. Clin. Med.* **2016**, *5*, 39. [CrossRef]
143. Aguirre-Ghiso, J.A. Models, mechanisms, and clinical evidence for cancer dormancy. *Nat. Rev. Cancer* **2007**, *7*, 834–846. [CrossRef]
144. Braun, S.; Naume, B. Circulating and disseminated tumor cells. *J. Clin. Oncol.* **2005**, *23*, 1623–1626. [CrossRef]
145. Braun, S.; Vogl, F.D.; Naume, B.; Janni, W.; Osborne, M.P.; Coombes, R.C.; Schlimok, G.; Diel, I.J.; Gerber, B.; Gebauer, G.; et al. A pooled analysis of bone marrow micrometastasis in breast cancer. *N. Engl. J. Med.* **2005**, *353*, 793–802. [CrossRef]
146. Quayle, L.; Ottewell, P.D.; Holen, I. Bone metastasis: Molecular mechanisms implicated in tumour cell dormancy in breast and prostate cancer. *Curr. Cancer Drug Targets* **2015**, *15*, 469–480. [CrossRef]
147. Scimeca, M.; Antonacci, C.; Toschi, N.; Giannini, E.; Bonfiglio, R.; Buonomo, C.O.; Pistolese, C.A.; Tarantino, U.; Bonanno, E. Breast osteoblast-like cells: A reliable early marker for bone metastases from breast cancer. *Clin. Breast Cancer* **2018**, *18*, e659–e669. [CrossRef]
148. Bodenstine, T.M.; Beck, B.H.; Cao, X.; Cook, L.M.; Ismail, A.; Powers, S.J.K.; Powers, J.K.; Mastro, A.M.; Welch, D. Pre-osteoblastic MC3T3-E1 cells promote breast cancer growth in bone in a murine xenograft model. *Chin. J. Cancer* **2011**, *30*, 189–196. [CrossRef] [PubMed]
149. Yumoto, K.; Eber, M.R.; Wang, J.; Cackowski, F.C.; Decker, A.M.; Lee, E.; Nobre, A.R.; Aguirre-Ghiso, J.A.; Jung, Y.; Taichman, R.S. Axl is required for TGF-beta2-induced dormancy of prostate cancer cells in the bone marrow. *Sci. Rep.* **2016**, *6*, 36520. [CrossRef] [PubMed]
150. Kobayashi, A.; Okuda, H.; Xing, F.; Pandey, P.R.; Watabe, M.; Hirota, S.; Pai, S.K.; Liu, W.; Fukuda, K.; Chambers, C.; et al. Bone morphogenetic protein 7 in dormancy and metastasis of prostate cancer stem-like cells in bone. *J. Exp. Med.* **2011**, *208*, 2641–2655. [CrossRef] [PubMed]
151. Weigelt, B.; Ghajar, C.M.; Bissell, M.J. The need for complex 3D culture models to unravel novel pathways and identify accurate biomarkers in breast cancer. *Adv. Drug Deliv. Rev.* **2014**, *69-70*, 42–51. [CrossRef]
152. Gómez-Cuadrado, L.; Tracey, N.; Ma, R.; Qian, B.; Brunton, V.G. Mouse models of metastasis: Progress and prospects. *Dis. Model. Mech.* **2017**, *10*, 1061–1074. [CrossRef]
153. Saxena, M.; Christofori, G. Rebuilding cancer metastasis in the mouse. *Mol. Oncol.* **2013**, *7*, 283–296. [CrossRef]
154. Ghajar, C.M.; Peinado, H.; Mori, H.; Matei, I.R.; Evason, K.J.; Brazier, H.; De Almeida, D.L.; Koller, A.; Hajjar, K.A.; Stainier, D.Y.R.; et al. The perivascular niche regulates breast tumour dormancy. *Nat. Cell Biol.* **2013**, *15*, 807–817. [CrossRef]
155. Naumov, G.N.; Macdonald, I.C.; Weinmeister, P.M.; Kerkvliet, N.; Nadkarni, K.V.; Wilson, S.M.; Morris, V.L.; Groom, A.C.; Chambers, A.F. Persistence of solitary mammary carcinoma cells in a secondary site: A possible contributor to dormancy. *Cancer Res.* **2002**, *62*, 2162–2168.
156. Chen, Y.-C.; Mastro, A.M.; Sosnoski, D.M.; Norgard, R.J.; Grove, C.D.; Vogler, E.A. Abstract 4891: Dormancy and growth of metastatic breast cancer cells in a bone-like microenvironment. *Tumor Biol.* **2014**, *32*, 335–344. [CrossRef]
157. Angeloni, V.; Negrini, N.C.; De Marco, C.; Bertoldi, S.; Tanzi, M.C.; Daidone, M.G.; Farè, S. Polyurethane foam scaffold as in vitro model for breast cancer bone metastasis. *Acta Biomater.* **2017**, *63*, 306–316. [CrossRef]
158. Khanna, C.; Hunter, K. Modeling metastasis in vivo. *Carcinogenesis* **2005**, *26*, 513–523. [CrossRef] [PubMed]

159. Clarke, R.B. Human breast cancer cell line xenografts as models of breast cancer—The immunobiologies of recipient mice and the characteristics of several tumorigenic cell lines. *Breast Cancer Res. Treat.* **1996**, *39*, 69–86. [CrossRef] [PubMed]
160. Wright, L.E.; Ottewell, P.D.; Rucci, N.; Peyruchaud, O.; Pagnotti, G.M.; Chiechi, A.; Buijs, J.T.; Sterling, J.A. Murine models of breast cancer bone metastasis. *BoneKEy Rep.* **2016**, *5*, 804. [CrossRef] [PubMed]
161. Shupp, A.B.; Kolb, A.D.; Bussard, K.M. Novel techniques to study the bone-tumor microenvironment. *Adv. Exp. Med. Biol.* **2020**, *1225*, 1–18. [CrossRef] [PubMed]
162. Kang, Y.; Siegel, P.M.; Shu, W.; Drobnjak, M.; Kakonen, S.M.; Cordón-Cardo, C.; Guise, T.A.; Massagué, J. A multigenic program mediating breast cancer metastasis to bone. *Cancer Cell* **2003**, *3*, 537–549. [CrossRef]
163. Yoneda, T.; Williams, P.J.; Hiraga, T.; Niewolna, M.; Nishimura, R. A bone-seeking clone exhibits different biological properties from the MDA-MD-231 parental human breast cancer cells and a brain-seeking clone in vivo and in vitro. *J. Bone Miner Res.* **2001**, *16*, 1486–1495. [CrossRef]
164. Cassereau, L.; Miroshnikova, Y.A.; Ou, G.; Lakins, J.; Weaver, V.M. A 3D tension bioreactor platform to study the interplay between ECM stiffness and tumor phenotype. *J. Biotechnol.* **2015**, *193*, 66–69. [CrossRef]
165. Matei, I.; Rampersaud, S.; Lyden, D. Engineered niches model the onset of metastasis. *Nat. Biomed. Eng.* **2018**, *2*, 885–887. [CrossRef]
166. Vanderburgh, J.P.; Guelcher, S.A.; Sterling, J.A. 3D bone models to study the complex physical and cellular interactions between tumor and the bone microenvironment. *J. Cell. Biochem.* **2018**, *119*, 5053–5059. [CrossRef]
167. Caliari, S.R.; Burdick, J.A. A practical guide to hydrogels for cell culture. *Nat. Methods* **2016**, *13*, 405–414. [CrossRef]
168. Page, J.M.; Merkel, A.R.; Ruppender, N.S.; Guo, R.; Dadwal, U.C.; Cannonier, S.; Basu, S.; Guelcher, S.A.; Sterling, J.A. Matrix rigidity regulates the transition of tumor cells to a bone-destructive phenotype through integrin B3 and TGF-B receptor type II. *Biomaterials* **2015**, *64*, 33–44. [CrossRef] [PubMed]
169. Kimura, Y.; Matsugaki, A.; Sekita, A.; Nakano, T. Alteration of osteoblast arrangement via direct attack by cancer cells: New insights into bone metastasis. *Sci. Rep.* **2017**, *7*, 44824. [CrossRef]
170. Tibbitt, M.W.; Anseth, K.S. Hydrogels as extracellular matrix mimics for 3D cell culture. *Biotechnol. Bioeng.* **2009**, *103*, 655–663. [CrossRef] [PubMed]
171. Hung, B.P.; Naved, B.A.; Nyberg, E.L.; Dias, M.; Holmes, C.A.; Elisseeff, J.H.; Dorafshar, A.H.; Grayson, W. Three-Dimensional printing of bone extracellular matrix for craniofacial regeneration. *ACS Biomater. Sci. Eng.* **2016**, *2*, 1806–1816. [CrossRef] [PubMed]
172. Temple, J.P.; Hutton, D.L.; Hung, B.P.; Huri, P.Y.; Cook, C.A.; Kondragunta, R.; Jia, X.; Grayson, W. Engineering anatomically shaped vascularized bone grafts with hASCs and 3D-printed PCL scaffolds. *J. Biomed. Mater. Res. Part A* **2014**, *102*, 4317–4325. [CrossRef] [PubMed]
173. Carpenter, R.A.; Kwak, J.-G.; Peyton, S.R.; Lee, J. Implantable pre-metastatic niches for the study of the microenvironmental regulation of disseminated human tumour cells. *Nat. Biomed. Eng.* **2018**, *2*, 915–929. [CrossRef]
174. Patricio, T.; Domingos, M.; Gloria, A.; Bártolo, P. Characterisation of PCL and PCL/PLA scaffolds for tissue engineering. *Procedia CIRP* **2013**, *5*, 110–114. [CrossRef]
175. Kolb, A.D.; Shupp, A.B.; Bussard, K.M. *Unpublished Data*; Thomas Jefferson University: Philadelphia, PA, USA, 2021.
176. Dababneh, A.B.; Ozbolat, I.T. Bioprinting technology: A current state-of-the-art review. *J. Manuf. Sci. Eng.* **2014**, *136*, 061016. [CrossRef]
177. Cleversey, C.; Robinson, M.; Willerth, S.M. 3D printing breast tissue models: A review of past work and directions for future work. *Micromachines* **2019**, *10*, 501. [CrossRef]
178. Klebe, R.J. Cytoscribing: A method for micropositioning cells and the construction of two- and three-dimensional synthetic tissues. *Exp. Cell Res.* **1988**, *179*, 362–373. [CrossRef]
179. Ashammakhi, N.; Hasan, A.; Kaarela, O.; Byambaa, B.; Sheikhi, A.; Gaharwar, A.K.; Khademhosseini, A. Advancing frontiers in bone bioprinting. *Adv. Healthc. Mater.* **2019**, *8*, e1801048. [CrossRef] [PubMed]
180. Gao, G.; Hubbell, K.; Schilling, A.F.; Dai, G.; Cui, X. Bioprinting cartilage tissue from mesenchymal stem cells and PEG hydrogel. *Toxic. Assess.* **2017**, *1612*, 391–398. [CrossRef]
181. Gao, G.; Schilling, A.F.; Yonezawa, T.; Wang, J.; Dai, G.; Cui, X. Bioactive nanoparticles stimulate bone tissue formation in bioprinted three-dimensional scaffold and human mesenchymal stem cells. *Biotechnol. J.* **2014**, *9*, 1304–1311. [CrossRef] [PubMed]
182. Ling, K.; Huang, G.; Liu, J.; Zhang, X.; Ma, Y.; Lu, T.; Xu, F. Bioprinting-Based high-throughput fabrication of three-dimensional MCF-7 human breast cancer cellular spheroids. *Engineering* **2015**, *1*, 269–274. [CrossRef]
183. Mirani, B.; Pagan, E.; Shojaei, S.; Duchscherer, J.; Toyota, B.D.; Ghavami, S.; Akbari, M. A 3D bioprinted hydrogel mesh loaded with all-trans retinoic acid for treatment of glioblastoma. *Eur. J. Pharmacol.* **2019**, *854*, 201–212. [CrossRef] [PubMed]
184. Wei, X.; Liu, C.; Wang, Z.; Luo, Y. 3D printed core-shell hydrogel fiber scaffolds with NIR-triggered drug release for localized therapy of breast cancer. *Int. J. Pharm.* **2020**, *580*, 119219. [CrossRef] [PubMed]
185. Yu, Y.; Zhang, Y.; Ozbolat, I.T. A hybrid bioprinting approach for scale-up tissue fabrication. *J. Manuf. Sci. Eng.* **2014**, *136*, 061013. [CrossRef]
186. Gao, G.; Cui, X. Three-Dimensional bioprinting in tissue engineering and regenerative medicine. *Biotechnol. Lett.* **2015**, *38*, 203–211. [CrossRef]
187. Murphy, S.V.; Atala, A. 3D bioprinting of tissues and organs. *Nat. Biotechnol.* **2014**, *32*, 773–785.

188. Gudapati, H.; Dey, M.; Ozbolat, I.T. A comprehensive review on droplet-based bioprinting: Past, present and future. *Biomaterials* **2016**, *102*, 20–42. [CrossRef]
189. Ji, S.; Guvendiren, M. Recent advances in bioink design for 3D bioprinting of tissues and organs. *Front. Bioeng. Biotechnol.* **2017**, *5*, 23. [CrossRef] [PubMed]
190. Placone, J.K.; Engler, A.J. Recent advances in extrusion-based 3D printing for biomedical applications. *Adv. Healthc. Mater.* **2017**, *7*, e1701161. [CrossRef] [PubMed]
191. Colosi, C.; Shin, S.R.; Manoharan, V.; Massa, S.; Costantini, M.; Barbetta, A.; Dokmeci, M.R.; Dentini, M.; Khademhosseini, A. Microfluidic bioprinting of heterogeneous 3D tissue constructs using low-viscosity bioink. *Adv. Mater.* **2016**, *28*, 677–684. [CrossRef]
192. Ding, S.; Feng, L.; Wu, J.; Zhu, F.; Yao, R.; Tan, Z. Bioprinting of stem cells: Interplay of bioprinting process, bioinks, and stem cell properties. *ACS Biomater. Sci. Eng.* **2018**, *4*, 3108–3124. [CrossRef] [PubMed]
193. Skoog, S.A.; Goering, P.L.; Narayan, R.J. Stereolithography in tissue engineering. *J. Mater. Sci. Mater. Med.* **2014**, *25*, 845–856. [CrossRef]
194. Li, J.; Wu, C.; Chu, P.K.; Gelinsky, M. 3D printing of hydrogels: Rational design strategies and emerging biomedical applications. *Mater. Sci. Eng. Rep.* **2020**, *140*, 100543. [CrossRef]
195. Lee, A.; Hudson, A.; Shiwarski, D.J.; Tashman, J.W.; Hinton, T.J.; Yerneni, S.S.; Bliley, J.M.; Campbell, P.G.; Feinberg, A.W. 3D bioprinting of collagen to rebuild components of the human heart. *Science* **2019**, *365*, 482–487. [CrossRef]
196. Mirdamadi, E.; Tashman, J.W.; Shiwarski, D.J.; Palchesko, R.N.; Feinberg, A.W. Fresh 3D bioprinting a full-size model of the human heart. *ACS Biomater. Sci. Eng.* **2020**, *6*, 6453–6459. [CrossRef]
197. Kačarević, Z.P.; Rider, P.; Alkildani, S.; Retnasingh, S.; Smeets, R.; Jung, O.; Ivanišević, Z.; Barbeck, M. An introduction to 3D bioprinting: Possibilities, challenges and future aspects. *Materials* **2018**, *11*, 2199. [CrossRef]
198. Fedorovich, N.E.; Alblas, J.; De Wijn, J.R.; Hennink, W.E.; Verbout, A.J.; Dhert, W.J. Hydrogels as extracellular matrices for skeletal tissue engineering: State-of-the-Art and novel application in organ printing. *Tissue Eng.* **2007**, *13*, 1905–1925. [CrossRef]
199. Guvendiren, M.; Burdick, J.A. Engineering synthetic hydrogel microenvironments to instruct stem cells. *Curr. Opin. Biotechnol.* **2013**, *24*, 841–846. [CrossRef] [PubMed]
200. Dzobo, K.; Motaung, K.S.C.M.; Adesida, A. Recent trends in decellularized extracellular matrix bioinks for 3D printing: An updated review. *Int. J. Mol. Sci.* **2019**, *20*, 4628. [CrossRef] [PubMed]
201. Kabirian, F.; Mozafari, M. Decellularized ECM-derived bioinks: Prospects for the future. *Methods* **2020**, *171*, 108–118. [CrossRef] [PubMed]
202. Nam, S.Y.; Park, S.-H. *ECM-Based Bioink for Tissue Mimetic 3D Printing, in Biomimetic Medical Materials: From Nanotechnology to 3D Printing*; Noh, I., Ed.; Springer: Singapore, 2018; pp. 335–354.
203. Rickard, D.; Sullivan, T.; Shenker, B.; Leboy, P.; Kazhdan, I. Induction of rapid osteoblast differentiation in rat bone marrow stromal cell cultures by dexamethasone and BMP-2. *Dev. Biol.* **1994**, *161*, 218–228. [CrossRef] [PubMed]
204. Poldervaart, M.T.; Wang, H.; Van Der Stok, J.; Weinans, H.; Leeuwenburgh, S.C.G.; Öner, F.C.; Dhert, W.J.A.; Alblas, J. Sustained release of BMP-2 in bioprinted alginate for osteogenicity in mice and rats. *PLoS ONE* **2013**, *8*, e72610. [CrossRef] [PubMed]
205. Balasundaram, G.; Sato, M.; Webster, T.J. Using hydroxyapatite nanoparticles and decreased crystallinity to promote osteoblast adhesion similar to functionalizing with RGD. *Biomaterials* **2006**, *27*, 2798–2805. [CrossRef] [PubMed]
206. Kim, S.-S.; Park, M.S.; Jeon, O.; Choi, C.Y.; Kim, B.-S. Poly(lactide-co-glycolide)/hydroxyapatite composite scaffolds for bone tissue engineering. *Biomaterials* **2006**, *27*, 1399–1409. [CrossRef]
207. Zhou, X.; Zhu, W.; Nowicki, M.; Miao, S.; Cui, H.; Holmes, B.; Glazer, R.I.; Zhang, L.G. 3D bioprinting a cell-laden bone matrix for breast cancer metastasis study. *ACS Appl. Mater. Interfaces* **2016**, *8*, 30017–30026. [CrossRef]
208. Jakus, A.E.; Rutz, A.L.; Jordan, S.W.; Kannan, A.; Mitchell, S.M.; Yun, C.; Koube, K.D.; Yoo, S.C.; Whiteley, H.E.; Richter, C.-P.; et al. Hyperelastic "bone": A highly versatile, growth factor–free, osteoregenerative, scalable, and surgically friendly biomaterial. *Sci. Transl. Med.* **2016**, *8*, 358ra127. [CrossRef]
209. Huang, Y.-H.; Jakus, A.E.; Jordan, S.W.; Dumanian, Z.; Parker, K.; Zhao, L.; Patel, P.K.; Shah, R.N. Three-Dimensionally printed hyperelastic bone scaffolds accelerate bone regeneration in critical-size calvarial bone defects. *Plast. Reconstr. Surg.* **2019**, *143*, 1397–1407. [CrossRef]
210. Alluri, R.; Jakus, A.; Bougioukli, S.; Pannell, W.; Sugiyama, O.; Tang, A.; Shah, R.; Lieberman, J.R. 3D printed hyperelastic "bone" scaffolds and regional gene therapy: A novel approach to bone healing. *J. Biomed. Mater. Res. Part A* **2018**, *106*, 1104–1110. [CrossRef] [PubMed]
211. Plantz, M.A.; Hsu, W.K. Recent research advances in biologic bone graft materials for spine surgery. *Curr. Rev. Musculoskelet. Med.* **2020**, *13*, 318–325. [CrossRef] [PubMed]
212. Lewicki, J.; Bergman, J.; Kerins, C.; Hermanson, O. Optimization of 3D bioprinting of human neuroblastoma cells using sodium alginate hydrogel. *Bioprinting* **2019**, *16*, 00053. [CrossRef]
213. Ferreira, L.P.; Gaspar, V.M.; Mano, J.F. Decellularized extracellular matrix for bioengineering physiomimetic 3D in vitro tumor models. *Trends Biotechnol.* **2020**, *38*, 1397–1414. [CrossRef] [PubMed]
214. Zhao, Y.; Yao, R.; Ouyang, L.; Ding, H.; Zhang, T.; Zhang, K.; Cheng, S.; Sun, W. Three-Dimensional printing of Hela cells for cervical tumor model in vitro. *Biofabrication* **2014**, *6*, 035001. [CrossRef]

215. Zhao, S.; Zuo, W.-J.; Shao, Z.-M.; Jiang, Y.-Z. Molecular subtypes and precision treatment of triple-negative breast cancer. *Ann. Transl. Med.* **2020**, *8*, 499. [CrossRef]
216. Tang, M.; Xie, Q.; Gimple, R.C.; Prager, B.C.; Qiu, Z.; Schimelman, J.; Wang, P.; Lee, D.; Yu, A.; Miller, T.E.; et al. Abstract 320: 3D-bioprinting of biomimetic multicellular glioblastoma tissues enable modeling of tumor-immune interactions. *Mol. Cell. Biol. Genet.* **2020**, *80*, 320. [CrossRef]
217. Hakobyan, D.; Médina, C.; Dusserre, N.; Stachowicz, M.L.; Handschin, C.; Fricain, J.C.; Guillermet-Guibert, J.; Oliveira, H. Laser-Assisted 3D bioprinting of exocrine pancreas spheroid models for cancer initiation study. *Biofabrication* **2020**, *12*, 035001. [CrossRef]
218. Radhakrishnan, J.; Varadaraj, S.; Dash, S.K.; Sharma, A.; Verma, R.S. Organotypic cancer tissue models for drug screening: 3D constructs, bioprinting and microfluidic chips. *Drug Discov. Today* **2020**, *25*, 879–890. [CrossRef]
219. Zhu, W.; Castro, N.J.; Cui, H.; Zhou, X.; Boualam, B.; McGrane, R.; Glazer, R.I.; Zhang, L.G. A 3D printed nano bone matrix for characterization of breast cancer cell and osteoblast interactions. *Nanotechnology* **2016**, *27*, 315103. [CrossRef]
220. Kolb, A.D.; Bussard, K.M. The bone extracellular matrix as an ideal milieu for cancer cell metastases. *Cancers* **2019**, *11*, 1020. [CrossRef] [PubMed]
221. Ryan, D.G.; Murphy, M.P.; Frezza, C.; Prag, H.A.; Chouchani, E.T.; O'Neill, L.A.; Mills, E.L. Coupling Krebs cycle metabolites to signalling in immunity and cancer. *Nat. Metab.* **2019**, *1*, 16–33. [CrossRef] [PubMed]
222. Bahraminasab, M. Challenges on optimization of 3D-printed bone scaffolds. *Biomed. Eng. Online* **2020**, *19*, 1–33. [CrossRef]
223. Wang, L.; Xu, M.; Luo, L.; Zhou, Y.; Si, P. Iterative feedback bio-printing-derived cell-laden hydrogel scaffolds with optimal geometrical fidelity and cellular controllability. *Sci. Rep.* **2018**, *8*, 1–13. [CrossRef] [PubMed]
224. Wu, Y.; Heikal, L.; Ferns, G.A.; Ghezzi, P.; Nokhodchi, A.; Maniruzzaman, M. 3D bioprinting of novel biocompatible scaffolds for endothelial cell repair. *Polymers* **2019**, *11*, 1924. [CrossRef] [PubMed]
225. Koons, G.L.; Mikos, A.G. Progress in three-dimensional printing with growth factors. *J. Control. Release* **2019**, *295*, 50–59. [CrossRef] [PubMed]
226. Foresti, R.; Rossi, S.; Pinelli, S.; Alinovi, R.; Barozzi, M.; Sciancalepore, C.; Galetti, M.; Caffarra, C.; Lagonegro, P.; Scavia, G.; et al. Highly-Defined bioprinting of long-term vascularized scaffolds with Bio-Trap: Complex geometry functionalization and process parameters with computer aided tissue engineering. *Materialia* **2020**, *9*, 100560. [CrossRef]
227. Hanumantharao, S.N.; Que, C.A.; Vogl, B.J.; Rao, S. Engineered Three-Dimensional Scaffolds Modulating Fate of Breast Cancer Cells Using Stiffness and Morphology Related Cell Adhesion. *IEEE Open J. Eng. Med. Biol.* **2020**, *1*, 41–48. [CrossRef]
228. Freeman, F.E.; Pitacco, P.; Van Dommelen, L.H.A.; Nulty, J.; Browe, D.C.; Shin, J.-Y.; Alsberg, E.; Kelly, D.J. 3D bioprinting spatiotemporally defined patterns of growth factors to tightly control tissue regeneration. *Sci. Adv.* **2020**, *6*, eabb5093. [CrossRef]
229. Sun, B.; Lian, M.; Han, Y.; Mo, X.; Jiang, W.; Qiao, Z.; Dai, K. A 3D-Bioprinted dual growth factor-releasing intervertebral disc scaffold induces nucleus pulposus and annulus fibrosus reconstruction. *Bioact. Mater.* **2021**, *6*. [CrossRef]

cancers

MDPI

Review

Three-Dimensional Culture System of Cancer Cells Combined with Biomaterials for Drug Screening

Teruki Nii [1,2], Kimiko Makino [2,3] and Yasuhiko Tabata [1,*]

1 Laboratory of Biomaterials, Institute for Frontier Life and Medical Sciences, Kyoto University, 53 Kawara-cho Shogoin, Sakyo-ku, Kyoto 606-8507, Japan; nii.teruki.85x@st.kyoto-u.ac.jp or 3A18706@ed.tus.ac.jp
2 Faculty of Pharmaceutical Sciences, Tokyo University of Science, 2641 Yamazaki, Noda, Chiba 278-8510, Japan; makino@rs.noda.tus.ac.jp
3 Center for Drug Delivery Research, Tokyo University of Science, 2641 Yamazaki, Noda, Chiba 278-8510, Japan
* Correspondence: yasuhiko@infront.kyoto-u.ac.jp; Fax: +81-75-751-4646

Received: 26 August 2020; Accepted: 22 September 2020; Published: 24 September 2020

Simple Summary: For the research and development of drug discovery, it is of prime importance to construct the three-dimensional (3D) tissue models in vitro. To this end, the enhancement design of cell function and activity by making use of biomaterials is essential. In this review, 3D culture systems of cancer cells combined with several biomaterials for anticancer drug screening are introduced.

Abstract: Anticancer drug screening is one of the most important research and development processes to develop new drugs for cancer treatment. However, there is a problem resulting in gaps between the in vitro drug screening and preclinical or clinical study. This is mainly because the condition of cancer cell culture is quite different from that in vivo. As a trial to mimic the in vivo cancer environment, there has been some research on a three-dimensional (3D) culture system by making use of biomaterials. The 3D culture technologies enable us to give cancer cells an in vitro environment close to the in vivo condition. Cancer cells modified to replicate the in vivo cancer environment will promote the biological research or drug discovery of cancers. This review introduces the in vitro research of 3D cell culture systems with biomaterials in addition to a brief summary of the cancer environment.

Keywords: biomaterials; tissue engineering; 3D cell culture; cancer cells

1. Introduction

The basic concept of regenerative medicine is to achieve the regeneration and repairing of damaged or injured tissues by utilizing the natural healing potential of the body itself. Regenerative medicine consists of regenerative therapy and regenerative research. Regenerative therapy is to treat patients through the in vivo enhancement of cell activity. Regenerative research is positioned as the scientific support for the regeneration therapy of the next generation. Drug discovery is defined as regenerative research. The therapeutic efficacy, metabolism or toxicology of drugs are efficiently evaluated by taking advantage of activated cells. To enhance the cell activity, two methodologies have been recently noted. One is to utilize three-dimensional (3D) cell culture technologies. Cells are usually cultured in a two-dimensional (2D) system, with a plate or dish. However, the functions of cells cultured in the 2D system are lower than those of body cells because cells tend to interact with each other for the enhancement of their own activities in the body [1–4]. Due to the difference in the cell condition, the drug effect evaluated by the in vitro drug screening is not always the same as that in preclinical or clinical study, which leads to the failure of drug research and development [5,6] (Figure 1).

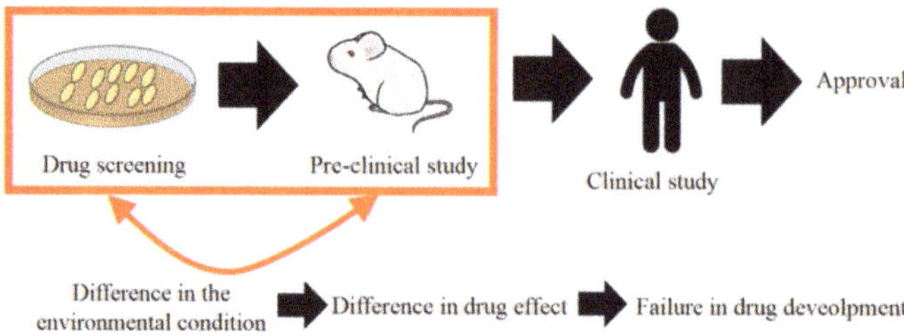

Figure 1. Research and development process of drug development. The difference in the environment condition between in vitro and in vivo leads to that in drug effects, which often causes a failure in drug development.

The comparison of cancer cell culture between 2D and 3D systems is shown in Table 1. There are merits or demerits between the two culture systems. Although the systems have been used depending on the purpose, the 3D culture is superior in terms of drug discovery which well reflects the in vivo cancer environment. The other methodology to enhance cell functions is the active utilization of biomaterials. Cell culture is often performed on the dish or plate which is mainly composed of polystyrene. This condition of an artificial environment is quite different from the in vivo body environment of cancer cells, and consequently, the drug effect or cytotoxicity evaluation is technologically limited. Biomaterials which consist of extracellular matrix (ECM) components are effective in enhancing the cell activity or functions. The interaction with biomaterials will enable cells to enhance their proliferation, differentiation, and biological functions, leading to the realization of cancer cell–environment interaction.

Table 1. Comparison of cancer cells culture between 2D and 3D systems.

Points Compared	Culture System	
	2D	3D
Cost	Low	High
Cell proliferation	High	Low
Cell differentiation	Low	High
Reproducibility	Good	Poor
In vivo imitation	Limited	Versatile
Cell–cell interaction	Low	High
Cell morphology change	Low	High
Diverse polarity	Loss	Diverse
ECM synthesis	Low	High
Drug sensitivity	High (in contrast to in vivo)	Low (Same as in vivo)

Anticancer drug screening is often performed by using the 2D culture system of cancer cells. As mentioned above, to mimic the cancer environment in the body, the combination of 3D cell culture technology and biomaterials is important. In addition to the technological methods, the interaction of cancer cells with stromal cells should be considered [7], because the cancer environment is composed of several stromal cells, such as cancer-associated fibroblasts (CAF) [8,9], tumor-associated macrophages (TAM) [10,11], mesenchymal stem cells (MSC) [12,13] or endothelial cells [14,15]. It has been demonstrated that cancer cells interact with stromal cells, leading to the promotion of cancer diseases [16] (Figure 2). Moreover, several humoral factors secreted from cells are also important to construct the cancer environment [17–19]. Therefore, to mimic the cancer environment or cancer diseases in vitro, a coculture system of cancer cells with stromal cells is essential.

Nowadays, to replicate the cancer environment and diseases in vitro, several studies have been reported on 3D cancer models combined with biomaterials. In this review, first, the important stromal cells and their characterization are briefly described. Second, we introduce 3D cancer models by making use of several biomaterials.

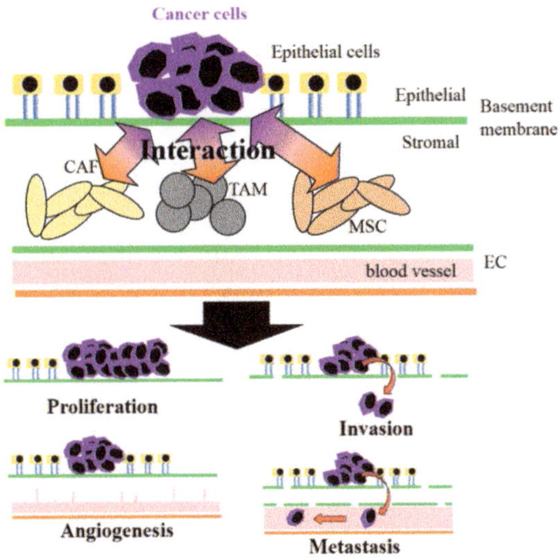

Figure 2. Cancer cells interact with various stromal cells of cancer-associated fibroblasts (CAF), tumor-associated macrophages (TAM), mesenchymal stem cells (MSC), and endothelial cells (EC), leading to the pathological maintenance and promotion of cancer characteristics.

2. Stromal Cells in Cancer Environment

There are four types of stromal cells which are composed of the cancer environment. The biological functions of stromal cells and the humoral factors secreted are briefly explained. Table 2 summarizes some key cytokines in the cancer environment.

Table 2. Cytokines secreted in cancer environment and the biological function.

Cytokines	Functions
Transforming growth factor-β (TGF-β)	Support of cancer cells proliferation Promotion of endothelial–mesenchymal transition (EMT) and the consequent invasion or metastasis Recruitment of fibroblasts Differentiation of fibroblasts or MSC into CAF Promotion of tumorigenicity Promotion of angiogenesis
Tumor necrosis factor-α (TNF-α)	Disruption of epithelial barrier Promotion of inflammatory cell infiltration Stimulation of TGF-β-induced EMT Induction of vascular endothelial growth factors (VEGF) secretion
Vascular endothelial growth factor (VEGF)	Promotion of angiogenesis ECM remodeling Promotion of inflammatory cytokine secretion Formation of tumor endothelial cells
Stromal derived factor-1 (SDF-1)	Promotion of angiogenesis by recruiting endothelial cell precursors Recruitment of MSC Promotion of cancer cells proliferation
Matrix metroproteinase (MMP)	ECM degradation and the consequent angiogenesis, invasion, and metastasis Promotion of tumorigenicity
Interuekin-6 (IL-6)	Stimulation of TGF-β-induced EMT Promotion of cancer cell proliferation Promotion of angiogenesis

2.1. Cancer-Associated Fibroblasts

Cancer-associated fibroblasts (CAF) are major stromal cells. CAF of a large-spindle shape are perpetually activated and never undergo apoptosis [8]. Although the origin of CAF is not completely clear, normal fibroblasts [20–22], mesenchymal stem cells (MSC) [23,24], or endothelial cells [25,26] are potential sources of CAF. As CAF markers, alpha-smooth muscle actin (α-SMA), fibroblast activation protein (FAP), and fibroblast specific protein-1 are well known [27]. In particular, approximately 90% of cancer cell types show the expression of FAP [28]. The interaction between cancer cells and CAF plays a key role in cancer diseases. An experimental trial to indicate the importance of CAF has been reported by Weinberg et al. Human CAF and breast cancer cells are injected to nude mice. It is demonstrated that cancer cells with CAF effectively proliferate compared with CAF-free cancer cells or cancer cells cocultured with normal fibroblasts groups. This proliferation enhancement was induced by stromal cell-derived factor-1 (SDF-1) secreted [29]. This study clearly indicates the importance of CAF existence for cancer cell activity. CAF not only promote cancer proliferation but also increase the invasion of cancer cells via the cancer–CAF interaction. The interaction also promotes the secretion of various matrix-degrading proteinases. Among them, matrix-metalloproteinase (MMP) has a key role in the cancer invasion or metastasis. MMP can degrade type IV collagen and laminin, which are major components of basement membrane [30–32]. In addition to SDF-1 and MMP, transforming growth factor-β1 (TGF-β1) [33,34] and interleukin (IL)-6 [35] are also important factors for the cancer–CAF interaction.

2.2. Tumor-Associated Macrophages

Macrophages are usually polarized to M1 or M2 phenotypes responding to the environment. M1 macrophages (proinflammatory) have a capacity of inflammation induction, chronic inflammation, and pathogen defense [36,37]. On the other hand, M2 macrophages (anti-inflammatory) are involved in noninflammatory response, wound healing, and tissue regeneration [37–39]. TAM are generally recognized as M2-type macrophages [40,41]. Due to the M2-type phenotype, CD163 and CD204 are well known as the TAM markers [42,43]. The stimulation of macrophages by lipopolysaccharide (LPS) and adenosines can induce TAM in vitro [44]. TAM play an important role in cancer progression. Grivennikov et al. indicate that IL-23 and IL-17 secreted from TAM promote the cancer proliferation [45]. Tumor-necrosis factor-α (TNF-α), vascular endothelial growth factor (VEGF), and TGF-β1 secreted from TAM can promote the cancer metastasis [46]. Taken together, TAM are recognized as important cells for cancer diseases. This promising TAM-targeted therapy has been investigated [47,48].

2.3. Cancer-Associated Fibroblasts and Tumor-Associated Macrophages for Different Cancer Types

CAF and TAM are major components of stromal cells in the cancer environment. However, their biological contribution and influence on cancer cells generally depend on the cancer regions. For example, in brain, liver, or kidney cancer, contribution of TAM is larger than that of CAF, while the effect of CAF on the lung or pancreatic cancer is high compared with that of TAM. This is mainly because of the existence ratio [49]. Therefore, the CAF/TAM contribution ratio should be considered to understand the characteristics of various cancer cell types.

2.4. Mesenchymal Stem Cells

Mesenchymal stem cells (MSC) have been noted in the field of tissue regeneration because MSC have a capacity of differentiation into bone, cartilage, or fat cells [50–52]. Therefore, MSC transplantation would be effective in regenerative medicine [53]. However, the differentiation capacity of MSC is unfavorable for cancer patients. For example, TGF-β1 secreted from several cells in the cancer environment can differentiate MSC into CAF [54]. Chowdhury et al. also report that exosomes secreted from cancer cells promote the differentiation MSC into CAF [55]. In addition to the differentiation into CAF, MSC also allow TAM to migrate into the cancer environment via C-C chemokine receptor type 2

(CCR2) [56]. Moreover, IL-6 and angiopoietin-1 secreted from primary human MSC can promote the angiogenesis [57]. Recently, it has been reported that MSC can polarize into a proinflammatory MSC-1 and an immunosuppressive MSC-2 phenotype. MSC-2 can enhance the cancer proliferation, spread, and promotion, while MSC-1 suppress the cancer proliferation [13,58,59]. The understanding of MSC roles at cancer sites would provide an important aspect for further cancer research and therapies.

2.5. Endothelial Cells

It is important for cancer cells to induce angiogenesis in terms of nutrient and oxygen supply, the elimination of waste products, invasion, and metastasis. However, since a vascularization suddenly advances at the cancer sites under a nonphysiological condition, it is well recognized that the blood vessels in the cancer environment are fragile and the wall is highly permeable. Enhanced permeation and retention effect (EPR effect) is a concept to symbolize this condition of cancer blood vessels [60]. Based on the EPR effect concept, a positive targeting of micelles containing anticancer drug to cancer has been reported [61,62]. Thus, there are some structural and functional differences between the cancer and normal blood vessels. To study cancer characteristics or therapeutic efficacy, the blood vessel properties and the cancer–endothelial cell interaction are important to consider. Some research has been reported to demonstrate that tumor endothelial cells (TEC) differ from normal endothelial cells in properties, such as the cell proliferation, the gene expression, the response to growth factors, or migration [63,64]. High metastatic tumor-derived TEC (HM-TEC) and low metastatic tumor-derived TEC (LM-TEC) can be isolated from mice. It is demonstrated that the secretion levels of VEGF, MMP-2, MMP-9, and SDF-1 from HM-TEC are higher than from that of LM-TEC [60,65]. It is reported that coculture with endothelial cells facilitates the in vitro culture of cancer cells [66].

3. 3D Culture System of Cancer Cells with Biomaterials

Biomaterials classify into natural biomaterials derived from animals or plants and synthetic biomaterials artificially prepared. Natural biomaterials are composed of polysaccharide (amylose, cellulose, alginate, chitosan, or hyaluronic acid), peptide (collagen or gelatin), nucleic acid, or polyhydroxyalkanoates. Since the degradative enzyme and metabolic system have already existed in the body, most natural biomaterials can enzymatically be degraded. Because the components constitute the cancer environment as the ECM and contribute to cancer diseases, natural biomaterials are often used to design the 3D culture system of cancer cells. Although natural biomaterials are of high biocompatible, there are some limitations of immunogenicity or homogeneity to use. To avoid the issues, synthetic biomaterials are used. Synthetic biomaterials are mainly degraded nonenzymatically based on simple hydrolysis. There are some merits of synthetic biomaterials, such as the characteristics control, the high stiffness, and the clarity of properties.

In this chapter, several 3D culture systems of cancer cells combined with biomaterials are introduced. To date, two types of biomaterials have been applied to the 3D culture system of cancer cells. One is the culture system of cancer cells with the biomaterials of a spherical shape. When incubated with microspheric hydrogels of biomaterial, cancer cells naturally form a cell aggregate of a tissue-like 3D structure, which mimics the cancer environment. The disadvantages of this system are the difficulty of cells separation from the cell-hydrogel aggregates, and consequently, the result is often of low repeatability. The other is the culture system of cancer cells with the biomaterials of nonspherical type, such as sponge shapes or nonwoven fabrics. In this system, cells effectively proliferate and migrate on the scaffold. This is suitable for immunohistochemical analysis. Table 3 summarizes the 3D culture systems of cancer cells combined with various types of biomaterials.

Table 3. 3D culture system of cancer cells combined with biomaterials.

Biomaterials	Characteristics	Types of Cancer Cells Cultured with Biomaterial Scaffolds of Spherical or Other Shapes		Stromal Cells Cocultured with Cancer Cells
		Spherical [(a)]	Other (Sponges Shapes or Nonwoven Fabrics) [(b)]	
Chitosan	Derived from crustacean shells Linear cationic polymer Formation of polyelectrolyte complexes with anionic polymers		Breast cancer [67] Liver cancer [68] Glioblastoma [69–72] Lung cancer [73,74] Prostate cancer [75–77]	MSC [73]
Alginate	Derived from seaweed Water-soluble Crosslinked by ions Easy cell encapsulation Nonadhesive nature to cells Easy stiffness control Thermally stable High water-holding capacity	Breast cancer [78,79] Liver cancer [80,81] Head and neck squamous cell carcinoma [82] Leukemia [83]	Liver cancer [68] Breast cancer [84,85] Glioblastoma [71,72] Prostate cancer [75,76] Oral squamous cell carcinoma [84] Lung cancer [84] Gastric cancer [84]	Fibroblasts [78,85] MSC [81]
Collagen	A major component of ECM Low inflammation High cell adhesion Biodegradability Affinity for integrin receptor	Breast cancer [86]	Breast cancer [85,87–91] Prostate cancer [92] Pancreatic cancer [93] Lung cancer [93–95]	CAF [89,93] Macrophages [94,95] Fibroblasts [85,93–95]
Gelatin	Denatured material of collagen Water-soluble Crosslinked by chemical or thermal methods Biodegradability High water-holding capacity Affinity for integrin receptor	Breast cancer [66,96–99] Lung cancer [96,100,101] Liver cancer [96] Pancreatic cancer [102]		CAF [96–102] TAM [96] Fibroblasts [66,97,102] Endothelial cells [66]
Hyaluronic acid	A major component of ECM Water-soluble Affinity for CD44 receptor High water-holding capacity High molecular weight affects the biological functions.		Glioblastoma [69,70,103] Lung cancer [73,74,104] Gastric cancer [103,104] Prostate cancer [103,105] Osteosarcoma [103] Liver cancer [103] Breast cancer [103] Glioblastoma [106] Endometrial adenocarcinoma [105]	MSC [73] Endometrial stromal sarcoma [105]
Matrigel	Alternative material of basement membrane Derived from mouse tumors Layer used for Boyden chamber Suitable for invasion assay	Breast cancer [79,86]	Breast cancer [107–113] Fibrosarcoma [109,114] Melanoma [109]	Fibroblasts [108,113] T_{REG} lymphocyte [111] NK cells [111] MSC [112] Endothelial cells [113]
Poly (lactic-*co*-glycolic acid)	Porosity morphology Biodegradability Hydrophobic property	Ovarian cancer [115] Breast cancer [116]	Breast cancer [117,118] Prostate cancer [118] Melanoma [118] Ovarian cancer [118] Lung cancer [118] Liver cancer [119]	
Polyethylene glycol	Chemical modification Water-holding capacity	Breast cancer [120–122] Lung cancer [123] Prostate cancer [122] Colon cancer [122]	Breast cancer [91,118,124] Lung cancer [118] Melanoma [118] Ovarian cancer [118] Prostate cancer [118,125] Fibrosarcoma [126] Glioblastoma [106]	Fibroblasts [123] Endothelial cells [123]

[(a)] 3D cell constructs are readily formed; [(b)] cells well proliferate and migrate on the scaffold.

3.1. Chitosan

Chitosan of poly (1, 4 D-glucosamine), a partially deacetylated derivative of chitin, is a natural cationic linear polysaccharide [127]. Chitin is known as primary structural polymers in arthropod exoskeletons. The antigenic response of chitosan is rather low among organonitrogen compounds, and the stiffness is also enough for the cell scaffold. Therefore, chitosan is used as a blood anticoagulant [128], a wound healing accelerator [129], and a surgical suture [130] and also for cardiac [131], neural [132], bone [133], or vein endothelial [134] tissue engineering. Chitosan is also an effective biomaterial for 3D culture of cancer cells because glycosaminoglycan (GAG), closely to the structure of chitosan, is one major component of ECM in the cancer environment [135]. A chitosan scaffold is reported for the 3D culture system of cancer cells. When human breast MCF-7 cancer cells were cultured on the chitosan scaffold, the cell attachment and proliferation were superior to the regular culture of plastic dish [67].

3.2. Alginate

Alginate, purified from seaweed, is a naturally-occurring anionic polysaccharide composed of α-L-guluronic acid and β-D-mannuronic acid [136]. As a pharmaceutical application, sodium alginate has already been used for the treatment of peptic ulcer [137]. One of the alginate merits is the quick gelation or cell encapsulation by ionic crosslinking using divalent metal ions of calcium or ferric ions [138,139]. Second, alginate is thermally stable [140]. The molecular structure of alginate is similar to that of polysaccharide in vivo [141]. Therefore, for the 3D culture system of cancer cells, there are many studies on the encapsulation of cancer cells by using alginate gels. Liu et al. prepare alginate gels to encapsulate head and neck squamous carcinoma cells. In addition, three types of gels with different stiffness are prepared by changing the alginate concentration. It is found that the tumorigenicity, the metastatic ability, and the drug resistance increased at the moderate stiffness [82]. The system is also applied to not only neck squamous cell carcinoma but also the hepatocellular carcinoma reaction [80]. In addition, it is reported that IL-8, inflammatory cytokines, secreted from cancer cells cultured within alginate gels under the hypoxia, was high compared with in 2D culture system [84]. Alginate is widely used as a material of cell encapsulation or scaffold for the 3D culture system of cancer cells.

3.3. Collagen

Collagen is the main protein of most tissues and contributes to the physical support of tissues [142]. Therefore, collagen is widely used as a material for nerve [143–145], bone [146–148], cartilage [149–152], tendon [153], ligament [154,155], or skin [156,157] tissue engineering. Chen et al. report that the expression of proangiogenic growth factors and the transcript of MMP of human breast MCF-7 cancer cells cultured on collagen sponges increased [158]. For the 3D cancer cell culture, collagen is often used to evaluate the invasion ability of breast cancer cells. This may be mainly because it has been reported that breast cancer cells prefer to migrate into collagen I [86]. When high-invasive breast MDA-MB-231 cancer cells were cultured on a collagen scaffold, the migration ability increased via the epithelial–mesenchymal transition (EMT) [88]. For the bone metastasis models, Bersini et al. prepared collagen hydrogels containing osteoblasts cells on a microfluidic device. Human breast MDA-MB-231 cancer cells were invaded into the collagen hydrogels embedding osteoblasts cells effectively via the CXCL5/CXCR2 system compared with the collagen hydrogel without cells [90]. It is demonstrated that the migration ability of breast cancer cells was induced by the degree of collagen fiber alignment or the fibril bending stiffness of the collagen matrix [87].

3.4. Hyaluronic Acid

Mucopolysaccharide, namely GAG, repeating units of amino acid and uronic acid, is a major ECM component in connective, epithelial, and neural tissues. Hyaluronic acid (HA) is a GAG family and is composed of D-glucuronic acid and D-N-acetylglucosamine [159,160]. The advantageous characteristic of HA is recognized by the CD44 surface receptor [161]. The interaction between HA and cells via the CD44 receptor affects the cell functions [162]. For cancer, the HA-CD44 interaction leads to the cancer invasion [163], MMP-2 secretion [164], RhoGTPase activation or c-Src phosphorylation [165], and the expression of TGF-β1 and basic fibroblast growth factor (b-FGF) [166]. Moreover, HA affects the stemness maintenance of cancer cells, leading to tumorigenesis, EMT, or drug resistance because CD44 is a major surface marker for stem cells [167,168]. It has been demonstrated that the higher expression of HA in the cancer environment increased the cancer progression, leading to the poor mortality rate [169]. In addition, the molecular weight of HA is also one of the most important factors for cell response. Rayahin et al. report that the molecular weight of HA affects the macrophage phenotypes. At a low molecular weight (5 kDa), the secretion of TNF-α and nitrite production increased. HA of high molecular weight (3 MDa) enhanced the alginase activity which is the characteristic of M2-type macrophages [170]. Therefore, when HA is selected for a 3D cell culture system, the molecular weight of HA should be sufficiently considered because macrophage phenotypes affect the characterization of cancer cells. David et al. report a 3D culture system of cancer cells by use of HA hydrogels crosslinked with adipic dihydrazide to evaluate the invasion ability of several cancer cell lines [103]. It is found by the same groups that the drug resistance enhanced on the same culture systems compared with that in the 2D culture [104].

3.5. Matrigel

Basement membrane (BM), a thin layer of ECM, is between the epithelial and stromal sites [171] (Figure 2). BM has a major role in tissue integrity, specificity, and separation [172]. The components of BM are collagen type IV, laminin, heparan sulfate proteoglycan, various growth factors, cytokines, and chemokines [173]. Although BM is an essential material for biological research, human BM of physiological integrity cannot be obtained. As an alternative, matrigel, an extract of Engelbreth–Holm–Swarm tumor derived from wild mice, is used in vitro and in vivo [173]. The major component of matrigel is laminin-111, and gelation is formed at 37 °C [174].

Kramer et al. report on the investigation method of human HT1080 fibrosarcoma cells by use of matrigel [114]. After that, matrigel is often used for cancer invasion assay [109,110]. Matrigel enables the evaluation of not only the cancer invasion ability but also morphology. High-invasive MDA-MB-231 breast cancer cells cultured on matrigel grew, forming a star-like appearance (invasive characterization), while near-sphere cell aggregates were formed when low-invasive breast MCF-7 cancer cells were cultured [107]. Nowadays, the Boyden chamber has been developed to widely investigate cancer invasion as a reliable method [175–177]. The two chambers are separated via matrigel-coated porous filter. Cancer cells are plated in the upper chamber, while the medium with or without invasion modulators are in the feeder chamber. When the high-invasion cancer cells are plated, the filter is degraded, leading to the migration of cancer cells and their localization on the feeder surface of filter. Cancer cells migrated are easily counted by the trypan blue stain or fluorescence intensity. The merit of this assay is not to take a long time (12–24 h) to evaluate [171]. The Boyden chamber is a powerful tool to evaluate the cancer invasion ability or perform a drug screening.

3.6. Poly (Lactic-Co-Glycolic Acid)

Poly (lactic-*co*-glycolic acid) (PLGA) of biodegradable lactic acid (LA) and glycolic acid (GA) copolymers are widely used for biomedical applications [178]. As an example, leuprolide-loaded PLGA microparticles are used for the treatment of breast or prostate cancer. The microparticles realize an extended release of leuprorelin, which enables once every few months [179]. The basic properties of PLGA are usually given by molecular weight and the LA/GA ratio. For example, PLGA7520 indicates a copolymer of 20,000 molecular weight, and 75 wt % PLA and 25 wt % PGA. Both the molecular weight and LA/GA ratio determine the crystallinity or glass transition temperature [180], which enables the control of the size, porosity, or stiffness of PLGA particles or scaffolds easily [178,181–184].

Due to the easiness of the functional control, PLGA particles or scaffolds are also used for the 3D culture system of cancer cells. Sahoo et al. prepare PLGA scaffolds for the human breast MCF-7 cancer cell line by a solvent evaporation method. Since the PLGA scaffolds are hydrophobic, the difficulty of wetting and swelling in the culture medium is often a problem. The incorporation of poly (vinyl alcohol) (PVA) into the scaffolds enhanced the hydrophilic nature, leading to improved cell adherence and proliferation [116]. Besides breast cancer cells, several PLGA sponges have been prepared for a cell line of human liver Hep3B cancer by changing the LA/GA ratio. The sponges were prepared by a supercritical CO_2 gas-foaming method. The growth, mitochondrial activity, DNA amounts, hepatic function, and invasion ability of Hep3B cells on the sponges became maximum at the ratio of 85/15 [119]. In addition, PLGA porous microparticles have been prepared for ovarian HO-8910 cancer cell growth [115].

3.7. Polyethylene Glycol

Polyethylene glycol (PEG) is widely used for chemical modification in the field of drug delivery system or biomaterials [185]. PEG-based hydrogels are studied for the 3D cell culture system to investigate the migration of human fibrosarcoma HT-1080 cell line [126] or to mimic the prostate cancer environment [125]. PEG scaffolds in a layer-by-layer fashion with tunable stiffness are reported to evaluate the cell mortality [124]. In addition, Yang et al. report that the mouse breast 4T1 cancer cells are encapsulated in inert PEG hydrogels. The PEG hydrogels enabled cancer cells to form tumorspheres and maintain the cancer stemness [120].

4. 3D Culture System of Cancer Cells with Combination of Several Biomaterials

Considering unique properties and functions of each biomaterial, different biomaterials are often combined to use for 3D culture system of cancer cells. In this chapter, the 3D culture systems of cancer cells with combined biomaterials are introduced.

4.1. Chitosan–Alginate

Chitosan forms insoluble ionic complexes with alginate to improve the mechanical strength or replicate cancer environment [186–188]. Chitosan and alginate (CA) hybrid materials are used to create a 3D material with an interconnected and porous structure. The CA materials have a mechanical strength and shape maintenance significantly improved as compared with chitosan only. This is due to the electrostatic interaction between the amine groups of chitosan and the carboxyl groups of alginate [189]. When human liver HepG2 cancer cells were cultured on the CA scaffolds, both the malignancy and drug resistance increased [68]. The CA scaffolds can be applied not only for hepatocellular carcinoma cells, but also for human glioblastoma U-87 MG and U-118 MG cell lines. The expression levels of genes involved in EMT or cancer stem cells were rapidly promoted [71,72].

4.2. Chitosan–Hyaluronic Acid

The mixed hydrogel of chitosan and hyaluronic acid (CH) is often used as a nonadhesive material for spheroids formation. The CH has an ability to maintain the stemness of MSC spheroids through the Rho/Rock activation. A short time of spheroid formation and the enlargement of spheroid size were achieved compared with the conventional culture system [190]. When the 3D spheroids of human nonsmall cell lung cancer cells were prepared on the CH membrane, the expression level of EMT marker, the stemness, or the drug resistance increased compared with those of cells in the 2D culture system [74]. In addition, upon culturing on the CH scaffolds, the expression of stem cell marker and drug resistance of 3D human glioblastoma cancer stem cells was enhanced [70]. A porous CH scaffold promoted the formation of cancer spheroids and their stemness [69].

4.3. Matrigel–Collagen or Alginate

Nguyen-Ngoc et al. formulate matrigel hydrogels embedding human breast cancer cell aggregates. Cancer cells are individually dissociated from aggregates to promote their invasion nature because matrigel gives cancer cells a suitable environment. Moreover, the addition of collagen type I into the matrigel increased further cancer invasion [86]. It is reported that the mixed alginate matrigel hydrogel (a mixing ratio of 50:50) enabled human breast cancer cells incorporated to replicate the cancer invasion [79].

4.4. Polyethylene Glycol–Other Biomaterials

For the formation of cancer cell scaffolds, PEG is often conjugated with various biomaterials of collagen [91], HA [106], PLGA [118], fibrin [123], and fibrinogen [121,122,185]. PEG/collagen hydrogels of interpenetrating network are prepared to investigate the functions of human breast cancer cells, such as their proliferation, viability, or migration [91]. PEG/HA hydrogels with different stiffness are prepared by changing the PEG concentration to investigate the behavior of brain cancer cells embedded into the hydrogels [106]. Lipke groups have intensively studied the function of cancer cells cultured with PEG/fibrinogen materials [121,122]. Fibrinogen is one of the ECM components and has an important role in the polymerization or deposition of collagen [191]. Breast cancers [121] and colon or prostate cancer cells [122] are embedded in the 3D PEG/fibrinogen hydrogel to experimentally confirm the possibility of a long-time culture. Girard et al. culture several cancer cells on the 3D nanofibers of PLGA-PLA-PEG. Tight irregular aggregates were formed similarly to those of cancers in vivo, and the EMT was induced [118].

5. 3D Coculture System of Cancer and Stromal Cells Combined with Biomaterials

5.1. Alginate

Coculture of cancer cells and stromal cells with alginate has been investigated. Alginate hydrogels encapsulating human breast MCF-7 cancer cell aggregates were cocultured with human fibroblasts. The oestrogen receptor and the membrane E-cadherin expression increased, the polarity was lost, and the cell migration and angiogenesis increased, in contrast to the monoculture of MCF-7 cells [78]. These phenotypic alterations are important at the advanced stage of cancer. Liu et al. embed hepatocellular carcinoma in the algiante hydrogels, and then, the hydrogels are cocultured with MSC. In this culture system, efficient induction of EMT and the metastasis of cancer cells via TGF-β were observed [81].

5.2. Collagen

Nikkhah groups prepare a 3D microengineered cancer model composed of breast cancer cells and CAF embedded into collagen hydrogels. This culture system enabled cancer cells and CAF to achieve their interaction in vitro, which leads to better evaluation of invasion level of cancer cells, MMP secretion, and drug resistance [89]. 3D lung or pancreatic cancer cell aggregates embedded in collagen hydrogels are cocultured with CAF. Cancer cells were attached to CAF and quickly migrated on the CAF protrusions, while CAF-free cancer cells hardly invaded into the matrix [93].

5.3. Gelatin

Collagen of one ECM components is often used in the research field of 3D cell culture. However, collagen is water-insoluble and has biological activities, such as blood coagulation and a specific affinity for humoral factors. Considered as a material to design the cell culture system, the inherent properties are sometimes not suitable. Gelatin, a denatured form of collagen, is a cell friendly (high cell adhesion and low inflammation induction) material and is water-soluble [192,193]. In addition, it is technologically easy to prepare gelatin with various physicochemical properties by changing the preparation process from collagen [194,195]. Hydrogel formulations of water-insoluble gelatin can be freely prepared by the physical or chemical crosslinking methods, while the degradation profile can be modified as well [194,196]. The gelatin material is used for a coculture system of cancer cells and stromal cells. Netti groups have extensively investigated cancer microtissues by use of gelatin porous microbeads (GPM). Gelatin scaffolds with interconnected pores of about 20 μm diameter are designed for a 3D culture system, and the microtissues of cancer are formulated [197]. 3D CAF microtissues with GPM showed the higher deposition of collagen, fibronectin, and hyaluronic acid than that of GPM-free 3D CAF. GPM are effective materials to replicate the 3D cancer-stroma condition in vitro [97]. Moreover, human MCF-7 breast cancer and CAF microtissues with GPM are prepared to mimic the cancer microenvironment. The diffusion coefficient of anticancer drugs and the drug action for the 3D MCF-7-CAF microtissues with GPM were higher than those for the GPM-free 3D MCF-7-CAF. In addition, there was a good correlation of the expression of some cancer biomarkers related to cell junctions between the 3D MCF-7-CAF microtissues combined with GPM and in vivo cancer site [98]. The combination of endothelial cells with the culture system is reported [66].

5.4. Hyaluronic Acid

As a coculture system of cancer cells and stromal cells with HA, a multilayer system of high-invasive prostate C4–2B cancer cells, or endometrial Ishikawa cancer cells and stromal cells with HA hydrogels is reported. This culture system enables the evaluation of the cytotoxicity of compounds used clinically for both prostate and endometrial cancer cells in vitro. In addition, it is technically possible to anticipate and identify drugs that fail in clinical trials [105]. Han et al. prepare multicellular spheroids of human cell lung carcinoma cell line A549 and human MSC isolated from adipose tissue on CH coating plates. It is found that the gene expression levels of tumorigenicity markers in cancer cells associated with cancer stemness, EMT property, and cell mobility were up-regulated in the MSC-tumor multicellular spheroids [73].

5.5. Matrigel

There are several reports on matrigel-assisted coculture systems with stromal cells, such as fibroblasts [108], regulatory T lymphocyte (T_{REG} lymphocyte) or natural killer cells (NK cells) [111], and MSC [112]. Augustine et al. culture both T_{REG} lymphocytes and NK cells with luminal phenotype MCF-7 and basal phenotype MDA-MB-231 to study the immune reaction of breast cancer progression. Cancer morphology, the expression of biomarkers, and CC-chemokine 4 (CCL4) secretion were influenced by the phenotype of breast cancer cells and their immune stimulation [111]. MSC are cocultured with estrogen receptor-positive breast cancer cells embedded in matrigel. Cancer cells rapidly proliferated compared with the MSC-free cells [112].

5.6. Collagen–Alginate

Mixed hydrogels of collagen and alginate are investigated to form the multicellular spheroids of human breast cancer cells and fibroblasts. The hydrogel system developed in this study enables the control of the stiffness without altering the major gel components, since the concentration of alginate and collagen in the hydrogel remains constant. The change in the degree of calcium crosslinking does not affect the cell adhesion on the collagen network [85]. Alginate has been extensively used as a material whose stiffness can be readily regulated.

An increase in ECM stiffness is involved in the cancer progression [198]. In addition, there have been reports on the relationship between the stiffness and drug resistance [199,200]. Based on these findings, it is important to design the 3D culture system of cancer cells by making use of biomaterials of which the stiffness can be changed. It has been recently reported that the stiffness of biomaterials affects the characteristics of cancer cells, such as drug resistance [80,201–203]. It is promising for the 3D coculture of cancer and stromal cells to use biomaterials of the right material for the right place.

6. 3D Coculture System of Cancer and Stromal Cells Combined with Biomaterials of Drug Delivery System

The drug delivery system (DDS) is defined as a technology and methodology to enhance the biological activities of drugs or reduce the adverse effects by appropriately combining with biomaterials. To date, the DDS has been mainly used for in vivo cancer therapy through drug delivery [62,204,205]. However, the technology and methodology are also applicable for drug screening because cancer–environmental normal cell interaction is biologically supported by humoral factors secreted from the cells [8,12,13,16,19,27,33]. The combination of humoral factors in the DDS will enable the enhancement of the interaction between cancer and stromal cells which physiologically takes place in the body.

Gelatin hydrogel microspheres (GM) for regenerative medicine have been explored. GM can incorporate various growth factors, such as b-FGF [206–209], TGF-β1 [100,210,211], insulin-like growth factor-1 [212,213], or SDF-1 [214] for controlled release. Growth factors and gelatin molecules effectively interact by physicochemical interaction (e.g., ionic or hydrogen interaction) [194]. Due to the interaction, the mechanism of gelatin matrix-degradation-driven drug release is achievable. This is different from the conventional release system where the drug is usually released from release matrices by the drug diffusion. In addition, GM are in vivo and in vitro enzymatically degraded with time, and finally disappear. The characteristic behavior of GM disappearance is essential as a material for drug release used for tissue regeneration. To repair the damaged tissues, cells should migrate, proliferate, and differentiate. If drug release materials remain for a long time period after drug release is completed, the material remaining will cause the physical impairment of tissue regeneration. The speed of tissue regeneration should be synchronized to that of material degradation. Taken together, the growth factor release as the result of GM degradation with time is effective in realizing tissue regeneration based on the cell activity enhancement for natural healing potential [101,193,215–219]. In addition, a water phase of GM matrices is a pathway to permeate oxygen or nutrients [220]. This permeability is very important considering the 3D cell culture because cells in cell aggregates easily die because of the lack of oxygen or nutrients [221–223]. As a trial to break through the issue and culture 3D cell aggregates for a long time period, GM incorporation into the aggregates has been attempted [224–226]. Moreover, to enhance the cell activity, drugs to activate the cell function can be impregnated into GM for sustained release. Incorporation of GM containing drugs in cell aggregates is useful to give cells cultured in the 3D system a better condition. It is reported that CAF aggregates incorporating GM containing TGF-β1 (3D CAF-GM-TGF-β1) showed an activated function of CAF. When the activated CAF aggregates and cancer cells were cocultured via a model basement membrane, the invasion rate of cancer cells through the membrane was significantly higher than that of 2D cultured CAF (Figure 3) [100]. The findings indicate that the combination of 3D cell culture and DDS technology is promising to enhance the activity of cancer cells in the 3D culture system. TAM aggregates incorporating GM containing adenosines (3D TAM-GM-adenosines) were formulated to activate and maintain TAM functions. It is found that a 3D cancer cell coculture system of combined 3D CAF-GM-TGF-β1 and 3D TAM-GM-adenosines enabled the effective evaluation of the in vitro invasion of various cancer cells [96].

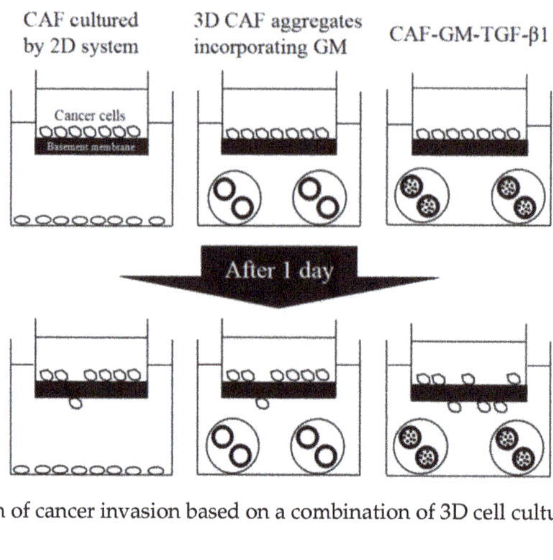

Figure 3. Illustration of cancer invasion based on a combination of 3D cell culture and drug delivery system technology.

The body tissue fundamentally consists of cells and the surrounding environment. The environment generally is made of ECM and nutrients for cells. In the case that the two factors of cell environment were not biologically sufficient, the functions of cells would rapidly decrease. The gelatin hydrogel microspheres (GM) function not only as the cell scaffold, but also as the release carrier of TGF-β1 and adenosines of nutrients for CAF and TAM.

7. Future Prospective and Conclusion

Biomaterials can assist the 3D culture system of cancer cells through the biological induction of ECM components. Several studies have reported on 3D culture systems by taking advantage of biomaterials. For further development of the 3D culture system of cancer cells, several biomaterials should be combined considering their unique properties and functions. In addition, substantial and close interaction between tissue engineering and the biological research of cancer cells or cancer environment would bring about further development of the 3D cell culture system for anticancer drug screening. In future, patient-derived cancer cells or stromal cells should be combined with biomaterials selected to allow the culture system to approach a more realistic cancer environment. The 3D culture system with biomaterials is a promising tool for cancer research and anticancer drug screening.

Author Contributions: Conceptualization, T.N., K.M., and Y.T.; writing—original draft preparation, T.N. and Y.T.; writing—review and editing, T.N. and Y.T. All authors have read and agreed to the published version of the manuscript.

Funding: This research received no external funding.

Conflicts of Interest: The authors declare no conflict of interest.

References

1. Fukuda, J.; Sakai, Y.; Nakazawa, K. Novel hepatocyte culture system developed using microfabrication and collagen/polyethylene glycol microcontact printing. *Biomaterials* **2006**, *27*, 1061–1070. [CrossRef] [PubMed]
2. Rodriguez-Enriquez, S.; Gallardo-Perez, J.C.; Aviles-Salas, A.; Marin-Hernandez, A.; Carreno-Fuentes, L.; Maldonado-Lagunas, V.; Moreno-Sanchez, R. Energy metabolism transition in multi-cellular human tumor spheroids. *J. Cell Physiol.* **2008**, *216*, 189–197. [CrossRef] [PubMed]
3. Kurosawa, H. Methods for inducing embryoid body formation: In vitro differentiation system of embryonic stem cells. *J. Biosci. Bioeng.* **2007**, *103*, 389–398. [CrossRef] [PubMed]
4. Lin, R.Z.; Chang, H.Y. Recent advances in three-dimensional multicellular spheroid culture for biomedical research. *Biotechnol. J.* **2008**, *3*, 1172–1184. [CrossRef]
5. Breslin, S.; O'Driscoll, L. Three-dimensional cell culture: The missing link in drug discovery. *Drug Discov. Today* **2013**, *18*, 240–249. [CrossRef]
6. Hait, W.N. Anticancer drug development: The grand challenges. *Nat. Rev. Drug Discov.* **2010**, *9*, 253–254. [CrossRef]
7. Burdett, E.; Kasper, F.K.; Mikos, A.G.; Ludwig, J.A. Engineering Tumors: A Tissue Engineering Perspective in Cancer Biology. *Tissue Eng. Part B Rev.* **2010**, *16*, 351–359. [CrossRef]
8. Shiga, K.; Hara, M.; Nagasaki, T.; Sato, T.; Takahashi, H.; Takeyama, H. Cancer-Associated Fibroblasts: Their Characteristics and Their Roles in Tumor Growth. *Cancers* **2015**, *7*, 2443–2458. [CrossRef]
9. Kalluri, R. The biology and function of fibroblasts in cancer. *Nat. Rev. Cancer* **2016**, *16*, 582–598. [CrossRef]
10. Kim, J.; Bae, J.S. Tumor-Associated Macrophages and Neutrophils in Tumor Microenvironment. *Mediat. Inflamm.* **2016**, *2016*, 6058147. [CrossRef]
11. Yang, L.; Zhang, Y. Tumor-associated macrophages: From basic research to clinical application. *J. Hematol. Oncol.* **2017**, *10*, 58. [CrossRef] [PubMed]
12. Melzer, C.; Yang, Y.; Hass, R. Interaction of MSC with tumor cells. *Cell Commun. Signal.* **2016**, *14*, 20. [CrossRef] [PubMed]
13. Barcellos-de-Souza, P.; Gori, V.; Bambi, F.; Chiarugi, P. Tumor microenvironment: Bone marrow-mesenchymal stem cells as key players. *Biochim. Biophys. Acta* **2013**, *1836*, 321–335. [CrossRef] [PubMed]
14. Baluk, P.; Hashizume, H.; McDonald, D.M. Cellular abnormalities of blood vessels as targets in cancer. *Curr. Opin. Genet. Dev.* **2005**, *15*, 102–111. [CrossRef] [PubMed]
15. Matsuda, K.; Ohga, N.; Hida, Y.; Muraki, C.; Tsuchiya, K.; Kurosu, T.; Akino, T.; Shih, S.C.; Totsuka, Y.; Klagsbrun, M.; et al. Isolated tumor endothelial cells maintain specific character during long-term culture. *Biochem. Biophys. Res. Commun.* **2010**, *394*, 947–954. [CrossRef]
16. Mbeunkui, F.; Johann, D.J., Jr. Cancer and the tumor microenvironment: A review of an essential relationship. *Cancer Chemother. Pharmacol.* **2009**, *63*, 571–582. [CrossRef]

17. Kessenbrock, K.; Plaks, V.; Werb, Z. Matrix metalloproteinases: Regulators of the tumor microenvironment. *Cell* **2010**, *141*, 52–67. [CrossRef]
18. Whiteside, T.L. The tumor microenvironment and its role in promoting tumor growth. *Oncogene* **2008**, *27*, 5904–5912. [CrossRef]
19. Chanmee, T.; Ontong, P.; Konno, K.; Itano, N. Tumor-associated macrophages as major players in the tumor microenvironment. *Cancers* **2014**, *6*, 1670–1690. [CrossRef]
20. Mitra, A.K.; Zillhardt, M.; Hua, Y.; Tiwari, P.; Murmann, A.E.; Peter, M.E.; Lengyel, E. MicroRNAs reprogram normal fibroblasts into cancer-associated fibroblasts in ovarian cancer. *Cancer Discov.* **2012**, *2*, 1100–1108. [CrossRef]
21. Kojima, Y.; Acar, A.; Eaton, E.N.; Mellody, K.T.; Scheel, C.; Ben-Porath, I.; Onder, T.T.; Wang, Z.C.; Richardson, A.L.; Weinberg, R.A.; et al. Autocrine TGF-beta and stromal cell-derived factor-1 (SDF-1) signaling drives the evolution of tumor-promoting mammary stromal myofibroblasts. *Proc. Natl. Acad. Sci. USA* **2010**, *107*, 20009–20014. [PubMed]
22. Wen, S.; Niu, Y.; Yeh, S.; Chang, C. BM-MSCs promote prostate cancer progression via the conversion of normal fibroblasts to cancer-associated fibroblasts. *Int. J. Oncol.* **2015**, *47*, 719–727. [PubMed]
23. Quante, M.; Tu, S.P.; Tomita, H.; Gonda, T.; Wang, S.S.; Takashi, S.; Baik, G.H.; Shibata, W.; Diprete, B.; Betz, K.S.; et al. Bone marrow-derived myofibroblasts contribute to the mesenchymal stem cell niche and promote tumor growth. *Cancer Cell* **2011**, *19*, 257–272. [PubMed]
24. Direkze, N.C.; Hodivala-Dilke, K.; Jeffery, R.; Hunt, T.; Poulsom, R.; Oukrif, D.; Alison, M.R.; Wright, N.A. Bone marrow contribution to tumor-associated myofibroblasts and fibroblasts. *Cancer Res.* **2004**, *64*, 8492–8495.
25. Zeisberg, E.M.; Tarnavski, O.; Zeisberg, M.; Dorfman, A.L.; McMullen, J.R.; Gustafsson, E.; Chandraker, A.; Yuan, X.; Pu, W.T.; Roberts, A.B.; et al. Endothelial-to-mesenchymal transition contributes to cardiac fibrosis. *Nat. Med.* **2007**, *13*, 952–961.
26. Lin, F.; Wang, N.; Zhang, T.C. The role of endothelial-mesenchymal transition in development and pathological process. *IUBMB Life* **2012**, *64*, 717–723.
27. Orimo, A.; Weinberg, R.A. Heterogeneity of stromal fibroblasts in tumors. *Cancer Biol. Ther.* **2007**, *6*, 618–619.
28. Garin-Chesa, P.; Old, L.J.; Rettig, W.J. Cell surface glycoprotein of reactive stromal fibroblasts as a potential antibody target in human epithelial cancers. *Proc. Natl. Acad. Sci. USA* **1990**, *87*, 7235–7239.
29. Orimo, A.; Gupta, P.B.; Sgroi, D.C.; Arenzana-Seisdedos, F.; Delaunay, T.; Naeem, R.; Carey, V.J.; Richardson, A.L.; Weinberg, R.A. Stromal fibroblasts present in invasive human breast carcinomas promote tumor growth and angiogenesis through elevated SDF-1/CXCL12 secretion. *Cell* **2005**, *121*, 335–348.
30. Boire, A.; Covic, L.; Agarwal, A.; Jacques, S.; Sherifi, S.; Kuliopulos, A. PAR1 is a matrix metalloprotease-1 receptor that promotes invasion and tumorigenesis of breast cancer cells. *Cell* **2005**, *120*, 303–313.
31. Koontongkaew, S.; Amornphimoltham, P.; Monthanpisut, P.; Saensuk, T.; Leelakriangsak, M. Fibroblasts and extracellular matrix differently modulate MMP activation by primary and metastatic head and neck cancer cells. *Med. Oncol.* **2012**, *29*, 690–703. [CrossRef] [PubMed]
32. Takahashi, M.; Fukami, S.; Iwata, N.; Inoue, K.; Itohara, S.; Itoh, H.; Haraoka, J.; Saido, T. In vivo glioma growth requires host-derived matrix metalloproteinase 2 for maintenance of angioarchitecture. *Pharmacol. Res.* **2002**, *46*, 155–163. [CrossRef]
33. Casey, T.M.; Eneman, J.; Crocker, A.; White, J.; Tessitore, J.; Stanley, M.; Harlow, S.; Bunn, J.Y.; Weaver, D.; Muss, H.; et al. Cancer associated fibroblasts stimulated by transforming growth factor beta1 (TGF-beta 1) increase invasion rate of tumor cells: A population study. *Breast Cancer Res. Treat.* **2008**, *110*, 39–49. [CrossRef]
34. Yu, Y.; Xiao, C.H.; Tan, L.D.; Wang, Q.S.; Li, X.Q.; Feng, Y.M. Cancer-associated fibroblasts induce epithelial-mesenchymal transition of breast cancer cells through paracrine TGF-beta signalling. *Br. J. Cancer* **2014**, *110*, 724–732. [CrossRef] [PubMed]
35. Shintani, Y.; Fujiwara, A.; Kimura, T.; Kawamura, T.; Funaki, S.; Minami, M.; Okumura, M. IL-6 Secreted from Cancer-Associated Fibroblasts Mediates Chemoresistance in NSCLC by Increasing Epithelial-Mesenchymal Transition Signaling. *J. Thorac. Oncol.* **2016**, *11*, 1482–1492. [CrossRef]
36. Gordon, S.; Taylor, P.R. Monocyte and macrophage heterogeneity. *Nat. Rev. Immunol.* **2005**, *5*, 953–964. [CrossRef]

37. Yoshimoto, Y.; Jo, J.I.; Tabata, Y. Preparation of antibody-immobilized gelatin nanospheres incorporating a molecular beacon to visualize the biological function of macrophages. *Regen. Ther.* **2020**, *14*, 11–18. [CrossRef]
38. Gordon, S. Alternative activation of macrophages. *Nat. Rev. Immunol.* **2003**, *3*, 23–35. [CrossRef]
39. Brown, L.F.; Yeo, K.T.; Berse, B.; Yeo, T.K.; Senger, D.R.; Dvorak, H.F.; van de Water, L. Expression of vascular permeability factor (vascular endothelial growth factor) by epidermal keratinocytes during wound healing. *J. Exp. Med.* **1992**, *176*, 1375–1379. [CrossRef]
40. Mantovani, A.; Sozzani, S.; Locati, M.; Allavena, P.; Sica, A. Macrophage polarization: Tumor-associated macrophages as a paradigm for polarized M2 mononuclear phagocytes. *Trends Immunol.* **2002**, *23*, 549–555. [CrossRef]
41. Shu, Y.; Qin, M.; Song, Y.; Tang, Q.; Huang, Y.; Shen, P.; Lu, Y. M2 polarization of tumor-associated macrophages is dependent on integrin beta3 via peroxisome proliferator-activated receptor-gamma up-regulation in breast cancer. *Immunology* **2020**. [CrossRef] [PubMed]
42. Hu, J.M.; Liu, K.; Liu, J.H.; Jiang, X.L.; Wang, X.L.; Chen, Y.Z.; Li, S.G.; Zou, H.; Pang, L.J.; Liu, C.X.; et al. CD163 as a marker of M2 macrophage, contribute to predicte aggressiveness and prognosis of Kazakh esophageal squamous cell carcinoma. *Oncotarget* **2017**, *8*, 21526–21538. [CrossRef] [PubMed]
43. Kawachi, A.; Yoshida, H.; Kitano, S.; Ino, Y.; Kato, T.; Hiraoka, N. Tumor-associated CD204(+) M2 macrophages are unfavorable prognostic indicators in uterine cervical adenocarcinoma. *Cancer Sci.* **2018**, *109*, 863–870. [CrossRef]
44. Huang, X.; Li, Y.; Fu, M.; Xin, H.B. Polarizing Macrophages In Vitro. *Methods Mol. Biol.* **2018**, *1784*, 119–126.
45. Grivennikov, S.I.; Wang, K.; Mucida, D.; Stewart, C.A.; Schnabl, B.; Jauch, D.; Taniguchi, K.; Yu, G.Y.; Osterreicher, C.H.; Hung, K.E.; et al. Adenoma-linked barrier defects and microbial products drive IL-23/IL-17-mediated tumour growth. *Nature* **2012**, *491*, 254–258. [CrossRef] [PubMed]
46. Tomita, T.; Sakurai, Y.; Ishibashi, S.; Maru, Y. Imbalance of Clara cell-mediated homeostatic inflammation is involved in lung metastasis. *Oncogene* **2011**, *30*, 3429–3439. [CrossRef]
47. Dong, X.; Huang, X.; Yao, Z.; Wu, Y.; Chen, D.; Tan, C.; Lin, J.; Zhang, D.; Hu, Y.; Wu, J.; et al. Tumour-associated macrophages as a novel target of VEGI-251 in cancer therapy. *J. Cell Mol. Med.* **2020**, *24*, 7884–7895. [CrossRef]
48. Evrard, D.; Szturz, P.; Tijeras-Raballand, A.; Astorgues-Xerri, L.; Abitbol, C.; Paradis, V.; Raymond, E.; Albert, S.; Barry, B.; Faivre, S. Macrophages in the microenvironment of head and neck cancer: Potential targets for cancer therapy. *Oral. Oncol.* **2019**, *88*, 29–38. [CrossRef]
49. Komohara, Y.; Takeya, M. CAFs and TAMs: Maestros of the tumour microenvironment. *J. Pathol.* **2017**, *241*, 313–315. [CrossRef]
50. Sasaki, M.; Abe, R.; Fujita, Y.; Ando, S.; Inokuma, D.; Shimizu, H. Mesenchymal stem cells are recruited into wounded skin and contribute to wound repair by transdifferentiation into multiple skin cell type. *J. Immunol.* **2008**, *180*, 2581–2587. [CrossRef]
51. Kassem, M.; Abdallah, B.M.; Saeed, H. Osteoblastic cells: Differentiation and trans-differentiation. *Arch. Biochem. Biophys.* **2008**, *473*, 183–187. [CrossRef] [PubMed]
52. Song, L.; Tuan, R.S. Transdifferentiation potential of human mesenchymal stem cells derived from bone marrow. *FASEB J.* **2004**, *18*, 980–982. [CrossRef] [PubMed]
53. Murphy, M.B.; Moncivais, K.; Caplan, A.I. Mesenchymal stem cells: Environmentally responsive therapeutics for regenerative medicine. *Exp. Mol. Med.* **2013**, *45*, e54. [PubMed]
54. Barcellos-de-Souza, P.; Comito, G.; Pons-Segura, C.; Taddei, M.L.; Gori, V.; Becherucci, V.; Bambi, F.; Margheri, F.; Laurenzana, A.; Del Rosso, M.; et al. Mesenchymal Stem Cells are Recruited and Activated into Carcinoma-Associated Fibroblasts by Prostate Cancer Microenvironment-Derived TGF-beta1. *Stem Cells* **2016**, *34*, 2536–2547. [CrossRef] [PubMed]
55. Chowdhury, R.; Webber, J.P.; Gurney, M.; Mason, M.D.; Tabi, Z.; Clayton, A. Cancer exosomes trigger mesenchymal stem cell differentiation into pro-angiogenic and pro-invasive myofibroblasts. *Oncotarget* **2015**, *6*, 715–731. [CrossRef]
56. Mantovani, A. MSCs, macrophages, and cancer: A dangerous menage-a-trois. *Cell Stem Cell* **2012**, *11*, 730–732. [CrossRef]
57. Papaccio, F.; Paino, F.; Regad, T.; Papaccio, G.; Desiderio, V.; Tirino, V. Concise Review: Cancer Cells, Cancer Stem Cells, and Mesenchymal Stem Cells: Influence in Cancer Development. *Stem Cells Transl. Med.* **2017**, *6*, 2115–2125.

58. Waterman, R.S.; Tomchuck, S.L.; Henkle, S.L.; Betancourt, A.M. A new mesenchymal stem cell (MSC) paradigm: Polarization into a pro-inflammatory MSC1 or an Immunosuppressive MSC2 phenotype. *PLoS ONE* **2010**, *5*, e10088.
59. Waterman, R.S.; Henkle, S.L.; Betancourt, A.M. Mesenchymal stem cell 1 (MSC1)-based therapy attenuates tumor growth whereas MSC2-treatment promotes tumor growth and metastasis. *PLoS ONE* **2012**, *7*, e45590.
60. Hida, K.; Ohga, N.; Akiyama, K.; Maishi, N.; Hida, Y. Heterogeneity of tumor endothelial cells. *Cancer Sci.* **2013**, *104*, 1391–1395.
61. Cabral, H.; Kataoka, K. Progress of drug-loaded polymeric micelles into clinical studies. *J. Control Release* **2014**, *190*, 465–476. [CrossRef]
62. Matsumura, Y.; Kataoka, K. Preclinical and clinical studies of anticancer agent-incorporating polymer micelles. *Cancer Sci.* **2009**, *100*, 572–579. [CrossRef] [PubMed]
63. Akino, T.; Hida, K.; Hida, Y.; Tsuchiya, K.; Freedman, D.; Muraki, C.; Ohga, N.; Matsuda, K.; Akiyama, K.; Harabayashi, T.; et al. Cytogenetic abnormalities of tumor-associated endothelial cells in human malignant tumors. *Am. J. Pathol.* **2009**, *175*, 2657–2667. [CrossRef] [PubMed]
64. Amin, D.N.; Hida, K.; Bielenberg, D.R.; Klagsbrun, M. Tumor endothelial cells express epidermal growth factor receptor (EGFR) but not ErbB3 and are responsive to EGF and to EGFR kinase inhibitors. *Cancer Res.* **2006**, *66*, 2173–2180. [CrossRef] [PubMed]
65. Kurosu, T.; Ohga, N.; Hida, Y.; Maishi, N.; Akiyama, K.; Kakuguchi, W.; Kuroshima, T.; Kondo, M.; Akino, T.; Totsuka, Y.; et al. HuR keeps an angiogenic switch on by stabilising mRNA of VEGF and COX-2 in tumour endothelium. *Br. J. Cancer* **2011**, *104*, 819–829. [CrossRef]
66. Mazio, C.; Casale, C.; Imparato, G.; Urciuolo, F.; Netti, P.A. Recapitulating spatiotemporal tumor heterogeneity in vitro through engineered breast cancer microtissues. *Acta Biomater.* **2018**, *73*, 236–249. [CrossRef]
67. Dhiman, H.K.; Ray, A.R.; Panda, A.K. Characterization and evaluation of chitosan matrix for in vitro growth of MCF-7 breast cancer cell lines. *Biomaterials* **2004**, *25*, 5147–5154. [CrossRef] [PubMed]
68. Leung, M.; Kievit, F.M.; Florczyk, S.J.; Veiseh, O.; Wu, J.; Park, J.O.; Zhang, M. Chitosan-alginate scaffold culture system for hepatocellular carcinoma increases malignancy and drug resistance. *Pharm. Res.* **2010**, *27*, 1939–1948. [CrossRef] [PubMed]
69. Florczyk, S.J.; Wang, K.; Jana, S.; Wood, D.L.; Sytsma, S.K.; Sham, J.; Kievit, F.M.; Zhang, M. Porous chitosan-hyaluronic acid scaffolds as a mimic of glioblastoma microenvironment ECM. *Biomaterials* **2013**, *34*, 10143–10150. [CrossRef]
70. Wang, K.; Kievit, F.M.; Erickson, A.E.; Silber, J.R.; Ellenbogen, R.G.; Zhang, M. Culture on 3D Chitosan-Hyaluronic Acid Scaffolds Enhances Stem Cell Marker Expression and Drug Resistance in Human Glioblastoma Cancer Stem Cells. *Adv. Healthc. Mater.* **2016**, *5*, 3173–3181. [CrossRef]
71. Kievit, F.M.; Florczyk, S.J.; Leung, M.C.; Veiseh, O.; Park, J.O.; Disis, M.L.; Zhang, M. Chitosan-alginate 3D scaffolds as a mimic of the glioma tumor microenvironment. *Biomaterials* **2010**, *31*, 5903–5910. [CrossRef] [PubMed]
72. Kievit, F.M.; Florczyk, S.J.; Leung, M.C.; Wang, K.; Wu, J.D.; Silber, J.R.; Ellenbogen, R.G.; Lee, J.S.; Zhang, M. Proliferation and enrichment of CD133(+) glioblastoma cancer stem cells on 3D chitosan-alginate scaffolds. *Biomaterials* **2014**, *35*, 9137–9143. [CrossRef] [PubMed]
73. Han, H.W.; Hsu, S.H. Chitosan-hyaluronan based 3D co-culture platform for studying the crosstalk of lung cancer cells and mesenchymal stem cells. *Acta Biomater.* **2016**, *42*, 157–167. [CrossRef] [PubMed]
74. Huang, Y.J.; Hsu, S.H. Acquisition of epithelial-mesenchymal transition and cancer stem-like phenotypes within chitosan-hyaluronan membrane-derived 3D tumor spheroids. *Biomaterials* **2014**, *35*, 10070–10079. [CrossRef] [PubMed]
75. Wang, K.; Kievit, F.M.; Florczyk, S.J.; Stephen, Z.R.; Zhang, M. 3D Porous Chitosan-Alginate Scaffolds as an In Vitro Model for Evaluating Nanoparticle-Mediated Tumor Targeting and Gene Delivery to Prostate Cancer. *Biomacromolecules* **2015**, *16*, 3362–3372. [CrossRef]
76. Xu, K.; Ganapathy, K.; Andl, T.; Wang, Z.; Copland, J.A.; Chakrabarti, R.; Florczyk, S.J. 3D porous chitosan-alginate scaffold stiffness promotes differential responses in prostate cancer cell lines. *Biomaterials* **2019**, *217*, 119311. [CrossRef]
77. Xu, K.; Wang, Z.; Copland, J.A.; Chakrabarti, R.; Florczyk, S.J. 3D porous chitosan-chondroitin sulfate scaffolds promote epithelial to mesenchymal transition in prostate cancer cells. *Biomaterials* **2020**, *254*, 120126. [CrossRef]

78. Estrada, M.F.; Rebelo, S.P.; Davies, E.J.; Pinto, M.T.; Pereira, H.; Santo, V.E.; Smalley, M.J.; Barry, S.T.; Gualda, E.J.; Alves, P.M.; et al. Modelling the tumour microenvironment in long-term microencapsulated 3D co-cultures recapitulates phenotypic features of disease progression. *Biomaterials* **2016**, *78*, 50–61. [CrossRef]
79. Cavo, M.; Caria, M.; Pulsoni, I.; Beltrame, F.; Fato, M.; Scaglione, S. A new cell-laden 3D Alginate-Matrigel hydrogel resembles human breast cancer cell malignant morphology, spread and invasion capability observed "in vivo". *Sci. Rep.* **2018**, *8*, 5333. [CrossRef]
80. Liu, C.; Liu, Y.; Xie, H.G.; Zhao, S.; Xu, X.X.; Fan, L.X.; Guo, X.; Lu, T.; Sun, G.W.; Ma, X.J. Role of three-dimensional matrix stiffness in regulating the chemoresistance of hepatocellular carcinoma cells. *Biotechnol. Appl. Biochem.* **2015**, *62*, 556–562. [CrossRef]
81. Liu, C.; Liu, Y.; Xu, X.X.; Guo, X.; Sun, G.W.; Ma, X.J. Mesenchymal stem cells enhance the metastasis of 3D-cultured hepatocellular carcinoma cells. *BMC Cancer* **2016**, *16*, 566.
82. Liu, C.; Liu, Y.; Xu, X.X.; Wu, H.; Xie, H.G.; Chen, L.; Lu, T.; Yang, L.; Guo, X.; Sun, G.W.; et al. Potential effect of matrix stiffness on the enrichment of tumor initiating cells under three-dimensional culture conditions. *Exp. Cell Res.* **2015**, *330*, 123–134. [PubMed]
83. Vu, T.T.; Lim, C.; Lim, M. Characterization of leukemic cell behaviors in a soft marrow mimetic alginate hydrogel. *J. Biomed. Mater. Res. B Appl. Biomater.* **2012**, *100*, 1980–1988. [PubMed]
84. DelNero, P.; Lane, M.; Verbridge, S.S.; Kwee, B.; Kermani, P.; Hempstead, B.; Stroock, A.; Fischbach, C. 3D culture broadly regulates tumor cell hypoxia response and angiogenesis via pro-inflammatory pathways. *Biomaterials* **2015**, *55*, 110–118. [PubMed]
85. Liu, C.; Lewin Mejia, D.; Chiang, B.; Luker, K.E.; Luker, G.D. Hybrid collagen alginate hydrogel as a platform for 3D tumor spheroid invasion. *Acta Biomater.* **2018**, *75*, 213–225. [PubMed]
86. Nguyen-Ngoc, K.V.; Cheung, K.J.; Brenot, A.; Shamir, E.R.; Gray, R.S.; Hines, W.C.; Yaswen, P.; Werb, Z.; Ewald, A.J. ECM microenvironment regulates collective migration and local dissemination in normal and malignant mammary epithelium. *Proc. Natl. Acad. Sci. USA* **2012**, *109*, E2595–E2604. [PubMed]
87. Sapudom, J.; Kalbitzer, L.; Wu, X.; Martin, S.; Kroy, K.; Pompe, T. Fibril bending stiffness of 3D collagen matrices instructs spreading and clustering of invasive and non-invasive breast cancer cells. *Biomaterials* **2019**, *193*, 47–57.
88. Campbell, J.J.; Husmann, A.; Hume, R.D.; Watson, C.J.; Cameron, R.E. Development of three-dimensional collagen scaffolds with controlled architecture for cell migration studies using breast cancer cell lines. *Biomaterials* **2017**, *114*, 34–43.
89. Saini, H.; Eliato, K.R.; Silva, C.; Allam, M.; Mouneimne, G.; Ros, R.; Nikkhah, M. The Role of Desmoplasia and Stromal Fibroblasts on Anti-cancer Drug Resistance in a Microengineered Tumor Model. *Cell. Mol. Bioeng.* **2018**, *11*, 419–433.
90. Bersini, S.; Jeon, J.S.; Dubini, G.; Arrigoni, C.; Chung, S.; Charest, J.L.; Moretti, M.; Kamm, R.D. A microfluidic 3D in vitro model for specificity of breast cancer metastasis to bone. *Biomaterials* **2014**, *35*, 2454–2461.
91. Reynolds, D.S.; Bougher, K.M.; Letendre, J.H.; Fitzgerald, S.F.; Gisladottir, U.O.; Grinstaff, M.W.; Zaman, M.H. Mechanical confinement via a PEG/Collagen interpenetrating network inhibits behavior characteristic of malignant cells in the triple negative breast cancer cell line MDA.MB.231. *Acta Biomater.* **2018**, *77*, 85–95. [PubMed]
92. Fitzgerald, K.A.; Guo, J.F.; Tierney, E.G.; Curtin, C.M.; Malhotra, M.; Darcy, R.; O'Brien, F.J.; O'Driscoll, C.M. The use of collagen-based scaffolds to simulate prostate cancer bone metastases with potential for evaluating delivery of nanoparticulate gene therapeutics. *Biomaterials* **2015**, *66*, 53–66. [CrossRef] [PubMed]
93. Miyazaki, K.; Oyanagi, J.; Hoshino, D.; Togo, S.; Kumagai, H.; Miyagi, Y. Cancer cell migration on elongate protrusions of fibroblasts in collagen matrix. *Sci. Rep.* **2019**, *9*, 292. [CrossRef] [PubMed]
94. Liu, W.; Song, J.; Du, X.; Zhou, Y.; Li, Y.; Li, R.; Lyu, L.; He, Y.; Hao, J.; Ben, J.; et al. AKR1B10 (Aldo-keto reductase family 1 B10) promotes brain metastasis of lung cancer cells in a multi-organ microfluidic chip model. *Acta Biomater.* **2019**, *91*, 195–208.
95. Xu, Z.; Li, E.; Guo, Z.; Yu, R.; Hao, H.; Xu, Y.; Sun, Z.; Li, X.; Lyu, J.; Wang, Q. Design and Construction of a Multi-Organ Microfluidic Chip Mimicking the in vivo Microenvironment of Lung Cancer Metastasis. *ACS Appl. Mater. Interfaces* **2016**, *8*, 25840–25847. [CrossRef]
96. Nii, T.; Kuwahara, T.; Makino, K.; Tabata, Y. A co-culture system of three-dimensional tumor-associated macrophages and three-dimensional cancer-associated fibroblasts combined with biomolecule release for cancer cell migration. *Tissue Eng. Part A* **2020**. [CrossRef]

97. Brancato, V.; Garziano, A.; Gioiella, F.; Urciuolo, F.; Imparato, G.; Panzetta, V.; Fusco, S.; Netti, P.A. 3D is not enough: Building up a cell instructive microenvironment for tumoral stroma microtissues. *Acta Biomater.* **2017**, *47*, 1–13. [CrossRef]
98. Brancato, V.; Gioiella, F.; Imparato, G.; Guarnieri, D.; Urciuolo, F.; Netti, P.A. 3D breast cancer microtissue reveals the role of tumor microenvironment on the transport and efficacy of free-doxorubicin in vitro. *Acta Biomater.* **2018**, *75*, 200–212. [CrossRef]
99. Brancato, V.; Gioiella, F.; Profeta, M.; Imparato, G.; Guarnieri, D.; Urciuolo, F.; Melone, P.; Netti, P.A. 3D tumor microtissues as an in vitro testing platform for microenvironmentally-triggered drug delivery systems. *Acta Biomater.* **2017**, *57*, 47–58. [CrossRef]
100. Nii, T.; Makino, K.; Tabata, Y. A cancer invasion model of cancer-associated fibroblasts aggregates combined with TGF-beta1 release system. *Regen. Ther.* **2020**, *14*, 196–204.
101. Nii, T.; Makino, K.; Tabata, Y. A Cancer Invasion Model Combined with Cancer-Associated Fibroblasts Aggregates Incorporating Gelatin Hydrogel Microspheres Containing a p53 Inhibitor. *Tissue Eng. Part C Methods* **2019**, *25*, 711–720. [CrossRef] [PubMed]
102. Brancato, V.; Comunanza, V.; Imparato, G.; Cora, D.; Urciuolo, F.; Noghero, A.; Bussolino, F.; Netti, P.A. Bioengineered tumoral microtissues recapitulate desmoplastic reaction of pancreatic cancer. *Acta Biomater.* **2017**, *49*, 152–166. [CrossRef] [PubMed]
103. David, L.; Dulong, V.; Le Cerf, D.; Chauzy, C.; Norris, V.; Delpech, B.; Lamacz, M.; Vannier, J.P. Reticulated hyaluronan hydrogels: A model for examining cancer cell invasion in 3D. *Matrix Biol.* **2004**, *23*, 183–193. [CrossRef] [PubMed]
104. David, L.; Dulong, V.; Le Cerf, D.; Cazin, L.; Lamacz, M.; Vannier, J.P. Hyaluronan hydrogel: An appropriate three-dimensional model for evaluation of anticancer drug sensitivity. *Acta Biomater.* **2008**, *4*, 256–263. [CrossRef] [PubMed]
105. Engel, B.J.; Constantinou, P.E.; Sablatura, L.K.; Doty, N.J.; Carson, D.D.; Farach-Carson, M.C.; Harrington, D.A.; Zarembinski, T.I. Multilayered, Hyaluronic Acid-Based Hydrogel Formulations Suitable for Automated 3D High Throughput Drug Screening of Cancer-Stromal Cell Cocultures. *Adv. Healthc. Mater.* **2015**, *4*, 1664–1674. [CrossRef]
106. Wang, C.; Tong, X.; Yang, F. Bioengineered 3D brain tumor model to elucidate the effects of matrix stiffness on glioblastoma cell behavior using PEG-based hydrogels. *Mol. Pharm.* **2014**, *11*, 2115–2125. [CrossRef]
107. Kenny, P.A.; Lee, G.Y.; Myers, C.A.; Neve, R.M.; Semeiks, J.R.; Spellman, P.T.; Lorenz, K.; Lee, E.H.; Barcellos-Hoff, M.H.; Petersen, O.W.; et al. The morphologies of breast cancer cell lines in three-dimensional assays correlate with their profiles of gene expression. *Mol. Oncol.* **2007**, *1*, 84–96. [CrossRef]
108. Olsen, C.J.; Moreira, J.; Lukanidin, E.M.; Ambartsumian, N.S. Human mammary fibroblasts stimulate invasion of breast cancer cells in a three-dimensional culture and increase stroma development in mouse xenografts. *BMC Cancer* **2010**, *10*, 444. [CrossRef]
109. Yu, X.; Machesky, L.M. Cells assemble invadopodia-like structures and invade into matrigel in a matrix metalloprotease dependent manner in the circular invasion assay. *PLoS ONE* **2012**, *7*, e30605. [CrossRef]
110. Zhu, J.; Liang, L.; Jiao, Y.; Liu, L.; Alliance, U.S.-C.P.S.-O. Enhanced invasion of metastatic cancer cells via extracellular matrix interface. *PLoS ONE* **2015**, *10*, e0118058. [CrossRef]
111. Augustine, T.N.; Dix-Peek, T.; Duarte, R.; Candy, G.P. Establishment of a heterotypic 3D culture system to evaluate the interaction of TREG lymphocytes and NK cells with breast cancer. *J. Immunol. Methods* **2015**, *426*, 1–13. [CrossRef] [PubMed]
112. Sasser, A.K.; Mundy, B.L.; Smith, K.M.; Studebaker, A.W.; Axel, A.E.; Haidet, A.M.; Fernandez, S.A.; Hall, B.M. Human bone marrow stromal cells enhance breast cancer cell growth rates in a cell line-dependent manner when evaluated in 3D tumor environments. *Cancer Lett.* **2007**, *254*, 255–264. [CrossRef] [PubMed]
113. Pinto, M.P.; Dye, W.W.; Jacobsen, B.M.; Horwitz, K.B. Malignant stroma increases luminal breast cancer cell proliferation and angiogenesis through platelet-derived growth factor signaling. *BMC Cancer* **2014**, *14*, 735. [CrossRef] [PubMed]
114. Kramer, R.H.; Bensch, K.G.; Wong, J. Invasion of reconstituted basement membrane matrix by metastatic human tumor cells. *Cancer Res.* **1986**, *46*, 1980–1989.
115. Zhang, T.Z.; Zhang, Q.Y.; Chen, J.S.; Fang, K.; Dou, J.; Gu, N. The controllable preparation of porous PLGA microspheres by the oil/water emulsion method and its application in 3D culture of ovarian cancer cells. *Colloids Surfaces Physicochem. Eng. Asp.* **2014**, *452*, 115–124. [CrossRef]

116. Sahoo, S.K.; Panda, A.K.; Labhasetwar, V. Characterization of porous PLGA/PLA microparticles as a scaffold for three dimensional growth of breast cancer cells. *Biomacromolecules* **2005**, *6*, 1132–1139. [CrossRef]
117. Luo, H.L.; Zhang, Y.; Gan, D.Q.; Yang, Z.W.; Ao, H.Y.; Zhang, Q.C.; Yao, F.L.; Wan, Y.Z. Incorporation of hydroxyapatite into nanofibrous PLGA scaffold towards improved breast cancer cell behavior. *Mater. Chem. Phys.* **2019**, *226*, 177–183. [CrossRef]
118. Girard, Y.K.; Wang, C.; Ravi, S.; Howell, M.C.; Mallela, J.; Alibrahim, M.; Green, R.; Hellermann, G.; Mohapatra, S.S.; Mohapatra, S. A 3D fibrous scaffold inducing tumoroids: A platform for anticancer drug development. *PLoS ONE* **2013**, *8*, e75345. [CrossRef]
119. Zhu, X.H.; Lee, L.Y.; Jackson, J.S.; Tong, Y.W.; Wang, C.H. Characterization of porous poly(D,L-lactic-co-glycolic acid) sponges fabricated by supercritical CO2 gas-foaming method as a scaffold for three-dimensional growth of Hep3B cells. *Biotechnol. Bioeng.* **2008**, *100*, 998–1009. [CrossRef]
120. Yang, X.; Sarvestani, S.K.; Moeinzadeh, S.; He, X.; Jabbari, E. Three-dimensional-engineered matrix to study cancer stem cells and tumorsphere formation: Effect of matrix modulus. *Tissue Eng. Part A* **2013**, *19*, 669–684. [CrossRef]
121. Pradhan, S.; Hassani, I.; Seeto, W.J.; Lipke, E.A. PEG-fibrinogen hydrogels for three-dimensional breast cancer cell culture. *J. Biomed. Mater. Res. Part A* **2017**, *105*, 236–252. [CrossRef] [PubMed]
122. Pradhan, S.; Clary, J.M.; Seliktar, D.; Lipke, E.A. A three-dimensional spheroidal cancer model based on PEG-fibrinogen hydrogel microspheres. *Biomaterials* **2017**, *115*, 141–154. [CrossRef] [PubMed]
123. Del Bufalo, F.; Manzo, T.; Hoyos, V.; Yagyu, S.; Caruana, I.; Jacot, J.; Benavides, O.; Rosen, D.; Brenner, M.K. 3D modeling of human cancer: A PEG-fibrin hydrogel system to study the role of tumor microenvironment and recapitulate the in vivo effect of oncolytic adenovirus. *Biomaterials* **2016**, *84*, 76–85. [CrossRef]
124. Soman, P.; Kelber, J.A.; Lee, J.W.; Wright, T.N.; Vecchio, K.S.; Klemke, R.L.; Chen, S. Cancer cell migration within 3D layer-by-layer microfabricated photocrosslinked PEG scaffolds with tunable stiffness. *Biomaterials* **2012**, *33*, 7064–7070. [CrossRef] [PubMed]
125. Sieh, S.; Taubenberger, A.V.; Rizzi, S.C.; Sadowski, M.; Lehman, M.L.; Rockstroh, A.; An, J.Y.; Clements, J.A.; Nelson, C.C.; Hutmacher, D.W. Phenotypic Characterization of Prostate Cancer LNCaP Cells Cultured within a Bioengineered Microenvironment. *PLoS ONE* **2012**, *7*, e40217. [CrossRef]
126. Singh, S.P.; Schwartz, M.P.; Lee, J.Y.; Fairbanks, B.D.; Anseth, K.S. A peptide functionalized poly(ethylene glycol) (PEG) hydrogel for investigating the influence of biochemical and biophysical matrix properties on tumor cell migration. *Biomater. Sci.* **2014**, *2*, 1024–1034. [CrossRef]
127. Healy, K.E.; Lom, B.; Hockberger, P.E. Spatial distribution of mammalian cells dictated by material surface chemistry. *Biotechnol. Bioeng.* **1994**, *43*, 792–800. [CrossRef]
128. Jayakumar, R.; Nwe, N.; Tokura, S.; Tamura, H. Sulfated chitin and chitosan as novel biomaterials. *Int. J. Biol. Macromol.* **2007**, *40*, 175–181. [CrossRef]
129. Ueno, H.; Mori, T.; Fujinaga, T. Topical formulations and wound healing applications of chitosan. *Adv. Drug Deliv. Rev.* **2001**, *52*, 105–115. [CrossRef]
130. Miyazaki, S.; Ishii, K.; Nadai, T. The use of chitin and chitosan as drug carriers. *Chem. Pharm. Bull.* **1981**, *29*, 3067–3069. [CrossRef]
131. Hussain, A.; Collins, G.; Yip, D.; Cho, C.H. Functional 3-D cardiac co-culture model using bioactive chitosan nanofiber scaffolds. *Biotechnol. Bioeng.* **2013**, *110*, 637–647. [CrossRef] [PubMed]
132. Crompton, K.E.; Goud, J.D.; Bellamkonda, R.V.; Gengenbach, T.R.; Finkelstein, D.I.; Horne, M.K.; Forsythe, J.S. Polylysine-functionalised thermoresponsive chitosan hydrogel for neural tissue engineering. *Biomaterials* **2007**, *28*, 441–449. [CrossRef] [PubMed]
133. Muzzarelli, R.A.A.; Biagini, G.; Bellardini, M.; Simonelli, L.; Castaldini, C.; Fratto, G. Osteoconduction Exerted by Methylpyrrolidinone Chitosan Used in Dental Surgery. *Biomaterials* **1993**, *14*, 39–43. [CrossRef]
134. Wei, Y.N.; Wang, Q.Q.; Gao, T.T.; Kong, M.; Yang, K.K.; An, Y.; Jiang, S.Y.; Li, J.; Cheng, X.J.; Chen, X.G. 3-D culture of human umbilical vein endothelial cells with reversible thermosensitive hydroxybutyl chitosan hydrogel. *J. Mater. Sci. Mater. Med.* **2013**, *24*, 1781–1787. [CrossRef] [PubMed]
135. Croisier, F.; Jerome, C. Chitosan-based biomaterials for tissue engineering. *Eur. Polym. J.* **2013**, *49*, 780–792. [CrossRef]
136. Miller, I.J. Alginate composition of some New Zealand brown seaweeds. *Phytochemistry* **1996**, *41*, 1315–1317. [CrossRef]

137. Shishu; Gupta, N.; Aggarwal, N. Stomach-specific drug delivery of 5-fluorouracil using floating alginate beads. *AAPS PharmSciTech* **2007**, *8*, 48.
138. Sun, J.C.; Tan, H.P. Alginate-Based Biomaterials for Regenerative Medicine Applications. *Materials* **2013**, *6*, 1285–1309. [CrossRef]
139. Jejurikar, A.; Seow, X.T.; Lawrie, G.; Martin, D.; Jayakrishnan, A.; Grondahl, L. Degradable alginate hydrogels crosslinked by the macromolecular crosslinker alginate dialdehyde. *J. Mater. Chem.* **2012**, *22*, 9751–9758. [CrossRef]
140. Zia, K.M.; Zia, F.; Zuber, M.; Rehman, S.; Ahmad, M.N. Alginate based polyurethanes: A review of recent advances and perspective. *Int. J. Biol. Macromol.* **2015**, *79*, 377–387. [CrossRef]
141. Scott, J.E. Supramolecular organization of extracellular matrix glycosaminoglycans, in vitro and in the tissues. *FASEB J.* **1992**, *6*, 2639–2645. [CrossRef]
142. Gelse, K.; Poschl, E.; Aigner, T. Collagens—Structure, function, and biosynthesis. *Adv. Drug Deliv. Rev.* **2003**, *55*, 1531–1546. [CrossRef] [PubMed]
143. Lee, Y.S.; Arinzeh, T.L. Electrospun Nanofibrous Materials for Neural Tissue Engineering. *Polymers* **2011**, *3*, 413–426. [CrossRef]
144. Liu, T.; Houle, J.D.; Xu, J.Y.; Chan, B.P.; Chew, S.Y. Nanofibrous Collagen Nerve Conduits for Spinal Cord Repair. *Tissue Eng. Part A* **2012**, *18*, 1057–1066. [CrossRef] [PubMed]
145. Boecker, A.H.; van Neerven, S.G.; Scheffel, J.; Tank, J.; Altinova, H.; Seidensticker, K.; Deumens, R.; Tolba, R.; Weis, J.; Brook, G.A.; et al. Pre-differentiation of mesenchymal stromal cells in combination with a microstructured nerve guide supports peripheral nerve regeneration in the rat sciatic nerve model. *Eur. J. Neurosci.* **2016**, *43*, 404–416. [CrossRef]
146. Inzana, J.A.; Olvera, D.; Fuller, S.M.; Kelly, J.P.; Graeve, O.A.; Schwarz, E.M.; Kates, S.L.; Awad, H.A. 3D printing of composite calcium phosphate and collagen scaffolds for bone regeneration. *Biomaterials* **2014**, *35*, 4026–4034. [CrossRef]
147. Chen, G.; Dong, C.; Yang, L.; Lv, Y. 3D Scaffolds with Different Stiffness but the Same Microstructure for Bone Tissue Engineering. *ACS Appl. Mater. Interfaces* **2015**, *7*, 15790–15802. [CrossRef]
148. Chen, G.; Yang, L.; Lv, Y. Cell-free scaffolds with different stiffness but same microstructure promote bone regeneration in rabbit large bone defect model. *J. Biomed. Mater. Res. A* **2016**, *104*, 833–841. [CrossRef]
149. Zheng, X.; Wang, W.; Liu, S.; Wu, J.; Li, F.; Cao, L.; Liu, X.D.; Mo, X.; Fan, C. Enhancement of chondrogenic differentiation of rabbit mesenchymal stem cells by oriented nanofiber yarn-collagen type I/hyaluronate hybrid. *Mater. Sci. Eng. C Mater. Biol. Appl.* **2016**, *58*, 1071–1076. [CrossRef]
150. Muhonen, V.; Salonius, E.; Haaparanta, A.M.; Jarvinen, E.; Paatela, T.; Meller, A.; Hannula, M.; Bjorkman, M.; Pyhalto, T.; Ella, V.; et al. Articular cartilage repair with recombinant human type II collagen/polylactide scaffold in a preliminary porcine study. *J. Orthop. Res.* **2016**, *34*, 745–753. [CrossRef]
151. Griffin, D.J.; Bonnevie, E.D.; Lachowsky, D.J.; Hart, J.C.; Sparks, H.D.; Moran, N.; Matthews, G.; Nixon, A.J.; Cohen, I.; Bonassar, L.J. Mechanical characterization of matrix-induced autologous chondrocyte implantation (MACI(R)) grafts in an equine model at 53 weeks. *J. Biomech.* **2015**, *48*, 1944–1949. [CrossRef] [PubMed]
152. Basad, E.; Wissing, F.R.; Fehrenbach, P.; Rickert, M.; Steinmeyer, J.; Ishaque, B. Matrix-induced autologous chondrocyte implantation (MACI) in the knee: Clinical outcomes and challenges. *Knee Surg. Sports Traumatol. Arthrosc.* **2015**, *23*, 3729–3735. [CrossRef]
153. Gigante, A.; Busilacchi, A.; Lonzi, B.; Cecconi, S.; Manzotti, S.; Renghini, C.; Giuliani, A.; Mattioli-Belmonte, M. Purified collagen I oriented membrane for tendon repair: An ex vivo morphological study. *J. Orthop. Res.* **2013**, *31*, 738–745. [CrossRef] [PubMed]
154. Cardwell, R.D.; Dahlgren, L.A.; Goldstein, A.S. Electrospun fibre diameter, not alignment, affects mesenchymal stem cell differentiation into the tendon/ligament lineage. *J. Tissue Eng. Regen. Med.* **2014**, *8*, 937–945. [CrossRef]
155. Yunoki, S.; Hatayama, H.; Ebisawa, M.; Kondo, E.; Yasuda, K. A novel fabrication method to create a thick collagen bundle composed of uniaxially aligned fibrils: An essential technology for the development of artificial tendon/ligament matrices. *J. Biomed. Mater. Res. A* **2015**, *103*, 3054–3065. [CrossRef] [PubMed]
156. Ma, L.; Gao, C.; Mao, Z.; Zhou, J.; Shen, J.; Hu, X.; Han, C. Collagen/chitosan porous scaffolds with improved biostability for skin tissue engineering. *Biomaterials* **2003**, *24*, 4833–4841. [CrossRef]

157. Rho, K.S.; Jeong, L.; Lee, G.; Seo, B.M.; Park, Y.J.; Hong, S.D.; Roh, S.; Cho, J.J.; Park, W.H.; Min, B.M. Electrospinning of collagen nanofibers: Effects on the behavior of normal human keratinocytes and early-stage wound healing. *Biomaterials* **2006**, *27*, 1452–1461. [CrossRef]
158. Chen, L.; Xiao, Z.; Meng, Y.; Zhao, Y.; Han, J.; Su, G.; Chen, B.; Dai, J. The enhancement of cancer stem cell properties of MCF-7 cells in 3D collagen scaffolds for modeling of cancer and anti-cancer drugs. *Biomaterials* **2012**, *33*, 1437–1444. [CrossRef]
159. Laurent, T.C.; Fraser, J.R. Hyaluronan. *FASEB J.* **1992**, *6*, 2397–2404. [CrossRef]
160. Necas, J.; Bartosikova, L.; Brauner, P.; Kolar, J. Hyaluronic acid (hyaluronan): A review. *Vet. Med.* **2008**, *53*, 397–411. [CrossRef]
161. Knudson, W.; Chow, G.; Knudson, C.B. CD44-mediated uptake and degradation of hyaluronan. *Matrix Biol.* **2002**, *21*, 15–23. [CrossRef]
162. Jakobsen, R.B.; Shahdadfar, A.; Reinholt, F.P.; Brinchmann, J.E. Chondrogenesis in a hyaluronic acid scaffold: Comparison between chondrocytes and MSC from bone marrow and adipose tissue. *Knee Surg. Sports Traumatol. Arthrosc.* **2010**, *18*, 1407–1416. [CrossRef] [PubMed]
163. Bourguignon, L.Y.; Wong, G.; Earle, C.; Krueger, K.; Spevak, C.C. Hyaluronan-CD44 interaction promotes c-Src-mediated twist signaling, microRNA-10b expression, and RhoA/RhoC up-regulation, leading to Rho-kinase-associated cytoskeleton activation and breast tumor cell invasion. *J. Biol. Chem.* **2010**, *285*, 36721–36735. [CrossRef] [PubMed]
164. Zhang, Y.; Thant, A.A.; Machida, K.; Ichigotani, Y.; Naito, Y.; Hiraiwa, Y.; Senga, T.; Sohara, Y.; Matsuda, S.; Hamaguchi, M. Hyaluronan-CD44s signaling regulates matrix metalloproteinase-2 secretion in a human lung carcinoma cell line QG90. *Cancer Res.* **2002**, *62*, 3962–3965. [PubMed]
165. Bourguignon, L.Y.; Zhu, H.; Shao, L.; Chen, Y.W. CD44 interaction with c-Src kinase promotes cortactin-mediated cytoskeleton function and hyaluronic acid-dependent ovarian tumor cell migration. *J. Biol. Chem.* **2001**, *276*, 7327–7336. [CrossRef] [PubMed]
166. Ahrens, T.; Assmann, V.; Fieber, C.; Termeer, C.; Herrlich, P.; Hofmann, M.; Simon, J.C. CD44 is the principal mediator of hyaluronic-acid-induced melanoma cell proliferation. *J. Invest. Dermatol.* **2001**, *116*, 93–101.
167. Zhu, H.; Mitsuhashi, N.; Klein, A.; Barsky, L.W.; Weinberg, K.; Barr, M.L.; Demetriou, A.; Wu, G.D. The role of the hyaluronan receptor CD44 in mesenchymal stem cell migration in the extracellular matrix. *Stem Cells* **2006**, *24*, 928–935. [CrossRef]
168. T, L.R.; Sanchez-Abarca, L.I.; Muntion, S.; Preciado, S.; Puig, N.; Lopez-Ruano, G.; Hernandez-Hernandez, A.; Redondo, A.; Ortega, R.; Rodriguez, C.; et al. MSC surface markers (CD44, CD73, and CD90) can identify human MSC-derived extracellular vesicles by conventional flow cytometry. *Cell Commun. Signal.* **2016**, *14*, 2.
169. Auvinen, P.; Tammi, R.; Parkkinen, J.; Tammi, M.; Agren, U.; Johansson, R.; Hirvikoski, P.; Eskelinen, M.; Kosma, V.M. Hyaluronan in peritumoral stroma and malignant cells associates with breast cancer spreading and predicts survival. *Am. J. Pathol.* **2000**, *156*, 529–536. [CrossRef]
170. Rayahin, J.E.; Buhrman, J.S.; Zhang, Y.; Koh, T.J.; Gemeinhart, R.A. High and low molecular weight hyaluronic acid differentially influence macrophage activation. *ACS Biomater. Sci. Eng.* **2015**, *1*, 481–493. [CrossRef]
171. Benton, G.; Arnaoutova, I.; George, J.; Kleinman, H.K.; Koblinski, J. Matrigel: From discovery and ECM mimicry to assays and models for cancer research. *Adv. Drug Deliv. Rev.* **2014**, *79–80*, 3–18. [CrossRef]
172. Yousif, L.F.; Di Russo, J.; Sorokin, L. Laminin isoforms in endothelial and perivascular basement membranes. *Cell Adh. Migr.* **2013**, *7*, 101–110. [CrossRef] [PubMed]
173. Kleinman, H.K.; Martin, G.R. Matrigel: Basement membrane matrix with biological activity. *Semin. Cancer Biol.* **2005**, *15*, 378–386. [CrossRef] [PubMed]
174. Kikkawa, Y.; Hozumi, K.; Katagiri, F.; Nomizu, M.; Kleinman, H.K.; Koblinski, J.E. Laminin-111-derived peptides and cancer. *Cell Adh. Migr.* **2013**, *7*, 150–256. [CrossRef] [PubMed]
175. Albini, A.; Iwamoto, Y.; Kleinman, H.K.; Martin, G.R.; Aaronson, S.A.; Kozlowski, J.M.; McEwan, R.N. A rapid in vitro assay for quantitating the invasive potential of tumor cells. *Cancer Res.* **1987**, *47*, 3239–3245. [PubMed]
176. Albini, A.; Noonan, D.M. The 'chemoinvasion' assay, 25 years and still going strong: The use of reconstituted basement membranes to study cell invasion and angiogenesis. *Curr. Opin. Cell Biol.* **2010**, *22*, 677–689. [CrossRef] [PubMed]

177. Tu, Y.; Gao, X.; Li, G.; Fu, H.; Cui, D.; Liu, H.; Jin, W.; Zhang, Y. MicroRNA-218 inhibits glioma invasion, migration, proliferation, and cancer stem-like cell self-renewal by targeting the polycomb group gene Bmi1. *Cancer Res.* **2013**, *73*, 6046–6055. [CrossRef]
178. Gentile, P.; Chiono, V.; Carmagnola, I.; Hatton, P.V. An overview of poly(lactic-co-glycolic) acid (PLGA)-based biomaterials for bone tissue engineering. *Int. J. Mol. Sci.* **2014**, *15*, 3640–3659. [CrossRef]
179. Enayati, M.; Mobedi, H.; Hojjati-Emami, S.; Mirzadeh, H.; Jafari-Nodoushan, M. In situ forming PLGA implant for 90days controlled release of leuprolide acetate for treatment of prostate cancer. *Polym. Adv. Technol.* **2017**, *28*, 867–875. [CrossRef]
180. Tracy, M.A.; Ward, K.L.; Firouzabadian, L.; Wang, Y.; Dong, N.; Qian, R.; Zhang, Y. Factors affecting the degradation rate of poly(lactide-co-glycolide) microspheres in vivo and in vitro. *Biomaterials* **1999**, *20*, 1057–1062. [CrossRef]
181. Ravivarapu, H.B.; Burton, K.; DeLuca, P.P. Polymer and microsphere blending to alter the release of a peptide from PLGA microspheres. *Eur. J. Pharm. Biopharm.* **2000**, *50*, 263–270. [CrossRef]
182. Nii, T.; Takeuchi, I.; Kimura, Y.; Makino, K. Effects of the conformation of PLGA molecules in the organic solvent on the aerodynamic diameter of spray dried microparticles. *Colloids Surfaces a-Physicochem. Eng. Asp.* **2018**, *539*, 347–353. [CrossRef]
183. Pan, Z.; Ding, J. Poly(lactide-co-glycolide) porous scaffolds for tissue engineering and regenerative medicine. *Interface Focus* **2012**, *2*, 366–377. [CrossRef] [PubMed]
184. Lanao, R.P.F.; Jonker, A.M.; Wolke, J.G.C.; Jansen, J.A.; Van Hest, J.C.M.; Leeuwenburgh, S.C.G. Physicochemical Properties and Applications of Poly(lactic-co-glycolic acid) for Use in Bone Regeneration. *Tissue Eng. Part B Rev.* **2013**, *19*, 380–390. [CrossRef]
185. Kolate, A.; Baradia, D.; Patil, S.; Vhora, I.; Kore, G.; Misra, A. PEG—A versatile conjugating ligand for drugs and drug delivery systems. *J. Control Release* **2014**, *192*, 67–81. [CrossRef]
186. Kulig, D.; Zimoch-Korzycka, A.; Jarmoluk, A.; Marycz, K. Study on Alginate-Chitosan Complex Formed with Different Polymers Ratio. *Polymers* **2016**, *8*, 167. [CrossRef]
187. Shaari, N.; Kamarudin, S.K. Chitosan and alginate types of bio-membrane in fuel cell application: An overview. *J. Power Sources* **2015**, *289*, 71–80. [CrossRef]
188. Krayukhina, M.A.; Samoilova, N.A.; Yamskov, I.A. Polyelectrolyte Complexes of Chitosan: Formation, Properties, and Applications. *Uspekhi Khimii* **2008**, *77*, 854–869. [CrossRef]
189. Li, Z.; Ramay, H.R.; Hauch, K.D.; Xiao, D.; Zhang, M. Chitosan-alginate hybrid scaffolds for bone tissue engineering. *Biomaterials* **2005**, *26*, 3919–3928. [CrossRef]
190. Huang, G.S.; Dai, L.G.; Yen, B.L.; Hsu, S.H. Spheroid formation of mesenchymal stem cells on chitosan and chitosan-hyaluronan membranes. *Biomaterials* **2011**, *32*, 6929–6945. [CrossRef]
191. Velling, T.; Risteli, J.; Wennerberg, K.; Mosher, D.F.; Johansson, S. Polymerization of type I and III collagens is dependent on fibronectin and enhanced by integrins alpha(11)beta(1) and alpha(2)beta(1). *J. Biol. Chem.* **2002**, *277*, 37377–37381. [CrossRef]
192. Takahashi, Y.; Yamamoto, M.; Tabata, Y. Osteogenic differentiation of mesenchymal stem cells in biodegradable sponges composed of gelatin and beta-tricalcium phosphate. *Biomaterials* **2005**, *26*, 3587–3596. [CrossRef] [PubMed]
193. Narita, A.; Takahara, M.; Ogino, T.; Fukushima, S.; Kimura, Y.; Tabata, Y. Effect of gelatin hydrogel incorporating fibroblast growth factor 2 on human meniscal cells in an organ culture model. *Knee* **2009**, *16*, 285–289. [CrossRef] [PubMed]
194. Ikada, Y.; Tabata, Y. Protein release from gelatin matrices. *Adv. Drug Deliv. Rev.* **1998**, *31*, 287–301. [PubMed]
195. Fujii, T. The effect of amines added to an alkali-pretreatment on the solubilisation of collagen and on the properties of gelatin. *Hoppe Seylers Z Physiol. Chem.* **1969**, *350*, 1257–1265. [CrossRef] [PubMed]
196. Tabata, Y. Tissue regeneration based on growth factor release. *Tissue Eng.* **2003**, *9* (Suppl. 1), S5–S15. [CrossRef] [PubMed]
197. Imparato, G.; Urciuolo, F.; Casale, C.; Netti, P.A. The role of microscaffold properties in controlling the collagen assembly in 3D dermis equivalent using modular tissue engineering. *Biomaterials* **2013**, *34*, 7851–7861. [CrossRef]
198. Zhao, G.; Cui, J.; Qin, Q.; Zhang, J.; Liu, L.; Deng, S.; Wu, C.; Yang, M.; Li, S.; Wang, C. Mechanical stiffness of liver tissues in relation to integrin beta1 expression may influence the development of hepatic cirrhosis and hepatocellular carcinoma. *J. Surg. Oncol.* **2010**, *102*, 482–489. [CrossRef]

199. Sethi, T.; Rintoul, R.C.; Moore, S.M.; MacKinnon, A.C.; Salter, D.; Choo, C.; Chilvers, E.R.; Dransfield, I.; Donnelly, S.C.; Strieter, R.; et al. Extracellular matrix proteins protect small cell lung cancer cells against apoptosis: A mechanism for small cell lung cancer growth and drug resistance in vivo. *Nat. Med.* **1999**, *5*, 662–668. [CrossRef]
200. Conti, J.A.; Kendall, T.J.; Bateman, A.; Armstrong, T.A.; Papa-Adams, A.; Xu, Q.; Packham, G.; Primrose, J.N.; Benyon, R.C.; Iredale, J.P. The desmoplastic reaction surrounding hepatic colorectal adenocarcinoma metastases aids tumor growth and survival via alphav integrin ligation. *Clin. Cancer Res.* **2008**, *14*, 6405–6413. [CrossRef]
201. Feng, J.; Tang, Y.; Xu, Y.; Sun, Q.; Liao, F.; Han, D. Substrate stiffness influences the outcome of antitumor drug screening in vitro. *Clin. Hemorheol. Microcirc.* **2013**, *55*, 121–131. [CrossRef] [PubMed]
202. Rehfeldt, F.; Engler, A.J.; Eckhardt, A.; Ahmed, F.; Discher, D.E. Cell responses to the mechanochemical microenvironment–implications for regenerative medicine and drug delivery. *Adv. Drug Deliv. Rev.* **2007**, *59*, 1329–1339. [CrossRef] [PubMed]
203. Rohwer, N.; Cramer, T. Hypoxia-mediated drug resistance: Novel insights on the functional interaction of HIFs and cell death pathways. *Drug Resist. Updat.* **2011**, *14*, 191–201. [CrossRef] [PubMed]
204. Plummer, R.; Wilson, R.H.; Calvert, H.; Boddy, A.V.; Griffin, M.; Sludden, J.; Tilby, M.J.; Eatock, M.; Pearson, D.G.; Ottley, C.J.; et al. A Phase I clinical study of cisplatin-incorporated polymeric micelles (NC-6004) in patients with solid tumours. *Br. J. Cancer* **2011**, *104*, 593–598. [CrossRef] [PubMed]
205. Nishiyama, N.; Okazaki, S.; Cabral, H.; Miyamoto, M.; Kato, Y.; Sugiyama, Y.; Nishio, K.; Matsumura, Y.; Kataoka, K. Novel cisplatin-incorporated polymeric micelles can eradicate solid tumors in mice. *Cancer Res.* **2003**, *63*, 8977–8983.
206. Tabata, Y.; Ikada, Y. Vascularization effect of basic fibroblast growth factor released from gelatin hydrogels with different biodegradabilities. *Biomaterials* **1999**, *20*, 2169–2175. [CrossRef]
207. Tabata, Y.; Nagano, A.; Ikada, Y. Biodegradation of hydrogel carrier incorporating fibroblast growth factor. *Tissue Eng.* **1999**, *5*, 127–138. [CrossRef]
208. Tabata, Y.; Hijikata, S.; Muniruzzaman, M.; Ikada, Y. Neovascularization effect of biodegradable gelatin microspheres incorporating basic fibroblast growth factor. *J. Biomater. Sci. Polym. Ed.* **1999**, *10*, 79–94. [CrossRef]
209. Tabata, Y.; Nagano, A.; Muniruzzaman, M.; Ikada, Y. In vitro sorption and desorption of basic fibroblast growth factor from biodegradable hydrogels. *Biomaterials* **1998**, *19*, 1781–1789. [CrossRef]
210. Ogawa, T.; Akazawa, T.; Tabata, Y. In vitro proliferation and chondrogenic differentiation of rat bone marrow stem cells cultured with gelatin hydrogel microspheres for TGF-beta1 release. *J. Biomater. Sci. Polym. Ed.* **2010**, *21*, 609–621. [CrossRef]
211. Yamamoto, M.; Ikada, Y.; Tabata, Y. Controlled release of growth factors based on biodegradation of gelatin hydrogel. *J. Biomater. Sci. Polym. Ed.* **2001**, *12*, 77–88. [CrossRef] [PubMed]
212. Lee, K.Y.; Nakagawa, T.; Okano, T.; Hori, R.; Ono, K.; Tabata, Y.; Lee, S.H.; Ito, J. Novel therapy for hearing loss: Delivery of insulin-like growth factor 1 to the cochlea using gelatin hydrogel. *Otol. Neurotol.* **2007**, *28*, 976–981. [CrossRef] [PubMed]
213. Nakagawa, T.; Sakamoto, T.; Hiraumi, H.; Kikkawa, Y.S.; Yamamoto, N.; Hamaguchi, K.; Ono, K.; Yamamoto, M.; Tabata, Y.; Teramukai, S.; et al. Topical insulin-like growth factor 1 treatment using gelatin hydrogels for glucocorticoid-resistant sudden sensorineural hearing loss: A prospective clinical trial. *BMC Med.* **2010**, *8*, 76. [CrossRef] [PubMed]
214. Kimura, Y.; Tabata, Y. Controlled release of stromal-cell-derived factor-1 from gelatin hydrogels enhances angiogenesis. *J. Biomater. Sci. Polym. Ed.* **2010**, *21*, 37–51. [CrossRef] [PubMed]
215. Hori, Y.; Inoue, S.; Hirano, Y.; Tabata, Y. Effect of culture substrates and fibroblast growth factor addition on the proliferation and differentiation of rat bone marrow stromal cells. *Tissue Eng.* **2004**, *10*, 995–1005. [CrossRef] [PubMed]
216. Akagawa, Y.; Kubo, T.; Koretake, K.; Hayashi, K.; Doi, K.; Matsuura, A.; Morita, K.; Takeshita, R.; Yuan, Q.; Tabata, Y. Initial bone regeneration around fenestrated implants in Beagle dogs using basic fibroblast growth factor-gelatin hydrogel complex with varying biodegradation rates. *J. Prosthodont. Res.* **2009**, *53*, 41–47. [CrossRef] [PubMed]

217. Hiraoka, Y.; Yamashiro, H.; Yasuda, K.; Kimura, Y.; Inamoto, T.; Tabata, Y. In situ regeneration of adipose tissue in rat fat pad by combining a collagen scaffold with gelatin microspheres containing basic fibroblast growth factor. *Tissue Eng.* **2006**, *12*, 1475–1487. [CrossRef]
218. Ozeki, M.; Tabata, Y. In vivo degradability of hydrogels prepared from different gelatins by various cross-linking methods. *J. Biomater. Sci. Polym. Ed.* **2005**, *16*, 549–561. [CrossRef]
219. Kikuchi, T.; Kubota, S.; Asaumi, K.; Kawaki, H.; Nishida, T.; Kawata, K.; Mitani, S.; Tabata, Y.; Ozaki, T.; Takigawa, M. Promotion of bone regeneration by CCN2 incorporated into gelatin hydrogel. *Tissue Eng. Part A* **2008**, *14*, 1089–1098. [CrossRef]
220. Nii, T.; Makino, K.; Tabata, Y. Influence of shaking culture on the biological functions of cell aggregates incorporating gelatin hydrogel microspheres. *J. Biosci. Bioeng.* **2019**, *128*, 606–612. [CrossRef]
221. Kellner, K.; Liebsch, G.; Klimant, I.; Wolfbeis, O.S.; Blunk, T.; Schulz, M.B.; Gopferich, A. Determination of oxygen gradients in engineered tissue using a fluorescent sensor. *Biotechnol. Bioeng.* **2002**, *80*, 73–83. [CrossRef] [PubMed]
222. Bradford, A. The role of hypoxia and platelets in air travel-related venous thromboembolism. *Curr. Pharm. Des.* **2007**, *13*, 2668–2672. [CrossRef]
223. Tajima, S.; Tabata, Y. Preparation and functional evaluation of cell aggregates incorporating gelatin microspheres with different degradabilities. *J. Tissue Eng. Regen. Med.* **2013**, *7*, 801–811. [CrossRef] [PubMed]
224. Hayashi, K.; Tabata, Y. Preparation of stem cell aggregates with gelatin microspheres to enhance biological functions. *Acta Biomater.* **2011**, *7*, 2797–2803. [CrossRef] [PubMed]
225. Inoo, K.; Yamamoto, M.; Tabata, Y. Preparation of cell aggregates incorporating gelatin hydrogel microspheres of sugar-responsive water solubilization. *J. Tissue Eng. Regen. Med.* **2020**, *14*, 1050–1062. [CrossRef]
226. Tajima, S.; Tabata, Y. Preparation of cell aggregates incorporating gelatin hydrogel microspheres containing bone morphogenic protein-2 with different degradabilities. *J. Biomater. Sci. Polym. Ed.* **2018**, *29*, 775–792. [CrossRef]

cancers

MDPI

Article

A Systematic Comparative Assessment of the Response of Ovarian Cancer Cells to the Chemotherapeutic Cisplatin in 3D Models of Various Structural and Biochemical Configurations—Does One Model Type Fit All?

Priyanka Gupta [1,2], Aline Miller [3], Adedamola Olayanju [3], Thumuluru Kavitha Madhuri [4,5] and Eirini Velliou [1,2,*]

1 Centre for 3D Models of Health and Disease, Division of Surgery and Interventional Science, University College London, London W1W 7TY, UK; priyanka.g.gupta@ucl.ac.uk
2 Bioprocess and Biochemical Engineering Group (BioProChem), Department of Chemical and Process Engineering, University of Surrey, Surrey GU2 7XH, UK
3 Manchester BIOGEL, 19F4, Mereside, Alderley Park, Alderley Edge, Chesire SK10 4TG, UK; a.miller@manchesterbiogel.com (A.M.); a.olayanju@manchesterbiogel.com (A.O.)
4 Department of Gynaecological Oncology Royal Surrey NHS Foundation Trust, Egerton Road, Guildford GU2 7XX, UK; docmadhuri231@gmail.com
5 Honorary Senior Lecturer in Cancer Research, School of Applied Sciences, University of Brighton, Huxley Building, Lewes Road, Brighton BN2 4GJ, UK
* Correspondence: e.velliou@ucl.ac.uk

Simple Summary: Epithelial Ovarian Cancer is considered to be a 'silent killer' and a challenge for gynaecological health across the world due to its asymptotic nature in the early stages, its late-stage diagnosis, high recurrence rate and resistance to currently available treatment methods (chemotherapy). These disheartening figures highlight the need for extensive in vitro studies to better understand this disease. A number of in vitro 3D models are currently available to aid in the study of ovarian cancer and its response to therapeutic methods. In this work, we report, for the first time, a comprehensive comparative study of three widely used 3D in vitro models for ovarian cancer, along with chemotherapy assessment of primary and metastatic cells. Our study highlights the importance of selecting an appropriate 3D in vitro platform, which is based on multiple factors including the origin of cells used, experimental time period and experimental design, even for one specific disease.

Abstract: Epithelial Ovarian Cancer (EOC) is a silent, deadly and aggressive gynaecological disease with a relatively low survival rate. This has been attributed, to some extent, to EOC's high recurrence rate and resistance to currently available platinum-based chemotherapeutic treatment methods. Multiple groups have studied and reported the effect of chemotherapeutic agents on various EOC 3D in vitro models. However, there are very few studies wherein a direct comparative study has been carried out between the different in vitro 3D models of EOC and the effect of chemotherapy within them. Herein, we report, for the first time, a direct comprehensive systematic comparative study of three different 3D in vitro platforms, namely (i) spheroids, (ii) synthetic PeptiGels/hydrogels of various chemical configurations and (iii) polymeric scaffolds with coatings of various extracellular matrices (ECMs) on the cell growth and response to the chemotherapeutic (Cisplatin) for ovary-derived (A2780) and metastatic (SK-OV-3) EOC cell lines. We report that all three 3D models are able to support the growth of EOC, but for different time periods (varying from 7 days to 4 weeks). We have also reported that chemoresistance to Cisplatin, in vitro, observed especially for metastatic EOC cells, is platform-dependent, in terms of both the structural and biochemical composition of the model/platform. Our study highlights the importance of selecting an appropriate 3D platform for in vitro tumour model development. We have demonstrated that the selection of the best platform for producing in vitro tumour models depends on the cancer/cell type, the experimental time period and the application for which the model is intended.

Keywords: epithelial ovarian cancer; tissue engineering; 3D in vitro model; chemotherapy; Cisplatin; spheroids; hydrogels; polymeric scaffolds; A2780; SK-OV-3

Citation: Gupta, P.; Miller, A.; Olayanju, A.; Madhuri, T.K.; Velliou, E. A Systematic Comparative Assessment of the Response of Ovarian Cancer Cells to the Chemotherapeutic Cisplatin in 3D Models of Various Structural and Biochemical Configurations—Does One Model Type Fit All? *Cancers* **2022**, *14*, 1274. https://doi.org/10.3390/cancers14051274

Academic Editor: Donatella Aldinucci

Received: 13 January 2022
Accepted: 25 February 2022
Published: 1 March 2022

1. Introduction

Epithelial Ovarian Cancer (EOC) is the seventh most common cancer amongst women across the world and the third most common gynaecological cancer, ranked after cervical and uterine [1]. Even with extensive advancements in the field of cancer diagnosis and treatment, the 5-year survival rate of ovarian cancer is only around 30% [2] and it is the most common cause of gynaecological cancer-related deaths worldwide [3,4]. The relatively high mortality rate of ovarian cancer is attributed to its asymptotic nature, delayed onset and recognition of symptoms, lack of proper screening, high recurrence rate along with resistance to available chemotherapeutic methods of treatment [5,6]. The current gold standard for EOC treatment involves 'debulking' via reductive surgery in combination with the use of platinum-based chemotherapy involving chemotherapeutic agents such as Cisplatin and Carboplatin [7–9]. However, it has been observed that over 80% of patients have a relapse post chemotherapy, along with the development of a platinum-resistant aggressive form of EOC, all of which can be attributed, to a large extent, to EOC's complex tumour microenvironment (TME) [10–13]. The latter is a cocktail of different cellular, structural, biochemical (extracellular matrix (ECM) protein composition), biophysical and biomechanical features, all of which interact in complex and sometimes unknown ways with the tumour, leading to its progression, resistance to treatment and metastasis. Overall, these disheartening data suggest that there is an unmet need for ex vivo models of EOC in order to better understand the disease, EOC's complex TME and its unique mode of metastasis, as well as to predict patient-specific, personalised drug responses.

Similar to other cancers, traditionally, EOC studies including treatment screening are carried out either in (i) 2D in vitro systems such as T-Flasks and petri dishes [14–18] or (ii) in vivo, in animal models such as mice, rats and hens [19–21]. Firstly, 2D tumour models are easy to use, reproducible and are generally responsive to most therapeutic methods [22–24]. However, they are unable to capture key properties of the in vivo TME, including cell–cell interactions, cell–ECM interactions, structure, stiffness, spatial orientation and various environmental gradients [25,26] that are typically formed in a 3D tumour during growth and progression. In vivo animal models are currently considered to be the gold standard for therapeutic assessment, as they are more realistic in terms of capturing the in vivo organ complexity in comparison to 2D systems [27–30]. However, these models are expensive, difficult to reproduce and time-consuming. Additionally, there is evidence suggesting that they undergo genetic changes that differ from the evolutionary course of human diseases, raising further concern regarding their validity as models for personalised treatment [31–34].

Thus, 3D in vitro models are slowly emerging to tide over these issues associated with 2D as well as animal models. To date, 3D models used for ovarian cancer in vitro research include (i) cell spheroids, (ii) hydrogels and (iii) natural or synthetic biomaterial-based polymeric scaffolds. Spheroids are the most commonly used 3D in vitro models for ovarian cancer [24,35–42]. The earliest known spheroid model of EOC was reported in 1995 by Griffon et al. [43], wherein they developed cellular aggregates/spheroids from patient samples and exposed them to photon radiation (0–8 Gy), followed by analysis after 7 days post-treatment. They reported that the effect of radiation on the spheroids was dependent on the spheroids' size and proposed that spheroids are able to mimic the patient-specific radio-response to a large extent [43]. Since then, many groups have reported the feasibility of using EOC spheroids developed via different methods of fabrication, as suitable 3D models for obtaining an understanding of EOC and for predicting therapeutic outcomes in vitro [35–38,40,41,44–49]. In general, these therapeutic assessment studies report observing higher chemoresistance in spheroid models in comparison to 2D monolayer models for different established cell lines as well as patient samples. For example, recently, Gunay et al. (2020), in their comparative study between OVCAR-3 and OVCAR-8 cell lines, reported that different cell lines have different responses to Taxol and Cisplatin [44]. Similarly, Raghavan et al. (2017) reported that different patient samples responded differently to the chemotherapeutic treatment in spheroid-based 3D models, highlighting the feasibility of using the model for the personalised screening of therapy [35]. An extensive comparison of 3D spheroid models using 16 different commercially available EOC cell lines was carried out by Heredia-Soto et al. (2018). Similarly to other groups, they reported that 3D models for the different cell lines showed higher resistance to Cisplatin treatment in comparison to 2D monolayer models [38]. The study also reported that, for most cell lines, there was increased expression of master EMT regulators in 3D models in comparison to 2D, showing the advantage of 3D models in capturing the biochemical 3D features of the EOC TME.

Although simple cell-based spheroids are proven to be feasible models for therapeutic assessment and very promising tools especially for fast drug screening, they come with certain inherent constraints, including limited culture time; the formation of unrealistically high gradients of nutrients and oxygen, which can lead to the formation of extreme/extended necrotic cores at the spheroid centre; a lack of robust ECM mimicry and a lack of spatial architecture and structural orientation [26,32,50,51]. However, 3D models based on hydrogels and polymeric scaffolds can solve some of the issues associated with simple cell-based spheroid models. More specifically, they can sustain a longer culture period (several weeks), they enable better diffusion of biochemical reagents and they allow for the presence of specific ECM proteins, spatial orientation and a relatively defined and tuneable architecture [50,51]. Several groups have developed and used both natural and synthetic material-based hydrogels as 3D models of EOC for the purpose of therapeutic assessment [52–62]. For example, Yang and Zhao (2011) carried out a comparative study between a collagen I hydrogel and a RADA16-I peptide hydrogel with different cell lines (A2780, A2780/DDP and SK-OV-3) using three different chemotherapeutic agents (5-FU, Paclitaxel and curcumin). They reported that synthetic hydrogels with RADA16-I peptide were able to maintain all three cell lines in culture and, similarly to spheroids, hydrogel-based 3D models of EOC showed approximately 2- to 5-fold higher chemoresistance in comparison to 2D monoculture [53]. Liu et al. (2018) reported the growth of patient-derived platinum-sensitive and platinum-resistant cell lines within a collagen I hydrogel, for 7 days, wherein they observed the overexpression of mesenchymal markers (N-cadherin, vimentin and fibronectin) and transcriptional factors (snail and slug) along with higher chemoresistance in comparison to 2D monolayer systems [61]. Chen et al. (2014) used a commercially available Basement Membrane Extract (BME) hydrogel to compare the characteristics of a $CD44^{+}/CD117^{+}$ double-positive EOC cancer stem cell (CSC) population between 3D (the hydrogel) and a 2D culture system [59]. They attributed the increased chemoresistance of the CSC population in the 3D hydrogel to the increased expression of ABCG2, ABCB1, MMP-2 and MMP-9 as compared to 2D. The longest study for EOC using hydrogels as 3D models was conducted by Loessner et al. (2010), wherein EOC cell lines OV-MZ-6 and SK-OV-3 were cultured in PEG hydrogels biofunctionalised with RGD, Gln and MMP-sensitive sites for 14 days, followed by treatment using Paclitaxel and 7-day post-treatment analysis [58].

Currently, there are very few reported studies wherein polymeric scaffolds have been used for the development of ovarian cancer 3D models [63–66]. However, to the best of our knowledge, there are currently no reported studies wherein polymeric scaffold-based 3D models of EOC have been used for therapeutic assessment. Girard et al. (2013) developed a nanofibrous polymeric scaffold composed of PLGA and mPEG-PLA polymers (3P scaffold), via electrospinning, to culture BG-1 ovarian cancer cell lines for up to 5 days, wherein the EOC cells were shown to be growing as cell aggregates [63]. Ul-Islam et al. (2019) developed a 3D model of EOC using A-2780 ovarian cancer cell lines and a chitosan and bacterial cellulose-based polymeric scaffold and maintained it for 7 days [65].

In addition to these static monocellular models of EOC (containing cancer cells only), efforts have also been made to study the effect of shear stress and fluid flow, as well as the presence of other stromal cells, e.g., mesothelial cells and adipocytes, in EOC 3D models [62,66–73].

Despite extensive advancement in the field of EOC in vitro 3D models, there are very few publications available for a direct comparative study between the various types of in vitro 3D models, i.e., simple cell spheroids vs. hydrogels vs. polymeric scaffolds, and most of them compare either different types of hydrogels or spheroid models prepared via different fabrication methods [52,74]. For example, Zheng et al. (2014), carried out a comparative study between hydrogels from collagen I, Matrigel, alginate and agarose using SK-OV-3 cell lines, injected in vivo in a nude mice model for a period of 4 weeks [52]. They reported that tumour formation by SK-OV-3 cells was best supported by collagen, followed by Matrigel, alginate, control (cell suspension only) and agarose in vivo, along with increased MMP activity and upregulated expression of laminin, fibronectin, HIF-1α and VEGF-A in collagen I hydrogels. They concluded that the bioactive and biomimetic hydrogels were superior to 'inert' (i.e., lacking in native ligands that allow mammalian cells to attach) hydrogels at promoting tumour regeneration/growth. A comparative study between different fabrication methods for spheroid formation (polydimethylsiloxane-based microfluidic chips, ultra-low-attachment plates and hanging drop method) and their effect on drug sensitivity for carboplatin was carried out by Patra et al. (2020), wherein they highlighted the challenges of choosing appropriate preclinical models for drug testing [74].

In this current study, we have taken a step further to systematically compare the chemotherapeutic (Cisplatin) response (cell viability and apoptosis analysis) between spheroid, hydrogel and polymeric scaffold-based 3D in vitro models of epithelial ovarian cancer. For such a comparative study, we have used two different cell lines, namely (i) A2780, which is derived from the ovary (primary tumour), and (ii) SK-OV-3, which is derived from the ascites fluid (metastatic site), to identify the impact of cell origin on the growth and drug response in various 3D systems.

2. Materials and Methods

2.1. Cell Culture

Human Epithelial Ovarian Cancer (EOC) cell lines A2780 (Merck, Gillingham, UK) and SK-OV-3 (HTB 77, ATCC, Teddington, UK) were cultured in RPMI-1640 (Thermo Fisher Scientific, Loughborough, UK) and McCoy's 5a (Thermo Fisher Scientific, Loughborough, UK) media, respectively, supplemented with 10% Foetal Bovine Serum (FBS, Thermo Fisher Scientific, Loughborough, UK), 2 mM glutamine (Merck, Gillingham, UK) and 1% antibiotic–antimycotic (Thermo Fisher Scientific, Loughborough, UK) in a humidified incubator at 37 °C and 5% CO_2. Both cell lines were passaged regularly upon reaching 75–80% confluency with Typle E (Thermo Fisher Scientific, Loughborough, UK) till the required cell densities were obtained. The two cell lines were selected to reflect two different stages of ovarian cancer. A 2780 cell line was derived from the ovarian tumour of an untreated patient while SK-OV-3 was derived from the ascites fluid post-metastasis.

2.2. 3D Cell Culture and Chemotherapeutic Treatment on Polymeric Scaffolds

Polyurethane (PU) polymeric scaffolds were prepared via the Thermal-Induced Phase Separation (TIPS) method, sterilised and coated (surface modified with physisorption) with the ECM proteins fibronectin and collagen I, as previously described [75–77]. Both proteins are important elements of the EOC TME [78,79]. The scaffolds were highly porous (85% porosity) and they were microporous, with interconnected pores with an average pore diameter of 100–150 μm. A2780 and SK-OV-3 cell lines were seeded in the scaffolds ($5 \times 5 \times 5$ mm^3) at a seeding density of 0.5×10^6 cells/scaffold and cultured for 28 days. Post 28 days (4 weeks) of culture, the chemotherapeutic agent Cisplatin (Merck, Gillingham, UK) was added to the culture at a concentration of 50 μM for 1 feeding cycle (48 h) and removed thereafter. The concentration of Cisplatin was selected based on published IC_{50} data for A2780 and SK-OV-3 in 3D spheroids [38]. The scaffolds were then characterised 24 h post-treatment with sectioning, staining, microscopy and image analysis.

2.3. 3D Cell Culture and Chemotherapeutic Treatment on Synthetic Hydrogels

Synthetic PeptiGels (Manchester BIOGEL, Manchester, UK) were also used for our study with A2780 and SK-OV-3 cells. An initial fast screening was carried out between 4 different hydrogels of different stiffness and charge: α1 (5 kPa, neutral charge), α2 (10 kPa, medium charge), α3 (5 kPa, low charge) and α4 (1 kPa, high charge) [60,80]. Based on this preliminary screening (data not shown), the α4 PeptiGel was selected for further long-term studies as it led to optimal cell proliferation and longer hydrogel chemical stability. In order to incorporate ECM mimicry, α4 hydrogel conjugated with RGD (cell binding sequence found in fibronectin) and GFOGER (integrin binding site withing collagen I) were also used. The incorporation of these ECM protein conjugates enabled a more biomimetic synthetic system and also allowed us to compare the hydrogels with our protein-coated PU scaffolds (see Section 2.4 below). PeptiGels were used as per manufacturer's instructions. Briefly, both A2780 and SK-OV-3 cells were encapsulated via physical mixing of 100 μL of cell suspension into 1 mL of hydrogel, providing a final cell concentration of 0.5×10^6 cells/mL. Thereafter, cells were mixed to ensure a homogeneous solution. Aliquots of 200 μL of hydrogels with cells were pipetted into 24-well cell culture inserts with 1 mL of cell culture media added to each well and 200 μL on top of each gel within the inserts. Thereafter, the cell culture plates were incubated at 37 °C and 5%CO_2. Cell culture medium was changed every 20 min for the first hour to calibrate the hydrogel to neutral pH and every 2 days thereafter and cultured for a period of 3 weeks. Post 3 weeks of culture, the chemotherapeutic agent Cisplatin (Merck, Gillingham, UK) was added to the culture at a concentration of 50 μM for 1 feeding cycle (48 h) and removed thereafter. Hydrogels were then assessed 24 h post-treatment via staining, microscopy and image analysis (see following sections).

2.4. 3D Cell Culture and Chemotherapeutic Treatment on Cellular Spheroids

Spheroids of A2780 and SK-OV-3 cell lines were fabricated using specialised 96-well round-bottom plates, provided by faCellitate (Manheim, Germany). More specifically, to fabricate the spheroids, 200 μL of cell suspension containing 25,000 cells was seeded in each well and cultured for 1 week with media change every 2 days. Post 1 week of culture, chemotherapeutic agent Cisplatin (Merck, Gillingham, UK) was added to the culture at a concentration of 50 μM for 1 feeding cycle (48 h) and removed thereafter. Spheroids were then assessed 24 h post-treatment via staining, microscopy and image analysis (Cytation 5, BioTek, Agilent Technologies, Stockport, UK).

2.5. Spatial Evaluation of Live and Dead Cells via Imaging

To visualise the spatial distribution of live and dead cells pre- and post-treatment, model-specific methods were used. More specifically, PU scaffolds were collected at appropriate time points, snap-frozen in liquid nitrogen for 15 min and then preserved at −80 °C for further analysis, as previously described [75–77]. This method has been widely used in the field of tissue engineering for sample preservation without harming the cells [81–83]. Prior to analysis, scaffolds were sectioned and washed twice with PBS. For live/dead cell analysis, a Live/Dead Viability/Cytotoxicity Kit was used (Molecular Probes, Thermo Scientific, Loughborough, UK). Scaffold sections were stained with 2 μM of Calcein-AM (4 mM stock) and 4 μM of ethidium homodimer (2 mM stock) and were then incubated at 37 °C for 1 h. The solution was then removed, and the samples were washed twice in PBS, followed by imaging using a Nikon Ti-Eclipse inverted confocal microscope (Nikon Instruments, Surbiton, UK).

PeptiGels and spheroids were stained and imaged live at appropriate time points. Unlike the PU polymeric scaffolds, snap-freezing the spheroids and the hydrogels was not feasible as this process would have destroyed their native structure. Hydrogels and spheroids were stained with 2 μM of calcein-AM (4 mM stock) and 4 μM of ethidium homodimer (2 mM stock) and were incubated at 37 °C for 2 h for proper penetration of the dyes. The solution was then removed, and the samples were washed twice in PBS followed by imaging. PeptiGels were imaged whole using a Nikon Ti-Eclipse inverted confocal microscope (Nikon Instruments, Surbiton, UK). Spheroids were imaged using Cytation 5 Cell Imaging Reader (BioTek, Agilent Technologies, Stockport, UK).

2.6. Spatial Evaluation of Apoptotic Cells (Caspase 3/7 Activity) via Imaging

The caspase 3/7 activity was visualised and quantified in situ to assess the induction of cellular apoptosis after different treatments. As described in Section 2.5, model-specific methods were used to stain and image the different 3D models. Post processing the different 3D models similarly to Section 2.5, samples were incubated in culture medium containing (i) the Cell Event Caspase-3/7 green detection reagent (Fisher Scientific, Loughborough, UK) and (ii) DAPI, 1:200 (Fisher Scientific, Loughborough, UK) for 1 h (PU scaffolds) and 2 h (PeptiGels and spheroids) at 37 °C. The presence of caspase 3/7 positive cells (green) was immediately evaluated with a Nikon Ti-Eclipse inverted confocal microscope (Nikon Instruments, Surbiton, UK) for PU scaffolds and PeptiGels. Spheroids were imaged using Cytation 5 Cell Imaging Reader (Biotek, Agilent Technologies, Stockport, UK).

2.7. Advanced Microscopy Imaging

Immunofluorescent samples of PU scaffolds and PeptiGels (prepared as described in Sections 2.5 and 2.6 above) were imaged on a Nikon Ti-Eclipse inverted confocal microscope (Nikon Instruments, Surbiton, UK) and processed with the NIS-Elements software, using 405, 488 and 561 nm lasers for DAPI (blue), green fluorescence (calcein and caspase 3/7) and ethidium homodimer (red) staining, respectively. Confocal images were captured using a 10× objective and a 5–10 μm Z-stack distance. The same acquisition conditions were used for the positive controls. Cytation 5 Cell imaging Reader (BioTek, Agilent Technologies, Stockport, UK) was used with similar lasers and 10× magnification with Z-stacking and montage creation to image the spheroid models, using 405, 488 and 561 nm lasers for DAPI (blue), green fluorescence (calcein and caspase 3/7) and ethidium homodimer (red) staining, respectively. Imaging was carried out using 10× objective along with Z-stack and the Montage feature of the instrument in order to image the complete spheroid construct.

Multiple samples as well as multiple areas and multiple sections per sample were imaged for all models under study to ensure reproducibility. Representative images are presented in this manuscript.

2.8. Image Analysis and Quantification

Image-based quantification was carried out in a model-specific manner. Within PU scaffolds and PeptiGels, for the quantitative evaluation of (i) live (green) and dead (red) populations, as well as (ii) caspase-positive/apoptotic (green) and non-apoptotic (blue) populations of each image, the percentage of green vs. red (live/dead) or green vs. blue (caspase-positive/caspase-negative) areas of each image was calculated using Image J® software (Wayne Rasband, NIH, Bethesda, MD, USA). The particle analyser macro (Image J®, Wayne Rasband, NIH, Bethesda, MD, USA) was used in each individual channel (green or read for live/dead and green or blue for caspase 3/7-DAPI, respectively). Multiple samples ($n \geq 3$) were imaged and analysed for statistical relevance.

In the spheroid model, the average mean grey values for calcein (Live-Dead) and caspase 3/7 (apoptotic) were calculated using the particle measure macro in Image J® software (Wayne Rasband, NIH, Bethesda, MD, USA). Multiple samples ($n \geq 3$) were imaged and analysed for statistical relevance.

2.9. Statistical Analysis

Statistical analysis was performed for at least 3 independent experiments with at least 3 replicates per time point ($n \geq 3$, $n \geq 3$). Analysis of variance (one-way ANOVA) followed by the Bonferroni's multiple comparison test or T-Test (depending on sample) using the GraphPad Prism® software (version 8.00 for Windows) were carried out depending on samples, in order to find statistically significant differences between data ($p < 0.05$). Untreated samples were considered as controls in all cases. The error bars in the graphs represent the standard error of mean.

3. Results

3.1. Assessment of the Impact of the Chemotherapeutic Cisplatin on EOC Cells in a 3D Spheroid Model

As mentioned before, spheroids are the first in vitro 3D culture system established in tissue engineering and they have been extensively used for the therapeutic assessment of EOC [35,43,44]. In this study, Cisplatin, a widely used chemotherapeutic agent for EOC [35,44], was introduced to the culture medium for both A2780 and SK-OV-3 EOC spheroids on day 7 of culture [40,44] at a concentration of 50 μM for 48 h [38] (Section 2.4). Thereafter, Cisplatin-containing medium was replaced with fresh medium and the spheroids were maintained for 24 h, followed by post-treatment analysis of viability and apoptosis. More specifically, spheroid staining, imaging and image processing were carried out as described in Sections 2.5–2.8, for spatial assessment and the quantification of the impact of Cisplatin on the cell viability and cell apoptosis for A2780 and SK-OV-3 spheroids (Figures 1 and 2). A2780 spheroids were larger in size (2194 $\pm$ 130 μm) in comparison to SK-OV-3 spheroids (1305 $\pm$ 203 μm). Post treatment with Cisplatin, aggregate loosing was observed for both cell lines, with a resulting increase in spheroid mean diameter (A2780 = 2477 $\pm$ 168 μm; SK-OV-3 = 2093 $\pm$ 37 μm).

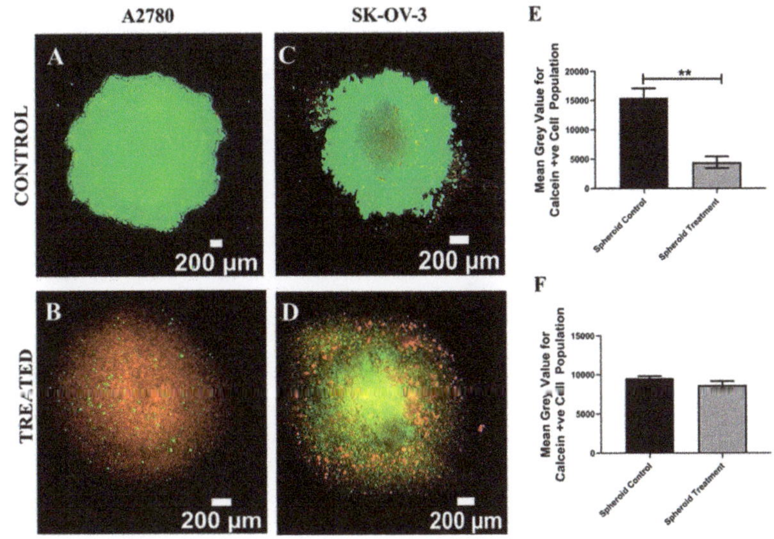

Figure 1. Effect of the chemotherapeutic Cisplatin on the viability EOC spheroids 24 h post-treatment (**A–D**): Representative images for live–dead (green–red) staining for untreated (control) and Cisplatin-treated A2780 and SK-OV-3 spheroids. (**E**) Image analysis-based quantification of live (green) image areas for A2780-treated and untreated spheroids. (**F**) Image analysis-based quantification of live (green) image areas for SK-OV-3-treated and untreated spheroids. Scale bar = 200 μm. Quantitative data represent mean $\pm$ SEM for multiple images ($\geq$3) and multiple spheroids ($\geq$3). ** $p \leq 0.01$.

Figure 1A–D show representative images for live–dead staining for A2780 and SK-OV-3 spheroids, for both Cisplatin-treated and untreated (control) spheroids. Figure 1E,F show the equivalent quantification of the percentage of live areas from image-based analysis.

As can be seen in Figure 1, 24 h post Cisplatin treatment, A2780 spheroids show a statistically significant decrease in cell viability as compared to untreated spheroids (Figure 1A,B,E). In contrast, no significant cell death was observed for SK-OV-3 cell lines 24 h post-treatment. This suggests that 50 μM of Cisplatin had an immediate and extremely damaging effect on A2780 spheroids, while its effect on SK-OV-3 viability in spheroids was much less, indicating some degree of resistance to Cisplatin.

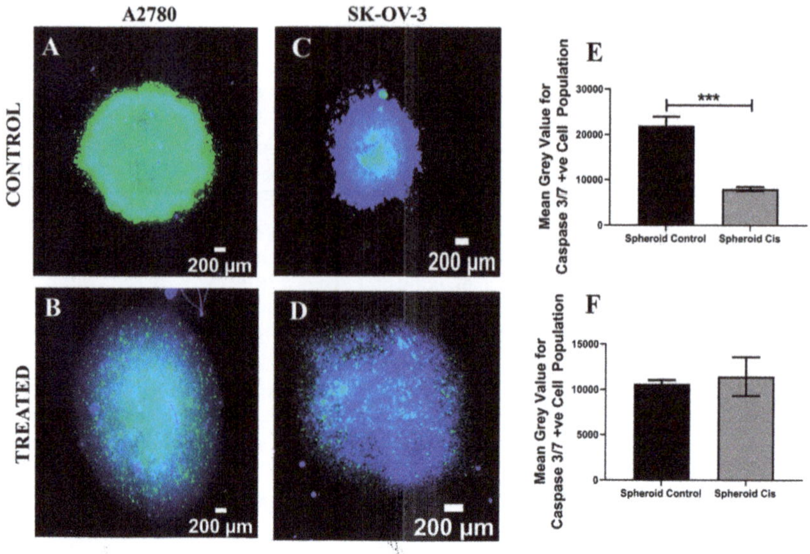

Figure 2. Effect of the chemotherapeutic Cisplatin on the apoptosis of EOC spheroids 24 h post-treatment (**A–D**): Representative images for caspase 3/7 (apoptosis)–DAPI (green–blue) staining for untreated (control) and Cisplatin-treated A2780 and SK-OV-3 spheroids. (**E**) Image analysis-based quantification of apoptotic (green) image areas for A2780-treated and untreated spheroids. (**F**) Image analysis-based quantification of apoptotic (green) image areas for SK-OV-3-treated and untreated spheroids. Scale bar = 200 µm. Quantitative data represent mean ± SEM for multiple images (≥3) and multiple spheroids (≥3). *** $p \leq 0.001$.

In terms of apoptosis induction, we noticed a high amount of caspase 3/7 positive (apoptotic) cells present even in the untreated spheroids for both the cell lines, with A2780 control spheroids having a significantly higher apoptotic cell number in comparison to the Cisplatin-treated samples (Figure 2A,B,E). For SK-OV-3, both treated and untreated spheroids showed a high amount of apoptotic cells (Figure 2C,D,F). This suggests that both EOC cell lines had started undergoing programmed cell death post 10 days of culture in a spheroid-based culture, most likely due to diffusional and spatial limitations that are inherent to this 3D culture method.

3.2. *Assessment of the Impact of the Chemotherapeutic Cisplatin on EOC Cells in a Synthetic Peptide Hydrogel 3D Model*

As mentioned in Section 2.3, peptide-based synthetic hydrogels were used for long-term culture (3 weeks) and Cisplatin chemotherapeutic assessment of A2780 and SK-OV-3 ovarian cancer cell lines. At the first stage, a preliminary screening was carried out to compare between hydrogels and different stiffness and charge combinations, and α4 PeptiGel was selected for long-term study as it was the hydrogel supporting the best viability and hydrogel stability for the timeframe of our experiments, i.e., 3 weeks (see also Section 2.3). As described in Section 2.3, in order to incorporate ECM mimicry, which has a crucial effect on cancer development and treatment response [84–86], ECM matrix-conjugated α4 PeptiGels were also tested. More specifically, RGD- and GFOGER-conjugated hydrogels were tested, mimicking fibronectin and collagen, respectively. Cisplatin was introduced to the culture medium for both A2780 and SK-OV-3 hydrogels at 3 weeks [58] of culture and at a concentration of 50 µM for 48 h [38] (Section 2.3). Thereafter, Cisplatin-containing medium was replaced with fresh medium and the hydrogels were maintained in culture for 24 h followed by post-treatment analysis.

Figures 3–6 show image-based spatial assessment and quantification of the impact of Cisplatin on cell viability and apoptosis induction for A2780 and SK-OV-3 hydrogels of all three configurations. More specifically, Figures 3A and 4A show representative images of live–dead staining for A2780- and SK-OV-3-treated and untreated PeptiGels, respectively, while Figures 3B and 4B show the equivalent quantification (% live areas) from image-based analysis. As observed in Figures 3 and 4, both A 2780 and SK-OV-3 cells were able to attach and proliferate for all three PeptiGel configurations, i.e., α4, α4 + RGD, α4 + GFOGER. However, post Cisplatin treatment, we observed differences in cell viability between the two cell lines. More specifically, a significant loss of cell viability was observed in all three hydrogel configurations for A2780 cells 24 h post-treatment (Figure 3). In contrast, for SK-OV-3, only cells in the α4 hydrogel (without ECM inspired conjugated motifs) showed a significant decrease in cell viability, while SK-OV-3 cells cultured within the biomimetic hydrogels (α4 + RGD and α4 + GFOGER) did not show any significant decrease in cell viability in Cisplatin-treated PeptiGels as compared to untreated controls (Figure 4).

Analysis of the apoptotic marker caspase 3/7, for both cell lines, in the hydrogels shows a similar trend (Figures 5 and 6). More specifically, for both A2780 and SK-OV-3 cells, an increase in the number of apoptotic cells post Cisplatin treatment was observed in all three hydrogel configurations, but SK-OV-3 had overall less apoptotic cells/less induction of apoptosis in comparison to A2780 for all three hydrogel configurations (Figures 5B and 6B). Taken together, these data suggest that the metastatic cell line SK-OV-3 shows some degree of resistance to Cisplatin treatment in synthetic PeptiGels and that the chemoresistance is substantially increased in the presence of an ECM-inspired peptide conjugation within the hydrogels (Figure 6). In contrast, no such peptide-related Cisplatin resistance is observed for A2780 cells in the PeptiGels (Figure 5).

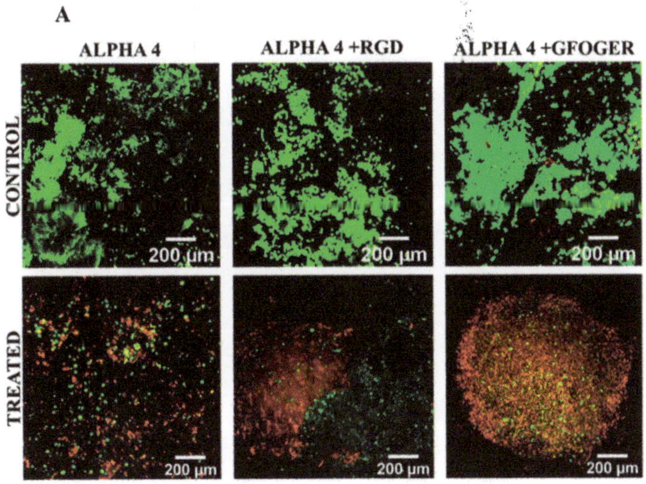

Figure 3. *Cont.*

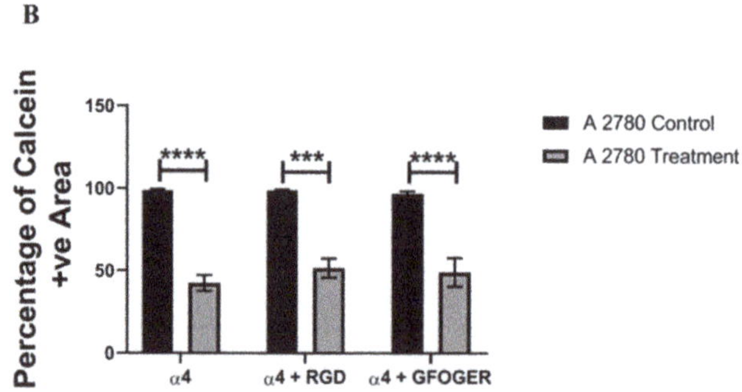

Figure 3. Effect of the chemotherapeutic Cisplatin on the viability of A2780 EOC cells grown in different synthetic PeptiGels, 24 h post-treatment (**A**): Representative images for live–dead (green–red) staining for both treated and untreated (control) A2780 PeptiGels. (**B**) Image analysis-based quantification of live (green) image areas for A2780 cells grown in the peptides. Scale bar = 200 μm. Quantitative data represent mean ± SEM for multiple images (≥3) and multiple hydrogels (≥3). *** $p \leq 0.001$, **** $p \leq 0.0001$.

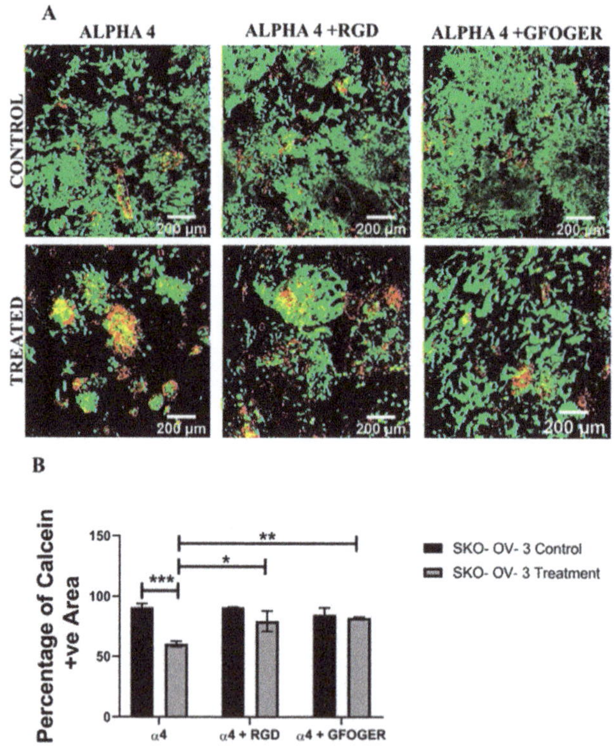

Figure 4. Effect of the chemotherapeutic Cisplatin on the viability of SK-OV-3 EOC cells grown in different PeptiGels, 24 h post-treatment (**A**): Representative images for live–dead (green–red) staining for both treated and untreated (control) SK-OV-3 PeptiGels. (**B**) Image analysis-based quantification of live (green) image areas for SK-OV-3 cells grown in the PeptiGels. Scale bar = 200 μm. Quantitative data represent mean ± SEM for multiple images (≥3) and multiple hydrogels (≥3). * $p \leq 0.05$, ** $p \leq 0.01$, *** $p \leq 0.001$.

3.3. Assessment of the Impact of the Chemotherapeutic Cisplatin on EOC Cells in an ECM Protein-Coated PU Polymeric Scaffold 3D Model

As previously mentioned, fibronectin- and collagen I-coated polymeric (PU) scaffolds were used for the development and maintenance of long-term (4 weeks) 3D in vitro models of EOC (using A2780 and SK-OV-3 cells). Cisplatin was introduced to the culture medium for both A2780 and SK-OV-3 cell lines 4 weeks into culture at a concentration of 50 μM for 48 h (Section 2.1). Thereafter, similarly to the spheroid (Section 3.1) and PeptiGel (Section 3.2) 3D models, Cisplatin-containing medium was replaced with fresh medium and the polymeric scaffolds were maintained in culture for 24 h, followed by post-treatment analysis of both Cisplatin-treated and untreated A2780 and SK-OV-3 scaffolds of various ECM coatings.

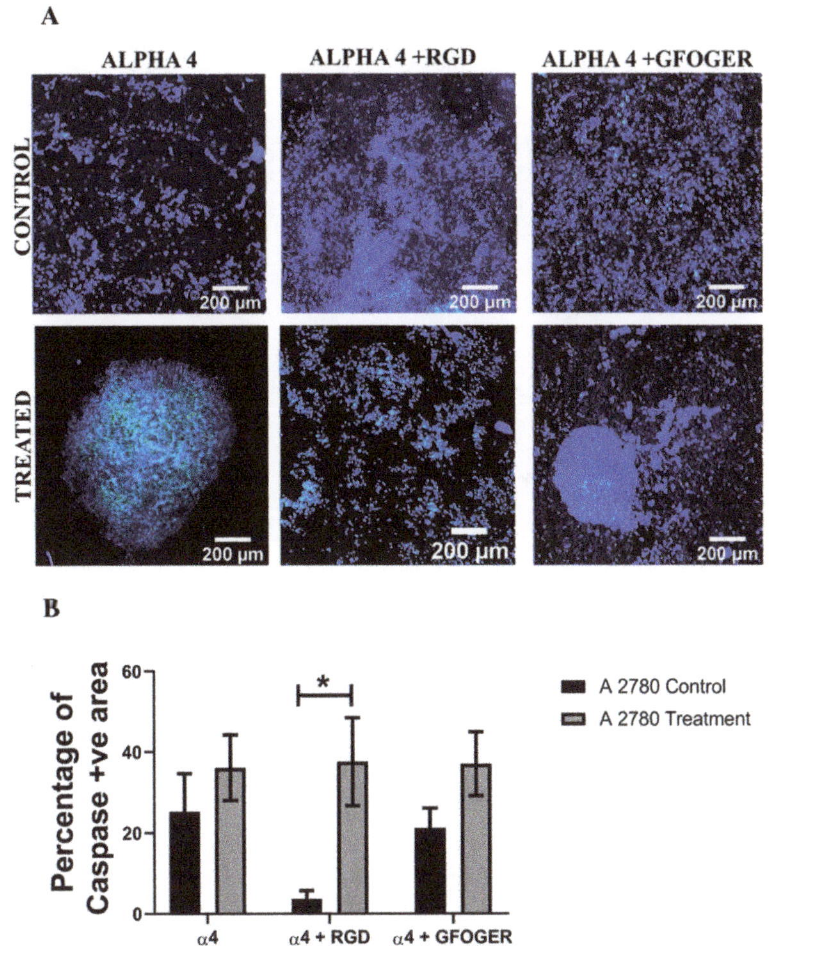

Figure 5. Effect of the chemotherapeutic Cisplatin on the apoptosis of A2780 EOC cells grown in different PeptiGels, 24 h post-treatment (**A**): Representative images for caspase 3/7 (apoptosis)–DAPI (green–blue) staining for both treated and untreated (control) A2780 PeptiGels. (**B**) Image analysis-based quantification of apoptotic (green) image areas for A2780 cells grown in the PeptiGels. Scale bar = 200 μm. Quantitative data represent mean ± SEM for multiple images (≥3) and multiple hydrogels (≥3). * $p \leq 0.05$.

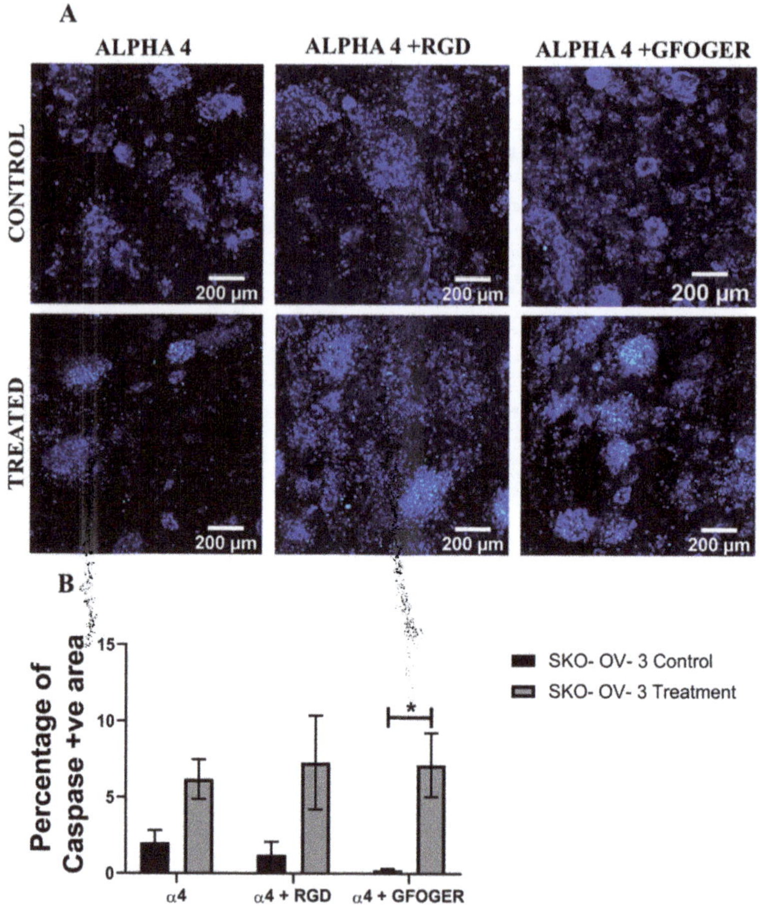

Figure 6. Effect of the chemotherapeutic Cisplatin on the apoptosis of SK-OV-3 EOC cells grown in different PeptiGels, 24 h post-treatment (**A**): Representative images for caspase 3/7 (apoptosis)–DAPI (green–blue) staining for both treated and untreated (control) SK-OV-3 PeptiGels. (**B**) Image analysis-based quantification of apoptotic (green) image areas for SK-OV-3 cells grown in the PeptiGels. Scale bar = 200 µm. Quantitative data represent mean ± SEM for multiple images (≥3) and multiple hydrogels (≥3). * $p \leq 0.05$.

Figures 7–10 show image-based spatial assessment and quantification of the impact of Cisplatin on the cell viability and apoptosis induction for A2780 and SK-OV-3 cells within fibronectin- and collagen I-coated PU scaffolds. More specifically, Figures 7A and 8A show representative images of live–dead analysis/staining for A2780 and SK-OV-3 scaffolds, respectively. Figures 7B,C and 8B,C show the equivalent quantification of the % of live cell population from image-based analysis on collagen I- and fibronectin-coated scaffolds for A2780 and SK-OV-3, respectively. As observed, PU-based scaffolds were able to maintain a long-term (4 weeks) viable culture for both A2780 and SK-OV-3 cells. Post application of Cisplatin, i.e., 24 h post-treatment, a significant decrease in cell viability was observed for both cell lines, irrespective of the coating (fibronectin or collagen I) of the PU scaffolds (Figures 7B,C and 8B,C).

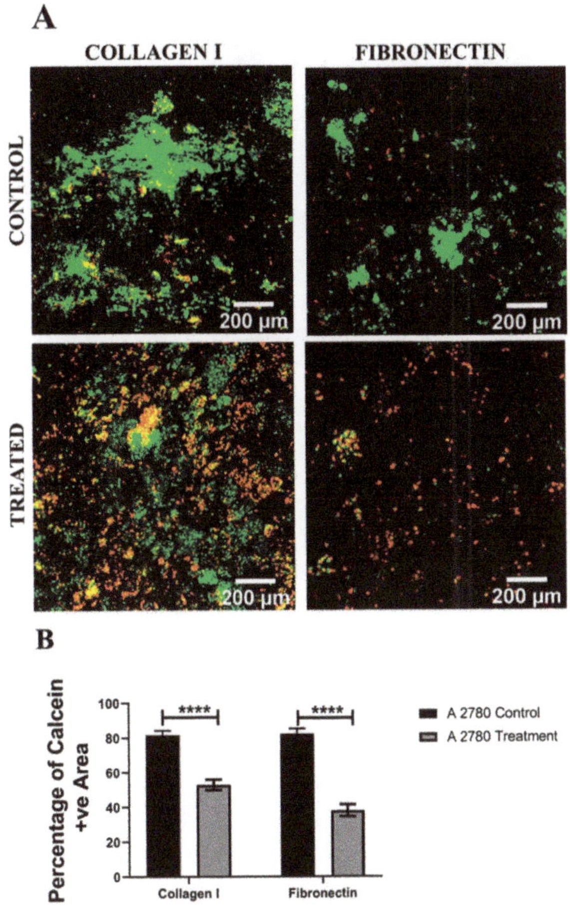

Figure 7. Effect of the chemotherapeutic Cisplatin on the viability of A2780 EOC cells grown in PU scaffolds of various protein coatings, 24 h post-treatment (**A**): Representative images for live–dead (green–red) staining in collagen I- and fibronectin-coated treated and untreated (control) A2780 polymeric scaffolds. (**B**) Image analysis-based quantification of live (green) image areas for A2780 cells grown in collagen I and fibronectin coated polymeric scaffolds. Scale bar = 200 μm. Quantitative data represent mean ± SEM for multiple images (≥3) and multiple scaffolds (≥3). **** $p \leq 0.0001$.

Further to cell viability, the induction of cellular apoptosis post Cisplatin treatment in the PU scaffolds was also assessed via caspase 3/7 staining. Figures 9 and 10 show representative images and image-based quantification of apoptosis for Cisplatin-treated and untreated (control) scaffolds for both A2780 and SK-OV-3 cells and both scaffold coatings, i.e., collagen I and fibronectin coating. It was observed that both cell lines showed an increase in the number of apoptotic cells within the scaffolds post Cisplatin treatment. Furthermore, both A2780 and SK-OV-3 showed significantly higher cellular apoptosis in fibronectin-coated scaffolds post-Cisplatin treatment (Figures 9C and 10C). No significant chemoresistance was observed for either of the cell lines within the PU scaffolds, irrespective of the ECM scaffold coating.

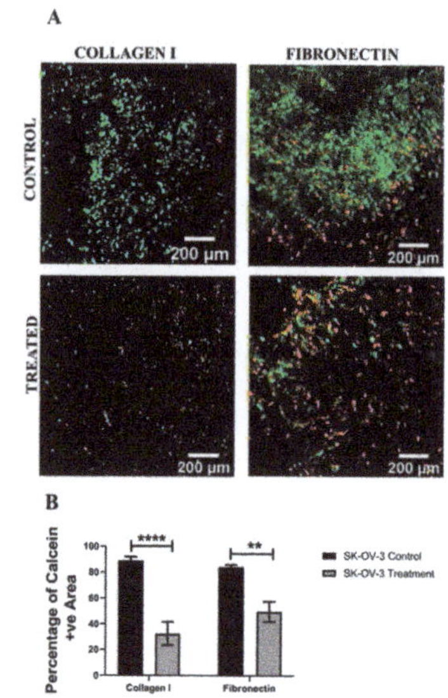

Figure 8. Effect of the chemotherapeutic Cisplatin on the viability of SK-OV-3 EOC cells grown in PU scaffolds of various protein coatings, 24 h post-treatment (**A**): Representative images for live–dead (green–red) staining in collagen I- and fibronectin-coated treated and untreated (control) SK-OV-3 polymeric scaffolds. (**B**) Image analysis-based quantification of live (green) image areas for SK-OV-3 cells grown in collagen I and fibronectin coated polymeric scaffolds. Scale bar = 200 µm. Quantitative data represent mean ± SEM for multiple images (≥3) and multiple scaffolds (≥3). ** $p \leq 0.01$, **** $p \leq 0.0001$.

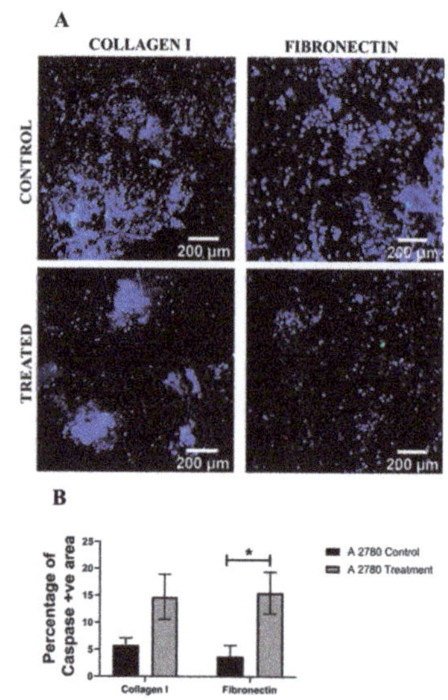

Figure 9. Effect of the chemotherapeutic Cisplatin on the apoptosis A2780 cells grown in PU scaffolds of various protein coatings, 24 h post-treatment (**A**): Representative images for caspase 3/7 (apoptosis)–DAPI (green–blue) staining in collagen I- and fibronectin-coated treated and untreated (control) A2780 polymeric scaffolds (**B**) Image analysis-based quantification of apoptotic (green) image areas for A2780 cells grown in collagen I and fibronectin coated polymeric scaffolds. Scale bar = 200 µm. Quantitative data represent mean ± SEM for multiple images (≥3) and multiple scaffolds (≥3). * $p \leq 0.05$.

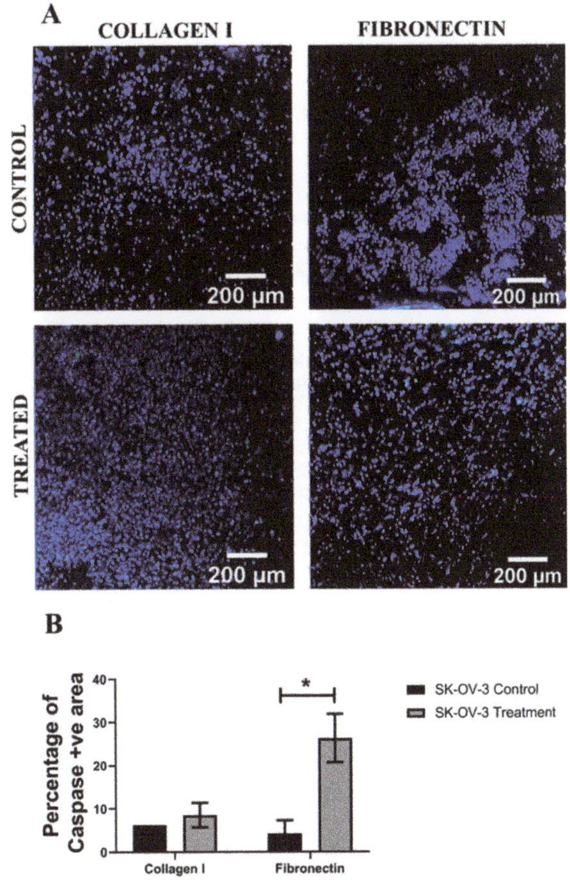

Figure 10. Effect of the chemotherapeutic Cisplatin on the apoptosis of SK-OV-3 cells grown in PU scaffolds of various protein coatings, 24 h post-treatment (**A**): Representative images for caspase 3/7 (apoptosis)–DAPI (green–blue) staining in collagen I- and fibronectin-coated treated and untreated (control) SK-OV-3 polymeric scaffolds. (**B**) Image analysis-based quantification of apoptotic (green) image areas for SK-OV-3 cells grown in collagen I and fibronectin coated polymeric scaffolds. Scale bar = 200 µm. Quantitative data represent mean ± SEM for multiple images (≥3) and multiple scaffolds (≥3). * $p \leq 0.05$.

4. Discussion

In this work, we have carried out a systematic comparative study to assess the effect of the type of 3D in vitro model/platform on the response of primary and epithelial ovarian cancer cells to the application of chemotherapy (Cisplatin). The 3D models used for this study were (i) simple cell spheroids, (ii) synthetic hydrogels/PeptiGels and (iii) polymeric scaffolds. Two different cell lines were used to assess the effect of the 'site of cell line origin', i.e., A2780, which is derived from the ovary (primary site), and SK-OV-3, which is derived from the ascites fluid (metastatic site). Due to the inherent structural differences between the three 3D models, they can be maintained in culture for different time periods. For example, the average time for which spheroids have been maintained in culture is between 6 and 12 days for EOC [35,43,44], while, for hydrogels, it is between 5 and 21 days [53,56,58,61]. There are currently very few studies involving polymeric scaffolds (PLGA–mPEG–PLA, bacterial cellulose–chitosan) and EOC and the average culture duration in these publications was 5–8 days only [63,65,66]. Based on these data, in our systematic study/comparison of different EOC 3D models, we selected different time points for the application of the chemotherapeutic agent Cisplatin. More specifically, Cisplatin was added to spheroids 7 days into culture, while, for hydrogels, it was added 21 days into the culture. Although not with EOC, our group has previously demonstrated that polymeric polyurethane (PU)-based scaffolds are able to support the long-term culture (4–5 weeks) of other cancer cells, including pancreatic cancer and melanoma, along with successful therapeutic assessment [75–77,87,88]. Hence, for the PU scaffold-based EOC model, Cisplatin was introduced at the end of 28 days in culture. Currently, there are very few studies where a comparative evaluation of different 3D in vitro models has been carried out for EOC, and they are restricted to comparing either different fabrication methods of the same 3D model, e.g., hanging drop or ultra-low plate for spheroids [74], or different materials such as collagen and alginate for the synthesis of hydrogels [52]. To the best of our knowledge, there are currently no reported studies that compare completely different types of 3D in vitro models for EOC and, with our current study, we are addressing this gap.

4.1. Spheroid EOC Model

As described in Section 2.2 (Methods and Materials), A2780 and SK-OV-3 spheroids were prepared using specialised round-bottom plates and maintained in culture for 7 days, followed by 48 h of Cisplatin (50 µM) treatment and analysis, i.e., viability and apoptosis (live–dead, caspase 3/7) 24 h post-treatment. For both cell lines, we observed morphological differences between untreated controls and Cisplatin-treated spheroids, with the Cisplatin-treated aggregates being less compact than the control spheroids (Figures 1 and 2). This is similar to observations made by Gunay et al. (2020), wherein they reported that the addition of Cisplatin at a concentration of 100 µM disrupted the morphology of EOC spheroids for OVCAR- 3 and OVCAR-8 cell lines [3]. Through live–dead and caspase 3/7 staining, along with image-based quantification (Figures 1 and 2), we observed a cell line-dependent response to the application of Cisplatin within our spheroids. More specifically, A2780 cells that originated from the ovary showed higher cell death 24 h post-chemotherapy in comparison to SK-OV-3 cells, which are ascites-derived (metastatic) (Figure 1). This suggested that SK-OV-3 were more resistant to Cisplatin in comparison to A2780 within our spheroid system. Such a cell line-dependent response for EOC to chemotherapeutic agents including Cisplatin has also been reported by other groups, including Raghavan et al. (2015) and Heredia-Soto et al. (2018) [38,40]. We also observed a certain degree of diffusion limitation within our spheroid models for calcein–ethidium homodimer as well as DAPI–caspase, especially for the compact untreated (control) spheroids (Figures 1A,C and 2A,C). It is well documented that diffusion limitation for nutrients, oxygen and even therapeutic agents is observed in spheroids with diameters higher than 200 µm [89–92]. Loessner et al. (2010) hypothesise that the phenomenon of cell spheroids displaying elevated chemoresistance to chemotherapeutic agents can be attributed to a number of mechanisms, including decreased penetration of the drugs, increased pro-survival signalling and/or upregulation of genes conferring drug resistance [58]. Although not for EOC, spheroid cultures' ability to display chemoresistance has also been attributed to the decreased penetration of chemotherapeutic reagents for other cancers (breast, lung and prostrate) by Stock et al. (2016) [92]. Finally, we also observed a high degree of cellular apoptosis for our control spheroids at day 10 of culture (Figure 2A,C), suggesting that the EOC cells within the spheroids had started undergoing programmed cell death, most likely due to a lack of structural integrity and increased diffusional limitations of nutrients and oxygen. This suggests that spheroid models are not suited for long-term culture and that they are more suitable models for rapid therapeutic assessment.

4.2. Hydrogel-Based EOC Models

PeptiGels, commercially available synthetic peptide-based materials (Manchester BIOGEL, Manchester, UK) were used to develop our EOC hydrogel models (Section 2.3,

Methods and Materials). More specifically, α4 PeptiGel with a stiffness of 1 kPa was used, along with its RGD- and GFOGER-conjugated versions, to compare between a purely synthetic hydrogel and its ECM biomimetic versions, i.e., with RGD mimicking fibronectin and GFOGER mimicking collagen. A2780 and SK-OV-3 EOC cells were grown in all three hydrogel configurations for 21 days, followed by a 48 h Cisplatin (50 μM) treatment and a 24 h post-treatment analysis, i.e., viability, apoptosis (live–dead, caspase 3/7). Both cell lines were able to attach and proliferate and were viable in all three hydrogel configurations for the entire duration of the experiment (Figures 3 and 4). Both cell lines showed a fairly uniform spread and growth within the hydrogels and no aggregates were observed. However, on application of Cisplatin, cellular aggregation within the matrix was observed, which was more pronounced for the A2780 cell line. Similarly to the spheroid model, we again observed a cell line-dependent response to Cisplatin within the hydrogels. More specifically, A2780 showed a significant decrease in cell viability irrespective of the gel type (Figure 3) and a corresponding increase in cellular apoptosis (Figure 5) post-treatment. In contrast, SK-OV-3 showed a significant decrease in cell viability on application of Cisplatin only within the pure α4 synthetic hydrogel. In the presence of RGD and GFOGER motifs, there was very little change in cell viability on Cisplatin application (Figure 4), although an increase in apoptotic cell numbers was observed post-treatment (Figure 6). These data suggest that SK-OV-3 cells show some degree of chemoresistance to Cisplatin within the biomimetic PeptiGels and highlights the importance of ECM proteins in therapeutic resistance for EOC. Although not directly highlighted, as with spheroid models, such cell line-dependent responses to various chemotherapeutic agents for EOC in hydrogel systems have also been reported by other groups in the form of differing IC_{50} [53,61]. For example, Liu et al. (2018) reported that the patient-derived OV-NC cell line had an IC_{50} of 92 ± 3.1 μmol/L in a collagen I gel for carboplatin, while, for the OV-206 patient-derived cell line, it was 154 ± 5.9 for the same drug [61]. They promote two key hypotheses: (i) the chemoresistance observed within the hydrogels can be attributed to the limited delivery of drugs into the core of the tumour model and increased cell survival and (ii) the presence of the collagen matrix can limit the effect of chemotherapy by activating specific signalling pathways, contributing towards chemoresistance via epithelial–mesenchymal transition (EMT) [61]. These hypotheses justify our observations within the PeptiGels and highlight the role of ECM proteins in chemoresistance for EOC. The role of the RGD motif in conferring chemoresistance to EOC has also been reported by Bondong et al. (2012) in their analysis of an ovarian cancer patient tumour and ascites fluid. They reported that the overexpression of the L1 cell adhesion molecule (L1CAM) is linked to poor prognosis in patients. Specifically, L1CAM, present in the ascites fluid, contains an RGD motif and is linked to the development of chemoresistance of EOC amongst patients [93]. Their observation is in line with our data where we see increased chemoresistance for the ascites-derived cell line SK-OV-3 (metastatic) within RGD-conjugated PeptiGels (Figure 4). Similarly, the role of collagen in the chemoresistance of EOC has also been reported by some other researchers [84,94]. For example, Januchowski et al. (2016) reported the overexpression of a number of different types of collagen (COL1A1, COL5A2, COL1A2, COL15A1, COL3A1, etc.) in chemotherapy-resistant versions of different EOC cell lines. including A2780 and SK-OV-3 [84]. They attribute this to cell adhesion-mediated drug resistance (CAM-DR) and suggest that the interaction of ECM components, including collagen, with the cancer cells results in chemoresistance. These interactions can even change the apoptosis sensitivity and increase the drug resistance of cancer cells [85,86]. Our observation of chemoresistance for SK-OV-3 (Figures 4 and 6) in RGD- and collagen I (GFOGER)-conjugated PeptiGels could also be due to the CAM-DR phenomenon, although further studies are needed to validate this.

4.3. Polymeric Scaffold-Based EOC Models

We have previously developed a highly porous and biocompatible PU scaffold via the TIPS method and have shown that a number of different cell types can be cultured long-term within these scaffolds [75–77,87]. Based on the timeframe of cell culture in the polymeric scaffolds reported in our previous publications, EOC cell lines in this study were grown within the polymeric scaffolds for 28 days, followed by 48 h of Cisplatin (50 μM) treatment and a 24 h post-treatment analysis, i.e., for viability and apoptosis. Collagen I- and fibronectin-coated scaffolds were used in line with our previous observations on the importance of ECM mimicry within these scaffolds for both healthy and diseased (cancerous) cells [76,77]. The choice of collagen I and fibronectin also allowed us to compare between the ECM protein-coated scaffolds and the peptide-conjugated PeptiGels to some extent. As observed in Figures 7 and 8, both A2780 and SK-OV-3 cell lines were able to grow for 28 days within our polymeric scaffolds, irrespective of the coating protein. On application of Cisplatin, both cell lines showed a significant drop in cell viability within the scaffolds, irrespective of the ECM protein present (Figures 7 and 8). Similarly, the number of apoptotic cells also increased post-treatment within the scaffolds for both cell lines (Figures 9 and 10). For A2780, the cell response to Cisplatin within the PU scaffolds was similar to that observed within our spheroids and hydrogel models (Figures 1–3 and Figure 5) However, our results with SK-OV-3 in the scaffolds contradict our observations within the spheroid and hydrogel models, wherein the SK-OV-3 cell line showed some degree of Cisplatin resistance, which was particularly pronounced in the presence of ECM protein mimicry (RGD, GFOGER) within the hydrogel models (Figures 3 and 4). To the best of our knowledge, there is currently no available literature for the chemotherapeutic assessment of ovarian cancer within polymeric scaffolds to enable us to compare our data. It is possible that the relatively large pore size (100–150 μm) and highly interconnected pores of the PU scaffolds allowed for the extensive diffusion of Cisplatin, which affected A2780 and SK-OV-3′s response to chemotherapy. This is a feasible theory since diffusion limitation of therapeutic agents within spheroids and hydrogels has been considered to be one of the reasons for the chemoresistance of EOC cells observed within these systems [58,61]. Another reason for this difference in SK-OV-3′s response to Cisplatin can be the difference in stiffness between the models. The elastic modulus for our PU scaffolds is around 28 ± 3 kPa [77,95], while the elastic modulus of α4 hydrogel is only 1 kPa. SK-OV-3′s preferential chemoresistance on softer matrices has been reported by other groups too [96,97]. For example, Fan et al. (2021) cultured SK-OV-3 cell lines on glass sheets coated with hydrogel substrates of varying stiffness (0.5–25 kPa). They observed that SK-OV-3 showed higher chemoresistance to Cisplatin and Paclitaxel on softer substrates and linked it to the overexpression of ABC transporters ABCB1 and ABCB4 on the soft substrates, which are genes linked to the development of multidrug resistance [96]. McGrail et al. (2014) carried out a comparative study between SK-OV-3 (highly metastatic) and OVCAR-3 (less metastatic) cell lines on Polyacrylamide substrates coated with equal densities of collagen I with two different stiffness types, i.e., 2.83 kPa mimicking adipocytes and 34.88 kPa mimicking osteoblasts [97]. They reported that the SK-OV-3 cell line was more mechanosensitive than OVCAR-3, resulting in a display of higher malignancy, a mesenchymal phenotype and higher resistance to Carboplatin on a softer substrate [97]. The findings of these studies are in line with our observation of chemoresistance by the SK-OV-3 cell line on the softer spheroid and hydrogel models as compared to the stiffer PU polymeric scaffold model.

Overall, in this work, we have conducted a novel, systematic comparative study of the response of EOC to a chemotherapeutic (Cisplatin) in three different in vitro 3D models: (i) spheroids, (ii) synthetic PeptiGels/hydrogels and (iii) polymeric scaffolds with various ECM coatings. Two different cell lines (A2780 and SK-OV-3) were used to understand the impact of the 'site of origin' of the cells, as well as to assess platform versatility. We have reported that all three platforms were able to support EOC in vitro 3D models with both cell lines, albeit for different culture time points. Polymeric scaffolds and hydrogels were maintained for 4 weeks and 3 weeks, respectively (Figures 3, 4, 7 and 8), highlighting that they are suitable models for the long-term 3D culture of EOC cell lines. Spheroids were able to survive for a shorter time period of around 7 days (Figure 1). On application of

Cisplatin, A2780 cells (primary) showed a decrease in viable cell population across all three platforms (Figures 1, 3 and 7). In comparison, SK-OV-3 cells (metastatic) showed decreased cell viability on chemotherapeutic application only on the PU polymeric scaffold (Figure 8) but showed resistance to Cisplatin when grown as spheroids (Figure 1) or within hydrogels, with the latter being peptide conjugation-dependent (Figure 4). More specifically, SK-OV-3′s Cisplatin resistance within hydrogels was observed in the presence of conjugated ECM protein motifs (RGD and GFOGER), highlighting the importance of ECM proteins in the chemoresistance of metastatic EOC. Our data also show that the response to chemotherapy is dependent on the cell site/location of origin of the cells. To date, most comparative studies for different 3D models have usually used the same platform system and focused on either different materials or on different fabrication methods of the models. To the best of our knowledge, this is the first time that such a comparative study for three completely different 3D in vitro models, across different culture time points and with different cell lines, has been carried out.

5. Conclusions

The goal of the present study was to carry out, for the first time, a robust comparative study of the growth and chemotherapy response of EOC in three widely different in vitro 3D models: (i) cell spheroids, (ii) synthetic PeptiGels/hydrogels and (iii) polymeric scaffolds of various protein coatings. We have shown the feasibility of using all three models for the culture of EOC cell lines A2780 and SK-OV-3, representing primary and metastatic disease, respectively, and assessed the impact of chemotherapy (Cisplatin) on the cell viability and apoptosis within these models. Our study highlights that the selection of a 3D in vitro platform depends on (i) the planned experimental/assessment time period, (ii) the type of cell to be studied, (iii) the site of cell origin in vivo and (iv) the question that needs to be answered. For example, a rapid screening analysis may benefit from the use of simple cell spheroid models; however, the need to study the effect of ECM proteins on cell growth and chemotherapy response long-term will benefit from structured hydrogel or polymeric scaffolds. Similarly, softer tumours or tumours originating from soft tissues such as the ovary or the omentum may prefer less stiff 3D platforms such as spheroids or hydrogels as compared to stiffer polymeric scaffolds. In conclusion, our study highlights that, as with most tissue engineering applications, there is no 'one-size-fits-all model' [92] and the selection of an appropriate model requires careful assessment of the available input variables and the expected outputs.

Author Contributions: Conceptualisation, P.G. and E.V.; data collection, P.G.; writing—original draft preparation, P.G.; writing and editing, P.G., E.V., A.M., A.O. and T.K.M.; funding acquisition, E.V., P.G., A.M. and T.K.M.; supervision, E.V. All authors have read and agreed to the published version of the manuscript.

Funding: P.G. and E.V. have received funding from the 3DbioNet. E.V. is grateful to the Royal Academy of Engineering for an Industrial Fellowship and to the Medical Research Council UK for a New Investigator Research Grant (MR/V028553/1), which also financially supports P.G.

Institutional Review Board Statement: Not Applicable.

Informed Consent Statement: Not Applicable.

Data Availability Statement: The data presented in this study are available on request from the corresponding author.

Conflicts of Interest: The authors declare no conflict of interest.

References

1. Augustine, R.; Kalva, S.N.; Ahmad, R.; Zahid, A.A.; Hasan, S.; Nayeem, A.; McClements, L.; Hasan, A. 3D Bioprinted cancer models: Revolutionizing personalized cancer therapy. *Transl. Oncol.* **2021**, *14*, 101015. [CrossRef] [PubMed]
2. Trinidad, C.V.; Tetlow, A.L.; Bantis, L.E.; Godwin, A.K. Reducing Ovarian Cancer Mortality Through Early Detection: Approaches Using Circulating Biomarkers. *Cancer Prev. Res.* **2020**, *13*, 241–252. [CrossRef] [PubMed]

3. Ahmed, N.; Kadife, E.; Raza, A.; Short, M.; Jubinsky, P.T.; Kannourakis, G. Ovarian Cancer, Cancer Stem Cells and Current Treatment Strategies: A Potential Role of Magmas in the Current Treatment Methods. *Cells* **2020**, *9*, 719. [CrossRef]
4. Nowacka, M.; Sterzynska, K.; Andrzejewska, M.; Nowicki, M.; Januchowski, R. Drug resistance evaluation in novel 3D in vitro model. *Biomed. Pharmacother.* **2021**, *138*, 111536. [CrossRef] [PubMed]
5. Peña, C.M.; Skipper, T.; Hsu, J.; Schechter, I.; Ghosh, D.; Dawson, M. Development of a Novel 3D Model to Investigate the Role of Heterogeneity in Ovarian Cancer Chemoresistance. *FASEB J.* **2021**, *35*. [CrossRef]
6. Momenimovahed, Z.; Tiznobaik, A.; Taheri, S.; Salehiniya, H. Ovarian cancer in the world: Epidemiology and risk factors. *Int. J. Women's Health* **2019**, *11*, 287–299. [CrossRef] [PubMed]
7. Rosen, B.; Laframboise, S.; Ferguson, S.; Dodge, J.; Bernardini, M.; Murphy, J.; Segev, Y.; Sun, P.; Narod, S.A. The impacts of neoadjuvant chemotherapy and of debulking surgery on survival from advanced ovarian cancer. *Gynecol. Oncol.* **2014**, *134*, 462–467. [CrossRef]
8. Redman, C.W.E.; Warwick, J.; Luesley, D.M.; Varma, R.; Lawton, F.G.; Blackledge, G.R.P. Intervention debulking surgery in advanced epithelial ovarian cancer. *BJOG Int. J. Obstet. Gynaecol.* **1994**, *101*, 142–146. [CrossRef]
9. Vergote, I.; Amant, F.; Kristensen, G.; Ehlen, T.; Reed, N.S.; Casado, A. Primary surgery or neoadjuvant chemotherapy followed by interval debulking surgery in advanced ovarian cancer. *Eur. J. Cancer* **2011**, *47*, S88–S92. [CrossRef]
10. Board, P.A.T.E. Ovarian Epithelial, Fallopian Tube, and Primary Peritoneal Cancer Treatment (PDQ®): Patient Version. In *PDQ Cancer Information Summaries*; Bethesda: Rockville, MI, USA, 2021.
11. Baci, D.; Bosi, A.; Gallazzi, M.; Rizzi, M.; Noonan, D.M.; Poggi, A.; Bruno, A.; Mortara, L. The Ovarian Cancer Tumor Immune Microenvironment (TIME) as Target for Therapy: A Focus on Innate Immunity Cells as Therapeutic Effectors. *Int. J. Mol. Sci.* **2020**, *21*, 3125. [CrossRef]
12. Westergaard, M.C.W.; Milne, K.; Pedersen, M.; Hasselager, T.; Olsen, L.R.; Anglesio, M.S.; Borch, T.H.; Kennedy, M.; Briggs, G.; LeDoux, S.; et al. Changes in the Tumor Immune Microenvironment during Disease Progression in Patients with Ovarian Cancer. *Cancers* **2020**, *12*, 3828. [CrossRef] [PubMed]
13. Jiang, Y.; Wang, C.; Zhou, S. Targeting tumor microenvironment in ovarian cancer: Premise and promise. *Biochim. Biophys. Acta* **2020**, *1873*, 188361. [CrossRef] [PubMed]
14. Beaufort, C.M.; Helmijr, J.C.A.; Piskorz, A.M.; Hoogstraat, M.; Ruigrok-Ritstier, K.; Besselink, N.; Murtaza, M.; van Ijcken, W.; Heine, A.; Smid, M.; et al. Ovarian Cancer Cell Line Panel (OCCP): Clinical Importance of In Vitro Morphological Subtypes. *PLoS ONE* **2014**, *9*, e103988. [CrossRef] [PubMed]
15. Buick, R.N.; Pullano, R.; Trent, J.M. Comparative properties of five human ovarian adenocarcinoma cell lines. *Cancer Res.* **1985**, *45*, 3668–3676. [PubMed]
16. Haley, J.; Tomar, S.; Pulliam, N.; Xiong, S.; Perkins, S.M.; Karpf, A.R.; Mitra, S.; Nephew, K.P.; Mitra, A.K. Functional characterization of a panel of high-grade serous ovarian cancer cell lines as representative experimental models of the disease. *Oncotarget* **2016**, *7*, 32810–32820. [CrossRef] [PubMed]
17. Hernandez, L.; Kim, M.K.; Lyle, L.T.; Bunch, K.P.; House, C.D.; Ning, F.; Noonan, A.M.; Annunziata, C.M. Characterization of ovarian cancer cell lines as in vivo models for preclinical studies. *Gynecol. Oncol.* **2016**, *142*, 332–340. [CrossRef]
18. Havrilesky, L.J.; Elbendary, A.; Hurteau, J.A.; Whitaker, R.S.; Rodriguez, G.C.; Berchuck, A.W. Chemotherapy-induced apoptosis in epithelial ovarian cancers. *Obstet. Gynecol.* **1995**, *85*, 1007–1010. [CrossRef]
19. Johnson, P.A.; Giles, J.R. The hen as a model of ovarian cancer. *Nat. Cancer* **2013**, *13*, 432–436. [CrossRef]
20. Roby, K.F.; Taylor, C.C.; Sweetwood, J.P.; Cheng, Y.; Pace, J.L.; Tawfik, O.; Persons, D.L.; Smith, P.; Terranova, P.F. Development of a syngeneic mouse model for events related to ovarian cancer. *Carcinogenesis* **2000**, *21*, 585–591. [CrossRef]
21. Stakleff, K.S.; Rouse, A.; Ryan, A.; Haller, N.; Von Gruenigen, V. A novel early-stage orthotopic model for ovarian cancer in the Fischer 344 rat. *Int. J. Gynecol. Cancer* **2005**, *15*, 246–254. [CrossRef]
22. Brodeur, M.N.; Simeone, K.; Leclerc-Deslauniers, K.; Fleury, H.; Carmona, E.; Provencher, D.M.; Mes-Masson, A.-M. Carboplatin response in preclinical models for ovarian cancer: Comparison of 2D monolayers, spheroids, ex vivo tumors and in vivo models. *Sci. Rep.* **2021**, *11*, 18183. [CrossRef] [PubMed]
23. Hadi, L.M.; Yaghini, E.; MacRobert, A.J.; Loizidou, M. Synergy between Photodynamic Therapy and Dactinomycin Chemotherapy in 2D and 3D Ovarian Cancer Cell Cultures. *Int. J. Mol. Sci.* **2020**, *21*, 3203. [CrossRef] [PubMed]
24. Sonoda, T.; Kobayashi, H.; Kaku, T.; Hirakawa, T.; Nakano, H. Expression of angiogenesis factors in monolayer culture, multicellular spheroid and in vivo transplanted tumor by human ovarian cancer cell lines. *Cancer Lett.* **2003**, *196*, 229–237. [CrossRef]
25. Chim, L.K.; Mikos, A.G. Biomechanical forces in tissue engineered tumor models. *Curr. Opin. Biomed. Eng.* **2018**, *6*, 42–50. [CrossRef]
26. Totti, S.; Vernardis, S.; Meira, L.; Pérez-Mancera, P.A.; Costello, E.; Greenhalf, W.; Palmer, D.; Neoptolemos, J.; Mantalaris, A.; Velliou, E.G. Designing a bio-inspired biomimetic in vitro system for the optimization of ex vivo studies of pancreatic cancer. *Drug Discov. Today* **2017**, *22*, 690–701. [CrossRef]
27. Goff, B.; Blake, J.; Bamberg, M.; Hasan, T. Treatment of ovarian cancer with photodynamic therapy and immunoconjugates in a murine ovarian cancer model. *Br. J. Cancer* **1996**, *74*, 1194–1198. [CrossRef]
28. Konstantinopoulos, P.A.; Matulonis, U.A. Current Status and Evolution of Preclinical Drug Development Models of Epithelial Ovarian Cancer. *Front. Oncol.* **2013**, *3*, 296. [CrossRef]

29. Magnotti, E.; Marasco, W.A. The latest animal models of ovarian cancer for novel drug discovery. *Expert Opin. Drug Discov.* **2018**, *13*, 249–257. [CrossRef]
30. McCloskey, C.W.; Rodriguez, G.M.; Galpin, K.J.C.; Vanderhyden, B.C. Ovarian Cancer Immunotherapy: Preclinical Models and Emerging Therapeutics. *Cancers* **2018**, *10*, 244. [CrossRef]
31. Erstad, D.J.; Sojoodi, M.; Taylor, M.; Ghoshal, S.; Razavi, A.A.; Graham-O'Regan, K.A.; Bardeesy, N.; Ferrone, C.R.; Lanuti, M.; Caravan, P.; et al. Orthotopic and heterotopic murine models of pancreatic cancer and their different responses to FOLFIRINOX chemotherapy. *Dis. Model. Mech.* **2018**, *11*, dmm034793. [CrossRef]
32. Nyga, A.; Cheema, U.; Loizidou, M. 3D tumour models: Novel in vitro approaches to cancer studies. *J. Cell Commun. Signal.* **2011**, *5*, 239–248. [CrossRef] [PubMed]
33. Tentler, J.J.; Tan, A.C.; Weekes, C.D.; Jimeno, A.; Leong, S.; Pitts, T.M.; Arcaroli, J.J.; Messersmith, W.A.; Eckhardt, S.G. Patient-derived tumour xenografts as models for oncology drug development. *Nat. Rev. Clin. Oncol.* **2012**, *9*, 338–350. [CrossRef] [PubMed]
34. Johnson, J.; Decker, S.; Zaharevitz, D.; Rubinstein, L.V.; Venditti, J.M.; Schepartz, S.; Kalyandrug, S.; Christian, M.; Arbuck, S.; Hollingshead, M.; et al. Relationships between drug activity in NCI preclinical in vitro and in vivo models and early clinical trials. *Br. J. Cancer* **2001**, *84*, 1424–1431. [CrossRef]
35. Raghavan, S.; Mehta, P.; Ward, M.R.; Bregenzer, M.E.; Fleck, E.M.A.; Tan, L.; McLean, K.; Buckanovich, R.J.; Mehta, G. Personalized Medicine–Based Approach to Model Patterns of Chemoresistance and Tumor Recurrence Using Ovarian Cancer Stem Cell Spheroids. *Clin. Cancer Res.* **2017**, *23*, 6934–6945. [CrossRef] [PubMed]
36. Liao, J.; Qian, F.; Tchabo, N.; Mhawech-Fauceglia, P.; Beck, A.; Qian, Z.; Wang, X.; Huss, W.J.; Lele, S.B.; Morrison, C.D.; et al. Ovarian Cancer Spheroid Cells with Stem Cell-Like Properties Contribute to Tumor Generation, Metastasis and Chemotherapy Resistance through Hypoxia-Resistant Metabolism. *PLoS ONE* **2014**, *9*, e84941. [CrossRef] [PubMed]
37. Lal-Nag, M.; McGee, L.; Titus, S.A.; Brimacombe, K.; Michael, S.; Sittampalam, G.; Ferrer, M. Exploring Drug Dosing Regimens In Vitro Using Real-Time 3D Spheroid Tumor Growth Assays. *SLAS Discov. Adv. Sci. Drug Discov.* **2017**, *22*, 537–546. [CrossRef] [PubMed]
38. Heredia-Soto, V.; Redondo, A.; Berjón, A.; Miguel-Martín, M.; Díaz, E.; Crespo, R.; Hernández, A.; Yébenes, L.; Gallego, A.; Feliu, J.; et al. High-throughput 3-dimensional culture of epithelial ovarian cancer cells as preclinical model of disease. *Oncotarget* **2018**, *9*, 21893–21903. [CrossRef] [PubMed]
39. Shuford, S.; Wilhelm, C.; Rayner, M.; Elrod, A.; Millard, M.; Mattingly, C.; Lotstein, A.; Smith, A.M.; Guo, Q.J.; O'Donnell, L.; et al. Prospective Validation of an Ex Vivo, Patient-Derived 3D Spheroid Model for Response Predictions in Newly Diagnosed Ovarian Cancer. *Sci. Rep.* **2019**, *9*, 11153. [CrossRef] [PubMed]
40. Raghavan, S.; Ward, M.R.; Rowley, K.R.; Wold, R.M.; Takayama, S.; Buckanovich, R.J.; Mehta, G. Formation of stable small cell number three-dimensional ovarian cancer spheroids using hanging drop arrays for preclinical drug sensitivity assays. *Gynecol. Oncol.* **2015**, *138*, 181–189. [CrossRef] [PubMed]
41. Rafehi, S.; Valdes, Y.R.; Bertrand, M.; McGee, J.; Préfontaine, M.; Sugimoto, A.; DiMattia, G.; Shepherd, T.G. TGFβ signaling regulates epithelial–mesenchymal plasticity in ovarian cancer ascites-derived spheroids. *Endocr.-Relat. Cancer* **2015**, *23*, 147–159. [CrossRef]
42. Puiffe, M.-L.; Le Page, C.; Filali-Mouhim, A.; Zietarska, M.; Ouellet, V.; Tonin, P.N.; Chevrette, M.; Provencher, D.M.; Mes-Masson, A.-M. Characterization of Ovarian Cancer Ascites on Cell Invasion, Proliferation, Spheroid Formation, Gene Expression in an In Vitro Model of Epithelial Ovarian Cancer. *Neoplasia* **2007**, *9*, 820-IN8. [CrossRef] [PubMed]
43. Griffon, G.; Marchal, C.; Merlin, J.-L.; Parache, R.; Bey, P. Radiosensitivity of multicellular tumour spheroids obtained from human ovarian cancers. *Eur. J. Cancer* **1995**, *31*, 85–91. [CrossRef]
44. Gunay, G.; Kirit, H.A.; Kamatar, A.; Baghdasaryan, O.; Hamsici, S.; Acar, H. The effects of size and shape of the ovarian cancer spheroids on the drug resistance and migration. *Gynecol. Oncol.* **2020**, *159*, 563–572. [CrossRef] [PubMed]
45. Zhang, S.; Balch, C.; Chan, M.; Lai, H.-C.; Matei, D.; Schilder, J.M.; Yan, P.S.; Huang, T.H.-M.; Nephew, K.P. Identification and Characterization of Ovarian Cancer-Initiating Cells from Primary Human Tumors. *Cancer Res.* **2008**, *68*, 4311–4320. [CrossRef] [PubMed]
46. Shield, K.; Ackland, L.; Ahmed, N.; Rice, G. Multicellular spheroids in ovarian cancer metastases: Biology and pathology. *Gynecol. Oncol.* **2009**, *113*, 143–148. [CrossRef]
47. Masiello, T.; Dhall, A.; Hemachandra, L.P.M.; Tokranova, N.; Melendez, J.A.; Castracane, J. A Dynamic Culture Method to Produce Ovarian Cancer Spheroids under Physiologically-Relevant Shear Stress. *Cells* **2018**, *7*, 277. [CrossRef]
48. Li, S.-S.; Ip, C.K.; Tang, M.Y.H.; Sy, S.K.H.; Yung, S.; Chan, T.-M.; Yang, M.; Shum, H.C.; Wong, A.S. Modeling Ovarian Cancer Multicellular Spheroid Behavior in a Dynamic 3D Peritoneal Microdevice. *J. Vis. Exp.* **2017**, e55337. [CrossRef]
49. Lawrenson, K.; Mhawech-Fauceglia, P.; Worthington, J.; Spindler, T.J.; O'Brien, D.; Lee, J.M.; Spain, G.; Sharifian, M.; Wang, G.; Darcy, K.M. Identification of novel candidate biomarkers of epithelial ovarian cancer by profiling the secretomes of three-dimensional genetic models of ovarian carcinogenesis. *Int. J. Cancer* **2015**, *137*, 1806–1817. [CrossRef]
50. Kapałczyńska, M.; Kolenda, T.; Przybyła, W.; Zajączkowska, M.; Teresiak, A.; Filas, V.; Ibbs, M.; Bliźniak, R.; Łuczewski, L.; Lamperska, K. 2D and 3D cell cultures—A comparison of different types of cancer cell cultures. *Arch. Med. Sci.* **2018**, *14*, 910–919. [CrossRef]

51. Fang, Y.; Eglen, R.M. Three-Dimensional Cell Cultures in Drug Discovery and Development. *SLAS Discov. Adv. Sci. Drug Discov.* **2017**, *22*, 456–472. [CrossRef]
52. Zheng, L.; Hu, X.; Huang, Y.; Xu, G.; Yang, J.; Li, L. In vivo bioengineered ovarian tumors based on collagen, matrigel, alginate and agarose hydrogels: A comparative study. *Biomed. Mater.* **2015**, *10*, 15016. [CrossRef] [PubMed]
53. Yang, Z.; Zhao, X. A 3D model of ovarian cancer cell lines on peptide nanofiber scaffold to explore the cell–scaffold interaction and chemotherapeutic resistance of anticancer drugs. *Int. J. Nanomed.* **2011**, *6*, 303–310. [CrossRef] [PubMed]
54. Xu, G.; Yin, F.; Wu, H.; Hu, X.; Zheng, L.; Zhao, J. In vitro ovarian cancer model based on three-dimensional agarose hydrogel. *J. Tissue Eng.* **2014**, *5*. [CrossRef] [PubMed]
55. Xu, F.; Celli, J.; Rizvi, I.; Moon, S.; Hasan, T.; Demirci, U. A three-dimensional in vitro ovarian cancer coculture model using a high-throughput cell patterning platform. *Biotechnol. J.* **2011**, *6*, 204–212. [CrossRef] [PubMed]
56. Hedegaard, C.L.; Redondo-Gómez, C.; Tan, B.Y.; Ng, K.W.; Loessner, D.; Mata, A. Peptide-protein coassembling matrices as a biomimetic 3D model of ovarian cancer. *Sci. Adv.* **2020**, *6*, eabb3298. [CrossRef] [PubMed]
57. Sodek, K.L.; Brown, T.J.; Ringuette, M.J. Collagen I but not Matrigel matrices provide an MMP-dependent barrier to ovarian cancer cell penetration. *BMC Cancer* **2008**, *8*, 223. [CrossRef]
58. Loessner, D.; Stok, K.S.; Lutolf, M.P.; Hutmacher, D.W.; Clements, J.A.; Rizzi, S.C. Bioengineered 3D platform to explore cell–ECM interactions and drug resistance of epithelial ovarian cancer cells. *Biomaterials* **2010**, *31*, 8494–8506. [CrossRef]
59. Chen, J.; Wang, J.; Zhang, Y.; Chen, D.; Yang, C.; Kai, C.; Wang, X.; Shi, F.; Dou, J. Observation of ovarian cancer stem cell behavior and investigation of potential mechanisms of drug resistance in three-dimensional cell culture. *J. Biosci. Bioeng.* **2014**, *118*, 214–222. [CrossRef]
60. Lachowski, D.; Matellan, C.; Cortes, E.; Saiani, A.; Miller, A.; Hernández, A.D.R. Self-Assembling Polypeptide Hydrogels as a Platform to Recapitulate the Tumor Microenvironment. *Cancers* **2021**, *13*, 3286. [CrossRef]
61. Liu, M.; Zhang, X.; Long, C.; Xu, H.; Cheng, X.; Chang, J.; Zhang, C.; Zhang, C.; Wang, X. Collagen-based three-dimensional culture microenvironment promotes epithelial to mesenchymal transition and drug resistance of human ovarian cancer in vitro. *RSC Adv.* **2018**, *8*, 8910–8919. [CrossRef]
62. Wan, X.; Ball, S.; Willenbrock, F.; Yeh, S.; Vlahov, N.; Koennig, D.; Green, M.; Brown, G.; Jeyaretna, D.; Delia, K.; et al. Perfused Three-dimensional Organotypic Culture of Human Cancer Cells for Therapeutic Evaluation. *Sci. Rep.* **2017**, *7*, 9408. [CrossRef] [PubMed]
63. Girard, Y.K.; Wang, C.; Ravi, S.; Howell, M.C.; Mallela, J.; Alibrahim, M.; Green, R.; Hellermann, G.; Mohapatra, S.S.; Mohapatra, S. A 3D Fibrous Scaffold Inducing Tumoroids: A Platform for Anticancer Drug Development. *PLoS ONE* **2013**, *8*, e75345. [CrossRef] [PubMed]
64. Alkmin, S.; Brodziski, R.; Simon, H.; Hinton, D.; Goldsmith, R.H.; Patankar, M.; Campagnola, P.J. Role of Collagen Fiber Morphology on Ovarian Cancer Cell Migration Using Image-Based Models of the Extracellular Matrix. *Cancers* **2020**, *12*, 1390. [CrossRef] [PubMed]
65. Ul-Islam, M.; Subhan, F.; Islam, S.U.; Khan, S.; Shah, N.; Manan, S.; Ullah, M.W.; Yang, G. Development of three-dimensional bacterial cellulose/chitosan scaffolds: Analysis of cell-scaffold interaction for potential application in the diagnosis of ovarian cancer. *Int. J. Biol. Macromol.* **2019**, *137*, 1050–1059. [CrossRef] [PubMed]
66. De Jaeghere, E.A.; De Vlieghere, E.; Van Hoorick, J.; Van Vlierberghe, S.; Wagemans, G.; Pieters, L.; Melsens, E.; Praet, M.; Van Dorpe, J.; Boone, M.N.; et al. Heterocellular 3D scaffolds as biomimetic to recapitulate the tumor microenvironment of peritoneal metastases in vitro and in vivo. *Biomaterials* **2018**, *158*, 95–105. [CrossRef] [PubMed]
67. Avraham-Chakim, L.; Elad, D.; Zaretsky, U.; Kloog, Y.; Jaffa, A.; Grisaru, D. Fluid-Flow Induced Wall Shear Stress and Epithelial Ovarian Cancer Peritoneal Spreading. *PLoS ONE* **2013**, *8*, e60965. [CrossRef]
68. Matte, I.; Legault, C.M.; Garde-Granger, P.; Laplante, C.; Bessette, P.; Rancourt, C.; Piché, A. Mesothelial cells interact with tumor cells for the formation of ovarian cancer multicellular spheroids in peritoneal effusions. *Clin. Exp. Metastasis* **2016**, *33*, 839–852. [CrossRef]
69. Kenny, H.A.; Krausz, T.; Yamada, S.D.; Lengyel, E. Use of a novel 3D culture model to elucidate the role of mesothelial cells, fibroblasts and extra-cellular matrices on adhesion and invasion of ovarian cancer cells to the omentum. *Int. J. Cancer* **2007**, *121*, 1463–1472. [CrossRef]
70. Ip, C.K.; Li, S.-S.; Tang, M.Y.H.; Sy, S.K.H.; Ren, Y.; Shum, H.C.; Wong, A.S.T. Stemness and chemoresistance in epithelial ovarian carcinoma cells under shear stress. *Sci. Rep.* **2016**, *6*, 26788. [CrossRef]
71. Hyler, A.R.; Baudoin, N.C.; Brown, M.S.; Stremler, M.; Cimini, D.; Davalos, R.V.; Schmelz, E.M. Fluid shear stress impacts ovarian cancer cell viability, subcellular organization, and promotes genomic instability. *PLoS ONE* **2018**, *13*, e0194170. [CrossRef]
72. Zhang, Y.; Tang, H.; Cai, J.; Zhang, T.; Guo, J.; Feng, D.; Wang, Z. Ovarian cancer-associated fibroblasts contribute to epithelial ovarian carcinoma metastasis by promoting angiogenesis, lymphangiogenesis and tumor cell invasion. *Cancer Lett.* **2011**, *303*, 47–55. [CrossRef] [PubMed]
73. Brooks, E.A.; Gencoglu, M.F.; Corbett, D.C.; Stevens, K.R.; Peyton, S.R. An omentum-inspired 3D PEG hydrogel for identifying ECM-drivers of drug resistant ovarian cancer. *APL Bioeng.* **2019**, *3*, 026106. [CrossRef] [PubMed]
74. Patra, B.; Lateef, M.A.; Brodeur, M.N.; Fleury, H.; Carmona, E.; Péant, B.; Provencher, D.; Mes-Masson, A.-M.; Gervais, T. Carboplatin sensitivity in epithelial ovarian cancer cell lines: The impact of model systems. *PLoS ONE* **2020**, *15*, e0244549. [CrossRef] [PubMed]

75. Gupta, P.; Totti, S.; Pérez-Mancera, P.A.; Dyke, E.; Nisbet, A.; Schettino, G.; Webb, R.; Velliou, E.G. Chemoradiotherapy screening in a novel biomimetic polymer based pancreatic cancer model. *RSC Adv.* **2019**, *9*, 41649–41663. [CrossRef]
76. Gupta, P.; Pérez-Mancera, P.A.; Kocher, H.; Nisbet, A.; Schettino, G.; Velliou, E.G. A Novel Scaffold-Based Hybrid Multicellular Model for Pancreatic Ductal Adenocarcinoma—Toward a Better Mimicry of the in vivo Tumor Microenvironment. *Front. Bioeng. Biotechnol.* **2020**, *8*, 290. [CrossRef]
77. Totti, S.; Allenby, M.C.; Dos Santos, S.B.; Mantalaris, A.; Velliou, E.G. A 3D bioinspired highly porous polymeric scaffolding system for in vitro simulation of pancreatic ductal adenocarcinoma. *RSC Adv.* **2018**, *8*, 20928–20940. [CrossRef]
78. Echo, A.; Howell, V.M.; Colvin, E.K. The Extracellular Matrix in Epithelial Ovarian Cancer—A Piece of a Puzzle. *Front. Oncol.* **2015**, *5*, 245. [CrossRef]
79. Kenny, H.A.; Lengyel, E. MMP-2 functions as an early response protein in ovarian cancer metastasis. *Cell Cycle* **2009**, *8*, 683–688. [CrossRef] [PubMed]
80. Kumar, D.; Workman, V.; O'Brien, M.; McLaren, J.; White, L.; Ragunath, K.; Rose, F.; Saiani, A.; Gough, J.E. Peptide Hydrogels-A Tissue Engineering Strategy for the Prevention of Oesophageal Strictures. *Adv. Funct. Mater.* **2017**, *27*. [CrossRef]
81. Allenby, M.C.; Misener, R.; Panoskaltsis, N.; Mantalaris, A. A Quantitative Three-Dimensional Image Analysis Tool for Maximal Acquisition of Spatial Heterogeneity Data. *Tissue Eng. Part C Methods* **2017**, *23*, 108–117. [CrossRef]
82. Allenby, M.C.; Panoskaltsis, N.; Tahlawi, A.; Dos Santos, S.B.; Mantalaris, A. Dynamic human erythropoiesis in a three-dimensional perfusion bone marrow biomimicry. *Biomaterials* **2018**, *188*, 24–37. [CrossRef] [PubMed]
83. Tahlawi, A.; Klontzas, M.E.; Allenby, M.C.; Morais, J.C.; Panoskaltsis, N.; Mantalaris, A. RGD-functionalized polyurethane scaffolds promote umbilical cord blood mesenchymal stem cell expansion and osteogenic differentiation. *J. Tissue Eng. Regen. Med.* **2018**, *13*, 232–243. [CrossRef] [PubMed]
84. Januchowski, R.; Świerczewska, M.; Sterzyńska, K.; Wojtowicz, K.; Nowicki, M.; Zabel, M. Increased Expression of Several Collagen Genes is Associated with Drug Resistance in Ovarian Cancer Cell Lines. *J. Cancer* **2016**, *7*, 1295–1310. [CrossRef] [PubMed]
85. Morin, P.J. Drug resistance and the microenvironment: Nature and nurture. *Drug Resist. Updat.* **2003**, *6*, 169–172. [CrossRef]
86. Croix, B.S.; Kerbel, R.S. Cell adhesion and drug resistance in cancer. *Curr. Opin. Oncol.* **1997**, *9*, 549–556. [CrossRef] [PubMed]
87. Totti, S.; Ng, K.W.; Dale, L.; Lian, G.; Chen, T.; Velliou, E.G. A novel versatile animal-free 3D tool for rapid low-cost assessment of immunodiagnostic microneedles. *Sens. Actuators B Chem.* **2019**, *296*, 126652. [CrossRef]
88. Wishart, G.; Gupta, P.; Nisbet, A.; Schettino, G.; Velliou, E. On the Evaluation of a Novel Hypoxic 3D Pancreatic Cancer Model as a Tool for Radiotherapy Treatment Screening. *Cancers* **2021**, *13*, 6080. [CrossRef]
89. Frankel, A.; Buckman, R.; Kerbel, R.S. Abrogation of taxol-induced G2-M arrest and apoptosis in human ovarian cancer cells grown as multicellular tumor spheroids. *Cancer Res.* **1997**, *57*, 2388–2393.
90. Mehta, G.; Hsiao, A.Y.; Ingram, M.; Luker, G.D.; Takayama, S. Opportunities and challenges for use of tumor spheroids as models to test drug delivery and efficacy. *J. Control Release* **2012**, *164*, 192–204. [CrossRef]
91. Folkman, J.; Hochberg, M. SELF-REGULATION OF GROWTH IN THREE DIMENSIONS. *J. Exp. Med.* **1973**, *138*, 745–753. [CrossRef]
92. Stock, K.; Estrada, M.; Vidic, S.; Gjerde, K.; Rudisch, A.; Santo, V.E.; Barbier, M.; Blom, S.; Arundkar, S.C.; Selvam, I.; et al. Capturing tumor complexity in vitro: Comparative analysis of 2D and 3D tumor models for drug discovery. *Sci. Rep.* **2016**, *6*, 28951. [CrossRef] [PubMed]
93. Bondong, S.; Kiefel, H.; Hielscher, T.; Zeimet, A.G.; Zeillinger, R.; Pils, D.; Schuster, E.; Castillo-Tong, D.C.; Cadron, I.; Vergote, I.; et al. Prognostic significance of L1CAM in ovarian cancer and its role in constitutive NF-κB activation. *Ann. Oncol.* **2012**, *23*, 1795–1802. [CrossRef] [PubMed]
94. Sterzyńska, K.; Klejewski, A.; Wojtowicz, K.; Świerczewska, M.; Nowacka, M.; Kaźmierczak, D.; Andrzejewska, M.; Rusek, D.; Brązert, M.; Brązert, J.; et al. Mutual Expression of ALDH1A1, LOX, and Collagens in Ovarian Cancer Cell Lines as Combined CSCs- and ECM-Related Models of Drug Resistance Development. *Int. J. Mol. Sci.* **2018**, *20*, 54. [CrossRef] [PubMed]
95. Safinia, L.; Mantalaris, A.; Bismarck, A. Nondestructive Technique for the Characterization of the Pore Size Distribution of Soft Porous Constructs for Tissue Engineering. *Langmuir* **2006**, *22*, 3235–3242. [CrossRef] [PubMed]
96. Fan, Y.; Sun, Q.; Li, X.; Feng, J.; Ao, Z.; Li, X.; Wang, J. Substrate Stiffness Modulates the Growth, Phenotype, and Chemoresistance of Ovarian Cancer Cells. *Front. Cell Dev. Biol.* **2021**, *9*, 718834. [CrossRef] [PubMed]
97. McGrail, D.J.; Kieu, Q.M.N.; Dawson, M.R. The malignancy of metastatic ovarian cancer cells is increased on soft matrices through a mechanosensitive Rho–ROCK pathway. *J. Cell Sci.* **2014**, *127*, 2621–2626. [CrossRef] [PubMed]

cancers

MDPI

Article

Erlotinib Promotes Ligand-Induced EGFR Degradation in 3D but Not 2D Cultures of Pancreatic Ductal Adenocarcinoma Cells

Nausika Betriu, Anna Andreeva and Carlos E. Semino *

Tissue Engineering Research Laboratory, Department of Bioengineering, IQS-School of Engineering, Ramon Llull University, 08017 Barcelona, Spain; nausikabetriur@iqs.url.edu (N.B.); annaandreeva@iqs.url.edu (A.A.)

* Correspondence: carlos.semino@iqs.url.edu; Tel.: +34-93-267-2107

Simple Summary: The EGFR is a tyrosine kinase receptor that responds to different stresses such as UV irradiation, hypoxia and drug treatment by internalizing into endosomal compartments. Receptor trafficking and degradation due to tyrosine kinase inhibitors has been widely studied in two- dimensional (2D) cell culture systems, but little is known about how cells respond to these types of drugs in more physiologically relevant models such as three-dimensional (3D) cultures, whose nanostructured properties allow cells to grow, proliferate, migrate and extend cellular processes in their 3D space. In this study, we show that EGFR suffers degradation in response to erlotinib treatment in 3D-cultured cancer cells but not in classic 2D culture systems, demonstrating that dimensionality strongly affects cell drug response. This 3D model may pave the way for the development of more physiological culture platforms to obtain mechanistic insights into how cells respond to chemotherapy.

Citation: Betriu, N.; Andreeva, A.; Semino, C.E. Erlotinib Promotes Ligand-Induced EGFR Degradation in 3D but Not 2D Cultures of Pancreatic Ductal Adenocarcinoma Cells. *Cancers* **2021**, *13*, 4504. https://doi.org/10.3390/cancers13184504

Academic Editors: Serena Danti, Nicola Contessi Negrini, Alessandro Franchi and Adam E. Frampton

Received: 26 July 2021
Accepted: 5 September 2021
Published: 7 September 2021

Abstract: The epithelial growth factor receptor (EGFR) is a tyrosine kinase receptor that participates in many biological processes such as cell proliferation. In addition, EGFR is overexpressed in many epithelial cancers and therefore is a target for cancer therapy. Moreover, EGFR responds to lots of stimuli by internalizing into endosomes from where it can be recycled to the membrane or further sorted into lysosomes where it undergoes degradation. Two-dimensional cell cultures have been classically used to study EGFR trafficking mechanisms in cancer cells. However, it has been widely demonstrated that in 2D cultures cells are exposed to a non-physiological environment as compared to 3D cultures that provide the normal cellular conformation, matrix dimensionality and stiffness, as well as molecular gradients. Therefore, the microenvironment of solid tumors is better recreated in 3D culture models, and this is why they are becoming a more physiological alternative to study cancer physiology. Here, we develop a new model of EGFR internalization and degradation upon erlotinib treatment in pancreatic ductal adenocarcinoma (PDAC) cells cultured in a 3D self-assembling peptide scaffold. In this work, we show that treatment with the tyrosine kinase inhibitor erlotinib promotes EGFR degradation in 3D cultures of PDAC cell lines but not in 2D cultures. We also show that this receptor degradation does not occur in normal fibroblast cells, regardless of culture dimensionality. In conclusion, we demonstrate not only that erlotinib has a distinct effect on tumor and normal cells but also that pancreatic ductal adenocarcinoma cells respond differently to drug treatment when cultured in a 3D microenvironment. This study highlights the importance of culture systems that can more accurately mimic the in vivo tumor physiology.

Keywords: EGFR; trafficking; degradation; self-assembling peptides; 3D culture; pancreatic ductal adenocarcinoma; PDAC; drug resistance

1. Introduction

The epithelial growth factor receptor (EGFR) is a tyrosine kinase receptor (TKR) that participates in many biological processes such as cell proliferation, differentiation and motility, under both physiological and pathological conditions. Overexpression and/or hyperactivation of the EGFR is a hallmark of many epithelial cancers such as breast, lung, colon and pancreatic cancer, and therefore EGFR is a target for cancer treatment. EGFR inhibitors can be classified as small molecules tyrosine kinase inhibitors (such as erlotinib, gefitinib and lapatinib) or monoclonal antibodies (such as cetuximab and panitumumab) [1].

Autophosphorylation and activation of the EGFR is triggered by ligand binding, which initiates signaling cascades at the plasma membrane. Complete activation of the EGFR as well as termination of its signaling depends on its internalization into endosomes and intracellular trafficking [2]. EGFR can also be internalized due to different stresses such as UV irradiation [3], hypoxia [4,5] and oxidative stress [6]. The internalization mechanism of EGFR as well as whether the receptor is subsequently degraded in lysosomes or recycled to the membrane will depend on the type of stimulus. For example, EGF binding induces ubiquitin-dependent lysosomal degradation of the receptor and a partial recycling to the membrane [7], while TGFα ligand induces endocytosis and a rapid recycling [7,8]. Other stresses such as UV radiation, serum starvation or cisplatin treatment trigger internalization and arrest in nondegradative endosomes [2]. Tyrosine kinase inhibitors (TKIs) have also been described to promote EGFR trafficking in cancer cells. For example, gefitinib induces EGFR endocytosis and non-degradative endosomal arrest [9] as well as mitochondrial translocation [10] in glioblastoma cell lines. However, EGFR internalization due to TKIs exposure may be cell type-dependent, since other works report that gefitinib inhibits endocytosis in non-small cell lung cancer cell lines [11]. In contrast, the monoclonal antibody cetuximab has been shown to induce EGFR degradation [12], its sorting to the endoplasmic reticulum and nucleus [13], and mitochondrial translocation of the truncated form EGFRvIII [12].

Two-dimensional cell cultures have been used for many years to study not only EGFR trafficking mechanisms in normal and cancer cells but also as a model to study cell physiology and pathophysiology in general. However, it has been widely demonstrated that 2D cultures do not recreate the microenvironment in which normal cells exist, nor the milieu of solid tumors. In such 2D conditions, cells are forced to adhere to a flat and stiff substrate, which results in morphological changes that ultimately modify cellular function [14]. Moreover, molecular gradients are not reproduced and cells along the 2D surface are exposed to the same nutrient, oxygen and drug levels, while cells within a tumor are exposed to a large gradient of concentration as molecules diffuse from blood vessels [15]. It is also important to note that ECM composition and configuration are strongly modified in 2D cultures, and consequently cells do not receive the proper signals that a normal ECM configuration provides [14]. Three-dimensional (3D) cancer cell models allow one to better mimic the tumor microenvironment and manipulate each component in order to study its implications in tumor progression [16,17]. In this sense, extracellular matrix analogs, also called scaffolds, have become very popular among researchers [18]. Biomaterial scaffolds permit not only cell–cell but also ECM–cell interactions and provide the chemical, physical and mechanical cues needed for cells to form tissue structures in vitro [14].

The selection of the type of scaffold is a key point when planning experiments, and different factors such as the application of the 3D model, the tumor etiology and the concrete step of tumor progression to be recreated should be considered. For example, scaffolds that have a natural origin, such as collagen, are typically used to study cancer cell migration and invasion [19–21]. On the other hand, polymeric scaffolds such as poly(vinyl alcohol), poly(ethylene oxide terephthalate) (PEOT) and poly(butylene terephthalate) (PBT) have been used to investigate the influence of the scaffold architecture on pancreatic cancer cell growth and behavior, thus permitting one to create stage-specific pancreatic cancer models [22]. Moreover, when working with 3D cell culture, it is important to interpret the obtained results in the context of each experimental design. For example, culture of PDAC cells in collagen matrices promotes epithelial-to-mesenchymal transition (EMT), while culturing the same cells in basement membrane extract gels at a matched stiffness promotes mesenchymal-to-epithelial transition (MET) [23]. Furthermore, not only the composition but also the stiffness of the matrix is an important parameter to adjust when culturing cells in three dimensions. Increased matrix stiffness is a hallmark of many cancers such as breast [24], colorectal [25] and pancreatic [26] cancers. PDAC is characterized by a dense and fibrotic stroma due to the production of abundant amounts of ECM (mostly collagens) by stromal pancreatic cells [27]. In consequence, PDAC tissue can be several folds stiffer than its healthy counterpart [26,28], and different 3D models using synthetic [29] and natural scaffolds [19] have been developed in order to study the effect of matrix stiffness on PDAC cells.

In this report, we present a new 3D cell model of EGFR trafficking and degradation in pancreatic ductal adenocarcinoma (PDAC), based on the synthetic self-assembling peptide RAD16-I as a biomaterial for cell culture. In a neutral pH, this peptide self-assembles into a nanofiber network (around 10 nm diameter and 50–200 nm pore size) that allows for the embedding of cells in a 3D environment [18]. The main advantage of self-assembling peptide scaffolds (SAPS) over natural matrices is that they do not suffer degradation in vitro and therefore allow for the maintenance of the same mechanical conditions (matrix stiffness) during culture time. Moreover, it allows one to establish both soft and stiff 3D environments by simply changing its final concentration. RAD16-I is a non-instructive matrix from the point of view of receptor recognition/activation, and therefore this synthetic matrix holds the cells in an inert 3D configuration until they produce ECM proteins and decorate their own physiological environment. RAD16-I has been widely used as a cell culture platform for different tissue engineering applications such as bone [30], cartilage [31], cardiac [32] and hair [33] tissue engineering, and it has also been used to develop 3D models of ovarian [34], breast [35] and pancreatic [36] cancers.

In this work, we describe for the first time that treatment with TKI erlotinib, together with EGF, promotes EGFR degradation in 3D cultures of pancreatic ductal adenocarcinoma (PDAC) cell lines but not in 2D cultures. Moreover, we show that EGFR degradation due to erlotinib treatment does not occur in normal fibroblast cells. This new 3D cell model may introduce new perspectives in the study of EGFR degradation and its implications in cancer therapy, in an environment that more accurately reproduces the in vivo conditions found in a tumor.

2. Materials and Methods

2.1. 2D Cell Culture

The human pancreatic ductal adenocarcinoma cell lines BxPC-3 (EP-CL-0042, Elabscience, Houston, TX, USA) and PANC-1 (CRL-1469, ATCC, Manassas, VA, USA), and primary human normal dermal fibroblasts (hNDF) (C-12302, Promocell, Heidelberg, Germany), were cultured at 10,000 cells/cm^2 for no more than 15 passages in DMEM (DMEM-HXA, Capricorn, Ebsdorfergrund, Germany) or RPMI (RPMI-XA, Capricorn) in the case of BxPC-3, supplemented with 10% Fetal Bovine Serum (FBS) (S1810; Biowest, Nuaillé, France), L-glutamine (X055, Biowest) and Penicillin/Streptomycin (P/S) (L0022, Biowest). Cultures were maintained at 37 °C and 5% CO_2 in a humidified atmosphere.

2.2. 3D Cell Culture in the Self-Assembling Peptide Scaffold RAD16-I

The protocol for cell encapsulation into self-assembling peptide scaffolds has been previously described in detail [37]. The peptide RAD16-I (commercially available at 1% in water, PuraMatrix^{TM}, 354250, Corning, New York, NY, USA) was diluted to a final concentration of 0.3% (v/v) in 10% (w/v) sucrose (S0389, Merck, St. Louis, MO, USA) or maintained at 1% (stock) and sonicated for 30 min. Meanwhile, cells were harvested by trypsinization and resuspended to $4{\cdot}10^6$ cells/mL in 10% (w/v) sucrose, which is an isotonic and non-ionic medium that avoids peptide spontaneous assembly during the encapsulation process. The cell suspension was then mixed with an equal volume of 0.3% or 1% RAD16-I peptide solution to obtain a mixture of $2{\cdot}10^6$ cells/mL and 0.15% (soft) or 0.5% (stiff) RAD16-I. Next, 40 µL of cell/peptide suspension (80,000 cells) was loaded into wells of a 48-well plate previously filled with 500 μL of culture medium, which induced the peptide spontaneous self-assembly. The plate was left in the flow cabinet for 20 min to let the peptide gel and then placed in the incubator for 1 h. Medium was changed twice to favor the leaching of sucrose. 3D cultures were maintained in DMEM or RPMI supplemented with 10% FBS, L-glutamine, P/S at 37 °C and 5% CO_2 in a humidified atmosphere, and medium was changed three times per week. Cells were cultured for 6 days to ensure adaptation to the 3D environment before being incubated with drugs and/or processed for protein extraction, MTT assay or immunofluorescence staining.

2.3. Drug Incubation

For 2D assays, cells were seeded at 10,000 cells/cm^2 in 48-well plates and drugs were added on the following day. For 3D assays, cells were cultured for 6 days before adding the drug. In both 2D and 3D conditions, cells were incubated with 50 μM erlotinib and 10 ng/mL EGF for 16 h. To inhibit lysosome and proteasome degradation, cells were pretreated for 8 h before erlotinib treatment with 600 nM bafilomycin A1 (SML1661, Merck) or 10 μM MG-132 (M7449, Merck), respectively.

2.4. MTT Assay for Cell Viability and Proliferation

MTT [3-(4,5-dimethylthiazol-2-yl)-2,5-diphenyltetrazolium bromide] (M5655, Merck) was used to assess cell viability in 2D and 3D cultures. To determine erlotinib IC_{50}, cells in 2D cultures were seeded at 10,000 cells/cm^2 and the drug was added the following day. For 3D cultures, cells were cultured for 6 days before adding erlotinib. In both cases (2D and 3D cultures), erlotinib IC_{50} was calculated at 72 h from drug addition using MTT assay. For that, cell culture medium was aspirated and 200 μL (for 2D cultures) or 500 μL (for 3D cultures) of MTT reagent were added to a final concentration of 0.5 mg/mL in culture medium. Samples were incubated for 2 h (2D cultures) or 3 h (3D cultures) at 37 °C and 5% CO_2 in a humidified atmosphere. MTT solution was then removed and cells were lysed with 200 μL of DMSO (D8418, Merck). Absorbance was read at 570 nm using a microplate reader (BiotekEpoch^{TM}, Biotek, Winooski, VT, USA).

2.5. Immunofluorescence

Cells in 2D and 3D cultures were fixed with 3.7% formaldehyde for 15 min and washed with 1× PBS. Cultures were blocked with 5% BSA/0.1% Triton X-100 in PBS for 1 h (for 2D cultures) or 2 h (3D cultures) and incubated overnight at 4°C with the following primary antibodies: anti-EGFR (700308, Invitrogen, Waltham, MA, USA) at 1:100, anti-integrin β1 (ab24693, abcam, Cambridge, UK) at 1:500, anti-LAMP1 (14-1079-80; Invitrogen) at 1:150 and anti-EEA1 (14-9114-80, Invitrogen) at 1:500 dilution in 1% BSA. Next, cells were washed with 1% BSA and incubated for 2 h with secondary antibodies conjugated with Alexa Fluor 488 (ab150105, abcam) and 647 (ab150079, abcam) at 1:500. Finally, cells were counterstained with Phalloidin-TRITC and DAPI for cytoskeleton and nuclei visualization. Pictures were acquired with Leica Thunder Imager widefield microscope (Leica Microsystems, Wetzlar, Germany) coupled to a Leica DFC9000 GTC sCMOS camera, using an APO 63× objective. Images were processed and analyzed with ImageJ software version 2017-05-30 (NIH, Bethesda, MD, USA) [38].

2.6. Image Analysis

For colocalization analysis, Manders' coefficients were calculated. Manders' coefficients are an overlapping parameter that describe the proportion of channel A signal coinciding with channel B over the total A intensity (M_1) and the proportion of channel B signal coinciding with channel A over the total B intensity (M_2) [39]. This coefficient ranges from 0 (no overlapping) to 1 (total overlapping). Manders' coefficients are very sensitive to noise, and for this reason, background needs to be set to zero. Moreover, to calculate Mander's coefficients it is important to establish a threshold for segmentation. The analysis was performed using ImageJ software [38]. Each channel was processed for background subtraction and filtered using Median and Gaussian Blur to reduce the presence of noise. Images were then segmented by thresholding using the Default option, and the resulting binary images were cleaned with the Erode and Open functions. Binary images were then used as masks to sample the denoised images using the Image Calculator function with the Min operator, creating a background-less image for each channel [40]. Finally, Manders' colocalization coefficients were calculated using JACoP (Just Another Colocalization Plugin) version 2.0 [41] by setting the threshold values to 1. Coefficients were obtained from at least 5 images ($n > 5$) containing 5–10 cells per colony in the case of tumor cells or single cells in hNDF.

2.7. Western Blot

2D cultures and 3D constructs were lysed with RIPA buffer (R0278, Merck) containing protease inhibitor cocktail (11836153001, Roche, Basel, Switzerland). Total protein content was quantified with a BCA protein assay kit (39228, Serva, Heidelberg, Germany) and 5 or 10 µg of protein were loaded into 8% polyacrylamide gels and run by applying 225 V for 40 min. Afterwards, proteins were transferred to a PVDF membrane (IPVH07850, Merck) by applying 40 V for 2 h. The membrane was then blocked for 1 h with 4% (w/v) nonfat powdered milk in 0.2% PBS-Tween. Next, the membrane was incubated with primary antibodies anti-EGFR (700308, Invitrogen) at 1:1000 and anti-GAPDH (649201, Biolegend, San Diego, CA, USA) at 1:2000 for 1 h at room temperature. The membrane was then washed and incubated with secondary antibodies anti rabbit-HRP (ab6721, abcam) and anti mouse-HRP (ab6820, abcam), both at 1:1000 for 1 h at RT. Finally, the membrane was revealed for HRP detection with a SuperSignal West Pico Chemiluminescent Substrate (34080, Thermo Fisher Scientific, Waltham, MA, USA). Chemiluminescent images were taken in the ImageQuant^{TM} LAS 4000 mini (GE HealthCare, Chicago, IL, USA). Protein bands were quantified using ImageJ software and expressed as a ratio between the protein of interest and the loading control. Each blot was repeated three times (N = 3).

2.8. Statistics

Data are presented as mean ± Standard Deviation. Conditions were tested in triplicate ($n = 3$) in three independent experiments (N = 3). Statistical differences were analyzed with GraphPad Prism 6 (San Diego, CA, USA) by one-way or two-way ANOVA followed by Tukey's multiple comparisons test. Statistical differences were indicated as * for $p < 0.05$, ** for $p < 0.01$, *** for $p < 0.001$ and **** for $p < 0.0001$.

3. Results

3.1. Cell Culture in RAD16-I Scaffold

In the present work, the pancreatic cancer cell lines BxPC-3 and PANC-1 were used, as well as primary human dermal fibroblasts (hNDF), as a control of healthy, non-tumor cells, to study the effect of the tyrosine kinase inhibitor (TKI) erlotinib on EGFR internalization and degradation in a three-dimensional (3D) environment. These two PDAC cell lines were chosen for three main reasons. First, they represent a model of epithelial phenotype (BxPC-3) and an intermediate epithelial-mesenchymal phenotype (PANC-1) [23]. Second, they present with differences regarding erlotinib sensitivity, with BxPC-3 being considered an erlotinib-sensitive line and PANC-1 an insensitive one [42]. Third, they present with different KRAS genetic signatures, an important gene involved in PDAC progression. BxPC-3 cells have wild type KRAS, while PANC-1 cells present with a G12C mutation that produces its constitutive activation. This is of vital importance since 95% of primary pancreatic tumors show mutations in the KRAS gene [43]. For three-dimensional cell culture, we used the self-assembling peptide scaffold RAD16-I, which has been previously used for different cancer cell line cultures [34–36]. Moreover, cultures in RAD16-I were prepared at two different peptide concentrations: 0.15% and 0.5%, which correspond to a stiffness of 120 Pa (namely, soft matrix) and 1500 Pa (namely, stiff matrix), respectively, previously measured by rheometry [44]. Results show that both pancreatic cancer cell lines presented good viability regardless of matrix stiffness (Figure 1a). Moreover, growth rate in 3D cultures decreased for both cell lines compared to 2D, partially recapitulating the growth behavior of in vivo cancer cells (Table 1). Dermal fibroblasts did not show proliferation in the soft RAD16-I matrix, as previously reported [36], but did proliferate in the stiff matrix (Table 1). This behavior was previously shown for fibroblasts cultured in 3D collagen gels [45]. We also determined cell viability after 72 h incubation with the TKI erlotinib. Under our culture conditions, we determined an IC_{50} of 10 μM for the erlotinib-sensitive cell line BxPC-3 and for normal fibroblasts, in both 2D and 3D cultures. For the erlotinib-insensitive cell line PANC-1, we determined an IC_{50} of 45 μM for 2D cultures and 100 μM for 3D cultures, in both matrix stiffness conditions (Table 1, Figure S1). Therefore, the 3D environment promoted these cells to be even more resistant to erlotinib treatment.

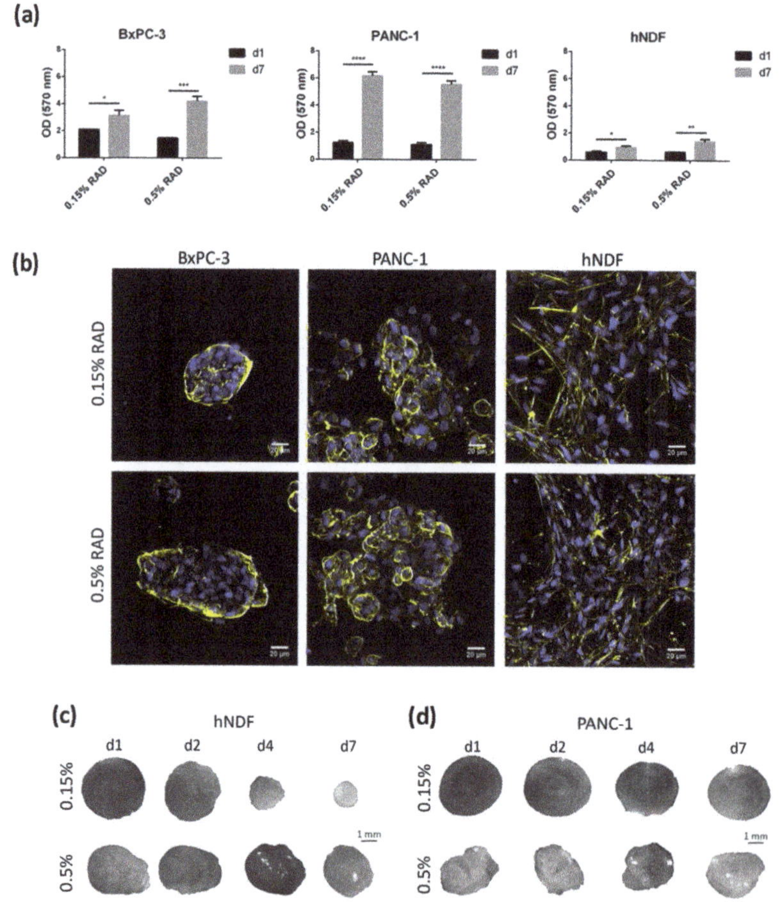

Figure 1. Cell culture in RAD16-I scaffold. (**a**) Viability of BxPC-3, PANC-1 and hNDF cultured in 0.15% and 0.5% RAD16-I scaffold measured by MTT assay at day 1 and day 7 of culture. Statistical differences are indicated as * for $p < 0.05$, ** for $p < 0.01$, *** for $p < 0.001$ and **** for $p < 0.0001$, two-way ANOVA, N = 2, $n = 3$); (**b**) Z-projection pictures of BxPC-3, PANC-1 and hNDF cells at day 6 of culture stained with Phalloidin (pseudo-colored in yellow) and DAPI (blue). Scale bars represent 20 μm; (**c**) macroscopic view of hNDF and (**d**) PANC-1 3D constructs in 0.15% and 0.5% RAD16-I scaffolds at different time points.

Table 1. Doubling time (h) and erlotinib sensitivity in 2D and 3D cultures.

	Condition	BxPC-3	PANC-1	hNDF
Doubling time (h)	2D	39	30	24.3
	0.15% RAD	ND	63.6	No proliferation
	0.5% RAD	95	62.3	65.3
Erlotinib IC_{50} (μM)	2D	10	45	10
	3D	10	100	10

ND: not determined.

Regarding cell phenotype, the epithelial cell line BxPC-3 grew forming round or oval clusters (Figure 1b, left), while PANC-1 cells formed more grape-like spheres (Figure 1b, middle). We did not find a correlation between matrix stiffness and cell phenotype or colony size in the tumor cells analyzed. Matrix stiffness had a great effect on dermal fibroblasts, which formed a highly interconnected network in the soft matrix, contracting the hydrogel and notably reducing its dimension in a few days (Figure 1c). In the stiff matrix, fibroblasts managed to interconnect to each other as well (Figure 1b, right) but required more time in culture to do so and contracted the matrix to a lesser extent (Figure 1c). This hydrogel-contraction behavior (also called matrix condensation) is characteristic of primary mesenchymal cells cultured in this kind of scaffold [33] but did not happen in tumor cells (Figure 1d).

3.2. EGFR Expression in 2D and RAD16-I 3D Cultures

We next analyzed the location and expression levels of the EGFR by immunofluorescence and western blot in both culture types. In 2D cultures of BxPC-3 cells, the EGFR displayed a strong staining in the cell periphery, colocalizing with β1-integrin, as well as a diffused cytoplasmatic staining (Figure 2a, top). In PANC-1 cells, the EGFR was found mainly in the cell periphery, presenting a strong staining that overlapped with β1-integrin (Figure 2a, middle). In fibroblasts instead, the EGFR showed a dotted staining all over the cell (Figure 2a, bottom). Similar to 2D cultures, BxPC-3 and PANC-1 in 3D cultures showed a peripheral staining of the EGFR, while hNDF displayed a dotted staining (Figure 2b). Western blot analysis revealed that total EGFR was downregulated in 3D cultures compared to 2D monolayer cultures in the three types of cells analyzed (Figure 2c), especially in dermal fibroblasts. Moreover, matrix stiffness in 3D cultures also influenced EGFR expression, being downregulated in stiff conditions compared to soft cultures in hNDF cells but not in BxPC-3 and PANC-1, which presented similar EGFR levels regardless of the stiffness (Figure 2d).

3.3. Effect of EGF and Erlotinib on the Location of the EGFR

We next analyzed the effect of EGF and acute erlotinib treatment on cells in 2D and 3D cultures. For that, cells were incubated with 10 ng/mL EGF or 50 μM erlotinib or a combination of both during 16 h. Under control conditions, EGFR in 2D-cultured PANC-1 cells was found predominantly in the cell membrane, showing a strong peripheral staining (Figure 3a,b) and colocalizing with β1-integrin (Figure 3c). The presence of EGF triggered ligand-induced endocytosis of the EGFR, as previously described [7], thus being internalized from the cell periphery (Figure 3a, white arrows) into the cytoplasm (Figure 3a, empty arrows), accumulating perinuclearly (Figure 3a,b). Treatment of cells with erlotinib did not induce EGFR internalization, as demonstrated by its peripheral staining (Figure 3a,b) and high colocalization degree with β1-integrin, similar to control conditions (Figure 3c). However, when we incubated the cells with erlotinib in combination with EGF, we detected both membrane and perinuclear EGFR staining (Figure 3a,b).

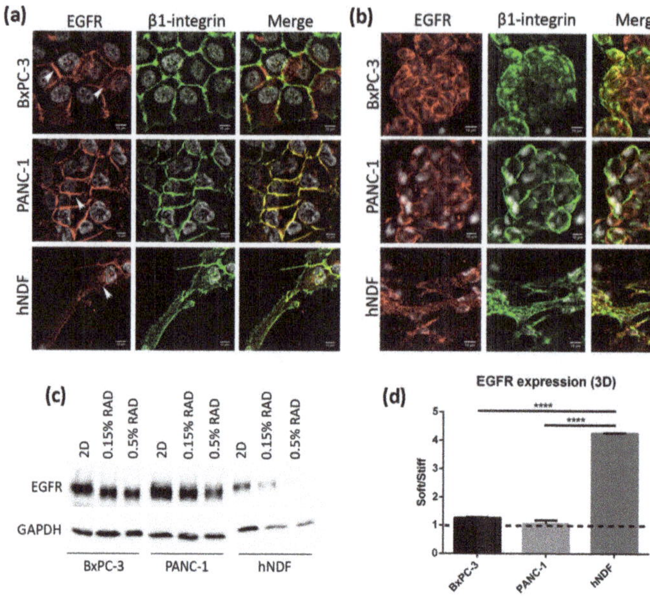

Figure 2. EGFR expression in 2D and 3D cultures. (**a**) EGFR (red) and β1-integrin (green) immunofluorescence counterstained with DAPI (grey) in BxPC-3, PANC-1 and hNDF cells cultured in 2D monolayer and (**b**) Z-projection of EGFR and β1-integrin immunofluorescence in BxPC-3, PANC-1 and hNDF cells in 3D RAD16-I scaffold at 0.15% peptide concentration. Scale bars represent 10 μm; (**c**) Western blot bands of EGFR in 2D and 3D cultures; (**d**) quantification of EGFR in 3D cultures represented as the ratio between soft and stiff cultures. GAPDH was used as loading control. One representative blot is shown. Experiments were repeated three times (N = 3), and statistical differences are indicated as **** for $p < 0.0001$.

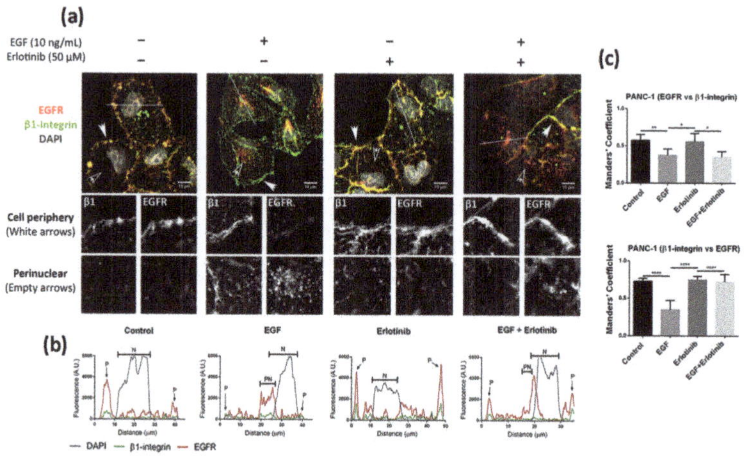

Figure 3. Immunofluorescence analysis of the EGFR in PANC-1 cells incubated with EGF, erlotinib or both in 2D cultures. (**a**) EGFR (red) and β1-integrin (green) immunofluorescence counterstained with DAPI (grey) in the presence of EGF, erlotinib or both in PANC-1 and close-up sections (gray pictures) of the cell periphery and the perinuclear area labeled with white and empty arrows, respectively. Scale bars represent 10 μm; (**b**) fluorescence intensity profiles corresponding to the white line in pictures from (**a**). Different cell regions are indicated as P for cell periphery, N for nucleus and PN for perinuclear area; (**c**) Manders' colocalization coefficients. Statistical differences are indicated as * for $p < 0.05$, ** for $p < 0.01$ and **** for $p < 0.0001$.

The location of EGFR in 2D cultures of BxPC-3 under control conditions was mainly peripheral and was associated with β1-integrin staining, and diffused expression was detected in the cytoplasm (Figure 4a). The presence of EGF induced a strong accumulation of the receptor in the cytoplasm and the perinuclear area (Figure 4a,b). In the presence of erlotinib, EGFR displayed a similar location as in the control (Figure 4a), but the fraction of β1-integrin colocalizing with EGFR increased compared to the control (Figure 4c). Finally, when BxPC-3 cells were incubated with erlotinib in combination with EGF, we detected strong perinuclear staining (Figure 4a,b) but also an increased colocalization degree with the membrane marker β1-integrin compared to EGF-incubated cells (Figure 4c), similar to what happened in PANC-1 cells. The cytoplasmatic staining of EGFR in BxPC-3 under control conditions suggests that basal levels of EGFR trafficking may exist, which could be for different reasons. First, even though EGFR mainly resides in the plasmatic membrane, it constantly undergoes trafficking through the endocytic system [2]. Second, it is important to note that in order to keep culture conditions between 2D and 3D as similar as possible, cells were not serum-starved prior to the experiments. In 2D cultures, FBS proteins are removed by simply changing the culture medium. However, protein release from RAD16-I hydrogels can take more than 50 h [46,47], and therefore complete serum depletion is not possible in the short term in our 3D culture system.

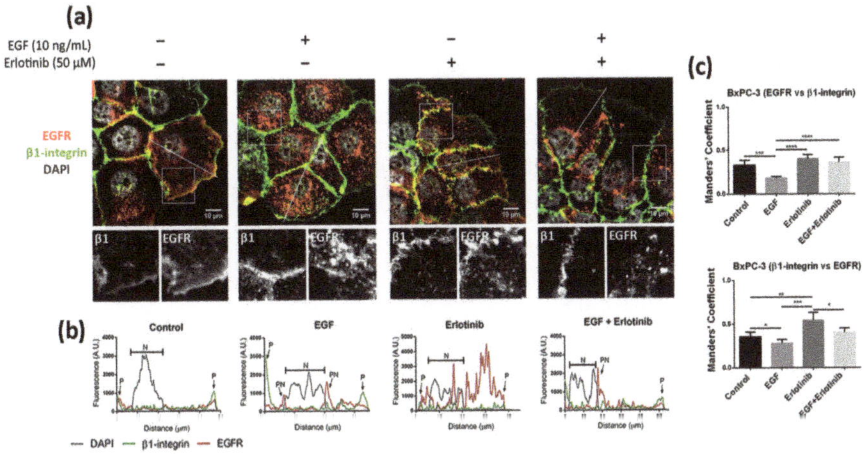

Figure 4. Immunofluorescence analysis of the EGFR in BxPC-3 cells incubated with EGF, erlotinib or both in 2D cultures. (**a**) EGFR (red) and β1-integrin (green) immunofluorescence counterstained with DAPI (grey) in the presence of EGF, erlotinib or both and close-up sections (gray pictures). Scale bars represent 10 µm; (**b**) fluorescence intensity profiles corresponding to the white line in pictures from (**a**). Different cell regions are indicated as: P for cell periphery, N for nucleus and PN for perinuclear area; (**c**) Manders' colocalization coefficients. Statistical differences are indicated as * for $p < 0.05$, ** for $p < 0.01$, *** for $p < 0.001$ and **** for $p < 0.0001$.

Finally, we analyzed the effect of EGF and erlotinib in 2D cultures of human normal dermal fibroblasts (hNDF). In this case, cells under control conditions showed a punctate staining of the EGFR distributed all over the cell and also some pericellular staining (Figure 5). The presence of EGF alone or combined with erlotinib induced the accumulation of the receptor mainly in the perinuclear area, probably in endosomal compartments, while erlotinib alone promoted the accumulation of the receptor along all the cell surface and pericellularly (Figure 5).

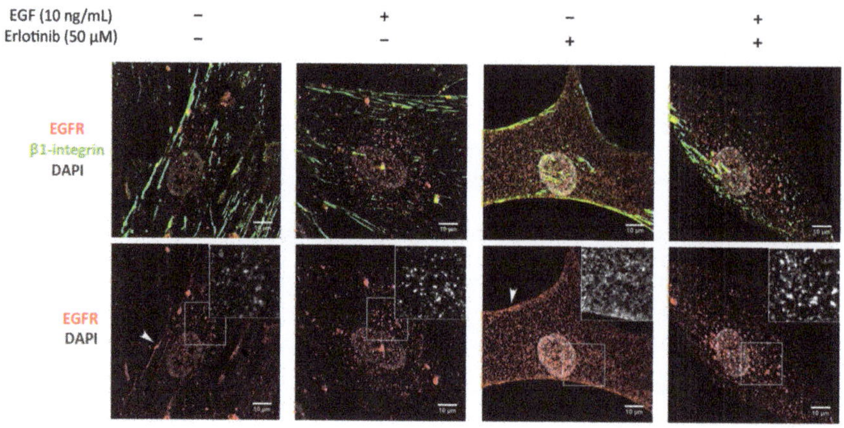

Figure 5. Immunofluorescence analysis of the EGFR in hNDF cells incubated with EGF, erlotinib or both in 2D cultures. Scale bars represent 10 µm. Insets represent high magnification images of the region indicated by a white square.

Altogether, these results suggest that erlotinib partially prevents EGFR internalization and trafficking to endosomal compartments upon EGF binding in 2D cultures, retaining part of the receptor in the plasmatic membrane, as previously described [48].

In 3D cultures, both tumor cell lines displayed peripheral staining of the EGFR, as happened in 2D cultures (Figure 6, top and middle). Under control conditions and in the presence of erlotinib alone, EGFR staining was peripheral and well-defined, but when EGF or EGF and erlotinib were added, EGFR signal in the membrane became more diffused and both pericellular and cytoplasmatic punctate staining were detected, which showed evidence of the internalization of the receptor (Figure 6, see arrows). In hNDF under control conditions, the EGFR displayed a punctate staining all over the cell while accumulating perinuclearly after EGF exposure (Figure 6, bottom, see arrows). In erlotinib-treated hNDF, the dotted EGFR expression found in the control was lost, becoming strongly concentrated all over the cell. When combining EGF and erlotinib treatment, both phenotypes could be detected.

3.4. Effect of EGF and Erlotinib on EGFR Degradation

It has been previously shown using 2D cultures that treatment of cells with monoclonal antibodies targeting the EGFR such as cetuximab [12] and Sym004 [48,49] resulted in overall decrease of EGFR levels due to protein degradation. In contrast, TKI treatment has not been shown to induce significant EGFR degradation in different 2D-cultured cancer cell lines [50–53]. Consistent with these studies we did not detect a relevant decrease in EGFR levels after 16 h erlotinib treatment (neither alone nor in combination with EGF) in any of the cells analyzed (Figure 7a,c, lanes 3–4, Figure S3a–c). On the other hand, EGF treatment alone induced strong EGFR degradation in hNDF but not in tumor cells (Figure 7a,c, lane 2), which express much higher EGFR levels than hNDF. Therefore, it is likely that the EGF dose used (10 ng/mL) is insufficient to promote EGFR degradation in these PDAC cell lines. Remarkably, the presence of erlotinib in hNDF treated with EGF rescued EGFR levels similar to those in untreated cells (Figure 7a,c, lane 4), reinforcing the hypothesis that erlotinib prevents EGFR internalization in 2D-cultered cells.

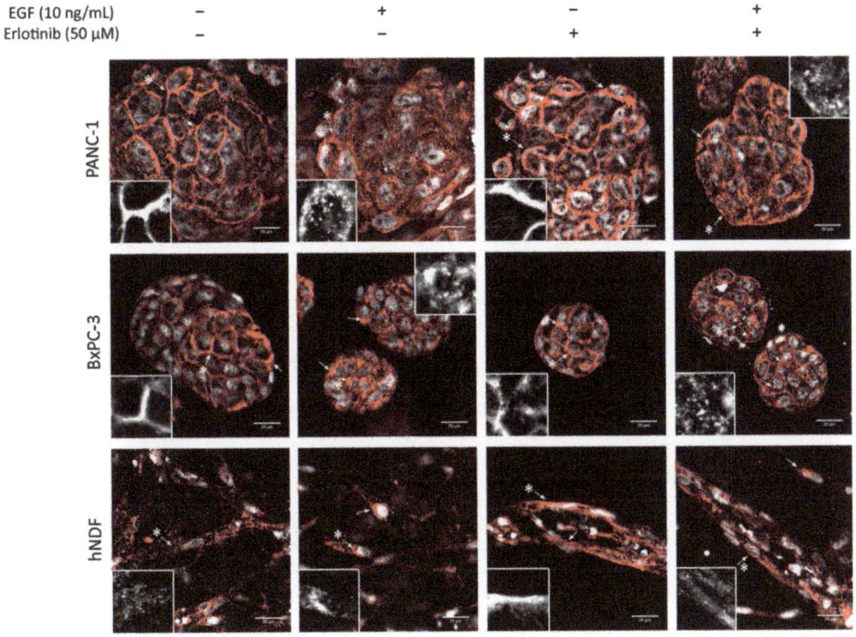

Figure 6. EGFR immunofluorescence in PANC-1, BxPC-3and hNDF cells incubated with EGF, erlotinib or both in 0.15% RAD16-I 3D cultures. One representative Z plane is shown. Scale bars represent 20 μm. Insets represent high magnification images of the region indicated by an asterisk (*).

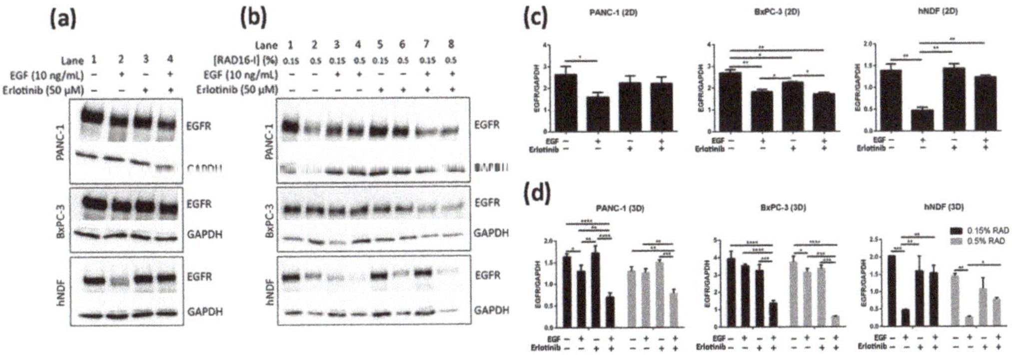

Figure 7. Western blot analysis of the EGFR in BxPC-3, PANC-1 and hNDF cells incubated with EGF or erlotinib or both in 2D cultures and soft and stiff 3D cultures. (**a**) Western blot bands of EGFR in 2D cultures; (**b**) Western blot bands of EGFR in 0.15% and 0.5% RAD16-I 3D cultures; (**c**) densitometry of bands shown in (**a**) for 2D cultures; (**d**) densitometry of bands shown in (**b**) for 3D cultures. GAPDH was used as an internal control. One representative blot is shown. Experiments were repeated three times (N = 3), and statistical differences are indicated as * for $p < 0.05$, ** for $p < 0.01$, *** for $p < 0.001$ and **** for $p < 0.0001$.

Similar to 2D cultures, EGF did not induce significant EGFR degradation in 3D-cultured tumor cells. On the contrary, EGF induced a dramatic EGFR degradation in 3D cultures of fibroblasts regardless of matrix stiffness (Figure 7b,d, lanes 1–4, Figure S3a–c). Interestingly, and contrary to what was found in 2D cultures, erlotinib promoted EGFR degradation when combined with EGF, but not alone, in PANC-1 and BxPC-3 cells cultured in 3D scaffolds under both stiffness conditions (Figure 7b,d, lanes 5–8). Moreover, the extent of EGFR degradation was erlotinib dose-dependent (Figures S2 and S3d).

Contrary to tumor cells, the presence of erlotinib in EGF-incubated fibroblasts in 3D not only induced the degradation of the receptor but prevented it (Figure 7b,d, lanes 5–8), as happened in 2D cultures. In conclusion, erlotinib treatment (combined with EGF) had a contrary effect depending on cell type and dimensionality, promoting EGFR degradation in PDAC cell lines cultured in 3D but preventing it in normal fibroblasts in both 2D and 3D cultures.

To confirm that the decrease of EGFR levels in 3D cultures of PANC-1 and BxPC-3 cells treated with EGF and erlotinib was actually due to protein degradation, we inhibited proteasomal and lysosomal degradation by pre-incubating the cells with MG-132 (proteasome inhibitor) and bafilomycin A1 (lysosome inhibitor). It has been extensively reported that EGF-induced degradation occurs via lysosomes [7], and therefore we questioned whether EGFR downregulation in erlotinib-treated cells (Figure 7b,d) was also due to lysosomal degradation.

Results show that lysosome inhibition in cells treated with erlotinib and EGF led to a notable increase in EGFR levels due to protein accumulation (Figure 8a,b, lane 8, Figure S3e,f). Bafilomycin A1 inhibits fusion between endosomes and lysosomes, which causes the accumulation of cargo unable to suffer degradation [54,55]. Moreover, MG-132 also led to EGFR accumulation upon EGF and erlotinib treatment, but to a lesser extent than bafilomycin A1 (Figure 8a,b, lane 7). Proteasomal inhibition has been reported to deplete the free ubiquitin pool within the cell, thus interfering with protein degradation [56]. Moreover, combinatorial treatment with both inhibitors (MG-132 and bafilomycin A1) also led to protein accumulation in PANC-1 but not in BxPC-3 cells, which presented with reduced EGFR levels compared to its respective control (Figure 8a,b, lanes 5 and 9). This could be explained by the fact that proteasome disruption leads to endoplasmic reticulum stress to which the cell responds by attenuating protein translation, thereby inhibiting global protein synthesis [57]. Altogether, these results confirm that EGFR degradation due to erlotinib and EGF treatment in our 3D cancer cell culture system occurs via lysosomes.

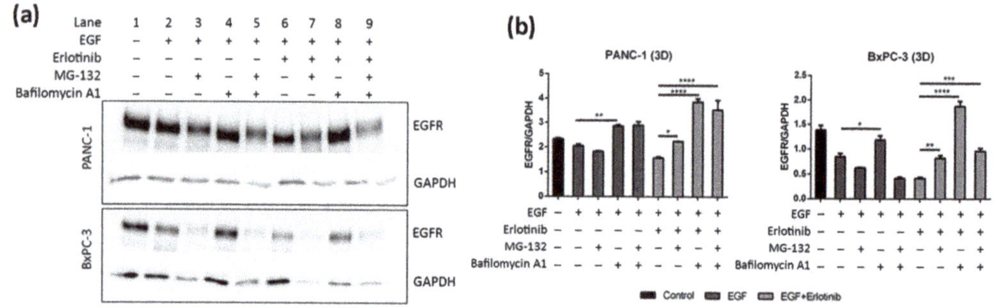

Figure 8. Western blot analysis of the EGFR in PANC-1 and BxPC-3 cells incubated with EGF and erlotinib in the presence of the proteasome (MG-132) and lysosomes (Bafilomycin A1) inhibitors in 0.15% RAD16-I 3D cultures. (**a**) Western blot bands of EGFR; (**b**) densitometry of bands shown in (**a**). GAPDH was used as an internal control. One representative blot is shown. Experiments were repeated three times (N = 3), and statistical differences are indicated as * for $p < 0.05$, ** for pv<0.01, *** for $p < 0.001$ and **** for $p < 0.0001$.

3.5. EGFR Trafficking to Early Endosomes and Lysosomes in 2D and 3D Cultures

It is well established that after stimulation with EGF, the EGFR is activated, internalized and sorted into endosomal compartments. Once in EEA1-positive early endosomes, a small fraction of the receptor is recycled to the plasmatic membrane, while most of it is sorted into late endosomes and subsequently degraded in lysosomes [7]. In 2D cultures, we found a low EGFR fraction colocalizing with EEA1-positive early endosomes in PANC-1

(Figure 9a,d) and BxPC-3 tumor cells (Figure 9b,d). After EGF exposure, the fraction of EGFR colocalizing with EEA1 increased for both tumor cell lines (Figure 9d). Moreover, and in concordance with the hypothesis that erlotinib partially prevents EGFR internalization (Figures 3–5), erlotinib-treated cells presented with the lowest colocalization values between EGFR and EEA1. Additionally, we found that in hNDF (Figure 9c) the EGFR fraction colocalizing with EEA1 was extremely low for all the conditions tested (Figure 9d), but on the contrary, Manders' coefficients obtained for the proportion of EEA1-positive early endosomes containing EGFR signal were almost 1 (Figure 9e), meaning that almost all EEA1-positive endosomes were carrying EGFR (see Section 2.6). These differences between both Manders' coefficients in hNDF exist because these are very large cells expressing EGFR all over the cellular milieu, while early endosomes are located mainly perinuclearly. Therefore, the EGFR signal coinciding with EEA1 signal over the total EGFR intensity (EGFR vs. EEA1) is much lower than the EEA1 signal coinciding with EGFR signal over the total EEA1 intensity (EEA1 vs. EGFR). Moreover, even colocalization between EEA1 and EGFR (Figure 9e) was similar between the control and the EGF-incubated cells, it significantly decreased in the presence of erlotinib alone, indicating a basal EGFR trafficking that was prevented by the presence of erlotinib.

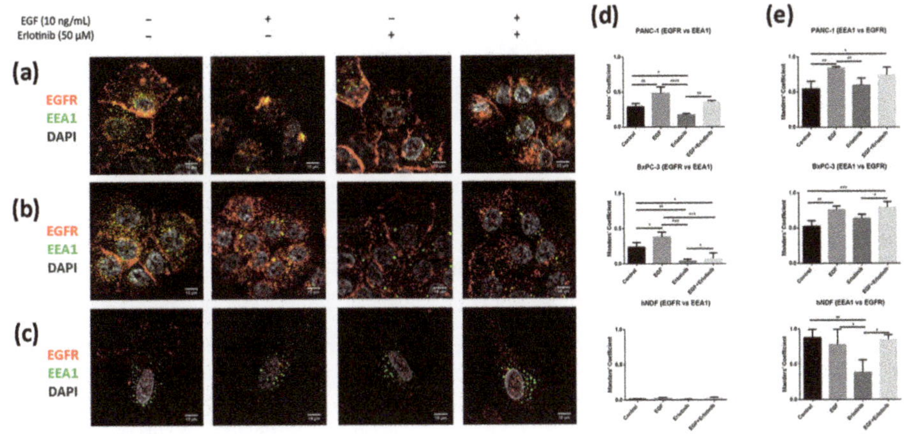

Figure 9. Colocalization analysis of the EGFR and EEA1 (early endosomes) in cells incubated with EGF, erlotinib or both in 2D cultures. EGFR (red) and EEA1 (green) immunofluorescence counterstained with DAPI (grey) in (**a**) PANC-1; (**b**) BxPC-3 and (**c**) hNDF cells. Scale bars represent 10 µm; (**d**) Manders' colocalization coefficients showing the proportion of EGFR overlapping with EEA1; (**e**) Manders' colocalization coefficients showing the proportion of EEA1 overlapping with EGFR. Statistical differences are indicated as * for $p < 0.05$, ** for $p < 0.01$, *** for $p < 0.001$ and **** for $p < 0.0001$.

We next analyzed colocalization between EGFR and LAMP1 (lysosomal-associated membrane protein 1), a well-known lysosomal marker. Lysosome distribution was mainly perinuclear in BxPC-1 and hNDF cells under all conditions tested (Figure 10b,c). However, in PANC-1 cells, lysosomes were distributed all over the cytoplasm in control conditions as well as when incubated with erlotinib but accumulated perinuclearly after EGF treatment (Figure 10a). LAMP1 staining showed significantly higher colocalization coefficients between lysosomes and the EGFR in cells treated with EGF compared to control in all the cell types analyzed (Figure 10d,e). Moreover, the presence of erlotinib alone produced significant differences only in BxPC-3 cells, in which Manders' coefficients (LAMP1 vs. EGFR) (Figure 10e) increased compared to the control.

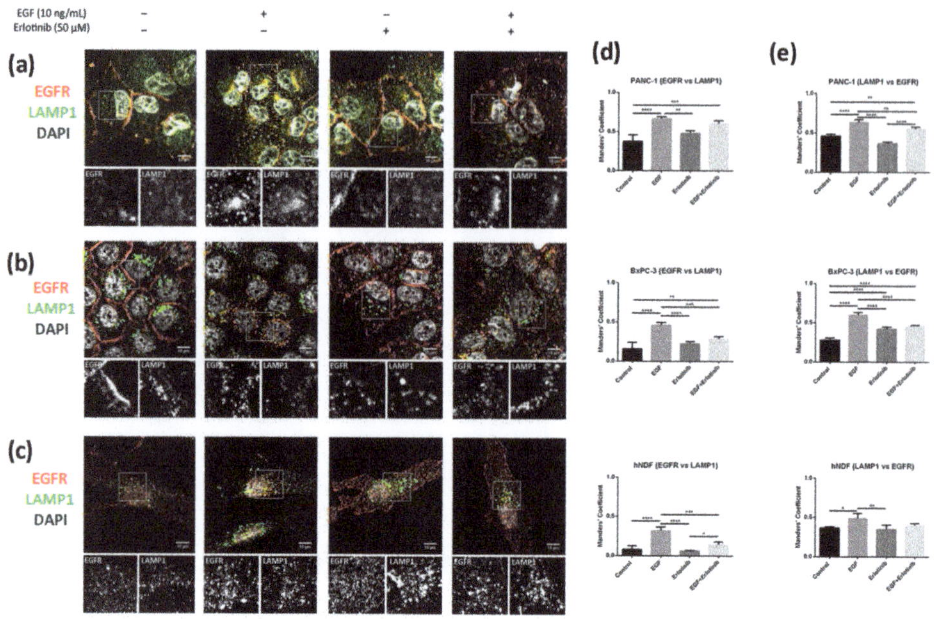

Figure 10. Colocalization analysis of the EGFR and LAMP1 (lysosomes) in cells incubated with EGF, erlotinib or both in 2D cultures. EGFR (red) and LAMP1 (green) immunofluorescence counterstained with DAPI (grey) in (**a**) PANC-1; (**b**) BxPC-3 and (**c**) hNDF cells. Scale bars represent 10 μm. Gray pictures show close-up sections of each marker; (**d**) Manders' colocalization coefficients showing the proportion of EGFR overlapping with LAMP1; (**e**) Manders' colocalization coefficients showing the proportion of LAMP1 overlapping with EGFR. Statistical differences are indicated as * for $p < 0.05$, ** for $p < 0.01$, *** for $p < 0.001$ and **** for $p < 0.0001$.

Finally, we analyzed EGFR sorting into early endosomes and lysosomes in 3D cultures. In BxPC-3 tumor cells, we detected EGFR trafficking to EEA1-positive early endosomes in all conditions tested, but colocalization was much more evident in cells treated with EGF or EGF and erlotinib (Figure 11a, see arrows), in which EGFR was undergoing endocytosis. A similar pattern was found in hNDF (Figure 11b, see arrows). Regarding EGFR sorting into lysosomes, we found that in PANC-1 (Figure 12a) and BxPC-3 cells (Figure 12b), colocalization between EGFR and LAMP1 was mostly detected in cells treated with EGF or EGF and erlotinib. Even though we detected certain colocalization between EGFR and LAMP1-positive lysosomes in 3D-cultured tumor cells incubated with EGF, degradation under these conditions was undetectable at a western blot level (Figure 7), suggesting that the presence of erlotinib exacerbated this degradation in EGF-treated cells.

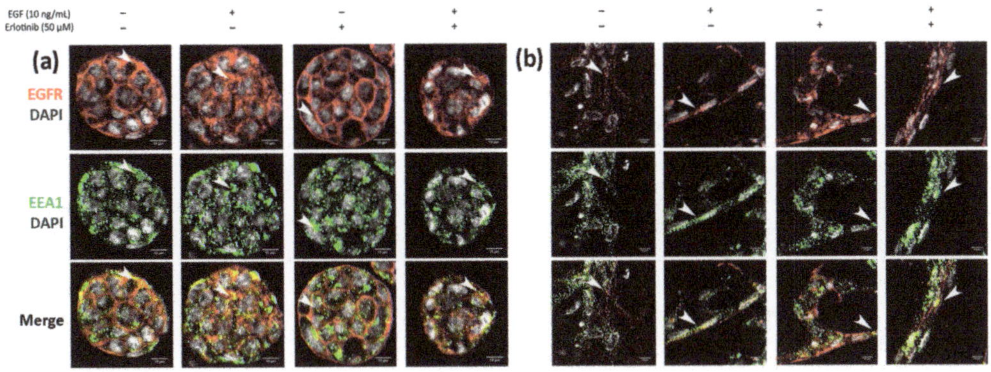

Figure 11. Immunofluorescence of EGFR and EEA1 in (**a**) BxPC-3 cells and (**b**) hNDF incubated with EGF, erlotinib or both in 0.15% RAD16-I 3D cultures. One representative Z plane is shown. Scale bars represent 10 µm. White arrows show EGFR colocalizing with early endosomes.

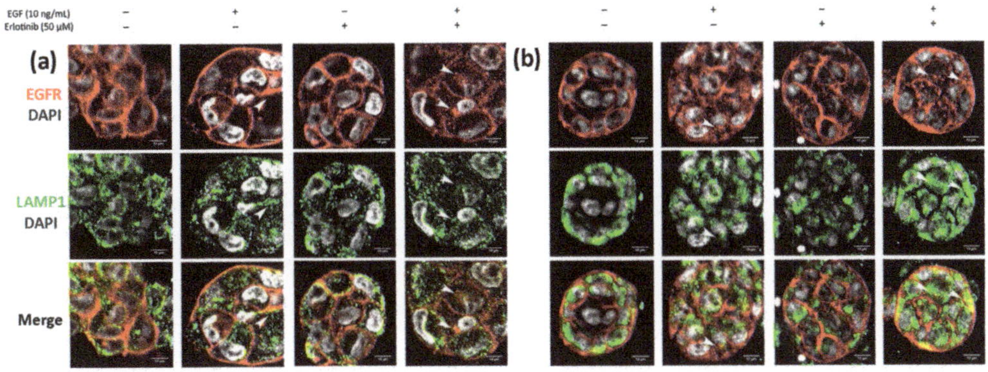

Figure 12. Immunofluorescence of EGFR and LAMP1 in (**a**) PANC-1 and (**b**) BxPC-3 cells incubated with EGF, erlotinib or both in 0.15% RAD16-I 3D cultures. One representative Z plane is shown. Scale bars represent 10 µm. White arrows show EGFR colocalizing with lysosomes.

4. Discussion

The EGFR is the unique TKR reported to respond to multiple stimuli by internalizing into endosomal compartments. Moreover, different stressors induce different trafficking pathways that can end in endosomal accumulation and recycling to the membrane or degradation. Two-dimensional cell cultures are generally used to study EGFR trafficking mechanisms because they are economically affordable and easier to handle as well as to analyze than 3D cultures. However, it is widely accepted that 2D cultures do not recreate the microenvironment of in vivo tissue cells nor can they predict therapy outcome as precisely as 3D cultures do. Growing cells in a 3D environment reveals a more realistic drug response, being that 3D-cultured cells are more resistant to chemotherapy when compared to the same cells grown in 2D monolayer [34,58,59]. Different mechanisms have been attributed to this enhanced drug resistance, but the most straightforward explanation is that the microenvironment provided by the 3D system protects the cells from drug penetration [60]. Besides, most drugs have rapidly dividing cells as a target, and 3D cultures have been described to decrease the proliferation rate of cancer cells compared to 2D cultures [36]. Moreover, quiescent cells that exist in the inner part of the 3D culture remain protected from drug effect [61]. In addition, some pancreatic cancer cell lines have shown an increased expression of diverse drug resistance genes in 3D culture compared to 2D culture [62].

For example, PDAC cells in 3D collagen hydrogels but not in 2D cultures present with gemcitabine resistance through the upregulation of the membrane type-I matrix metalloprotease (MT1-MMP) [63]. Therefore, 3D-cultured cells can recapitulate mechanisms of drug resistance found in tumors, thus offering the opportunity to analyze these mechanisms and test multidrug therapies in vitro in order to reduce animal experimentation.

In this paper, we report a new model for EGFR internalization and degradation due to tyrosine-kinase inhibitor treatment in a 3D cell culture of pancreatic ductal adenocarcinoma cells. Using the synthetic self-assembling peptide RAD16-I as a platform for 3D culture, we found that erlotinib treatment combined with EGF but not alone promoted EGFR degradation in 3D but not 2D cultures of both PDAC cell lines (Figure 7). These results suggest that in 3D culture, EGF is necessary to promote EGFR endocytosis in PDAC cell lines, and that upon internalization, erlotinib promotes its lysosomal degradation (Figure 8). Interestingly, this behavior was not detected in normal fibroblasts, in which erlotinib prevented EGF-induced EGFR internalization and degradation in both 2D and 3D cultures (Figures 5–7). In this sense, results can be controversial since previous in vitro studies demonstrate that TKI inhibitors such as erlotinib and gefitinib induce EGFR internalization and accumulation in non-degradative endosomes in different cell types [9], while others describe that the same TKI suppresses ligand-stimulated endocytosis [11]. Another work in which the authors develop an in vivo tumor model of human oral squamous carcinoma using xenografts reports that gefitinib treatment inhibits EGFR endocytosis [64], suggesting that receptor kinase activity is required for receptor internalization [65]. We indeed detected this kinase activity-dependence to internalize the receptor in fibroblasts (Figure 5) and in tumor cells, but to a lesser extent (Figures 3 and 4). One explanation could be that PDAC cell lines are more resistant to erlotinib and therefore the dose used could be not enough to inhibit the kinase activity of the receptor. However, we used an acute dose of 50 µM, corresponding to the IC_{50} of the erlotinib-resistant cell line PANC-1; this should be enough to inhibit the EGFR in the erlotinib-sensitive BxPC-3 cell line and in hNDF, which both presented an IC_{50} of 10 µM.

It is well established that receptor recycling to the plasmatic membrane after endocytosis leads to continuous signaling. By contrast, degradation in lysosomes is associated with signaling attenuation, and therefore endocytic downregulation could be associated with TKIs sensitization. For example, it has been reported that mutant-EGFR forms are internalized via clathrin-mediated endocytosis and sorted into endosomal compartments, where they continue signaling [66]. Instead, clathrin inhibition in mutant-EGFR-expressing cells induces micropinocytosis-dependent EGFR internalization followed by degradation in lysosomes, which results in signal extinction and apoptosis, and thus overcoming resistance to TKI [66]. Another recent study reports that TKI treatment induces intracellular accumulation of mutated-EGFR. The authors found a positive correlation between EGFR accumulation and clinical benefit in patients presenting with tumors harboring these EGFR mutations, suggesting that blocked EGFR-membrane recycling contributes to TKI sensitivity and a positive therapy outcome [67]. Our results demonstrate not only that erlotinib has a different effect on tumor and normal cells but also that PDAC cells respond differently to erlotinib treatment when cultured in a 3D microenvironment. In fact, PANC-1 cells increased their erlotinib resistance from 45 to 100 µM when cultured in 2D vs. 3D (Table 1), even though under these 3D conditions the EGFR was undergoing degradation. On the other hand, BxPC-3 cells did not show increased erlotinib resistance when cultured in 3D but also showed EGFR degradation. Therefore, under our experimental conditions, EGFR degradation did not sensitize the cells to erlotinib treatment. Given that this EGFR degradation was only detected in 3D cultures, it is possible that in this 3D environment PDAC cells acquire additional resistance mechanisms that allow them to survive independently of the EGFR signaling pathway.

Importantly, EGFR degradation in 3D cultures due to TKI was only detected in the presence of EGF, suggesting that ligand binding is necessary for endocytosis and that once the receptor is internalized, erlotinib treatment promotes its degradation. EGF can reach very high concentrations in different body fluids such as bile, urine and milk [68,69], and a normal epithelium avoids these fluids to reach EGF receptors, which are expressed basolaterally. However, as a result of a premalignant neoplasia, in which the tight junctions of the epithelium become leaky, high concentrations of EGF can reach and activate EGFRs [70]. Moreover, since EGFR ligands have been found in several tumors [71–73] it is likely that our 3D model could closely mimic the tumor in vivo scenario during erlotinib therapy.

5. Conclusions

We have developed a 3D cancer cell culture system to study EGFR trafficking and degradation due to TKI treatment. This study highlights not only that erlotinib has a distinct effect on tumor and normal cells but also that pancreatic ductal adenocarcinoma cells respond differently to erlotinib treatment when cultured in a 3D microenvironment. To our knowledge, this is the first time where EGFR trafficking is explored in a three-dimensional environment, and it is our hope that this report may encourage other researchers to introduce 3D cell culture systems in the study of receptor trafficking. Moreover, future research might be focused in studying the effect that ECM components present in PDAC, such as collagens and hyaluronic acid, could have on EGFR trafficking.

Supplementary Materials: The following are available online at https://www.mdpi.com/article/10.3390/cancers13184504/s1, Figure S1: Erlotinib dose-response curves in 2D and 3D cultures (0.15% and 0.5% RAD16-I) of BxPC-3, PANC-1 and hNDF cells. Figure S2: EGFR degradation with increasing concentrations of erlotinib in EGF-incubated BxPC-3 cells in 3D RAD16-I scaffolds. Figure S3: Uncropped western blot images.

Author Contributions: Conceptualization, N.B. and C.E.S.; methodology, N.B.; investigation, N.B. and A.A.; writing—original draft preparation, N.B. and A.A.; writing—review and editing, C.E.S.; supervision, C.E.S.; funding acquisition, C.E.S. All authors have read and agreed to the published version of the manuscript.

Funding: This research was funded by the Ministry of Science and Innovation (MICINN) of the Spanish Government, grant number RTI2018-096455-B-I00 to C.E.S. and by the Department of Bioengineering (IQS-School of Engineering, URL).

Institutional Review Board Statement: Not applicable.

Informed Consent Statement: Not applicable.

Data Availability Statement: Data is contained within the article or Supplementary Materials.

Acknowledgments: We thank Elena Rebollo from the Molecular Imaging Platform (MIP IBMB-PCB) for technical support during image acquisition. We are grateful to Claire Jarrosson Moral for creating the graphical abstract.

Conflicts of Interest: The authors declare no conflict of interest.

References

1. Mendelsohn, J.; Baselga, J. Epidermal Growth Factor Receptor Targeting in Cancer. *Semin. Oncol.* **2006**, *33*, 369–385. [CrossRef] [PubMed]
2. Tan, X.; Lambert, P.F.; Rapraeger, A.C.; Anderson, R.A. Stress-Induced EGFR Trafficking: Mechanisms, Functions, and Therapeutic Implications. *Trends Cell Biol.* **2016**, *26*, 352–366. [CrossRef]
3. Oksvold, M.P.; Hultfeldt, H.S.; Øtvold, A.C.; Skarpen, E. UV induces tyrosine kinase-independent internalisation and endosome arrest of the EGF receptor. *J. Cell Sci.* **2002**, *115*, 793–803. [CrossRef]
4. Shen, J.; Xia, W.; Khotskaya, Y.; Huo, L.; Nakanishi, K.; Lim, S.O.; Du, Y.; Wang, Y.; Chang, W.-C.; Chen, C.-H.; et al. EGFR modulates microRNA maturation in response to hypoxia through phosphorylation of AGO2. *Nature* **2013**, *497*, 383–387. [CrossRef] [PubMed]
5. Fang, Y.; Tan, J.; Zhang, Q. Signaling pathways and mechanisms of hypoxia-induced autophagy in the animal cells. *Cell Biol. Int.* **2015**, *39*, 891–898. [CrossRef] [PubMed]
6. Filomeni, G.; De Zio, D.; Cecconi, F. Oxidative stress and autophagy: The clash between damage and metabolic needs. *Cell Death Differ.* **2015**, *22*, 377–388. [CrossRef]

7. Roepstorff, K.; Grandal, M.V.; Henriksen, L.; Knudsen, S.L.J.; Lerdrup, M.; Grøvdal, L.; Willumsen, B.M.; van Deurs, B. Differential effects of EGFR ligands on endocytic sorting of the receptor. *Traffic* **2009**, *10*, 1115–1127. [CrossRef] [PubMed]
8. Henriksen, L.; Grandal, M.V.; Knudsen, S.L.J.; van Deurs, B.; Grøvdal, L.M. Internalization Mechanisms of the Epidermal Growth Factor Receptor after Activation with Different Ligands. *PLoS ONE* **2013**, *8*, e58148. [CrossRef] [PubMed]
9. Blandin, A.F.; Cruz Da Silva, E.; Mercier, M.C.; Glushonkov, O.; Didier, P.; Dedieu, S.; Schneider, C.; Devy, J.; Etienne-Selloum, N.; Dontenwill, M.; et al. Gefitinib induces EGFR and $\alpha5\beta1$ integrin co-endocytosis in glioblastoma cells. *Cell Mol. Life Sci.* **2020**, *78*, 2940–2962. [CrossRef]
10. Cao, X.; Zhu, H.; Ali-Osman, F.; Lo, H.W. EGFR and EGFRvIII undergo stress- and EGFR kinase inhibitor-induced mitochondrial translocalization: A potential mechanism of EGFR-driven antagonism of apoptosis. *Mol. Cancer* **2011**, *10*, 26. [CrossRef]
11. Nishimura, Y.; Bereczky, B.; Ono, M. The EGFR inhibitor gefitinib suppresses ligand-stimulated endocytosis of EGFR via the early/late endocytic pathway in non-small cell lung cancer cell lines. *Histochem. Cell Biol.* **2007**, *127*, 541–553. [CrossRef]
12. Dreier, A.; Barth, S.; Goswami, A.; Weis, J. Cetuximab induces mitochondrial translocalization of EGFRvIII, but not EGFR: Involvement of mitochondria in tumor drug resistance? *Tumor. Biol.* **2012**, *33*, 85–94. [CrossRef] [PubMed]
13. Liao, H.J.; Carpenter, G. Cetuximab/C225-induced intracellular trafficking of epidermal growth factor receptor. *Cancer Res.* **2009**, *69*, 6179–6183. [CrossRef]
14. Griffith, L.G.; Swartz, M.A. Capturing complex 3D tissue physiology in vitro. *Nat. Rev. Mol. Cell Biol.* **2006**, *7*, 211–224. [CrossRef]
15. Minchinton, A.I.; Tannock, I.F. Drug penetration in solid tumours. *Nat. Rev. Cancer* **2006**, *6*, 583–592. [CrossRef]
16. Thakuri, P.S.; Liu, C.; Luker, G.D.; Tavana, H. Biomaterials-Based Approaches to Tumor Spheroid and Organoid Modeling. *Adv. Healthc. Mater.* **2018**, *7*, 1–21. [CrossRef] [PubMed]
17. Cui, X.; Hartanto, Y.; Zhang, H. Advances in multicellular spheroids formation. *J. R. Soc. Interface* **2017**, *14*, 20160877. [CrossRef] [PubMed]
18. Alemany-Ribes, M.; Semino, C.E. Bioengineering 3D environments for cancer models. *Adv. Drug Deliv. Rev.* **2014**, *79*, 40–49. [CrossRef]
19. Liu, C.; Lewin, M.D.; Chiang, B.; Luker, K.E.; Luker, G.D. Hybrid Collagen Alginate Hydrogel as a Platform for 3D Tumor Spheroid Invasion. *Acta Biomater.* **2018**, *15*, 213–225. [CrossRef]
20. Miyazaki, K.; Oyanagi, J.; Hoshino, D.; Kumagai, H.; Miyagi, Y. Cancer cell migration on elongate protrusions of fibroblasts in collagen matrix. *Sci. Rep.* **2019**, *9*, 292. [CrossRef] [PubMed]
21. Kim, S.-K.; Jang, S.D.; Kim, H.; Chung, S.; Park, J.K.; Kuh, H.-J. Phenotypic Heterogeneity and Plasticity of Cancer Cell Migration in a Pancreatic Tumor Three-Dimensional Culture Model. *Cancers* **2020**, *12*, 1305. [CrossRef]
22. Ricci, C.; Mota, C.; Moscato, S.; Alessandro, D.D.; Sartoris, S.; Bronte, V.; Boggi, U.; Campani, D.; Funel, N.; Moroni, L.; et al. Interfacing polymeric scaffolds with primary pancreatic ductal adenocarcinoma cells to develop 3D cancer models Interfacing polymeric scaffolds with primary pancreatic ductal adenocarcinoma cells to develop 3D cancer models. *Biomatter* **2015**, *4*, e955386. [CrossRef] [PubMed]
23. Puls, T.J.; Tan, X.; Whittington, C.F.; Voytik-Harbin, S.L. 3D collagen fibrillar microstructure guides pancreatic cancer cell phenotype and serves as a critical design parameter for phenotypic models of EMT. *PLoS ONE* **2017**, *12*, e0188870. [CrossRef] [PubMed]
24. Paszek, M.J.; Zahir, N.; Johnson, K.R.; Lakins, J.N.; Rozenberg, G.I.; Gefen, A.; Reinhart-King, C.A.; Margulies, S.S.; Dembo, M.; Boettiger, D.; et al. Tensional homeostasis and the malignant phenotype. *Cancer cell* **2005**, *8*, 241–254. [CrossRef]
25. Kawano, S.; Kojima, M.; Higuchi, Y.; Sugimoto, M.; Ikeda, K.; Sakuyama, N.; Takahashi, S.; Hayashi, R.; Ochiai, A.; Saito, N. Assessment of elasticity of colorectal cancer tissue, clinical utility, pathological and phenotypical relevance. *Cancer Sci.* **2015**, *106*, 1232–1239. [CrossRef]
26. Rice, A.J.; Cortes, E.; Lachowski, D.; Cheung, B.C.H.; Karim, S.A.; Morton, J.P.; del Río Hernández, A. Matrix stiffness induces epithelial-mesenchymal transition and promotes chemoresistance in pancreatic cancer cells. *Oncogenesis* **2017**, *6*, e352. [CrossRef] [PubMed]
27. Apte, M.V.; Park, S.; Phillips, P.A.; Santucci, N.; Goldstein, D.; Kumar, R.K.; Ramm, G.A.; Buchler, M.; Friess, H.; McCarroll, J.A.; et al. Desmoplastic Reaction in Pancreatic Cancer: Role of Pancreatic Stellate Cells. *Pancreas* **2004**, *29*, 179–187. [CrossRef]
28. Itoh, Y.; Takehara, Y.; Kawase, T.; Terashima, K.; Ohkawa, Y.; Hirose, Y.; Koda, A.; Hyodo, N.; Ushio, T.; Hirai, Y.; et al. Feasibility of magnetic resonance elastography for the pancreas at 3T. *J. Magn. Reason. Imaging* **2016**, *43*, 384–390. [CrossRef]
29. Liu, H.Y.; Nguyen, H.D.; Lin, C.C. Dynamic PEG–Peptide Hydrogels via Visible Light and FMN-Induced Tyrosine Dimerization. *Adv. Healthc. Mater.* **2018**, *7*, 1800954. [CrossRef] [PubMed]
30. Marí-Buyé, N.; Luque, T.; Navajas, D.; Semino, C.E. Development of a Three-Dimensional Bone-Like Construct in a Soft Self-Assembling Peptide Matrix. *Tissue Eng. Part A* **2013**, *19*, 870–881. [CrossRef]
31. Recha-Sancho, L.; Semino, C.E. Heparin-based self-assembling peptide scaffold reestablish chondrogenic phenotype of expanded de-differentiated human chondrocytes. *J. Biomed. Mater. Res. Part A* **2016**, *104*, 1694–1706. [CrossRef] [PubMed]
32. Castells-Sala, C.; Recha-Sancho, L.; Llucià-Valldeperas, A.; Soler-Botija, C.; Bayes-Genis, A.; Semino, C.E. Three-Dimensional Cultures of Human Subcutaneous Adipose Tissue-Derived Progenitor Cells Based on RAD16-I Self-Assembling Peptide. *Tissue Eng. Part C Methods* **2016**, *22*, 113–124. [CrossRef]

33. Betriu, N.; Jarrosson-Moral, C.; Semino, C.E. Culture and Differentiation of Human Hair Follicle Dermal Papilla Cells in a Soft 3D Self-Assembling Peptide Scaffold. *Biomolecules* **2020**, *10*, 684. [CrossRef]
34. Song, H.; Cai, G.H.; Liang, J.; Ao, D.S.; Wang, H.; Yang, Z.H. Three-dimensional culture and clinical drug responses of a highly metastatic human ovarian cancer HO-8910PM cells in nanofibrous microenvironments of three hydrogel biomaterials. *J. Nanobiotechnol.* **2020**, *18*, 90. [CrossRef]
35. Miroshnikova, Y.A.; Jorgens, D.M.; Spirio, L.; Auer, M.; Sarang-Sieminski, A.L.; Weaver, V.M. Engineering strategies to recapitulate epithelial morphogenesis within synthetic three-dimensional extracellular matrix with tunable mechanical properties. *Phys. Biol.* **2011**, *8*, 026013. [CrossRef]
36. Betriu, N.; Semino, C.E. Development of a 3D Co-Culture System as a Cancer Model Using a Self-Assembling Peptide Scaffold. *Gels* **2018**, *4*, 65. [CrossRef] [PubMed]
37. Betriu, N.; Recha-sancho, L.; Semino, C.E. Culturing Mammalian Cells in Three-dimensional Peptide Scaffolds. *J. Vis. Exp.* **2018**, e57259. [CrossRef] [PubMed]
38. Schindelin, J.; Arganda-Carreras, I.; Frise, E.; Kaynig, V.; Longair, M.; Pietzsch, T.; Preibisch, S.; Rueden, C.; Saalfeld, S.; Schmid, B.; et al. Fiji: An open-source platform for biological-image analysis. *Nat. Methods* **2012**, *9*, 676–682. [CrossRef]
39. Manders, E.M.; Stap, J.; Brakenhoff, G.J.; Van Driel, R.; Aten, J.A. Dynamics of three-dimensional replication patterns during the S-phase, analysed by double labelling of DNA and confocal microscopy. *J. Cell Sci.* **1992**, *103*, 857–862. [CrossRef]
40. Sastre, D.; Estadella, I.; Bosch, M.; Felipe, A. Triple-Colocalization Approach to Assess Traffic Patterns and Their Modulation. In *Computer Optimized Microscopy*, 1st ed.; Rebollo, E., Bosch, M., Eds.; Springer: New York, NY, USA, 2019; pp. 215–233. [CrossRef]
41. Bolte, S.; Cordelières, F.P. A guided tour into subcellular colocalization analysis in light microscopy. *J. Microsc.* **2006**, *224*, 213–232. [CrossRef]
42. Buck, E.; Eyzaguirre, A.; Haley, J.D.; Gibson, N.W.; Cagnoni, P.; Iwata, K.K. Inactivation of Akt by the epidermal growth factor receptor inhibitor erlotinib is mediated by HER-3 in pancreatic and colorectal tumor cell lines and contributes to erlotinib sensitivity. *Mol. Cancer Ther.* **2006**, *5*, 2051–2059. [CrossRef]
43. Bryant, K.L.; Mancias, J.D.; Kimmelman, A.C.; Der, C.J. KRAS: Feeding pancreatic cancer proliferation. *Trends Biochem. Sci.* **2014**, *39*, 91–100. [CrossRef]
44. Sieminski, A.L.; Was, A.S.; Kim, G.; Gong, H.; Kamm, R.D. The stiffness of three-dimensional ionic self-assembling peptide gels affects the extent of capillary-like network formation. *Cell Biochem. Biophys.* **2007**, *49*, 73–83. [CrossRef]
45. Hadjipanayi, E.; Mudera, V.; Brown, A. Close dependence of fibroblast proliferation on collagen scaffold matrix stiffness. *J. Tissue Eng. Regen. Med.* **2008**, *3*, 77–84. [CrossRef]
46. Fernández-Muiños, T.; Recha-Sancho, L.; López-Chicón, P.; Castells-Sala, C.; Mata, A.; Semino, C.E. Bimolecular based heparin and self-assembling hydrogel for tissue engineering applications. *Acta Biomater.* **2015**, *16*, 35–48. [CrossRef]
47. Recha-Sancho, L.; Semino, C.E. Chondroitin sulfate- and decorin-based self-Assembling scaffolds for cartilage tissue engineering. *PLoS ONE* **2016**, *11*, e0157603. [CrossRef] [PubMed]
48. Jones, S.; King, P.J.; Antonescu, C.N.; Sugiyama, M.G.; Bhamra, A.; Surinova, S.; Angelopoulos, N.; Kragh, M.; Pedersen, M.W.; Hartley, J.A.; et al. Targeting of EGFR by a combination of antibodies mediates unconventional EGFR trafficking and degradation. *Sci. Rep.* **2020**, *10*, 663. [CrossRef] [PubMed]
49. Iida, M.; Brand, T.M.; Starr, M.M.; Li, C.; Huppert, E.J.; Luthar, N.; Pedersen, M.W.; Horak, I.D.; Kragh, M.; Wheeler, D.L. Sym004, a novel EGFR antibody mixture, can overcome acquired resistance to cetuximab. *Neoplasia* **2013**, *15*, 1196–1206. [CrossRef] [PubMed]
50. Sobhakumari, A.; Schickling, B.M.; Love-homan, L.; Raeburn, A.; Fletcher, E.V.M.; Case, A.J.; Domann, F.E.; Miller, F.J., Jr.; Simons, A.L. NOX4 mediates cytoprotective autophagy induced by the EGFR inhibitor erlotinib in head and neck cancer cells. *Toxicol. Appl. Pharmacol.* **2013**, *272*, 736–745. [CrossRef] [PubMed]
51. Zou, Y.; Ling, Y.; Sironi, J.; Schwartz, E.L. The Autophagy Inhibitor Chloroquine Overcomes the Innate Resistance of Wild-Type EGFR Non-Small-Cell Lung Cancer Cells to Erlotinib. *J. Thorac. Oncol.* **2013**, *8*, 693–702. [CrossRef]
52. Dragowska, W.H.; Weppler, S.A.; Wang, J.C.; Wong, L.Y.; Kapanen, A.I.; Rawji, J.S.; Warburton, C.; Qadir, M.A.; Donohue, E.; Roberge, M.; et al. Induction of Autophagy Is an Early Response to Gefitinib and a Potential Therapeutic Target in Breast Cancer. *PLoS ONE* **2013**, *8*, e76503. [CrossRef] [PubMed]
53. Sakuma, Y.; Matsukuma, S.; Nakamura, Y.; Yoshihara, M.; Koizume, S.; Sekiguchi, H.; Saito, H.; Nakayama, H.; Kameda, Y.; Yokose, T.; et al. Enhanced autophagy is required for survival in EGFR-independent EGFR-mutant lung adenocarcinoma cells. *Lab. Investig.* **2013**, *93*, 1137–1146. [CrossRef]
54. Yoshimori, T.; Yamamoto, A.; Moriyama, Y.; Futai, M.; Tashiro, Y. Bafilomycin A1, a specific inhibitor of vacuolar-type H+-ATPase, inhibits acidification and protein degradation in lysosomes of cultured cells. *J. Biol. Chem.* **1991**, *266*, 17707–17712. [CrossRef]
55. Mauvezin, C.; Nagy, P.; Juhász, G.; Neufeld, T.P. Autophagosome-lysosome fusion is independent of V-ATPase-mediated acidification. *Nat. Commun.* **2015**, *6*, 7007. [CrossRef] [PubMed]
56. Melikova, M.S.; Kondratov, K.A.; Kornilova, E.S. Two different stages of epidermal growth factor (EGF) receptor endocytosis are sensitive to free ubiquitin depletion produced by proteasome inhibitor MG132. *Cell Biol. Int.* **2006**, *30*, 31–43. [CrossRef]
57. Albornoz, N.; Bustamante, H.; Soza, A.; Burgos, P. Cellular responses to proteasome inhibition: Molecular mechanisms and beyond. *Int. J. Mol. Sci.* **2019**, *20*, 3379. [CrossRef] [PubMed]

58. Liu, M.; Zhang, X.; Long, C.; Xu, H.; Cheng, X.; Chang, J.; Zhang, C.; Zhang, C.; Wang, X. Collagen-based three-dimensional culture microenvironment promotes epithelial to mesenchymal transition and drug resistance of human ovarian cancer in vitro. *RSC Adv.* **2018**, *8*, 8910–8919. [CrossRef]
59. Firuzi, O.; Che, P.P.; El Hassouni, B.; Buijs, M.; Coppola, S.; Löhr, M.; Funel, N.; Heuchel, R.; Carnevale, I.; Schmidt, T.; et al. Role of c-MET inhibitors in overcoming drug resistance in spheroid models of primary human pancreatic cancer and stellate cells. *Cancers* **2019**, *11*, 638. [CrossRef]
60. Perche, F.; Torchilin, V.P. Cancer cell spheroids as a model to evaluate chemotherapy protocols. *Cancer Biol. Ther.* **2012**, *13*, 1205–1213. [CrossRef]
61. Chitcholtan, K.; Asselin, E.; Parent, S.; Sykes, P.H.; Evans, J.J. Differences in growth properties of endometrial cancer in three dimensional (3D) culture and 2D cell monolayer. *Exp. Cell Res.* **2013**, *319*, 75–87. [CrossRef] [PubMed]
62. Longati, P.; Jia, X.; Eimer, J.; Wagman, A.; Witt, M.-R.; Rehnmark, S.; Verbeke, C.; Toftgård, R.; Löhr, M.; Heuchel, R.L. 3D pancreatic carcinoma spheroids induce a matrix-rich, chemoresistant phenotype offering a better model for drug testing. *BMC Cancer* **2013**, *13*, 95. [CrossRef]
63. Dangi-Garimella, S.; Krantz, S.B.; Barron, M.R.; Shields, M.A.; Heiferman, M.J.; Grippo, P.J.; Bentrem, D.J.; Munshi, H.G. Three-dimensional collagen I promotes gemcitabine resistance in pancreatic cancer through MT1-MMP-mediated expression of HMGA2. *Cancer Res.* **2011**, *71*, 1019–1028. [CrossRef] [PubMed]
64. Pinilla-Macua, I.; Grassart, A.; Duvvuri, U.; Watkins, S.C.; Sorkin, A. EGF receptor signaling, phosphorylation, ubiquitylation and endocytosis in tumors in vivo. *eLife* **2017**, *6*, e31993. [CrossRef]
65. Heukers, R.; Vermeulen, J.F.; Fereidouni, F.; Bader, A.N.; Voortman, J.; Roovers, R.C.; Gerritsen, H.C.; van Bergen en Henegouwen, P.M.P. Endocytosis of EGFR requires its kinase activity and N-terminal transmembrane dimerization motif. *J. Cell Sci.* **2013**, *126*, 4900–4912. [CrossRef]
66. Ménard, L.; Floc'h, N.; Martin, M.J.; Cross, D.A.E. Reactivation of mutant-EGFR degradation through clathrin inhibition overcomes resistance to EGFR tyrosine kinase inhibitors. *Cancer Res.* **2018**, *78*, 3267–3279. [CrossRef]
67. De Wit, M.; Gao, Y.; Mercieca, D.; de Heer, I.; Valkenburg, B.; van Royen, M.E.; Aerts, J.; Smitt, P.S.; French, P. Mutation and drug-specific intracellular accumulation of EGFR predict clinical responses to tyrosine kinase inhibitors. *EBio Med.* **2020**, *56*, 102796. [CrossRef]
68. Beardmore, J.M.; Richards, R.C. Concentrations of epidermal growth factor in mouse milk throughout lactation. *J. Endocrinol.* **1983**, *96*, 287–292. [CrossRef]
69. Grau, M.; Rodríguez, C.; Soley, M.; Ramírez, I. Relationship between epidermal growth factor in mouse submandibular glands, plasma, and bile: Effects of catecholamines and fasting. *Endocrinology* **1994**, *135*, 1854–1862. [CrossRef]
70. Mullin, J.M. Epithelial Barriers, Compartmentation and Cancer. *Sci. Signal.* **2004**, *2004*, pe2. [CrossRef] [PubMed]
71. Thøgersen, V.B.; Sørensen, B.S.; Poulsen, S.S.; Ørntoft, T.F.; Wolf, H.; Nexo, E. A Subclass of HER1 Ligands Are Prognostic Markers for Survival in Bladder Cancer Patients. *Cancer Res.* **2001**, *61*, 6227–6233. [CrossRef] [PubMed]
72. Normanno, N.; Bianco, C.; Strizzi, L.; Mancino, M.; Maiello, M.R.; De Luca, A.; Caponigro, F.; Salomon, D.S. The ErbB Receptors and their Ligands in Cancer: An Overview. *Curr. Drug Targets* **2005**, *6*, 243–257. [CrossRef] [PubMed]
73. Révillion, F.; Lhotellier, V.; Hornez, L.; Bonneterre, J.; Peyrat, J.-P. ErbB/HER ligands in human breast cancer, and relationships with their receptors, the bio-pathological features and prognosis. *Ann. Oncol.* **2008**, *19*, 73–80. [CrossRef] [PubMed]

cancers

MDPI

Review

Tumor Microenvironment and Hydrogel-Based 3D Cancer Models for In Vitro Testing Immunotherapies

Chiara Vitale [1], Monica Marzagalli [2], Silvia Scaglione [2,3], Alessandra Dondero [1], Cristina Bottino [1,4,*] and Roberta Castriconi [1]

1 Department of Experimental Medicine (DIMES), University of Genova, 16132 Genova, Italy; chiara.vitale@edu.unige.it (C.V.); alessandra.dondero@unige.it (A.D.); roberta.castriconi@unige.it (R.C.)
2 React4life SRL, 16121 Genova, Italy; m.marzagalli@react4life.com (M.M.); s.scaglione@react4life.com (S.S.)
3 National Research Council of Italy, Institute of Electronics, Information Engineering and Telecommunications (IEIIT), 16149 Genova, Italy
4 IRCCS Istituto Giannina Gaslini, 16147 Genova, Italy
* Correspondence: cristina.bottino@unige.it

Simple Summary: Immunotherapies are emerging as promising strategies to cure cancer and extend patients' survival. Efforts should be focused, however, on the development of preclinical tools better able to predict the therapeutic benefits in individual patients. In this context, the availability of reliable preclinical models capable of recapitulating the tumor milieu while overcoming the limitations of traditional systems is mandatory. Here, we review the tumor immune responses, escape mechanisms, and the most recent 3D biomaterial-based cancer in vitro models useful for investigating the effects of the different immunotherapeutic approaches. The main challenges and possible future trends are also discussed.

Citation: Vitale, C.; Marzagalli, M.; Scaglione, S.; Dondero, A.; Bottino, C.; Castriconi, R. Tumor Microenvironment and Hydrogel-Based 3D Cancer Models for In Vitro Testing Immunotherapies. *Cancers* **2022**, *14*, 1013. https://doi.org/10.3390/cancers14041013

Academic Editors: Serena Danti, Nicola Contessi Negrini and Alessandro Franchi

Received: 14 January 2022
Accepted: 14 February 2022
Published: 17 February 2022

Abstract: In recent years, immunotherapy has emerged as a promising novel therapeutic strategy for cancer treatment. In a relevant percentage of patients, however, clinical benefits are lower than expected, pushing researchers to deeply analyze the immune responses against tumors and find more reliable and efficient tools to predict the individual response to therapy. Novel tissue engineering strategies can be adopted to realize in vitro fully humanized matrix-based models, as a compromise between standard two-dimensional (2D) cell cultures and animal tests, which are costly and hardly usable in personalized medicine. In this review, we describe the main mechanisms allowing cancer cells to escape the immune surveillance, which may play a significant role in the failure of immunotherapies. In particular, we discuss the role of the tumor microenvironment (TME) in the establishment of a milieu that greatly favors cancer malignant progression and impact on the interactions with immune cells. Then, we present an overview of the recent in vitro engineered preclinical three-dimensional (3D) models that have been adopted to resemble the interplays between cancer and immune cells and for testing current therapies and immunotherapeutic approaches. Specifically, we focus on 3D hydrogel-based tools based on different types of polymers, discussing the suitability of each of them in reproducing the TME key features based on their intrinsic or tunable characteristics. Finally, we introduce the possibility to combine the 3D models with technological fluid dynamics platforms, reproducing the dynamic complex interactions between tumor cells and immune effectors migrated in situ via the systemic circulation, pointing out the challenges that still have to be overcome for setting more predictive preclinical assays.

Keywords: 3D cancer models; immunotherapies; biomaterials; tumor escape mechanisms

1. Introduction

During the last decade, immunotherapy has emerged as a promising alternative to traditional anticancer treatments [1–4]. Harnessing the immune system represents a potent approach providing a patient-specific and durable strategy. The benefit and efficacy of immunotherapies, unleashing immune cell activity through antibodies specific for tumor antigens (TAs) or blocking immune checkpoint axes (ICB), have been documented in several clinical trials. However, in a significant portion of patients, resistance to immunotherapeutic procedures exists, whose causes are yet to be clarified. Indeed, immunotherapies, established according to results obtained from current preclinical models, resulted in only 20–40% of durable clinical responses [5]. In light of these considerations, there is an urgency to validate new platforms suitable to better set up anticancer treatments and predict their efficacy in individual patients. To this aim, it is imperative to develop highly predictive screening tools capable of resembling the complex structure of the human tumor microenvironment (TME), holding a heterogeneous population of cells and extracellular components, all involved in dynamic crosstalk. Then, ideally, such platforms might consider both the 3D contexture of the tissue and the complex interplay between different cell types.

Several cancer-related phenomena, such as metastasis, cell motility, and uncontrolled proliferation, are modulated by the surrounding extracellular matrix (ECM) [6]. Indeed, the ECM provides both mechanical support and biochemical signals, directly affecting cell activity in both physiological and pathological conditions [7]. In a solid tumor scenario, cancer cells are capable of shaping the ECM niche by changing its properties toward a pro-malignant phenotype [8]. Experimental evidence such as the histological analysis of tumor specimens isolated from animal models or patients strongly supports the concept that an altered ECM architecture and composition can play a pivotal role in the regulation of tumor onset, progression, immune evasion, and sensitivity to immunotherapeutic approaches [9,10]. ECM components, including glycoproteins, glycosaminoglycans (GAG), proteoglycans, soluble molecules, and fibrillar proteins, can directly interact with specific receptors on the tumor cell surface, regulating several aspects of their biology. In this context, particularly relevant is the CD44/hyaluronic acid (HA) axis [11], which is targeted in different therapeutic approaches, or the CXCR4/SDF1 axis driving bone marrow (BM) homing of tumor cells. ECM components can also contribute to the onset of an immunosuppressive milieu since many molecules such as the serine protease plasmin, matrix metalloprotease (MMP)-2 and MMP9, or thrombospondin-1 disrupt the latent form of tumor growth factor (TGF)-β (LAP–TGF-β) with the release of an active form of the cytokine [12,13]. Upon activation, TGF-β exerts potent immunomodulatory functions shaping the tumor immune landscape (see next paragraphs for more details). The ECM, depending on the relative abundance of the various components, can form "molecular sieves" able to regulate the migration of tumor cells and immune cells inside the ECM. In this context, it is relevant to stress that in tumor specimens from patients, components of the immune system such as natural killer (NK) cells are mainly observed in the tumor stroma rather than in tumor parenchyma [14,15], probably contributing to the low efficacy of the NK-mediated immune surveillance of solid tumors. Low tumor infiltration is also observed in the context of adoptive cell therapy based on the infusion of T cells, in vitro engineered to expressed TA-specific chimeric receptors (CAR-T) [16]. These effectors were shown to lose the expression of the enzyme heparanase (HPSE), which degrades heparan sulfate proteoglycans, the main components of the ECM [16]. Importantly, when CAR-T cells were also transduced with HPSE, they showed enhanced tumor infiltration and improved overall survival in xenograft tumor models. It is relevant to point out that the architecture of the extracellular fibers such as collagen also greatly influences the tumor biology and response to therapies [17]. In particular, it has been recently shown that the interaction of discoidin domain receptor 1 (DDR1), a collagen receptor expressed by tumor cells, promotes collagen fiber alignment, contributing to immune exclusion [18]. Ablation of DDR1 in tumors promotes the intra-tumoral infiltration of T cells. The described collagen remodeling requires the DDR1 extracellular (DDR1-ECD), but not the intracellular, kinase domain to be effective. Importantly, fibrillar components also impact the status of tumor cells in terms of quiescence or proliferation. In this context, "very dormant" cancer cells contribute to the establishment of type III collagen-enriched ECM niches that deeply sustain tumor dormancy. Histopathological analysis showed that tumor specimens from patients with lymph node-negative head and neck squamous cell carcinoma were enriched in type III collagen levels as compared to tumors from patients with tumor-infiltrated lymph nodes [19].

Besides the ECM components, tumor-infiltrating cells are key determinants of malignant advancement. Different cell types including regulatory T cells (Treg), cancer-associated fibroblasts (CAFs), tumor-associated macrophages (TAMs), mesenchymal stem cells, and endothelial cells can all contribute to tumor growth and escape from the host immune surveillance [20–22]. Therefore, to closely mimic in vitro the complex cancer dynamic environment, it is crucial to consider the tumor–stroma–immune cell interplays.

To date, most of the data available on the communication between the human immune system and cancer cells rely on 2D standard monolayers [23]. These flat systems are standardized, high-throughput, and cost-effective [24]. However, they are over-simplified tools that cannot replicate the complexity of the in vivo scenario, mainly due to the bi-dimensionality resulting in the lack of proper cell-to-cell and cell-to-ECM reciprocal interactions [23,25,26]. Moreover, although animal testing remains a gold standard in cancer research, it cannot faithfully reproduce the human TME [27]. In addition, animal models are costly, time-consuming, and hardly applicable on the road to large-scale personalized therapeutic approaches [28,29].

To solve these constraints, various 3D in vitro culture models have been realized to carry out preclinical experimental investigations under more physiological conditions. Initially, 3D multicellular tumor spheroids were proposed to imitate the human tumor native spatial arrangement, cellular reciprocal interplays, and diffusion gradients [30]. Such models revealed a closer resemblance to what occurs in vivo (cancer cell phenotype, proliferation rate, and drug resistance) when compared to 2D monolayers [31] and have allowed conducting systematic investigations in a reproducible manner that cannot be achieved with conventional models [32]. However, 3D in vitro culture models still have certain limitations, mainly due to the lack of an ECM favoring physical disintegration during their manipulation. Moreover, they do not allow obtaining information regarding cell interactions with the surrounding microenvironment. As a result, 3D scaffold-based cancer models have been more recently integrated with ECM components to better mimic pathophysiological features of native tumor tissues.

Merging the aforementioned considerations, here, we firstly describe the major tumor escape mechanisms and the therapeutic approaches potentiating the antitumor immune responses. Then, we review recent studies aimed at investigating tumor and immune system interactions and testing immunotherapeutic anticancer treatments by adopting 3D biomaterial-based cancer models. In particular, we focus on polymeric matrices used in the form of hydrogels. We present a schematic overview of the most important natural and synthetic biomaterials that have been adopted in this field, highlighting both benefits and limitations, and discussing how they can be optimized to fulfill some pivotal function of the tumor ECM.

Finally, we illustrate recent emerging microfluidic technologies that couple 3D hydrogel-based models with fluidical stimuli, thus mimicking the dynamic stimuli experienced by cells in vivo and affecting their physio-pathological behavior as well as their interplay with the immune system. Overall, these models could help to identify novel mechanisms making tumors resistant to immunotherapy, and to optimize innovative and personalized immunotherapeutic approaches, accelerating the clinical translation.

2. Mechanisms Allowing Cancer Immune Evasion, a Lesson from 2D Cultures and Animal Models

Several efforts have been made in recent years to modify the human TME to overcome the limitations related to the species-specific gaps existing in animal models. However, modeling the TME, from an immune point of view, is still challenging due to the highly complex relationships between cancer cells and immune cells. A plethora of immune cells can interact with tumor cells and the TME, and, depending on the nature of these interactions, immune effectors can acquire either a tumor-suppressive or tumor-promoting function. In addition, immune cells do not act alone but interact with each other, orchestrating tumor immune responses [33,34].

A huge amount of data indicates that a functional cancer immunosurveillance process exists. However, the relationship between cancer and immune cells is a complex dynamic process involving three phases, namely, Elimination, Equilibrium, and Escape, the so-called "3E's of cancer immune-editing". The Elimination phase is characterized by the successful activation of the immune system, leading to cancer cell recognition and death. In the Equilibrium phase, cancer cells adapt to the hostile environment established by the antitumor immune cells, enabling their survival and cohabitation. In the Escape phase, cancer cells, edited by the immune system, evade its aggression through mechanisms including the expression/upregulation of membrane-bound inhibitory axes, and the production of immunosuppressive soluble molecules [35]. The major tumor escape strategies are briefly described in the following paragraphs and illustrated in Figure 1.

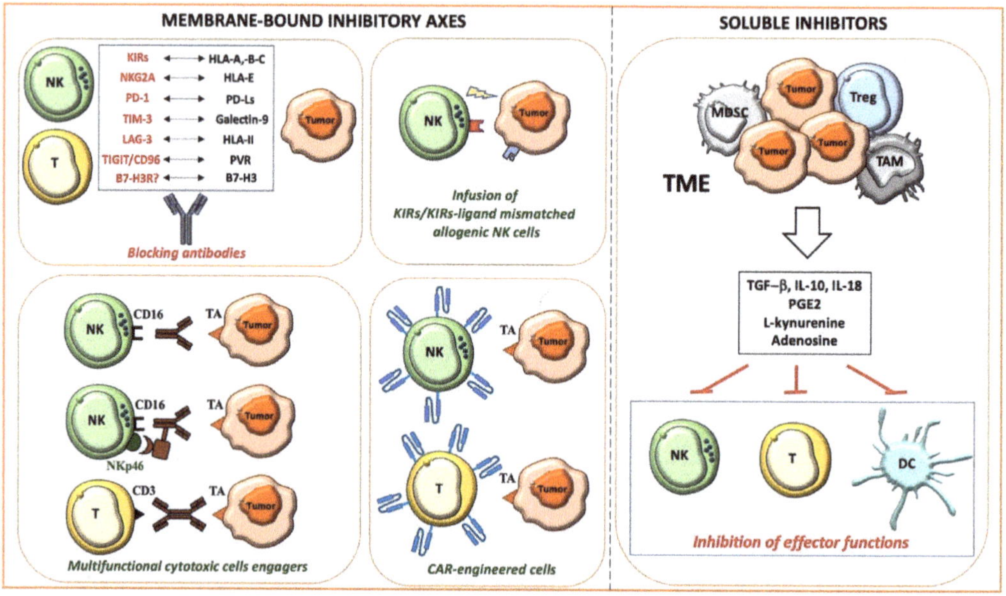

Figure 1. Main mechanisms allowing cancer immune evasion and therapeutic strategies.

2.1. Membrane-Bound Inhibitory Axes and Therapeutic Approaches

2.1.1. HLA-I-Related Axes and Inhibitory Immune Checkpoints

Relevant membrane-bound inhibitory axes, negatively impacting the antitumor activity of innate and adaptive cytotoxic cells, are those involving HLA class I (HLA-I) molecules on target cells, and specific inhibitory receptors on effector cells, such as killer Ig-like receptors (KIRs), NKG2A, and LIR-1 [36]. These interactions have physiological functions: for example, licensing NK cells to acquire a suitable cytolytic potential [37,38] or tuning the activity of triggering receptors such as NKp46, NKp30, and NKp44 (collectively termed natural cytotoxicity receptors, NCRs), mainly expressed by NK cells, NKG2D, and DNAM-1, also characterizing a significant population of T cells [36,39]. HLA-I^{high} autologous healthy cells are generally protected from the NK cell-mediated aggression since the strength of the inhibitory signals prevails over that of the activating signals; the activating signals overcome the inhibitory signals in pathological conditions including tumors, where cell transformation leads to partial or complete HLA-I expression, together with the upregulation or de novo expression of ligands for activating receptors [36,39,40]. In some instances, however, tumors can preserve high levels of protective HLA-I, as occurs in hematological malignancies [41], or upregulate HLA-I as an adaptive mechanism to the IFN-γ and TNF-α mediators released by cytotoxic cells during tumor aggression [35,42].

To potentiate the cytotoxic antitumor responses, different strategies breaking the inhibitory receptor/HLA-I axes have been planned including those blocking KIRs [43] or NKG2A [44] with specific antibodies or hemopoietic stem cell (HSC) transplant with selected allogenic donors, generally the patient's parents (haploidentical HSC, haplo-HSC) who can have NK cell populations with a KIR repertoire unable to recognize HLA-I alleles on the donors' (KIR-KIR Ligand-mismatched NK cells). A further advance in the transplant setting is represented by TCR αβ/CD19-depleted haplo-HSC transplant, where cells infused in the recipient contain CD34+ HSC cells and mature immune cells including γδ T cells and NK cells, which provide early and effective antitumor and antiviral activities acting before the immune cell reconstitution from CD34+ cells [45].

The antitumor activity of cytotoxic cells can also be negatively regulated by several non-HLA-I-specific co-inhibitory receptors such as PD-1, LAG-3, and TIM-3, expressed by T and NK cells, interacting with ligands on tumor cells [46–48]. As with HLA-I, their ligands, PD-Ls (-L1 and -L2), HLA-II, and galectin-9, can be upregulated/induced by INF-γ released during the immune responses. Interestingly, an opposite regulation by INF-γ has been observed for PVR (poliovirus receptor, CD155) [23], a ligand shared by the inhibitory checkpoints TIGIT and CD96, and the activating DNAM-1 receptor.

The PD-1/PD-Ls and CTLA/CD28 axes, whose discovery was awarded with the 2018 Nobel Prize for Medicine and Physiology, represent the prototypic immune checkpoints firstly targeted in cancer patients. In particular, the blockade of the PD-1/PD-Ls axis has revolutionized the treatment of many metastatic advanced tumors either as monotherapy or in combination with other therapeutic strategies. Several clinical trials combine PD-1/PD-Ls blockade with the infusion of antibodies specific for Tas, which unleash cytotoxicity and IFN-γ release by NK and T cells, promote phagocytosis, and complement activation. The use of antibodies as bullets reaching the right target, sparing normal cells, represents a strategy commonly used in tumors characterized by a high expression of the selected antigen that, conversely, shows a limited/low expression in normal tissues. These antibodies are often engineered to be more effective, for example, by mutating their Fc portion to reduce their binding with inhibitory or low-affinity FcγRs, or conjugating them with toxic drugs [49]. Recently, to further improve the cytotoxicity of NK cells, a strategy has been developed based on multifunctional engagers simultaneously targeting Tas and CD16 (FcγRIIIA) and NKp46 activating receptors in NK cells [50].

Additional tools to efficiently and specifically target both hematological malignancies and solid tumors are represented by T cells engineered with chimeric antigen receptors (CARs) specific for Tas, which have been optimized in recent years with the construction of more effective third-generation CARs, which also express an inducible suicide gene to induce, in case of adverse side effects, the rapid in vivo depletion of CAR-T. CARs deliver a cell activation signal that, in most instances, is strong enough to overcome the inhibitory axes. Recently, there has been an increasing interest in the generation of CAR-engineered NK cells, effectors that appear to be superior in terms of safety and that naturally express different receptors against tumor-associated molecules [51,52].

2.1.2. Novel Inhibitory Immune Checkpoints: B7-H3 and CD47

One of the most promising recently discovered tumor targets is B7-H3 (CD276) [53–55], highly expressed by several tumors and upregulated by IFN-γ [23]. Importantly, B7-H3 also shows a higher expression on the tumor-associated vasculature compared to normal vessels, whereas it is not expressed at significant levels on most normal tissues. B7-H3 represents an additional ligand of the growing list of immune checkpoint axes, physiologic mechanisms controlling the duration and the resolution of the immune responses [53,56–58]. Unfortunately, these inhibitory axes are "adopted" by tumors to escape immune surveillance.

B7-H3 acts on two sides, inhibiting the T and NK cell-mediated antitumor activity by reacting with a still unknown receptor, and favoring tumor progression by promoting migration, invasiveness, and drug resistance [52,59,60]. For these reasons, B7-H3 represents a consolidated negative prognostic marker in several adult and pediatric tumors including neuroblastoma (NB) [61]. In particular, in primary NB, high B7-H3 surface expression also correlates with poor survival in patients with localized disease, indicating that the analysis of its expression could improve patients' risk stratification [60,61].

Different therapeutic strategies targeting B7-H3 have been explored in preclinical studies with promising results [49,62–66]. Phase I clinical trials based on the infusion of humanized anti-B7-H3 monoclonal antibodies (mAbs) have been completed in adult and pediatric tumors including NB (NCT02982941), with results supporting the design of phase II and III clinical trials.

Whereas all the previously described molecules mainly impair lymphocyte-mediated immune surveillance, CD47 and its ligands, thrombospondin-1 and signal regulatory protein α (SIRPα), represent an inhibitory axis limiting phagocyte activity. Different from B7-H3, which can be considered a tumor-associated antigen, CD47 is overexpressed by many types of tumors but is also widely expressed in normal cells. The interaction of CD47 with SIRPα gives macrophages a "don't eat me" signal, inhibiting phagocytosis and allowing tumor cells to evade immune surveillance. The CD47/SIRPα axis is emerging as a key immune checkpoint in different cancers including hematological malignancies. This drives the development of immunotherapeutic strategies aimed to disrupt this brake. Importantly, however, due to the broad expression of CD47 on healthy cells, deep preclinical and clinical studies proving the safety of this therapeutic approach are required.

2.2. *Soluble Mediators and Therapeutic Approaches*

2.2.1. TGF-β and IL-10

Several soluble mediators are establishing an immunosuppressive milieu within the TME. Cytokines, growth factors, and metabolites, eventually packed into extracellular vesicles such as exosomes, play a central role in the intricate networking between cancer and immune cells, as well as between the different immune cell subsets.

Suppressive cytokines are either produced by tumor cells or immune cells having an immunosuppressive/pro-tumoral activity such as regulatory T cells (Tregs), myeloid-derived suppressor cells (MDSCs), and type 2 polarized tumor-associated neutrophils (N2, TANs) or macrophages (M2, TAMs). TAMs heavily contribute to tumor progression, exerting a suppressive and opposite role as compared to their proinflammatory M1 counterpart [67]. Among the cytokines involved in the generation of a suppressive microenvironment, TGF-β and IL-10 are known to play a central role. TGF-β has a direct pro-tumor effect on cancer cells and promotes the exhaustion of immune responses in different types of cancer [68]. In particular, TGF-β suppresses NK cells through multiple mechanisms. These include the direct inhibition of the mTOR pathway, impairing NK cell activation and function [69,70], the downregulation of the expression of different activating receptors including NKp30, NKG2D [71], DNAM-1, and CD16 [72,73], and the modulation of ligands on target cells [74,75]. TGF-β also modifies the chemokine receptor repertoire of NK cells, likely impacting their recruitment at the tumor site [76,77], and promotes the generation of NK cells with a low cytotoxic ILC1-like phenotype [78]. Interestingly, unlike other typical immunostimulatory cytokines such as IL2, IL-12, and IL-15, IL-18 potentiates rather than suppresses some of the TGF-β-mediated modulatory effects [79]. Besides the classical soluble form, recent findings show that the antitumor function of NK cells can also be suppressed via the contact with membrane-bound TGF-β expressed on metastasis-associated macrophages or Tregs [80,81].

Regarding the regulatory properties of TGF-β on T cells, the cytokine has been demonstrated to inhibit the differentiation of T cells toward the antitumor Th1 phenotype, inhibit their proliferation through IL-2 downregulation, and impair the cytotoxic effect of CD8+ T cells through the repression of granzyme B and IFN-γ. Moreover, TGF-β is involved in the upregulation of FoxP3 in CD4+ naïve T cells, inducing their differentiation toward Tregs [82–85], and, accordingly, TGF-β blockade results in Treg depletion in different cancers [86,87]. Importantly, TGF-β has also been correlated with resistance to the immune checkpoint blockade, as demonstrated by Hugo et al. through transcriptomic analysis on metastatic melanoma specimens [88]. Thus, TGF-β blockade can also be considered as a strategy to enhance the efficacy of therapies including the inhibition of immune checkpoints [89,90] and the adoptive transfer of CAR-engineered T cells [91,92]. An attractive approach is the combined targeting of immune checkpoint molecules and TGF-β within the same moiety, which has been demonstrated to be more effective in vivo than the single targeting [93]. Along this line, in a recent paper, Chen et al. engineered CAR-T cells secreting a bispecific trap protein binding PD-1 and TGF-β, demonstrating a significant improvement in effector T cell engagement, persistence, and expansion, preserving CAR-T cells from exhaustion, and leading to high antitumor efficacy and long-term remission in animal models [94].

IL-10 is a cytokine, mainly produced by Tregs, B cells, dendritic cells (DCs), and macrophages, suppressing the function of antigen-presenting cells (APCs) and CD4+ T cells [95]. The immunosuppressive role of IL-10 has been attributed to the downregulation of IFN-γ, the impairment of DC maturation, and the downregulation of HLA-I, on cancer cells, and HLA-II and costimulatory molecules (CD80 and CD86), on APCs [96–98]. Moreover, Ma et al. recently demonstrated that the over-production of IL-10 converts lymphoma-associated Th1 cells into FoxP3-negative/PD-1-overexpressing T regulatory type 1 cells, generating an immune escape signature [99]. IL-10 can also induce a pro-tumor phenotype in macrophages during the early phases of tumor formation, as demonstrated by Michielon et al. in 3D organotypic melanoma cultures [100]. Similar to TGF-β, IL-10 expression could be considered as a predictive biomarker of response to the blockade of immune checkpoints, especially when considering the IFN-γ/IL-10 ratio.

2.2.2. PGE2 and Metabolites

Specific classes of prostaglandins (PGs), molecules involved in inflammatory processes, have been associated with cancer development and progression. Namely, PGE2 contributes to the immunosuppressive tumor milieu. For example, produced by melanoma-associated fibroblasts, PGE2 negatively regulates the expression of NKp44 and NKp30 activating receptors in NK cells [101]. In breast cancer, it has been found to be associated with reduced CD80 expression on macrophages, thus hindering the antitumor immune response, and the administration of ibuprofen in vivo led to tumor shrinkage, active recruitment of T cells, and the reduction in immature monocytes [102,103]. Recently, the COX2/PGE2 pathway has been associated with M2 polarization of macrophages in hepatocellular carcinoma patients, and M2 macrophages were found to inhibit the production of IFN-γ and granzyme B from CD8+ T cells both in vitro and in vivo [104]. Therefore, the administration of PGE2-inhibiting drugs might help in the re-education of the TME. Studies performed on syngeneic mouse models revealed that the combinatory administration of a PGE2 receptor antagonist and PD1 blockade had a synergistic effect, leading to a massive reorganization of the tumor immune environment [105].

Studies in skin squamous cell carcinoma reported that PGE2 is associated with increased tumor cell migration and invasion, correlating with the staging [106]. This correlation has also been reported in gliomas, where PGE2 seems to be involved in the promotion of the tryptophan-2,3-dioxygenase pathway, known to mediate tolerogenic signaling through multiple mechanisms [107,108].

The accumulation of kynurenine due to tryptophan catabolism leads to its binding to AhR, further exerting an immunosuppressive pressure. The AhR nuclear translocation results in the upregulation of FoxP3 and IL-10 in T cells, driving the acquisition of a regulatory phenotype, reducing the immunogenic capacity of DCs [109–112], and upregulating PD-1 expression on effector T cells [113]. Importantly, the inhibitory effect of kynurenine has also been well documented in human NK cells [114]. In particular, l-kynurenine hampers the cytokine-mediated strengthening of the NK-cell-mediated killing, limiting the upregulation of NKp46 and NKG2D receptors. As a consequence, NK cells conditioned by l-kynurenine display a reduced ability to kill target cells mainly recognized via these receptors. Given all these observations, it is not surprising that different therapeutic approaches targeting the Trp-Kyn-AhR pathway are currently in preclinical development or clinical trials, in combination with standard therapies [115].

Another important metabolite exerting an immunosuppressive effect is adenosine, which can be generated within the TME due to the over-secretion by tumor cells of ATP and its catabolism by specific ectoenzymes. Intracellular ATP, produced by glycolytic or oxidative metabolism, can be released in the extracellular space through passive efflux or active secretion [116]. A "passive" release, due to a high intracellular concentration, can be associated with cytotoxicity, meaning that ATP represents a cell damage marker. The active secretion occurs through exocytosis or membrane transporters such as the ABC (ATP-binding cassette) proteins and is triggered by events such as hypoxia [117]. Importantly, hypoxia also induces the overexpression of CD39 and CD73 ectoenzymes, promoting the conversion of ATP to AMP and AMP to adenosine, respectively, as well as the downregulation of adenosine kinase, limiting the conversion of adenosine in its final metabolites and leading to the accumulation of adenosine in the extracellular space [118]. The ectonucleotidases can be expressed by tumor cells and different subsets of innate or adaptive immune cells [116,119,120]. Moreover, it has been reported that tumor cells and tumor-derived exosomes can carry CD39 and CD73 on their membranes, thus promoting ATP conversion and adenosine accumulation in the TME [120,121].

Adenosine also promotes the conversion of macrophages toward the immunosuppressive M2 phenotype, and the release of MMPs by tumor-associated neutrophils, thus favoring the invasive and metastatic process [122–125]. In Tregs, the activation of A2A receptors induces the proliferation, activation, and overexpression of the CTLA-4 and PD-1 immune checkpoints [126].

Finally, another molecule that is significantly involved in the generation of a pro-tumor setting is the vascular endothelial growth factor (VEGF), highly secreted by tumor cells and by pro-tumoral TAMs [127–130].

3. Three-Dimensional Culture Models: Moving from Spheroids to Next-Generation 3D Tools

Despite the significant information described in the previous sections and obtained by using 2D cultures and animal models, few tumor escape mechanisms have been addressed in 3D platforms, pointing out the need to move quickly towards these more reliable 3D culture systems.

These systems represent next-generation 3D tools that are slowly replacing spheroid-based strategies widely employed thus far to investigate tumor-immune system interactions in vitro and that still represent one of the gold standard 3D models to assess tumor–immune cell interactions. In particular, these tools have allowed us to gain insights into T and NK cell-related infiltration, cytotoxicity, and soluble factor release [131–138]. Moreover, spheroids have allowed deepening the understanding of the effects of the TME on macrophages' polarization and functions [139–141].

Despite these encouraging outcomes, the absence of an ECM limited the reliability of such spheroid-based systems and hampered the evaluation of the effects of chemical–physical properties of the surrounding microenvironment on cell activity [142,143]. Therefore, many researchers and material scientists are moving towards scaffold-based 3D platforms, which can also be integrated with different components of the TME. In the following sections, we will recapitulate the state of the art of 3D hydrogel-based in vitro models that have been adopted to study tumor and immune system interactions as well as novel immunotherapeutic approaches (summarized in Table 1), describing the different types of biomaterials that have been employed (schematically reported in Figure 2a).

Then, we will introduce recent emerging immune-on-chips that are paving the way for the assessment of more predictive 3D models.

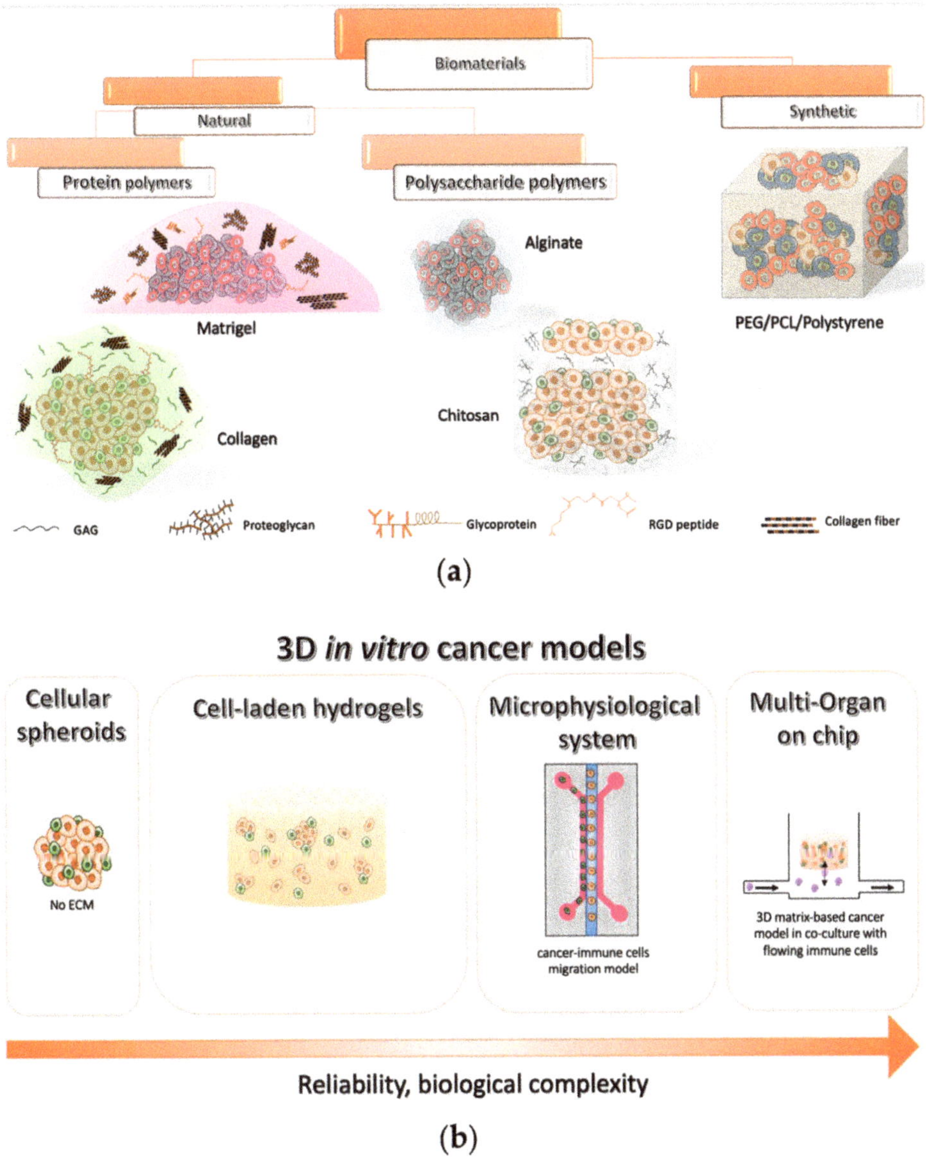

Figure 2. Overview of the current 3D in vitro cancer models (**a**) Classification of the most used polymeric biomaterials for 3D models and their schematic representation. (**b**) Evolution of 3D in vitro models for investigating cancer immunotherapies.

3.1. Natural Biomaterial Tools for 3D Tumor Modeling In Vitro

Biomaterials of natural origin are the most employed materials in several biomedical applications, due to their high biocompatibility, bioactivity, mechanical and biochemical properties similar to those of the ECM in vivo, and the presence of chemical cues promoting cell attachment and proliferation, reciprocal communication, and tumorigenesis, thus being a gold standard in cancer research [144,145]. They can be assigned to two main categories: (i) protein polymers, and (ii) polysaccharide polymers.

3.1.1. Protein-Based Polymers

Among biomaterials for 3D tumor modeling in vitro, the most adopted is Matrigel, which is an extract of the basement membrane matrix of Engelbreth Holm Swarm mouse sarcoma. This commercially available ECM, which generates a hydrogel at 24–37 °C, has a very similar content to the in vivo counterpart as it comprises various ECM macromolecules such as collagen IV, fibronectin, laminin, and proteoglycans, as well as different growth factors, chemokines, cytokines, and proteases [146,147]. Due to these constituents, Matrigel represents a biologically active platform able to promote the adhesion, migration, and differentiation of different cell types in vitro. Therefore, being easily available and versatile, and applicable with a wide variety of cellular phenotypes, it represents a standard support matrix for cell culture in several biomedical applications. In particular, it has been largely employed in 3D tumor modeling for the investigation of cancer progression, angiogenesis, metastasis, and drug efficacy [146]. Tumor cells are extremely proliferative in Matrigel-assisted cultures, differently from normal cells, showing an in vivo-like invasive profile. It has also been proved that 3D Matrigel matrices allow cells to express fundamental features related to their intrinsic malignancy [143]. For example, in the case of breast cancer, it is possible to discern poorly or highly aggressive cells when they are encapsulated within this hydrogel by examining their morphology. They usually organize into small aggregates (i.e., acini-like structures, luminal phenotype) or display an elongated shape with pronounced extensions (basal phenotype). Malignant cells are also capable of migrating through Matrigel matrices by enzymatic degradation, which is commonly studied through the Boyden chamber assay [143].

It is widely recognized that the cancer invasive profile also correlates with the ability of tumor cells to evade the immune surveillance as well as driving different types of immune cells to participate in cancer progression through cell–cell contacts and release of soluble factors [148,149]. For example, Ramirez et al. demonstrated that malignant cancer cells are capable of inducing macrophages to change the gene expression profile. Indeed, in a 3D Matrigel-based system, the interplay between the human macrophage U937 cell line and breast tumor cells caused, in U937, a significant upregulation of MMP1 and MMP9, both involved in tumor invasion via ECM degradation. Moreover, an upregulation of the inflammatory COX2 gene inducing the pro-tumoral factor PGE2 was observed. Such increments were significantly higher in the co-cultures of U937 with MDA-MB-231 cells, a highly aggressive triple-negative breast cancer cell line, than with MCF7, which has characteristics of a differentiated mammary epithelium [150]. The same group, in a later work, showed that primary breast cancer cells constitutively secrete high levels of CCL5, CCL2, and G-CSF, specifically involved in the attraction of circulating immune cells at the tumor site, while a remarkable increase in IL-1β, IL-8, MMP-1, MMP-2, and MMP-10 production was revealed when cancer cells were co-cultured with monocytes [151].

Taken together, these data support the idea that tumor aggressiveness is related to its capability to shape the inflammatory microenvironment by recruiting immune cell populations at the tumor site and instructing them to fulfill pro-tumoral functions. Therefore, it is evident that the interactions occurring within the TME between different cell types, including stromal cells, play a fundamental role in promoting disease progression [152,153].

Hence, tissues explanted during surgical resections or biopsies have been embedded in Matrigel to investigate immune cell populations infiltrating the tumors [154,155]. For instance, in slices derived from tissues of patients with colorectal and lung cancer, a great presence of myeloid-derived suppressor cells (MDSCs) (CD206+/CD33+/HLA-DR—) and CD4-/CD8-T cells, as well as a reduced number of NK cells and monocytes, has been observed [156]. Moreover, innovative organotypic cultures have been adopted by co-culturing organoids established from patient-derived cancer cells (due to their capability of retaining key pathophysiological and structural features of the original tumor in vitro [157]), with patient-matched stromal (e.g., CAFs) and immune components (e.g., T cells) [158]. These systems represent a valuable tool for studying the complex tumor–stroma–immune system communications in a highly reliable context, paving the way for the assessment of novel personalized immunotherapeutic strategies. To this end, more recently, Dijkstra et al. co-cultured autologous colorectal or non-small lung cancer tumor organoids with peripheral blood lymphocytes, with the intention of increasing the number of tumor-specific CD8+ T cells to be infused in patients [154]. Furthermore, other groups focused on testing novel engineered immune cell-mediated strategies. Among them, $\alpha\beta$T cells modified to express a tumor-specific $\gamma\delta$ TCR (TEGs) were used in primary myeloma cells grown within a 3D BM niche model [159]. Moreover, the CAR-NK-92 cell line was proposed as an effector against patient-derived colorectal cancer organoids by targeting the epidermal growth factor receptor variant III (EGFRvIII) [160], overexpressed in a wide variety of epithelial tumors [161]. Moreover, researchers are adopting such patient-derived preclinical platforms to evaluate different strategies targeting immune checkpoint axes, alone or in combination. In this latter context, an association of an anti-PD-L1 mAb (atezolizumab) with MEK inhibitors (selumetinib) led to a higher MHC-I expression on non-small lung cancer organoids, together with increased secretion of IFN-γ, IL-6, IL-1β, and TNF-α by immune cells [162].

However, despite all the encouraging results derived from in vitro and in vivo preclinical models, many patients do not respond to some promising therapies, even due to the great variety of mechanisms involved in cancer immune evasion that are still not completely understood. Furthermore, although Matrigel establishes a favorable TME [163], it is affected by several drawbacks that considerably limit its use. Firstly, because of its structural weakness, it is mainly adopted as a monolayer or a thin gel conformation, principally for short-term invasion assays [143]. Then, the applicability of Matrigel is severely hampered because of its variability in composition and structure, due to its natural origin (e.g., tumor sizes from which is extracted, prepared, etc.) [147]. Differences in mechanical and biochemical properties between the various batches and within a single batch negatively impact the experimental reproducibility [147,164]. These constraints, along with the fact that Matrigel is difficult to manipulate physically and biochemically, make comparisons between and within laboratories remarkably challenging [164,165]. Moreover, being an animal-derived ECM, the presence of xenogenic contaminants may hamper the use of Matrigel-based cell cultures as in vitro preclinical tool for screening effective immunotherapies. For instance, lactate dehydrogenase elevating virus (LDHV), a mouse virus capable of infecting macrophage cells, possibly influencing both the immune system and tumor behavior, was detected in multiple batches of Matrigel [166].

All these considerations should be kept in mind when interpreting results based on Matrigel-assisted cell cultures, to distinguish biological effects caused by controlled experimental conditions or variables from the hydrogel itself [164].

Collagen is another biomaterial belonging to this category that is largely employed as an ECM-supporting matrix for 3D models, as it contains fundamental cellular adhesion domains (i.e., arginine-glycine-aspartate (RGD) peptide) that favor cell growth in vitro. It is commonly deposited by different cancer types during malignant progression, thus being an important component of the TME. In particular, matrices made of collagen type I promote, in vitro, uncontrolled cancer cell growth, the establishment of hypoxic regions, and angiogenesis, thus being particularly suitable to resemble key environmental properties of tumors [167,168]. Considering this, several studies have been conducted to reproduce the complexity of the TME by including, in 3D collagen constructs, cancer cells with components of the tumor stroma as well as immune cells in close contact with each other. Cell-to-cell contact is notably critical when evaluating the anticancer activity of cytotoxic lymphocytes, which requires direct interactions with tumors to efficiently kill malignant cells [169]. Moreover, as discussed before, an immune-mediated pro-tumoral action is frequently observed within the TME, particularly due to the presence of TAMs supporting cancer progression and resistance to chemotherapies. For example, macrophages co-cultured with breast cancer cells in a more in vivo-like environment led to a significant increase in oxygen consumption as well as in the secretion of epidermal growth factor (EGF) and IL-10, suggesting a synergistic crosstalk between different types of cells and indicating a tumor-promoting activity of immune cells colonizing tumors such as M2-polarized macrophages [170]. It was demonstrated that macrophages' polarization towards an M2 phenotype is reached spontaneously in organotypic co-cultures including cancer cells and fibroblasts after three weeks, with a consequent reinforced proteolytic activity of the tumor cells through the increase in MMP2 and MMP9 production. Moreover, the same authors showed that organotypic co-cultures allow handling either M1 or M2 polarization via stimulation with IFN-γ and LPS or IL-4, respectively [171]. This can help to deeply elucidate the role of macrophages in the TME, where they can contemporarily show a tumor-promoting effect or exert an antitumor activity by attacking and eliminating cancer cells, depending on their polarized status [172]. Therefore, the importance of developing more reliable in vitro systems taking into account the complex reciprocal interactions occurring in vivo between malignant and non-malignant cells is evident. Recently, some platforms prepared the groundwork for the investigation of novel agents (e.g., immunotherapeutic antibodies) aimed at targeting the key cellular components of the TME (e.g., CAFs or TAMs) in a clinically relevant context [173].

Overall, collagen has been widely employed as an EMC-mimicking matrix in the field of cancer research, also due to its easy manipulation and low costs, making this biopolymer easily accessible to the scientific community [174]. Despite its intrinsic poor mechanical properties, it can be easily tuned by changing the concentration or adding synthetic crosslinking agents to finely tune its structure and stiffness based on the specific application [167,168,175]. However, because of its animal origin, as with Matrigel, it is affected by risks associated with biological materials, such as the batch-to-batch variability, that limit the reproducibility of the results [174].

3.1.2. Polysaccharide-Based Polymers

Polysaccharide-based biopolymers have been largely adopted as ECM-supporting matrices for in vitro cell culture since they are characterized by low immunogenicity as well as elevated biocompatibility [165]. Several biomaterials belonging to this group have been used to support cancer cells' interactions with the immune system, especially focusing on those mechanisms occurring within the TME that promote tumor growth and metastasis [176]. Among these polymers, alginate is one of the most employed. Alginate, derived from brown seaweeds, presents a molecular structure comparable to that of polysaccharides found in vivo [177]. It is particularly suitable for the formation of cell-laden microspheres, allowing for obtaining a high number of replicates due to its easy manipulation, fast gelation, thermal stability, and low cost [165,167,177,178].

In our recent publication, we selected alginate for developing a 3D model of NB. Both MYCN- and non-MYCN-amplified cell lines showed cellular proliferation, drug sensitivity, and immunophenotypic plasticity closer to those occurring in vivo, with respect to 2D models. Intriguingly, we observed molecular characteristics more similar to immunophenotypic variances occurring in vivo and not fully comprehended in traditional 2D culture conditions, such as the IFN-γ-induced negative regulation of PVR (CD155) expression on tumor cells after 7 days of 3D culture [23]. Moreover, it was possible to appreciate an IFN-γ-dependent upregulation of the immune checkpoint ligand B7-H3, a molecule deeply discussed above [179].

Alginate microencapsulation has also been used to explore the onset of either a proinflammatory or an immunosuppressive TME, especially focusing on the dynamic interactions occurring between the main cellular components that support the tumor malignant behavior [180]. In a 3D co-culture of non-small cell lung carcinoma cells with CAFs and monocytes, an accumulation of soluble factors (IL-4, IL-10, IL-13, CXCL1) was observed, promoting immune cell infiltration of the tumor and M2-like macrophage polarization.

This polarization was characterized by the expression of the CD68, CD163, and CD206 markers and the production of the CCL22 and CCL24 chemokines [181].

Chitosan is a linear polysaccharide derived from the partial deacetylation of chitin, which is abundantly available from different biological sources, being, for example, the main structural polymer of crustacean exoskeletons [177]. Due to its poor solubility in common solvents, the process of extraction of chitin is quite laborious, thus limiting its utilization. In general, chitosan offers a higher mechanical strength, and the possibility to be easily chemically modified, and to interact with other biomolecules due to the presence of reactive functional groups. Furthermore, it simply forms soft gels and crosslinks with other polymers [182,183]. Besides these characteristics, chitosan represents an effective alternative candidate for 3D cultures of cancer cells due to a structure similar to that of GAGs, one of the main constituents of the tumor ECM [177,184]. The chitosan and alginate (CA) combination has also been largely adopted to realize porous scaffolds that exhibit better mechanical strength and shape maintenance when compared to chitosan alone, because of the electrostatic contact between chitosan's amine groups and alginate's carboxyl groups [177,184]. Three-dimensional CA scaffolds provide a cost-effective feasible model to evaluate in vitro the interplays between tumors and the immune system in a clinically relevant context [185]. For example, these platforms can mimic the breast cancer TME. In this context, the inactivation of CAFs, which have been demonstrated to induce T cell suppression in breast tumor stroma [186], or combined gene therapies aimed at enhancing T cell infiltration and activation in the tumor milieu [187] may represent novel strategies for improving the efficacy of the current adoptive T cell therapies against breast cancer.

However, there are also different drawbacks associated with these types of biomaterials. For example, chitosan is characterized by poor mechanical properties [178], and alginate by a variable degradation rate. Moreover, the latter does not possess integrin-binding sites, thus often requiring chemical modification or conjugation with other bioactive polymers [167]. Indeed, extensive literature has been reported on the covalent functionalization of alginate with the RGDpeptide to favor cellular adhesion, proliferation, and migration [167,168,188,189].

In conclusion, natural polymers are highly suitable to recapitulate in vitro the main features of the native ECM. Nevertheless, they suffer from important limitations. Besides the aforementioned significant batch-to-batch variability (e.g., various mechanical and biochemical features, peptide or protein concentrations) and xenogeneic contaminations associated with polymers derived from an animal source, it is generally difficult to control scaffold degradation rates, possibly influencing cellular activity in unknown ways [144]. Moreover, natural polymers can be realized in a limited range of mechanical stiffness, porosity, or biochemical cues [145].

Therefore, the focus is shifting toward synthetic polymers that may mimic the biomimetic qualities of natural ones while providing more repeatability and control over the materials' physical and chemical properties.

3.2. Synthetic Biomaterial-Based 3D Tools

Synthetic polymer-based scaffolds represent a valid alternative to naturally derived ones. First, being free of xenogeneic and possible contaminants, they enable high reproducibility by reducing inter-batch variations, thus resulting in a greater consistency of the results. Moreover, they can be more easily manipulated to finely tune mechanical and chemical properties as well as degradation rates for specific cell culture applications [144,164,178,190]. Indeed, even though synthetic biomaterials are biologically inert, allowing, but not promoting, cellular activity, cell adhesion ligands and other bioactive molecules can be precisely introduced via covalent attachment, adsorption, or electrostatic interactions depending on the desirable environmental cues that need to be investigated [145,178,190,191]. Therefore, emerging studies have demonstrated the possibility to adopt these polymers for investigating the interactions between cancer cells and the immune system within the TME in vitro. For instance, 3D polystyrene-based scaffolds have been used to mimic T cell infiltration in non-small lung cancer and to explore the subset of inflammation proteins related to the co-cultivation of tumor cells with lymphocytes [192], while polycaprolactone (PCL) has been exploited for examining the capability of DCs to engulf dying colon cancer cells through the same mechanisms observed in the human body [193].

One of the most common synthetic polymers is polyethylene glycol (PEG), which has been largely used in the tissue engineering field both in vitro and in vivo, showing to be highly suitable as a model for ECM–cancer interaction studies [145]. It is biocompatible and fully hydrated, thus closely reproducing the soft tissues' characteristics, and particularly suitable for cell encapsulation due to the liquid-to-solid transition to form hydrogels encapsulating cells [184]. Even though it is biologically inert, it can be easily functionalized with protease-sensitive peptides to render the surrounding ECM enzymatically degradable by cells. Numerous studies have explored the inclusion, via crosslinking reactions, of different peptides sensitive to MMP-mediated cleavage, in order to evaluate cell migration and invasion [194,195]. The presence of MMP-cleavable sites in PEG hydrogels has also been shown to promote cell proliferation and differentiation [196–199]. Moreover, cellular adhesion and/or other molecules of interest (VEGF, TGF-β1, etc.) can been integrated through various non-toxic polymerization techniques [164,184,200]. Interestingly, in a recent study, the migration and function of NK-92 cells within a 3D RGD-functionalized PEG hydrogel containing either non-small lung cancer metastatic (H1299) or non-metastatic (A549) cell lines were investigated. The metastatic tumor model displayed a greater loss of stress ligands (ULBP1, MICA), downregulation of chemokine expression (MCP-1), and higher production of inhibitory soluble molecules (i.e., TGF-β, IL-6), as compared with a non-metastatic tumor model, more resembling the in vivo scenario. The NK cell migration toward cancer cells and their co-localization depended on the immunomodulatory profile of tumors, and NK-92 cells decreased the production of RANTES and MIP-1 α/β when incubated with H1299 cells. The study highlights the benefits of 3D cancer models that allow us to examine the effects of signals on NK cell migration. In addition to the release of soluble substances, immune cell infiltration might be influenced by the physical features of tumors. Nevertheless, the impact of matrix stiffness on NK cell migration is unknown, and more research is needed to fully understand the NK cell mechanotransduction pathways [201].

Despite recent promising outcomes, PEG and other synthetic biomaterials are still poorly adopted in cancer research. Despite the fact the raw materials for making PEG hydrogels are about half the price of Matrigel, the necessity for one or more synthetic peptides to provide the essential biochemical cues to drive cellular behavior can be prohibitively expensive for large-scale manufacturing. Furthermore, extensive adjustments to obtain the desired combination of physical and biochemical properties driving cellular behavior can be time-consuming, costly, and challenging [164], whereas degradation products are often non-biocompatible [202]. Finally, when compared to in vivo tumors, cells cultivated in completely synthetic platforms can proliferate without some tumor-like gene expression patterns, revealing inconsistent tumorigenicity and metastatic potential, or resistance to anticancer treatments. As one might expect, such difficulties have an impact on the creation of reliable tumor-mimicking 3D in vitro models [167].

To overcome these disadvantages and achieve more in vivo-like conditions, synthetic materials can be properly mixed with naturally generated biopolymers [202], in order to better address the physiological crosstalk between immune and cancer cells.

Table 1. Summary of 3D in vitro models based on different types of polymeric matrices to study cancer–immune interactions and immunotherapies.

3D Biomaterial	Cell Types	Main Objectives	Ref
Matrigel	Breast cancer cells, NK and Treg cells	To compare tumor biomarkers' expression and immune infiltration between luminal and basal tumor phenotypes	[149]
	Breast cancer cells, promonocytic cells/monocytes	To study tumor/immune cells' crosstalk	[150,151]
	Colorectal and non-small lung cancer spheroids, peripheral blood lymphocytes (PBLs)	To obtain patient-specific tumor-reactive T cells	[154]
	Colorectal or lung cancer tissues	To maintain primary cells in culture and study tumor-infiltrating immune cell populations	[156]
	Pancreatic cancer organoids, CAFs, PBLs	To analyze multiple cells' crosstalk	[158]
	Endothelial progenitor cells, multipotent mesenchymal stromal cells, CD138+ myeloma cells	To study engineered (to express a defined γδTCR) T cells' activity	[159]
	Colorectal cancer organoids	To study CAR-NK cells' activity	[160]
	Non-small lung cancer cells, peripheral blood mononuclear cells (PBMCs)	To establish an effective combined therapy based on MEK inhibitors and anti-PD-L1	[162]
Collagen	Pancreatic tumor spheroids, T cells	To monitor cancer invasive behavior and T cell cytotoxicity	[169]
	Breast cancer spheroids, macrophages	To investigate macrophages' polarization, localization, and function in the tumor mass	[170]
	Squamous carcinoma cells, fibroblasts, macrophages	TME-mediated regulation of macrophage polarization, both spontaneous and induced by exogenous factors	[171]
	Lung adenocarcinoma cells, fibroblasts, macrophages	To analyze multiple cells' crosstalk	[172]
	B lymphoma cells, fibroblasts, macrophages	To reproduce the lymphoma microenvironment to test therapeutic Abs	[173]
Agarose	Hepatocellular carcinoma cells, M2 macrophages	To investigate the impact of macrophages on cancer progression	[176]
Alginate	MYCN- and non-MYCN-amplified NB cells	To analyze tumor immunophenotype related to NK cell receptors	[23]
	Breast cancer cells, fibroblasts, and macrophages	To analyze multiple cells' crosstalk	[180,181]
	Non-small cell lung carcinoma cells, CAFs and monocytes	To analyze multiple cells' crosstalk	[180,181]

Table 1. *Cont.*

3D Biomaterial	Cell Types	Main Objectives	Ref
Alginate/ Chitosan	Prostate cancer cells, PBLs	To study tumor/immune cells' crosstalk	[185]
	Mammary carcinoma cells, CAFs, T cells	To evaluate the impact of CAFs on T cell function	[186]
	Mammary carcinoma cells, T cells	To explore how tumor CCL21 and IFN-γ expression affects T cell recruitment and activation	[187]
Polystyrene	Non-small lung cancer cells, T cells	To study tumor/immune cells' crosstalk	[192]
PCL	Colon cancer cells, DCs	To study tumor/immune cells' crosstalk	[193]
PEG	Non-small lung cancer cells, NK-92	To study NK cells' infiltration and function	[201]
PEG/Chitosan	Mammary carcinoma cells, T cells	To study the influence of TME on drug efficacy and immune resistance	[202]

4. Micro-Physiological Systems for Investigating Immune Cell–Tumor Dynamic Interactions

Cancer immunotherapy has shown many signs of progress in recent years, thanks to novel strategies aimed at enhancing the efficacy of immunomodulating agents and their patient-specific approaches. Furthermore, alternative in vitro platforms are in continuous development for overcoming some limitations of traditional preclinical models. As we discussed above, 3D tissue models allow us to better investigate tumor pathways in a tissue-like architecture than in simplistic cell culture monolayers, by introducing physiological barriers that mediate immune system–cancer cell crosstalk. However, the three-dimensionality itself cannot properly reproduce the complexity of the human tissues and organs, where dynamic stimuli (e.g., blood and lymphatic flow mechanical forces) shape the immune cell infiltration and their dynamic interactions with cancer cells. In this context, recent advancements in the realization of immunological tissues-on-chips have increased the relevance of in vitro models, leading to better knowledge of immune cell recruitment, selection, invasion, and activation within the tumor milieu [34,203].

Microfluidics is a rapidly growing technology using narrow channels, ranging in height/width from tens to hundreds of micrometers, to study cell migration by handling small fluid volumes, thus reducing the amounts of reagents and biological materials. Overall, microfluidic assays outperform other in vitro models in terms of physiological relevance because they allow local monitoring of the cellular, physical, and biochemical cues, making them a good compromise between in vivo and other types of in vitro systems [204,205]. They may also permit real-time imaging with fine control on the interplays between different cellular populations growing within the interconnected channels. This enables the investigation of spatiotemporal dynamics similar to those found in the TME, as well as the effects of environmental characteristics on cell behavior such as changes in pH and oxygen levels (i.e., acidification and tumor hypoxia) and cytokine/chemokine gradients [206]. Several studies focused on cancer and immune cell interactions as well as on cell-mediated immunotherapies used microfluidic devices coupled with 3D hydrogel-based models. The most recent and relevant publications are reported and summarized in Table 2.

Table 2. Recent immune-on-chips for studying immunotherapeutic strategies against 3D cancer models.

	Microfluidic Device				
Key Immune Cell/Checkpoint Axis	**3D Biomaterial**	**Cell Types**	**Method**	**Main Outcomes**	**Ref**
T cells	Collagen	Human hepatocellular cell line (HepG2); TCR-T cells	Tumor aggregates in a central gel region with tumor-specific T cell receptors added in the adjacent channels	Chemotactic migration, effective cytotoxicity	[207,208]
	Gelatin Methacrylate	Human epithelial ovarian cancer cell line (SKOV3); CAR-T cells	Cancer cells in a central gel-filled region flanked with two channels where CAR-T cells reside	Enhanced cancer cell killing within a hypoxic TME	[209]
NK cells	Collagen	Breast cancer cell line (MCF7); NK-92 cell line; endothelial cells (HUVECs)	3D matrix containing cancer spheroids and NK-92 cells, provided with two lateral endothelialized channels	Chemotactic migration and penetration; cytotoxicity	[210]
		Cervical cancer cell line (Hela cells); NK-92 cell line	Injection molded plastic array culture (CACI-IMPACT) patterning gel regions including cancer cells; NK cells deposited over hydrogel structures	3D ECM remarkably reduces NK cell migration	[211]
Monocytes/ Macrophages	Collagen	Human hepatocellular cell line (HepG2); monocytes; TCR-T cells	Immune and cancer cells embedded in the central part of the microfluidic device, while T cells added in the channel	Immunosuppressive potential of monocytes via PDL/PDL-1 signaling	[212]
		Lung adenocarcinoma cell line (A549); TAMs; HUVECs	Tumor aggregates and TAMs included in 3D hydrogel, in co-culture with an endothelial monolayer in an adjacent channel	Tumor cell migration, epithelial–mesenchymal transition	[213]
		Mouse glioblastoma cell lines (GL261 and CT2A); macrophages; HUVECs	Hydrogel cancer and immune cells crossed by two inner vessels lined with HUVECs	M2-like macrophage polarization, angiogenesis promoted via TGF-β1 and IL-10	[214]
		Mouse macrophage cell line (Raw 164.7); human metastatic breast cancer cell line (MDA-MB-231); prostate cancer cell line (PC3); melanoma cell line (MDA-MB-435S); monocytes	Immune and cancer cells co-embedded in a central gel region	Macrophages promote cancer cell migration by upregulating MMP expression of tumor and secreting TNF-α and TGF-β1	[215]
		Human metastatic breast cancer cell line (MDA-MB-231); monocytes	Cancer and immune cells embedded in 3D hydrogel within an endothelial channel	Tumor cell extravasation promoted via monocyte-derived MMP9	[216]
		Human metastatic breast cancer cell line (MDA-MB-231); monocytic cell line (U937); TAMs	Two separated adjacent hydrogel channels containing cancer cells and monocytes or TAMs	Monocyte conversion to TAMs, promoted cancer cell invasion	[217]
		Pancreatic ductal adenocarcinoma cell line (CRL-1469); macrophages	Cancer and immune cells cultured in separated gel channels	Macrophage migration	[218]

Table 2. *Cont.*

Microfluidic Device					
Key Immune Cell/Checkpoint Axis	**3D Biomaterial**	**Cell Types**	**Method**	**Main Outcomes**	**Ref**
	Collagen	Ovarian cancer cell line (OVCAR-3); neutrophils	Ovarian tumor spheroids embedded within hydrogel matrix with microfluidic channels carrying immune cells	Neutrophil extravasation, tumor cell migration	[219]
Neutrophils	Fibrin	Melanoma cancer cell line (A375-MA2), neutrophils; HUVECs	Cancer, immune, and endothelial cells co-embedded in a central hydrogel compartment	Increased tumor cell extravasation in an IL-8-dependent manner	[220]
DCs	Collagen	Colorectal cancer cell line (SW620); DCs	Cancer and immune cells cultured in a 3D chamber connected through microchannels to the immune compartments containing IFN-α-conditioned DCs	Crosstalk between dendritic and cancer cells	[221]
PD1/PDL-1	Collagen	Murine- and patient-derived melanoma cells; tumor-infiltrating lymphocytes	Organotypic tumor spheroids containing autologous immune cells embedded in a hydrogel	Effective response to PD-1 blockade treatment	[222]
ADCC	Collagen	Breast cancer cell line (BT474); CAFs; PBMCs; HUVECs	Central endothelial channel with two adjacent gel compartments including cancer cells, CAFs, and PBMCs	Trastuzumab antibody targeting the HER2 receptor promotes long cancer–immune interactions	[223]

Despite the benefits discussed above, some technical and biological limitations still affect these platforms. First, most of them are fabricated with polydimethylsiloxane (PDMS) that may cause toxicity due to the progressive release of non-crosslinked oligomers, and the retention of small hydrophobic molecules through adsorption, making some biochemical analyses difficult [204,206,208,212,213,215,220]. Furthermore, although the recovery of cells from the separate compartments of microfluidic devices is feasible, the low number of recovered cells makes it difficult to perform functional and phenotypic analysis [204]. For this reason, most data are derived from imaging analysis. Furthermore, in the case of immune cells loaded in microfluidic tools, monitoring the functional and phenotypic changes occurring over time without interrupting and altering the microfluidic flux is not feasible.

Moreover, the clinically relevant size of the tissue models hosted and cultured in such microfluidic devices still represents a challenge. Extended research in properly resembling the ECM complexity within these devices for the establishment of a tumor niche and the co-culture of different immune cell types (MDSCs, Th1 cells, Bregs, eosinophils, etc.) should be carried out [34,204].

Interestingly, at this time, the majority of these microfluidic devices are adopted for co-culturing tumor–immune cells in different compartments, physically separated by a polymeric gel, for evaluating the real-time cell migration under virtual static conditions, similar to the Boyden chamber-based assays [204,206,208,212,213,215,220]. However, these conditions do not properly recapitulate the circulatory flow in the human body, limiting the experience of blood flow-associated forces (e.g., shear stress) that influence the survival, escape, and activation of immune cells during their journey in the vascular network [142,224]. Along this line, the authors of this review are currently involved in the study of the NK-mediated cytotoxicity against NB cell-laden hydrogels [23] within the novel fluidic device MIVO® (Multi in Vitro Organ System), which is schematically represented in Figure 2b, capable of both mimicking the blood flow circulation in a highly reliable context [225–227], and culturing clinically relevant sized cancer tissues, as previously reported in recent works [226,227].

5. Conclusions and Future Directions

The recent assessment of innovative immunotherapies against cancer led to the urgent need for an increasing number of predictive preclinical models capable of reproducing the key features of the TME and properly testing the efficacy of immunomodulating agents. Up to now, the gold standard still relies on in vitro cell culture monolayers and animal testing, which, over the years, enabled us to take important steps forwards to the knowledge of tumor immunology. However, it is increasingly evident that novel platforms should be realized for overcoming the limitations that affect the traditional settings, which unfortunately often result in ruinous discrepancies between the benefits observed in preclinical and clinical studies.

In the present review, we have provided an overview of 3D in vitro models based on different types of biomaterials more commonly used to study tumor–immune cell interactions, and standard and innovative immunotherapeutic strategies. As discussed, the composition of the TME can play a pivotal role in malignant progression due to the evasion of the immune surveillance. Depending on their own characteristics, which may be more or less suitable for some specific applications, both natural and synthetic polymeric matrices are currently adopted to reproduce the key features of the tumor-surrounding ECM.

In general, typical ECM organic components, such as GAGs, proteoglycans, and glycoproteins interact with cancer, stromal, and immune cells and highly impact the hydration and stiffness of the ECM. Thus, to provide cells with such biological signals, bioactive polymers such as Matrigel or chitosan can be selected, while inert polymers such as alginate or PEG need to be functionalized to become biologically active. Other ECM elements, such fibrillary components, greatly impact tumor progression, also representing physical barriers both for tumor and immune cells. Thus, a polymer enriched in fiber elements may represent a good tool to investigate cancer cell invasiveness and metastasis as well as the mechanisms regulating immune cell recruitment in the tumor mass. It should be taken into consideration that the ECM can be shaped by the action of radio- and chemotherapy. In this context, in a recent study focused on lung cancer, chemotherapy-induced remodeling of the ECM occurred with the pivotal intervention of host T cells which, stimulated by paclitaxel, increased their release of lysyl oxidase, favoring the formation of tumor metastases [228].

It is important to highlight that the ECM composition and stiffness greatly vary among different types of tumors. This should be considered when setting an appropriate 3D model. Indeed, some solid tumors are characterized by a more rigid ECM, thus requiring hydrogels with high stiffness. In this case, an optimal choice may be represented by synthetic polymers that allow finely tuning their composition to reach a proper stiffness. Conversely, other materials such as collagen or chitosan form hydrogels characterized by poor mechanical properties. On the other hand, other malignancies arising, for example, in soft tissues may be modeled with low-stiffness and highly hydrated polymers such as alginate. Therefore, the choice of the most appropriate biomaterial-based model strictly depends on the type of tumor to be addressed and on the experimental aims.

Despite encouraging results, we are still far from the possibility to explore cancer immunotherapeutic approaches in a system that closely resembles the complexity and dynamic of the human body. Microfluidic tools represent a challenging approach in this sense, by introducing dynamical cues and the possibility to culture cells in a highly controllable environment that may help to elucidate the mechanisms exploited by immune cells to infiltrate tumors.

Some of the mentioned issues such as the experimental limits of microfluidic devices can be overcome by designing new fluidical platforms that can be adapted to the standard laboratories' tools commonly used for in vitro cell culture. Along this line, recent emerging technologies have inserted, in the fluidic devices, commercially available trans-wells that can be easily removed from the dynamic circuit and analyzed with conventional techniques. Other strategies take advantages of integrated biosensors able to measure some metabolic parameters and soluble factors, over time, without the interruption of the dynamic culture [229,230].

Nevertheless, such platforms may also remain too simplistic, due to the lack of a realistic vasculature where immune cells can flow and transmigrate through the endothelial wall, reaching the tumor. Moreover, it is important to highlight that most of the studies have based their works on the use of long-term cultured cell lines, known to be remarkably different from primary tumors or immune cells. Future implementations should focus on the use of patient-derived tumors and immune cells with stromal components, eventually in multi-organ on-chip platforms, where the contribution of different organs to immunotherapies can be considered (Figure 2b).

The new in vitro 3D preclinical approaches may lead to personalized medicine predicting response to immunotherapies with greater patient benefits.

Author Contributions: Writing—original draft preparation, C.V., M.M., A.D.; writing—review and editing, S.S., C.B., R.C. All authors have read and agreed to the published version of the manuscript.

Funding: This research was funded by the Italian Ministry of Health, "Ricerca Corrente 2021" and "5 per mille" (project 5M-2018-23680422), to C.B.

Conflicts of Interest: M.M. and S.S. are currently employees of React4Life Srl.

References

1. Sharma, P.; Siddiqui, B.A.; Anandhan, S.; Yadav, S.S.; Subudhi, S.K.; Gao, J.; Goswami, S.; Allison, J.P. The next decade of immune checkpoint therapy. *Cancer Discov.* **2021**, *11*, 838–857. [CrossRef] [PubMed]
2. Fritz, J.M.; Lenardo, M.J. Development of immune checkpoint therapy for cancer. *J. Exp. Med.* **2019**, *216*, 1244–1254. [CrossRef] [PubMed]
3. Wolchok, J. Putting the immunologic brakes on cancer. *Cell* **2018**, *175*, 1452–1454. [CrossRef] [PubMed]
4. Demaria, O.; Cornen, S.; Daëron, M.; Morel, Y.; Medzhitov, R.; Vivier, E. Harnessing innate immunity in cancer therapy. *Nature* **2019**, *574*, 45–56. [CrossRef] [PubMed]
5. Boucherit, N.; Gorvel, L.; Olive, D. 3D tumor models and their use for the testing of immunotherapies. *Front. Immunol.* **2020**, *11*, 3220. [CrossRef] [PubMed]
6. Cavo, M.; Fato, M.; Peñuela, L.; Beltrame, F.; Raiteri, R.; Scaglione, S. Microenvironment complexity and matrix stiffness regulate breast cancer cell activity in a 3D in vitro model. *Sci. Rep.* **2016**, *6*, 35367. [CrossRef]
7. Huang, G.; Wang, L.; Wang, S.; Han, Y.; Wu, J.; Zhang, Q.; Xu, F.; Lu, T.J. Engineering three-dimensional cell mechanical microenvironment with hydrogels. *Biofabrication* **2012**, *4*, 42001. [CrossRef]
8. Lu, P.; Weaver, V.M.; Werb, Z. The extracellular matrix: A dynamic niche in cancer progression. *J. Cell Biol.* **2012**, *196*, 395–406. [CrossRef]
9. Ulrich, T.A.; de Juan Pardo, E.M.; Kumar, S. The mechanical rigidity of the extracellular matrix regulates the structure, motility, and proliferation of glioma cells. *Cancer Res.* **2009**, *69*, 4167–4174. [CrossRef]
10. Suresh, S. Biomechanics and biophysics of cancer cells. *Acta Mater.* **2007**, *55*, 3989–4014. [CrossRef]
11. Hassn Mesrati, M.; Syafruddin, S.E.; Mohtar, M.A.; Syahir, A. CD44: A Multifunctional Mediator of Cancer Progression. *Biomolecules* **2021**, *11*, 1850. [CrossRef] [PubMed]
12. Annes, J.P.; Munger, J.S.; Rifkin, D.B. Making sense of latent TGFβ activation. *J. Cell Sci.* **2003**, *116*, 217–224. [CrossRef] [PubMed]
13. Cukkemane, A.; Seifert, R.; Kaupp, U.B. Cooperative and uncooperative cyclic-nucleotide-gated ion channels. *Trends Biochem. Sci.* **2011**, *36*, 55–64. [CrossRef] [PubMed]
14. Carrega, P.; Morandi, B.; Costa, R.; Frumento, G.; Forte, G.; Altavilla, G.; Ratto, G.B.; Mingari, M.C.; Moretta, L.; Ferlazzo, G. Natural killer cells infiltrating human nonsmall-cell lung cancer are enriched in $CD56^{bright}CD16^{-}$ cells and display an impaired capability to kill tumor cells. *Cancer* **2008**, *112*, 863–875. [CrossRef]
15. Delahaye, N.F.; Rusakiewicz, S.; Martins, I.; Ménard, C.; Roux, S.; Lyonnet, L.; Paul, P.; Sarabi, M.; Chaput, N.; Semeraro, M. Alternatively spliced NKp30 isoforms affect the prognosis of gastrointestinal stromal tumors. *Nat. Med.* **2011**, *17*, 700–707. [CrossRef]
16. Caruana, I.; Savoldo, B.; Hoyos, V.; Weber, G.; Liu, H.; Kim, E.S.; Ittmann, M.M.; Marchetti, D.; Dotti, G. Heparanase promotes tumor infiltration and antitumor activity of CAR-redirected T lymphocytes. *Nat. Med.* **2015**, *21*, 524–529. [CrossRef]
17. Leitinger, B. Pulling the strings of tumor collagen. *Nat. Cancer* **2022**, *3*, 9–10. [CrossRef]
18. Sun, X.; Wu, B.; Chiang, H.-C.; Deng, H.; Zhang, X.; Xiong, W.; Liu, J.; Rozeboom, A.M.; Harris, B.T.; Blommaert, E. Tumour DDR1 promotes collagen fibre alignment to instigate immune exclusion. *Nature* **2021**, *599*, 673–678. [CrossRef]

19. Di Martino, J.S.; Nobre, A.R.; Mondal, C.; Taha, I.; Farias, E.F.; Fertig, E.J.; Naba, A.; Aguirre-Ghiso, J.A.; Bravo-Cordero, J.J. A tumor-derived type III collagen-rich ECM niche regulates tumor cell dormancy. *Nat. Cancer* **2022**, *3*, 90–107. [CrossRef]
20. Opitz, F.V.; Haeberle, L.; Daum, A.; Esposito, I. Tumor Microenvironment in Pancreatic Intraepithelial Neoplasia. *Cancers* **2021**, *13*, 6188. [CrossRef]
21. Kochetkova, M.; Samuel, M.S. Differentiation of the tumor microenvironment: Are CAFs the Organizer? *Trends Cell Biol.* **2021**. [CrossRef] [PubMed]
22. Andón, F.T.; Digifico, E.; Maeda, A.; Erreni, M.; Mantovani, A.; Alonso, M.J.; Allavena, P. Targeting tumor associated macrophages: The new challenge for nanomedicine. *Semin. Immunol.* **2017**, *34*, 103–113. [CrossRef] [PubMed]
23. Marrella, A.; Dondero, A.; Aiello, M.; Casu, B.; Caluori, G.; Castriconi, R.; Scaglione, S. Cell-Laden Hydrogel as a Clinical-Relevant 3D Model for Analyzing Neuroblastoma Growth, Immunophenotype, and Susceptibility to Therapies. *Front. Immunol.* **2019**, *10*, 1876. [CrossRef] [PubMed]
24. Kapałczyńska, M.; Kolenda, T.; Przybyła, W.; Zajączkowska, M.; Teresiak, A.; Filas, V.; Ibbs, M.; Bliźniak, R.; Łuczewski, Ł.; Lamperska, K. 2D and 3D cell cultures—A comparison of different types of cancer cell cultures. *Arch. Med. Sci.* **2018**, *14*, 910–919. [CrossRef] [PubMed]
25. Mulhall, H.J.; Hughes, M.P.; Kazmi, B.; Lewis, M.P.; Labeed, F.H. Epithelial cancer cells exhibit different electrical properties when cultured in 2D and 3D environments. *Biochim. Biophys. Acta-Gen. Subj.* **2013**, *1830*, 5136–5141. [CrossRef] [PubMed]
26. Souza, A.G.; Ferreira, I.C.C. Advances in Cell Culture: More than a Century after Cultivating Cells. *J. Biotechnol. Biomater.* **2016**, *6*, 2–5. [CrossRef]
27. Doke, S.K.; Dhawale, S.C. Alternatives to animal testing: A review. *Saudi Pharm. J.* **2015**, *23*, 223–229. [CrossRef]
28. Meigs, L.; Smirnova, L.; Rovida, C.; Leist, M.; Hartung, T. Animal testing and its alternatives—The most important omics is economics. *ALTEX* **2018**, *35*, 275–305. [CrossRef]
29. Liguori, G.R.; Jeronimus, B.F.; De Aquinas Liguori, T.T.; Moreira, L.F.P.; Harmsen, M.C. Ethical Issues in the Use of Animal Models for Tissue Engineering: Reflections on Legal Aspects, Moral Theory, Three Rs Strategies, and Harm'Benefit Analysis. *Tissue Eng.-Part C Methods* **2017**, *23*, 850–862. [CrossRef]
30. Longati, P.; Jia, X.; Eimer, J.; Wagman, A.; Witt, M.R.; Rehnmark, S.; Verbeke, C.; Toftgård, R.; Löhr, M.; Heuchel, R.L. 3D pancreatic carcinoma spheroids induce a matrix-rich, chemoresistant phenotype offering a better model for drug testing. *BMC Cancer* **2013**, *13*, 95. [CrossRef] [PubMed]
31. Jacks, T.; Weinberg, R.A. Taking the study of cancer cell survival to a new dimension. *Cell* **2002**, *111*, 923–925. [CrossRef]
32. Gibbons, M.C.; Foley, M.A.; Cardinal, K.O. Thinking inside the box: Keeping tissue-engineered constructs in vitro for use as preclinical models. *Tissue Eng. Part B Rev.* **2013**, *19*, 14–30. [CrossRef] [PubMed]
33. Lee, J.Y.; Chaudhuri, O. Modeling the tumor immune microenvironment for drug discovery using 3D culture. *APL Bioeng.* **2021**, *5*, 010903. [CrossRef]
34. Shelton, S.E.; Nguyen, H.T.; Barbie, D.A.; Kamm, R.D. Engineering approaches for studying immune-tumor cell interactions and immunotherapy. *iScience* **2021**, *24*, 101985. [CrossRef] [PubMed]
35. Dunn, G.P.; Old, L.J.; Schreiber, R.D. The three Es of cancer immunoediting. *Annu. Rev. Immunol.* **2004**, *22*, 329–360. [CrossRef]
36. Bellora, F.; Castriconi, R.; Dondero, A.; Carrega, P.; Mantovani, A.; Ferlazzo, G.; Moretta, A.; Bottino, C. Human NK cells and NK receptors. *Immunol. Lett.* **2014**, *161*, 168–173. [CrossRef]
37. Thielens, A.; Vivier, E.; Romagné, F. NK cell MHC class I specific receptors (KIR): From biology to clinical intervention. *Curr. Opin. Immunol.* **2012**, *24*, 239–245. [CrossRef]
38. Kim, S.; Poursine-Laurent, J.; Truscott, S.M.; Lybarger, L.; Song, Y.-J.; Yang, L.; French, A.R.; Sunwoo, J.B.; Lemieux, S.; Hansen, T.H. Licensing of natural killer cells by host major histocompatibility complex class I molecules. *Nature* **2005**, *436*, 709–713. [CrossRef]
39. Guia, S.; Fenis, A.; Vivier, E.; Narni-Mancinelli, E. Activating and inhibitory receptors expressed on innate lymphoid cells. *Semin. Immunopathol.* **2018**, *40*, 331–341. [CrossRef]
40. Moretta, L.; Bottino, C.; Pende, D.; Castriconi, R.; Mingari, M.C.; Moretta, A. Surface NK receptors and their ligands on tumor cells. *Semin. Immunol.* **2006**, *18*, 151–158. [CrossRef]
41. Sivori, S.; Meazza, R.; Quintarelli, C.; Carlomagno, S.; Della Chiesa, M.; Falco, M.; Moretta, L.; Locatelli, F.; Pende, D. NK Cell-Based Immunotherapy for Hematological Malignancies. *J. Clin. Med.* **2019**, *8*, 1702. [CrossRef] [PubMed]
42. Balsamo, M.; Vermi, W.; Parodi, M.; Pietra, G.; Manzini, C.; Queirolo, P.; Lonardi, S.; Augugliaro, R.; Moretta, A.; Facchetti, F.; et al. Melanoma cells become resistant to NK-cell-mediated killing when exposed to NK-cell numbers compatible with NK-cell infiltration in the tumor. *Eur. J. Immunol.* **2012**, *42*, 1833–1842. [CrossRef] [PubMed]
43. Kohrt, H.E.; Thielens, A.; Marabelle, A.; Sagiv-Barfi, I.; Sola, C.; Chanuc, F.; Fuseri, N.; Bonnafous, C.; Czerwinski, D.; Rajapaksa, A. Anti-KIR antibody enhancement of anti-lymphoma activity of natural killer cells as monotherapy and in combination with anti-CD20 antibodies. *Blood J. Am. Soc. Hematol.* **2014**, *123*, 678–686. [CrossRef] [PubMed]
44. André, P.; Denis, C.; Soulas, C.; Bourbon-Caillet, C.; Lopez, J.; Arnoux, T.; Bléry, M.; Bonnafous, C.; Gauthier, L.; Morel, A. Anti-NKG2A mAb is a checkpoint inhibitor that promotes anti-tumor immunity by unleashing both T and NK cells. *Cell* **2018**, *175*, 1731–1743. [CrossRef] [PubMed]
45. Locatelli, F.; Pende, D.; Falco, M.; Della Chiesa, M.; Moretta, A.; Moretta, L. NK cells mediate a crucial graft-versus-leukemia effect in haploidentical-HSCT to cure high-risk acute leukemia. *Trends Immunol.* **2018**, *39*, 577–590. [CrossRef]

46. Bottino, C.; Dondero, A.; Castriconi, R. Inhibitory axes impacting on the activity and fate of Innate Lymphoid Cells. *Mol. Asp. Med.* **2021**, *80*, 100985. [CrossRef]
47. Chiossone, L.; Vienne, M.; Kerdiles, Y.M.; Vivier, E. Natural killer cell immunotherapies against cancer: Checkpoint inhibitors and more. *Semin. Immunol.* **2017**, *31*, 55–63. [CrossRef]
48. Sciumè, G.; Fionda, C.; Stabile, H.; Gismondi, A.; Santoni, A. Negative regulation of innate lymphoid cell responses in inflammation and cancer. *Immunol. Lett.* **2019**, *215*, 28–34. [CrossRef]
49. Scribner, J.A.; Brown, J.G.; Son, T.; Chiechi, M.; Li, P.; Sharma, S.; Li, H.; De Costa, A.; Li, Y.; Chen, Y. Preclinical Development of MGC018, a Duocarmycin-based Antibody–drug Conjugate Targeting B7-H3 for Solid Cancer. *Mol. Cancer Ther.* **2020**, *19*, 2235–2244. [CrossRef]
50. Gauthier, L.; Morel, A.; Anceriz, N.; Rossi, B.; Blanchard-Alvarez, A.; Grondin, G.; Trichard, S.; Cesari, C.; Sapet, M.; Bosco, F. Multifunctional natural killer cell engagers targeting NKp46 trigger protective tumor immunity. *Cell* **2019**, *177*, 1701–1713. [CrossRef]
51. Marofi, F.; Al-Awad, A.S.; Sulaiman Rahman, H.; Markov, A.; Abdelbasset, W.K.; Ivanovna Enina, Y.; Mahmoodi, M.; Hassanzadeh, A.; Yazdanifar, M.; Stanley Chartrand, M. CAR-NK cell: A new paradigm in tumor immunotherapy. *Front. Oncol.* **2021**, *11*, 2078. [CrossRef] [PubMed]
52. Bottino, C.; Dondero, A.; Bellora, F.; Moretta, L.; Locatelli, F.; Pistoia, V.; Moretta, A.; Castriconi, R. Natural killer cells and neuroblastoma: Tumor recognition, escape mechanisms, and possible novel immunotherapeutic approaches. *Front. Immunol.* **2014**, *5*, 56. [CrossRef] [PubMed]
53. Castriconi, R.; Dondero, A.; Augugliaro, R.; Cantoni, C.; Carnemolla, B.; Sementa, A.R.; Negri, F.; Conte, R.; Corrias, M.V.; Moretta, L.; et al. Identification of 4Ig-B7-H3 as a neuroblastoma-associated molecule that exerts a protective role from an NK cell-mediated lysis. *Proc. Natl. Acad. Sci. USA* **2004**, *101*, 12640–12645. [CrossRef] [PubMed]
54. Steinberger, P.; Majdic, O.; Derdak, S.V.; Pfistershammer, K.; Kirchberger, S.; Klauser, C.; Zlabinger, G.; Pickl, W.F.; Stöckl, J.; Knapp, W. Molecular Characterization of Human 4Ig-B7-H3, a Member of the B7 Family with Four Ig-Like Domains. *J. Immunol.* **2004**, *172*, 2352–2359. [CrossRef] [PubMed]
55. Chapoval, A.I.; Ni, J.; Lau, J.S.; Wilcox, R.A.; Flies, D.B.; Liu, D.; Dong, H.; Sica, G.L.; Zhu, G.; Tamada, K.; et al. B7-H3: A costimulatory molecule for T cell activation and IFN-γ production. *Nat. Immunol.* **2001**, *2*, 269–274. [CrossRef]
56. Lemke, D.; Pfenning, P.N.; Sahm, F.; Klein, A.C.; Kempf, T.; Warnken, U.; Schnölzer, M.; Tudoran, R.; Weller, M.; Platten, M.; et al. Costimulatory protein 4IgB7H3 drives the malignant phenotype of glioblastoma by mediating immune escape and invasiveness. *Clin. Cancer Res.* **2012**, *18*, 105–117. [CrossRef]
57. Prasad, D.V.R.; Nguyen, T.; Li, Z.; Yang, Y.; Duong, J.; Wang, Y.; Dong, C. Murine B7-H3 is a negative regulator of T cells. *J. Immunol.* **2004**, *173*, 2500–2506. [CrossRef]
58. Suh, W.-K.; Gajewska, B.U.; Okada, H.; Gronski, M.A.; Bertram, E.M.; Dawicki, W.; Duncan, G.S.; Bukczynski, J.; Plyte, S.; Elia, A. The B7 family member B7-H3 preferentially down-regulates T helper type 1–mediated immune responses. *Nat. Immunol.* **2003**, *4*, 899–906. [CrossRef]
59. Flem-Karlsen, K.; Fodstad, Ø.; Tan, M.; Nunes-Xavier, C.E. B7-H3 in cancer–beyond immune regulation. *Trends Cancer* **2018**, *4*, 401–404. [CrossRef]
60. Ni, L.; Dong, C. New B7 family checkpoints in human cancers. *Mol. Cancer Ther.* **2017**, *16*, 1203–1211. [CrossRef]
61. Gregorio, A.; Corrias, M.V.; Castriconi, R.; Dondero, A.; Mosconi, M.; Gambini, C.; Moretta, A.; Moretta, L.; Bottino, C. Small round blue cell tumours: Diagnostic and prognostic usefulness of the expression of B7-H3 surface molecule. *Histopathology* **2008**, *53*, 73–80. [CrossRef] [PubMed]
62. Vallera, D.A.; Ferrone, S.; Kodal, B.; Hinderlie, P.; Bendzick, L.; Ettestad, B.; Hallstrom, C.; Zorko, N.A.; Rao, A.; Fujioka, N. NK-cell-mediated targeting of various solid tumors using a B7-H3 Tri-specific killer engager in vitro and in vivo. *Cancers* **2020**, *12*, 2659. [CrossRef] [PubMed]
63. Kendsersky, N.M.; Lindsay, J.; Kolb, E.A.; Smith, M.A.; Teicher, B.A.; Erickson, S.W.; Earley, E.J.; Mosse, Y.P.; Martinez, D.; Pogoriler, J.; et al. The B7-H3–Targeting Antibody–Drug Conjugate m276-SL-PBD Is Potently Effective Against Pediatric Cancer Preclinical Solid Tumor Models. *Clin. Cancer Res.* **2019**, *27*, 221–237.e8. [CrossRef] [PubMed]
64. Majzner, R.G.; Theruvath, J.L.; Nellan, A.; Heitzeneder, S.; Cui, Y.; Mount, C.W.; Rietberg, S.P.; Linde, M.H.; Xu, P.; Rota, C.; et al. CAR T cells targeting B7-H3, a pan-cancer antigen, demonstrate potent preclinical activity against pediatric solid tumors and brain tumors. *Clin. Cancer Res.* **2019**, *25*, 2560–2574. [CrossRef] [PubMed]
65. Du, H.; Hirabayashi, K.; Ahn, S.; Kren, N.P.; Montgomery, S.A.; Wang, X.; Tiruthani, K.; Mirlekar, B.; Michaud, D.; Greene, K.; et al. Antitumor Responses in the Absence of Toxicity in Solid Tumors by Targeting B7-H3 via Chimeric Antigen Receptor T Cells. *Cancer Cell* **2019**, *35*, 221–237.e8. [CrossRef]
66. Park, J.A.; Cheung, N.-K. V Targets and antibody formats for immunotherapy of neuroblastoma. *J. Clin. Oncol.* **2020**, *38*, 1836. [CrossRef]
67. Locati, M.; Mantovani, A.; Sica, A. Macrophage activation and polarization as an adaptive component of innate immunity. *Adv. Immunol.* **2013**, *120*, 163–184. [CrossRef]
68. Kim, B.G.; Malek, E.; Choi, S.H.; Ignatz-Hoover, J.J.; Driscoll, J.J. Novel therapies emerging in oncology to target the TGF-β pathway. *J. Hematol. Oncol.* **2021**, *14*, 55. [CrossRef]

69. Viel, S.; Marçais, A.; Guimaraes, F.S.; Loftus, R.; Rabilloud, J.; Grau, M.; Degouve, S.; Djebali, S.; Sanlaville, A.; Charrier, E. TGF-β inhibits the activation and functions of NK cells by repressing the mTOR pathway. *Sci. Signal.* **2016**, *9*, ra19. [CrossRef]
70. Regis, S.; Dondero, A.; Caliendo, F.; Bottino, C.; Castriconi, R. NK Cell Function Regulation by TGF-β-Induced Epigenetic Mechanisms. *Front. Immunol.* **2020**, *11*, 311. [CrossRef]
71. Castriconi, R.; Cantoni, C.; Della Chiesa, M.; Vitale, M.; Marcenaro, E.; Conte, R.; Biassoni, R.; Bottino, C.; Moretta, L.; Moretta, A. Transforming growth factor β1 inhibits expression of NKP30 and NKG2d receptors: Consequences for the NK-mediated killing of dendritic cells. *Proc. Natl. Acad. Sci. USA* **2003**, *100*, 4120–4125. [CrossRef]
72. Mamessier, E.; Sylvain, A.; Thibult, M.-L.; Houvenaeghel, G.; Jacquemier, J.; Castellano, R.; Gonçalves, A.; André, P.; Romagné, F.; Thibault, G. Human breast cancer cells enhance self tolerance by promoting evasion from NK cell antitumor immunity. *J. Clin. Investig.* **2011**, *121*, 3609–3622. [CrossRef] [PubMed]
73. Allan, D.S.J.; Rybalov, B.; Awong, G.; Zúñiga-Pflücker, J.C.; Kopcow, H.D.; Carlyle, J.R.; Strominger, J.L. TGF-β affects development and differentiation of human natural killer cell subsets. *Eur. J. Immunol.* **2010**, *40*, 2289–2295. [CrossRef] [PubMed]
74. Song, H.; Kim, Y.; Park, G.; Kim, Y.S.; Kim, S.; Lee, H.K.; Chung, W.Y.; Park, S.J.; Han, S.Y.; Cho, D.; et al. Transforming growth factor-ß1 regulates human renal proximal tubular epithelial cell susceptibility to natural killer cells via modulation of the NKG2D ligands. *Int. J. Mol. Med.* **2015**, *36*, 1180–1188. [CrossRef] [PubMed]
75. Huang, C.H.; Liao, Y.J.; Chiou, T.J.; Huang, H.T.; Lin, Y.H.; Twu, Y.C. TGF-β regulated leukemia cell susceptibility against NK targeting through the down-regulation of the CD48 expression. *Immunobiology* **2019**, *224*, 649–658. [CrossRef]
76. Castriconi, R.; Carrega, P.; Dondero, A.; Bellora, F.; Casu, B.; Regis, S.; Ferlazzo, G.; Bottino, C. Molecular mechanisms directing migration and retention of natural killer cells in human tissues. *Front. Immunol.* **2018**, *9*, 2324. [CrossRef] [PubMed]
77. Regis, S.; Caliendo, F.; Dondero, A.; Casu, B.; Romano, F.; Loiacono, F.; Moretta, A.; Bottino, C.; Castriconi, R. TGF-ß1 downregulates the expression of CX3CR1 by inducing miR-27a-5p in primary human NK cells. *Front. Immunol.* **2017**, *8*, 868. [CrossRef] [PubMed]
78. Cortez, V.S.; Cervantes-Barragan, L.; Robinette, M.L.; Bando, J.K.; Wang, Y.; Geiger, T.L.; Gilfillan, S.; Fuchs, A.; Vivier, E.; Sun, J.C. Transforming growth factor-β signaling guides the differentiation of innate lymphoid cells in salivary glands. *Immunity* **2016**, *44*, 1127–1139. [CrossRef]
79. Casu, B.; Dondero, A.; Regis, S.; Caliendo, F.; Petretto, A.; Bartolucci, M.; Bellora, F.; Bottino, C.; Castriconi, R. Novel immunoregulatory functions of IL-18, an accomplice of TGF-β1. *Cancers* **2019**, *11*, 75. [CrossRef]
80. Brownlie, D.; Doughty-Shenton, D.; Yh Soong, D.; Nixon, C.; Carragher, N.O.; Carlin, L.M.; Kitamura, T. Metastasis-associated macrophages constrain antitumor capability of natural killer cells in the metastatic site at least partially by membrane bound transforming growth factor β. *J. Immunother. Cancer* **2021**, *9*, e001740. [CrossRef]
81. Han, Y.; Yang, Y.; Chen, Z.; Jiang, Z.; Gu, Y.; Liu, Y.; Xu, S.; Lin, C.; Pan, Z.; Zhou, W.; et al. Human hepatocellular carcinoma-infiltrating CD4+CD69 +Foxp3- regulatory T cell suppresses T cell response via membrane-bound TGF-β1. *J. Mol. Med.* **2014**, *92*, 539–550. [CrossRef] [PubMed]
82. Chen, W.J.; Jin, W.; Hardegen, N.; Lei, K.J.; Li, L.; Marinos, N.; McGrady, G.; Wahl, S.M. Conversion of Peripheral $CD4^+CD25^-$ Naive T Cells to $CD4^+CD25^+$ Regulatory T Cells by TGF-β Induction of Transcription Factor Foxp3. *J. Exp. Med.* **2003**, *198*, 1875–1886. [CrossRef] [PubMed]
83. Fantini, M.C.; Becker, C.; Monteleone, G.; Pallone, F.; Galle, P.R.; Neurath, M.F. Cutting Edge: TGF-β Induces a Regulatory Phenotype in $CD4^+CD25^-$ T Cells through Foxp3 Induction and Down-Regulation of Smad7. *J. Immunol.* **2004**, *172*, 5149–5153. [CrossRef] [PubMed]
84. Nakamura, S.; Yaguchi, T.; Kawamura, N.; Kobayashi, A.; Sakurai, T.; Higuchi, H.; Takaishi, H.; Hibi, T.; Kawakami, Y. TGF-β1 in tumor microenvironments induces immunosuppression in the tumors and sentinel lymph nodes and promotes tumor progression. *J. Immunother.* **2014**, *37*, 63–72. [CrossRef] [PubMed]
85. Wang, Y.; Liu, T.; Tang, W.; Deng, B.; Chen, Y.; Zhu, J.; Shen, X. Hepatocellular Carcinoma Cells Induce Regulatory T Cells and Lead to Poor Prognosis via Production of Transforming Growth Factor-β1. *Cell. Physiol. Biochem.* **2016**, *38*, 306–318. [CrossRef] [PubMed]
86. Soares, K.C.; Rucki, A.A.; Kim, V.; Foley, K.; Solt, S.; Wolfgang, C.L.; Jaffee, E.M.; Zheng, L. TGF-β blockade depletes T regulatory cells from metastatic pancreatic tumors in a vaccine dependent manner. *Oncotarget* **2015**, *6*, 43005–43015. [CrossRef]
87. Budhu, S.; Schaer, D.; Li, Y.; Houghton, A.; Silverstein, S.; Merghoub, T.; Wolchok, J. Blockade of surface bound TGF-β on regulatory T cells abrogates suppression of effector T cell function within the tumor microenvironment (TUM2P. 1015). *J. Immunol.* **2015**, *194*, 69.12.
88. Hugo, W.; Zaretsky, J.M.; Sun, L.; Song, C.; Moreno, B.H.; Hu-Lieskovan, S.; Berent-Maoz, B.; Pang, J.; Chmielowski, B.; Cherry, G.; et al. Genomic and Transcriptomic Features of Response to Anti-PD-1 Therapy in Metastatic Melanoma. *Cell* **2016**, *165*, 35–44. [CrossRef]
89. Bai, X.; Yi, M.; Jiao, Y.; Chu, Q.; Wu, K. Blocking tgf-β signaling to enhance the efficacy of immune checkpoint inhibitor. *Onco. Targets. Ther.* **2019**, *12*, 9527–9538. [CrossRef]
90. Martin, C.J.; Datta, A.; Littlefield, C.; Kalra, A.; Chapron, C.; Wawersik, S.; Dagbay, K.B.; Brueckner, C.T.; Nikiforov, A.; Danehy, F.T. Selective inhibition of TGFβ1 activation overcomes primary resistance to checkpoint blockade therapy by altering tumor immune landscape. *Sci. Transl. Med.* **2020**, *12*, eaay8456. [CrossRef]

91. Kloss, C.C.; Lee, J.; Zhang, A.; Chen, F.; Melenhorst, J.J.; Lacey, S.F.; Maus, M.V.; Fraietta, J.A.; Zhao, Y.; June, C.H. Dominant-Negative TGF-β Receptor Enhances PSMA-Targeted Human CAR T Cell Proliferation And Augments Prostate Cancer Eradication. *Mol. Ther.* **2018**, *26*, 1855–1866. [CrossRef] [PubMed]
92. Tang, N.; Cheng, C.; Zhang, X.; Qiao, M.; Li, N.; Mu, W.; Wei, X.F.; Han, W.; Wang, H. TGF-β inhibition via CRISPR promotes the long-term efficacy of CAR T cells against solid tumors. *JCI Insight* **2020**, *5*, e133977. [CrossRef] [PubMed]
93. Ravi, R.; Noonan, K.A.; Pham, V.; Bedi, R.; Zhavoronkov, A.; Ozerov, I.V.; Makarev, E.; Artemov, A.V.; Wysocki, P.T.; Mehra, R.; et al. Bifunctional immune checkpoint-targeted antibody-ligand traps that simultaneously disable TGFβ enhance the efficacy of cancer immunotherapy. *Nat. Commun.* **2018**, *9*, 741. [CrossRef] [PubMed]
94. Chen, X.; Yang, S.; Li, S.; Qu, Y.; Wang, H.Y.; Liu, J.; Dunn, Z.S.; Cinay, G.E.; MacMullan, M.A.; Hu, F.; et al. Secretion of bispecific protein of anti-PD-1 fused with TGF-β trap enhances antitumor efficacy of CAR-T cell therapy. *Mol. Ther.-Oncolytics* **2021**, *21*, 144–157. [CrossRef]
95. Li, L.; Yu, R.; Cai, T.; Chen, Z.; Lan, M.; Zou, T.; Wang, B.; Wang, Q.; Zhao, Y.; Cai, Y. Effects of immune cells and cytokines on inflammation and immunosuppression in the tumor microenvironment. *Int. Immunopharmacol.* **2020**, *88*, 106939. [CrossRef]
96. Vitale, M.; Cantoni, C.; Pietra, G.; Mingari, M.C.; Moretta, L. Effect of tumor cells and tumor microenvironment on NK-cell function. *Eur. J. Immunol.* **2014**, *44*, 1582–1592. [CrossRef]
97. Mannino, M.H.; Zhu, Z.; Xiao, H.; Bai, Q.; Wakefield, M.R.; Fang, Y. The paradoxical role of IL-10 in immunity and cancer. *Cancer Lett.* **2015**, *367*, 103–107. [CrossRef]
98. Konjević, G.M.; Vuletić, A.M.; Mirjačić Martinović, K.M.; Larsen, A.K.; Jurišić, V.B. The role of cytokines in the regulation of NK cells in the tumor environment. *Cytokine* **2019**, *117*, 30–40. [CrossRef]
99. Ma, Y.; Bauer, V.; Riedel, T.; Ahmetlić, F.; Hömberg, N.; Hofer, T.P.; Röcken, M.; Mocikat, R. Interleukin-10 counteracts T-helper type 1 responses in B-cell lymphoma and is a target for tumor immunotherapy. *Cancer Lett.* **2021**, *503*, 110–116. [CrossRef]
100. Michielon, E.; López González, M.; Burm, J.L.A.; Waaijman, T.; Jordanova, E.S.; de Gruijl, T.D.; Gibbs, S. Micro-environmental cross-talk in an organotypic human melanoma-in-skin model directs M2-like monocyte differentiation via IL-10. *Cancer Immunol. Immunother.* **2020**, *69*, 2319–2331. [CrossRef]
101. Balsamo, M.; Scordamaglia, F.; Pietra, G.; Manzini, C.; Cantoni, C.; Boitano, M.; Queirolo, P.; Vermi, W.; Facchetti, F.; Moretta, A.; et al. Melanoma-associated fibroblasts modulate NK cell phenotype and antitumor cytotoxicity. *Proc. Natl. Acad. Sci. USA* **2009**, *106*, 20847–20852. [CrossRef] [PubMed]
102. Olesch, C.; Sha, W.; Angioni, C.; Sha, L.K.; Açaf, E.; Patrignani, P.; Jakobsson, P.J.; Radeke, H.H.; Grösch, S.; Geisslinger, G.; et al. MPGES-1-derived PGE2 suppresses CD80 expression on tumorassociated phagocytes to inhibit anti-tumor immune responses in breast cancer. *Oncotarget* **2015**, *6*, 10284–10296. [CrossRef] [PubMed]
103. Pennock, N.D.; Martinson, H.A.; Guo, Q.; Betts, C.B.; Jindal, S.; Tsujikawa, T.; Coussens, L.M.; Borges, V.F.; Schedin, P. Ibuprofen supports macrophage differentiation, T cell recruitment, and tumor suppression in a model of postpartum breast cancer. *J. Immunother. Cancer* **2018**, *6*, 98. [CrossRef] [PubMed]
104. Xun, X.; Zhang, C.; Wang, S.; Hu, S.; Xiang, X.; Cheng, Q.; Li, Z.; Wang, Y.; Zhu, J. Cyclooxygenase-2 expressed hepatocellular carcinoma induces cytotoxic T lymphocytes exhaustion through M2 macrophage polarization. *Am. J. Transl. Res.* **2021**, *13*, 4360–4375. [PubMed]
105. Wang, Y.; Cui, L.; Georgiev, P.; Singh, L.; Zheng, Y.; Yu, Y.; Grein, J.; Zhang, C.; Muise, E.S.; Sloman, D.L.; et al. Combination of EP4 antagonist MF-766 and anti-PD-1 promotes anti-tumor efficacy by modulating both lymphocytes and myeloid cells. *Oncoimmunology* **2021**, *10*, 1896643. [CrossRef]
106. Lai, Y.H.; Liu, H.; Chiang, W.F.; Chen, T.W.; Chu, L.J.; Yu, J.S.; Chen, S.J.; Chen, H.C.; Tan, B.C.M. MiR-31-5p-ACOX1 axis enhances tumorigenic fitness in oral squamous cell carcinoma via the promigratory prostaglandin E2. *Theranostics* **2018**, *8*, 486–504. [CrossRef]
107. Munn, D.H.; Shafizadeh, E.; Attwood, J.T.; Bondarev, I.; Pashine, A.; Mellor, A.L. Inhibition of T cell proliferation by macrophage tryptophan catabolism. *J. Exp. Med.* **1999**, *189*, 1363–1372. [CrossRef]
108. Munn, D.H.; Sharma, M.D.; Baban, B.; Harding, H.P.; Zhang, Y.; Ron, D.; Mellor, A.L. GCN2 kinase in T cells mediates proliferative arrest and anergy induction in response to indoleamine 2,3-dioxygenase. *Immunity* **2005**, *22*, 633–642. [CrossRef]
109. Mezrich, J.D.; Fechner, J.H.; Zhang, X.; Johnson, B.P.; Burlingham, W.J.; Bradfield, C.A. An Interaction between Kynurenine and the Aryl Hydrocarbon Receptor Can Generate Regulatory T Cells. *J. Immunol.* **2010**, *185*, 3190–3198. [CrossRef]
110. Gandhi, R.; Kumar, D.; Burns, E.J.; Nadeau, M.; Dake, B.; Laroni, A.; Kozoriz, D.; Weiner, H.L.; Quintana, F.J. Activation of the aryl hydrocarbon receptor induces human type 1 regulatory T cell-like and Foxp3$^+$ regulatory T cells. *Nat. Immunol.* **2010**, *11*, 846–853. [CrossRef]
111. Nguyen, N.T.; Kimura, A.; Nakahama, T.; Chinen, I.; Masuda, K.; Nohara, K.; Fujii-Kuriyama, Y.; Kishimoto, T. Aryl hydrocarbon receptor negatively regulates dendritic cell immunogenicity via a kynurenine-dependent mechanism. *Proc. Natl. Acad. Sci. USA* **2010**, *107*, 19961–19966. [CrossRef] [PubMed]
112. Wagage, S.; John, B.; Krock, B.L.; Hall, A.O.; Randall, L.M.; Karp, C.L.; Simon, M.C.; Hunter, C.A. The Aryl Hydrocarbon Receptor Promotes IL-10 Production by NK Cells. *J. Immunol.* **2014**, *192*, 1661–1670. [CrossRef] [PubMed]
113. Liu, Y.; Liang, X.; Dong, W.; Fang, Y.; Lv, J.; Zhang, T.; Fiskesund, R.; Xie, J.; Liu, J.; Yin, X.; et al. Tumor-Repopulating Cells Induce PD-1 Expression in CD8$^+$ T Cells by Transferring Kynurenine and AhR Activation. *Cancer Cell* **2018**, *33*, 480–494.e7. [CrossRef] [PubMed]

114. Della Chiesa, M.; Carlomagno, S.; Frumento, G.; Balsamo, M.; Cantoni, C.; Conte, R.; Moretta, L.; Moretta, A.; Vitale, M. The tryptophan catabolite L-kynurenine inhibits the surface expression of NKp46- and NKG2D-activating receptors and regulates NK-cell function. *Blood* **2006**, *108*, 4118–4125. [CrossRef] [PubMed]
115. Labadie, B.W.; Bao, R.; Luke, J.J. Reimagining IDO pathway inhibition in cancer immunotherapy via downstream focus on the tryptophan–kynurenine–aryl hydrocarbon axis. *Clin. Cancer Res.* **2019**, *25*, 1462–1471. [CrossRef] [PubMed]
116. Vultaggio-Poma, V.; Sarti, A.C.; Di Virgilio, F. Extracellular ATP: A Feasible Target for Cancer Therapy. *Cells* **2020**, *9*, 2496. [CrossRef]
117. Dosch, M.; Gerber, J.; Jebbawi, F.; Beldi, G. Mechanisms of ATP release by inflammatory cells. *Int. J. Mol. Sci.* **2018**, *19*, 1222. [CrossRef]
118. Ohta, A. A metabolic immune checkpoint: Adenosine in Tumor Microenvironment. *Front. Immunol.* **2016**, *7*, 109. [CrossRef]
119. Antonioli, L.; Pacher, P.; Vizi, E.S.; Haskó, G. CD39 and CD73 in immunity and inflammation. *Trends Mol. Med.* **2013**, *19*, 355–367. [CrossRef]
120. Churov, A.; Zhulai, G. Targeting adenosine and regulatory T cells in cancer immunotherapy. *Hum. Immunol.* **2021**, *82*, 270–278. [CrossRef]
121. Clayton, A.; Al-Taei, S.; Webber, J.; Mason, M.D.; Tabi, Z. Cancer Exosomes Express CD39 and CD73, Which Suppress T Cells through Adenosine Production. *J. Immunol.* **2011**, *187*, 676–683. [CrossRef] [PubMed]
122. Cekic, C.; Day, Y.J.; Sag, D.; Linden, J. Myeloid expression of adenosine a2A receptor suppresses T and NK cell responses in the solid tumor microenvironment. *Cancer Res.* **2014**, *74*, 7250–7259. [CrossRef] [PubMed]
123. Vaupel, P.; Mayer, A. Hypoxia-driven adenosine accumulation: A crucial microenvironmental factor promoting tumor progression. In *Oxygen Transport to Tissue XXXVII*; Springer: New York, NY, USA, 2016; pp. 177–183. [CrossRef]
124. Antonioli, L.; Fornai, M.; Blandizzi, C.; Pacher, P.; Haskó, G. Adenosine signaling and the immune system: When a lot could be too much. *Immunol. Lett.* **2019**, *205*, 9–15. [CrossRef] [PubMed]
125. Chen, S.; Akdemir, I.; Fan, J.; Linden, J.; Zhang, B.; Cekic, C. The expression of adenosine A2B receptor on antigen-presenting cells suppresses $CD8^{+}$T-cell responses and promotes tumor growth. *Cancer Immunol. Res.* **2020**, *8*, 1064–1074. [CrossRef] [PubMed]
126. Ohta, A.; Kini, R.; Ohta, A.; Subramanian, M.; Madasu, M.; Sitkovsky, M. The development and immunosuppressive functions of $CD4^{+}CD25^{+}FoxP3^{+}$ regulatory T cells are under influence of the adenosine-A2A adenosine receptor pathway. *Front. Immunol.* **2012**, *3*, 190. [CrossRef] [PubMed]
127. Riabov, V.; Gudima, A.; Wang, N.; Mickley, A.; Orekhov, A.; Kzhyshkowska, J. Role of tumor associated macrophages in tumor angiogenesis and lymphangiogenesis. *Front. Physiol.* **2014**, *5*, 75. [CrossRef]
128. Balta, E.; Wabnitz, G.H.; Samstag, Y. Hijacked immune cells in the tumor microenvironment: Molecular mechanisms of immunosuppression and cues to improve t cell-based immunotherapy of solid tumors. *Int. J. Mol. Sci.* **2021**, *22*, 5736. [CrossRef]
129. Piccard, H.; Muschel, R.J.; Opdenakker, G. On the dual roles and polarized phenotypes of neutrophils in tumor development and progression. *Crit. Rev. Oncol. Hematol.* **2012**, *82*, 296–309. [CrossRef]
130. Masucci, M.T.; Minopoli, M.; Carriero, M.V. Tumor Associated Neutrophils. Their Role in Tumorigenesis, Metastasis, Prognosis and Therapy. *Front. Oncol.* **2019**, *9*, 1146. [CrossRef]
131. Hoh, A.; Dewerth, A.; Vogt, F.; Wenz, J.; Baeuerle, P.A.; Warmann, S.W. The activity of cd T cells against paediatric liver tumour cells and spheroids in cell culture. *Liver Int.* **2013**, *33*, 127–136. [CrossRef]
132. Varesano, S.; Zocchi, M.R.; Poggi, A. Zoledronate Triggers V δ 2 T Cells to Destroy and Kill Spheroids of Colon Carcinoma: Quantitative Image Analysis of Three-Dimensional Cultures. *Front. Immunol.* **2018**, *9*, 998. [CrossRef]
133. Zhou, S.; Zhu, M.; Meng, F.; Shao, J.; Xu, Q.; Wei, J.; Liu, B. Evaluation of PD-1 blockade using a multicellular tumor spheroid model. *Am. J. Transl. Res.* **2019**, *11*, 7471–7478.
134. Koeck, S.; Kern, J.; Zwierzina, M.; Gamerith, G.; Lorenz, E.; Sopper, S.; Zwierzina, H.; Amann, A.; Koeck, S.; Kern, J.; et al. The influence of stromal cells and tumor- microenvironment-derived cytokines and chemokines on CD3 CD8 tumor infiltrating lymphocyte subpopulations. *Oncoimmunology* **2017**, *6*, e1323617. [CrossRef] [PubMed]
135. Giannattasio, A.; Weil, S.; Kloess, S.; Ansari, N.; Stelzer, E.H.K.; Cerwenka, A.; Steinle, A.; Koehl, U.; Koch, J. Cytotoxicity and infiltration of human NK cells in in vivo- like tumor spheroids. *BMC Cancer* **2015**, *15*, 351. [CrossRef] [PubMed]
136. Evert, J.S.H.; Cany, J.; Van Den Brand, D.; Brock, R.; Torensma, R.; Bekkers, R.L.; Joop, H.; Massuger, L.F.; Dolstra, H.; Evert, J.S.H.; et al. Umbilical cord blood CD34 progenitor-derived NK cells efficiently kill ovarian cancer spheroids and intraperitoneal tumors in NOD/SCID/IL2Rg mice. *Oncoimmunology* **2017**, *6*, e1320630. [CrossRef] [PubMed]
137. Sherman, H.; Gitschier, H.J.; Rossi, A.E.; Rossi, A.E. A Novel three-dimensional Immune oncology Model for high-throughput testing of tumoricidal Activity. *Front. Immunol.* **2018**, *9*, 857. [CrossRef] [PubMed]
138. Courau, T.; Bonnereau, J.; Chicoteau, J.; Bottois, H.; Remark, R.; Miranda, L.A.; Toubert, A.; Blery, M.; Aparicio, T.; Allez, M.; et al. Cocultures of human colorectal tumor spheroids with immune cells reveal the therapeutic potential of MICA / B and NKG2A targeting for cancer treatment. *J. Immunother. Cancer* **2019**, *9*, 74. [CrossRef]
139. Ong, S.; Tan, Y.; Beretta, O.; Jiang, D.; Yeap, W.; Tai, J.J.Y.; Wong, W.; Yang, H.; Schwarz, H.; Lim, K.; et al. Macrophages in human colorectal cancer are pro-inflammatory and prime T cells towards an anti-tumour type-1 inflammatory response. *Eur. J. Immunol.* **2012**, *42*, 89–100. [CrossRef]
140. Kuen, J.; Darowski, D.; Kluge, T.; Majety, M. Pancreatic cancer cell / fibroblast co-culture induces M2 like macrophages that influence therapeutic response in a 3D model. *PLoS ONE* **2017**, *12*, e0182039. [CrossRef]

141. Raghavan, S.; Mehta, P.; Xie, Y.; Lei, Y.L.; Mehta, G. Ovarian cancer stem cells and macrophages reciprocally interact through the WNT pathway to promote pro-tumoral and malignant phenotypes in 3D engineered microenvironments. *J. Immunother. Cancer* **2019**, *1*, 190. [CrossRef]
142. Vitale, C.; Fedi, A.; Marrella, A.; Varani, G.; Fato, M.; Scaglione, S. 3D perfusable hydrogel recapitulating the cancer dynamic environment to in vitro investigate metastatic colonization. *Polymers* **2020**, *12*, 2467. [CrossRef] [PubMed]
143. Cavo, M.; Caria, M.; Pulsoni, I.; Beltrame, F.; Fato, M.; Scaglione, S. A new cell-laden 3D Alginate-Matrigel hydrogel resembles human breast cancer cell malignant morphology, spread and invasion capability observed "in vivo". *Sci. Rep.* **2018**, *8*, 5333. [CrossRef] [PubMed]
144. Sayd, T.; El Hamoui, O.; Alies, B.; Gaudin, K.; Lespes, G.; Battu, S. Biomaterials for Three-Dimensional Cell Culture: From Applications in Oncology to Nanotechnology. *Nanomaterials* **2021**, *11*, 481. [CrossRef] [PubMed]
145. Liu, Z.; Vunjak-novakovic, G. ScienceDirect Modeling tumor microenvironments using custom-designed biomaterial scaffolds. *Curr. Opin. Chem. Eng.* **2016**, *11*, 94–105. [CrossRef]
146. Monteiro, M.V.; Gaspar, V.M.; Ferreira, L.P.; Mano, J.F. Hydrogel 3D: In vitro tumor models for screening cell aggregation mediated drug response. *Biomater. Sci.* **2020**, *8*, 1855–1864. [CrossRef]
147. Benton, G.; Arnaoutova, I.; George, J.; Kleinman, H.K.; Koblinski, J. Matrigel: From discovery and ECM mimicry to assays and models for cancer research. *Adv. Drug Deliv. Rev.* **2014**, *79*, 3–18. [CrossRef]
148. West, N.R.; Kost, S.E.; Martin, S.D.; Milne, K.; Deleeuw, R.J.; Nelson, B.H.; Watson, P.H. Tumour-infiltrating FOXP3$^+$ lymphocytes are associated with cytotoxic immune responses and good clinical outcome in oestrogen receptor-negative breast cancer. *Br. J. Cancer* **2013**, *108*, 155–162. [CrossRef]
149. Augustine, T.N.; Dix-Peek, T.; Duarte, R.; Candy, G.P. Establishment of a heterotypic 3D culture system to evaluate the interaction of TREG lymphocytes and NK cells with breast cancer. *J. Immunol. Methods* **2015**, *426*, 1–13. [CrossRef]
150. Chimal-Ramírez, G.K.; Espinoza-Sánchez, N.A.; Utrera-Barillas, D.; Benítez-Bribiesca, L.; Velázquez, J.R.; Arriaga-Pizano, L.A.; Monroy-García, A.; Reyes-Maldonado, E.; Domínguez-López, M.L.; Piña-Sánchez, P.; et al. MMP1, MMP9, and COX2 expressions in promonocytes are induced by breast cancer cells and correlate with collagen degradation, transformation-like morphological changes in MCF-10A acini, and tumor aggressiveness. *BioMed Res. Int.* **2013**, *2013*, 279505. [CrossRef]
151. Espinoza-Sánchez, N.A.; Chimal-Ramírez, G.K.; Mantilla, A.; Fuentes-Pananá, E.M. IL-1β, IL-8, and Matrix Metalloproteinases-1, -2, and -10 Are Enriched upon Monocyte–Breast Cancer Cell Cocultivation in a Matrigel-Based Three-Dimensional System. *Front. Immunol.* **2017**, *8*, 205. [CrossRef]
152. Denton, A.E.; Roberts, E.W.; Fearon, D.T. Stromal cells in the tumor microenvironment. *Stromal Immunol.* **2018**, *1060*, 99–114. [CrossRef]
153. Bussard, K.M.; Mutkus, L.; Stumpf, K.; Gomez-Manzano, C.; Marini, F.C. Tumor-associated stromal cells as key contributors to the tumor microenvironment. *Breast Cancer Res.* **2016**, *18*, 84. [CrossRef]
154. Dijkstra, K.K.; Cattaneo, C.M.; Weeber, F.; Chalabi, M.; van de Haar, J.; Fanchi, L.F.; Slagter, M.; van der Velden, D.L.; Kaing, S.; Kelderman, S.; et al. Generation of Tumor-Reactive T Cells by Co-culture of Peripheral Blood Lymphocytes and Tumor Organoids. *Cell* **2018**, *174*, 1586–1598.e12. [CrossRef] [PubMed]
155. Kleinman, H.K.; Martin, G.R. Matrigel: Basement membrane matrix with biological activity. *Semin. Cancer Biol.* **2005**, *15*, 378–386. [CrossRef] [PubMed]
156. Finnberg, N.K.; Gokare, P.; Lev, A.; Grivennikov, S.I.; MacFarlane, A.W.; Campbell, K.S.; Winters, R.M.; Kaputa, K.; Farma, J.M.; Abbas, A.E.-S.; et al. Application of 3D tumoroid systems to define immune and cytotoxic therapeutic responses based on tumoroid and tissue slice culture molecular signatures. *Oncotarget* **2017**, *8*, 66747–66757. [CrossRef]
157. Shamir, E.R.; Ewald, A.J. Three-dimensional organotypic culture: Experimental models of mammalian biology and disease. *Nat. Rev. Mol. Cell Biol.* **2014**, *15*, 647–664. [CrossRef]
158. Tsai, S.; McOlash, L.; Palen, K.; Johnson, B.; Duris, C.; Yang, Q.; Dwinell, M.B.; Hunt, B.; Evans, D.B.; Gershan, J.; et al. Development of primary human pancreatic cancer organoids, matched stromal and immune cells and 3D tumor microenvironment models. *BMC Cancer* **2018**, *18*, 335. [CrossRef]
159. Braham, M.V.J.; Minnema, M.C.; Aarts, T.; Sebestyen, Z.; Straetemans, T.; Vyborova, A.; Kuball, J.; Öner, F.C.; Robin, C.; Alblas, J. Cellular immunotherapy on primary multiple myeloma expanded in a 3D bone marrow niche model. *Oncoimmunology* **2018**, *7*, e1434465. [CrossRef]
160. Schnalzger, T.E.; Groot, M.H.; Zhang, C.; Mosa, M.H.; Michels, B.E.; Röder, J.; Darvishi, T.; Wels, W.S.; Farin, H.F. 3D model for CAR -mediated cytotoxicity using patient-derived colorectal cancer organoids. *EMBO J.* **2019**, *38*, 1–15. [CrossRef]
161. Gan, H.K.; Cvrljevic, A.N.; Johns, T.G. The epidermal growth factor receptor variant III (EGFRvIII): Where wild things are altered. *FEBS J.* **2013**, *280*, 5350–5370. [CrossRef]
162. Della Corte, C.M.; Barra, G.; Ciaramella, V.; Di Liello, R.; Vicidomini, G.; Zappavigna, S.; Luce, A.; Abate, M.; Fiorelli, A.; Caraglia, M.; et al. Antitumor activity of dual blockade of PD-L1 and MEK in NSCLC patients derived three-dimensional spheroid cultures. *J. Exp. Clin. Cancer Res.* **2019**, *38*, 253. [CrossRef] [PubMed]
163. Edsparr, K.; Basse, P.H.; Goldfarb, R.H.; Albertsson, P. Matrix metalloproteinases in cytotoxic lymphocytes impact on tumour infiltration and immunomodulation. *Cancer Microenviron.* **2011**, *4*, 351–360. [CrossRef] [PubMed]
164. Aisenbrey, E.A.; Murphy, W.L. Synthetic alternatives to Matrigel. *Nat. Rev. Mater.* **2020**, *5*, 539–551. [CrossRef]

165. Carvalho, M.R.; Lima, D.; Reis, R.L.; Correlo, V.M.; Oliveira, J.M. Evaluating Biomaterial- and Microfluidic-Based 3D Tumor Models. *Trends Biotechnol.* **2015**, *33*, 667–678. [CrossRef] [PubMed]
166. Peterson, N.C. From bench to Cageside: Risk assessment for rodent pathogen contamination of cells and biologics. *ILAR J.* **2012**, *53*, 310–314. [CrossRef]
167. Ferreira, L.P.; Gaspar, V.M.; Mano, J.F. Design of spherically structured 3D in vitro tumor models -Advances and prospects. *Acta Biomater.* **2018**, *75*, 11–34. [CrossRef]
168. Utech, S.; Boccaccini, A.R. A review of hydrogel-based composites for biomedical applications: Enhancement of hydrogel properties by addition of rigid inorganic fillers. *J. Mater. Sci.* **2016**, *51*, 271–310. [CrossRef]
169. Lin, Y.N.; Nasir, A.; Camacho, S.; Berry, D.L.; Schmidt, M.O.; Pearson, G.W.; Riegel, A.T.; Wellstein, A. Monitoring cancer cell invasion and t-cell cytotoxicity in 3d culture. *J. Vis. Exp.* **2020**, *2020*, 61392. [CrossRef]
170. Tevis, K.M.; Cecchi, R.J.; Colson, Y.L.; Grinstaff, M.W. Mimicking the tumor microenvironment to regulate macrophage phenotype and assessing chemotherapeutic efficacy in embedded cancer cell/macrophage spheroid models. *Acta Biomater.* **2017**, *50*, 271–279. [CrossRef]
171. Linde, N.; Gutschalk, C.M.; Hoffmann, C.; Yilmaz, D.; Mueller, M.M. Integrating macrophages into organotypic co-cultures: A 3D in vitro model to study tumor-associated macrophages. *PLoS ONE* **2012**, *7*, e40058. [CrossRef]
172. Liu, X.Q.; Kiefl, R.; Roskopf, C.; Tian, F.; Huber, R.M. Interactions among lung cancer cells, fibroblasts, and macrophages in 3D co-cultures and the impact on MMP-1 and VEGF expression. *PLoS ONE* **2016**, *11*, e0156268. [CrossRef] [PubMed]
173. Foxall, R.; Narang, P.; Glaysher, B.; Hub, E.; Teal, E.; Coles, M.C.; Ashton-Key, M.; Beers, S.A.; Cragg, M.S. Developing a 3D B Cell Lymphoma Culture System to Model Antibody Therapy. *Front. Immunol.* **2021**, *11*, 605231. [CrossRef] [PubMed]
174. Ravi, M.; Paramesh, V.; Kaviya, S.R.; Anuradha, E.; Paul Solomon, F.D. 3D cell culture systems: Advantages and applications. *J. Cell. Physiol.* **2015**, *230*, 16–26. [CrossRef] [PubMed]
175. Szot, C.S.; Buchanan, C.F.; Freeman, J.W.; Rylander, M.N. 3D in vitro bioengineered tumors based on collagen I hydrogels. *Biomaterials* **2011**, *32*, 7905–7912. [CrossRef]
176. Lu, Y.; Li, S.; Ma, L.; Li, Y.; Zhang, X.; Peng, Q.; Cuiju, M.; Huang, L. Type conversion of secretomes in a 3D TAM2 and HCC cell co-culture system and functional importance of CXCL2 in HCC. *Sci. Rep.* **2016**, *6*, 24558. [CrossRef]
177. Nii, T.; Makino, K.; Tabata, Y. Three-Dimensional Culture System of Cancer Cells Combined with Biomaterials for Drug Screening. *Cancers* **2020**, *12*, 2754. [CrossRef]
178. Ahadian, S.; Civitarese, R.; Bannerman, D.; Mohammadi, M.H.; Lu, R.; Wang, E.; Davenport-huyer, L.; Lai, B.; Zhang, B.; Zhao, Y. Organ-On-A-Chip Platforms: A Convergence of Advanced Materials, Cells, and Microscale Technologies. *Adv. Healtch. Mater.* **2017**, *7*, 1700506. [CrossRef]
179. Dondero, A.; Morini, M.; Cangelosi, D.; Mazzocco, K.; Serra, M.; Spaggiari, G.M.; Rotta, G.; Tondo, A.; Locatelli, F.; Castellano, A.; et al. Multiparametric flow cytometry highlights B7-H3 as a novel diagnostic/therapeutic target in GD2neg/low neuroblastoma variants. *J. Immunother. Cancer* **2021**, *9*, e002293. [CrossRef]
180. Rama-Esendagli, D.; Esendagli, G.; Yilmaz, G.; Guc, D. Spheroid formation and invasion capacity are differentially influenced by co-cultures of fibroblast and macrophage cells in breast cancer. *Mol. Biol.* **2014**, *41*, 2885–2892. [CrossRef]
181. Rebelo, P.; Pinto, C.; Martins, T.R.; Harrer, N.; Alves, P.M.; Estrada, M.F.; Loza-alvarez, P.; Gualda, E.J.; Sommergruber, W.; Brito, C. Biomaterials 3D-3-culture: A tool to unveil macrophage plasticity in the tumour microenvironment. *Biomaterials* **2018**, *163*, 185–197. [CrossRef]
182. Joseph, S.M.; Krishnamoorthy, S.; Paranthaman, R.; Moses, J.A.; Anandharamakrishnan, C. A review on source-specific chemistry, functionality, and applications of chitin and chitosan. *Carbohydr. Polym. Technol. Appl.* **2021**, *2*, 100036. [CrossRef]
183. Ahmad, S.I.; Ahmad, R.; Khan, M.S.; Kant, R.; Shahid, S.; Gautam, L.; Hasan, G.M.; Hassan, M.I. Chitin and its derivatives: Structural properties and biomedical applications. *Int. J. Biol. Macromol.* **2020**, *164*, 526–539. [CrossRef]
184. Thakuri, P.S.; Liu, C.; Luker, G.D.; Tavana, H. Biomaterials-Based Approaches to Tumor Spheroid and Organoid Modeling. *Adv. Healtch. Mater.* **2018**, *7*, 1700980. [CrossRef]
185. Florczyk, S.J.; Liu, G.; Kievit, F.M.; Lewis, A.M.; Wu, J.D.; Zhang, M. 3D Porous Chitosan—Alginate Scaffolds: A New Matrix for Studying Prostate Cancer Cell—Lymphocyte Interactions In Vitro. *Adv. Healthc. Mater.* **2012**, *1*, 590–599. [CrossRef]
186. Phan-Lai, V.; Florczyk, S.J.; Kievit, F.M.; Wang, K.; Gad, E.; Disis, M.L.; Zhang, M. Three-Dimensional Scaffolds to Evaluate Tumor Associated Fibroblast-Mediated Suppression of Breast Tumor Specific T Cells. *Biomacromolecules* **2013**, *14*, 1330–1337. [CrossRef]
187. Phan-Lai, V.; Kievit, F.M.; Florczyk, S.J.; Wang, K.; Disis, M.L.; Zhang, M. CCL21 and IFNγ recruit and activate tumor specific T cells in 3D scaffold model of breast cancer. *Anti-Cancer Agents Med. Chem. (Former. Curr. Med. Chem. Agents)* **2014**, *14*, 204–210. [CrossRef]
188. Grigore, A.; Sarker, B.; Fabry, B.; Boccaccini, A.R.; Detsch, R. Behavior of encapsulated MG-63 cells in RGD and gelatine-modified alginate hydrogels. *Tissue Eng.-Part A* **2014**, *20*, 2140–2150. [CrossRef]
189. Sarker, B.; Singh, R.; Silva, R.; Roether, J.A.; Kaschta, J.; Detsch, R.; Schubert, D.W.; Cicha, I.; Boccaccini, A.R. Evaluation of fibroblasts adhesion and proliferation on alginate-gelatin crosslinked hydrogel. *PLoS ONE* **2014**, *9*, e0107952. [CrossRef]
190. Alemany-ribes, M.; Semino, C.E. Bioengineering 3D environments for cancer models ☆. *Adv. Drug Deliv. Rev.* **2014**, *79–80*, 40–49. [CrossRef]
191. Liaw, C.; Ji, S.; Guvendiren, M. Engineering 3D Hydrogels for Personalized In Vitro Human Tissue Models. *Adv. Healthc. Mater.* **2018**, *7*, 1701165. [CrossRef]

192. Alonso-Nocelo, M.; Abuín, C.; López-López, R.; De La Fuente, M. De Development and characterization of a three-dimensional co-culture model of tumor T cell in fi ltration. *Biofabrication* **2016**, *8*, 1–13. [CrossRef]
193. Kim, T.; Kim, G.; Kim, J.S.; Jin, S.; Yoon, S.; Bae, H.; Jeong, Y.H.; Young, H.J.; Kwak, J.-Y. Three-dimensional culture and interaction of cancer cells and dendritic cells in an electrospun nano-submicron hybrid fibrous scaffold. *Int. J. Nanomed.* **2016**, *11*, 823–835. [CrossRef]
194. Lutolf, M.P.; Lauer-Fields, J.L.; Schmoekel, H.G.; Metters, A.T.; Weber, F.E.; Fields, G.B.; Hubbell, J.A. Synthetic matrix metalloproteinase-sensitive hydrogels for the conduction of tissue regeneration: Engineering cell-invasion characteristics. *Proc. Natl. Acad. Sci. USA* **2003**, *100*, 5413–5418. [CrossRef]
195. Patterson, J.; Hubbell, J.A. Enhanced proteolytic degradation of molecularly engineered PEG hydrogels in response to MMP-1 and MMP-2. *Biomaterials* **2010**, *31*, 7836–7845. [CrossRef]
196. Rong, Y.; Zhang, Z.; He, C.L.; Chen, X.S. Matrix metalloproteinase-sensitive poly(ethylene glycol)/peptide hydrogels as an interactive platform conducive to cell proliferation during 3D cell culture. *Sci. China Technol. Sci.* **2021**, *64*, 1285–1294. [CrossRef]
197. Liu, S.; Cao, H.; Guo, R.; Li, H.; Lu, C.; Yang, G.; Nie, J.; Wang, F.; Dong, N.; Shi, J.; et al. Effects of the proportion of two different cross-linkers on the material and biological properties of enzymatically degradable PEG hydrogels. *Polym. Degrad. Stab.* **2020**, *172*, 109067. [CrossRef]
198. Aziz, A.H.; Bryant, S.J. A comparison of hMSC osteogenesis in PEG hydrogels as a function of MMP-sensitive crosslinker and crosslink density in chemically-defined medium. *Biotechnol. Bioeng.* **2020**, *116*, 1523–1536. [CrossRef]
199. Vedaraman, S.; Bernhagen, D.; Haraszti, T.; Licht, C.; Castro Nava, A.; Omidinia Anarkoli, A.; Timmerman, P.; De Laporte, L. Bicyclic RGD peptides enhance nerve growth in synthetic PEG-based Anisogels. *Biomater. Sci.* **2021**, *9*, 4329–4342. [CrossRef]
200. Infanger, D.W.; Lynch, M.E.; Fischbach, C. Engineered Culture Models for Studies of Tumor- Microenvironment Interactions. *Annu. Rev. Biomed. Eng.* **2013**, *15*, 29–53. [CrossRef]
201. Temples, M.N.; Adjei, I.M.; Nimocks, P.M.; Djeu, J.; Sharma, B. Engineered Three-Dimensional Tumor Models to Study Natural Killer Cell Suppression. *ACS Biomater. Sci. Eng.* **2020**, *6*, 4179–4199. [CrossRef]
202. Tsao, C.; Kievit, F.M.; Wang, K.; Erickson, A.E.; Ellenbogen, R.G. Chitosan-Based Thermoreversible Hydrogel as an in Vitro Tumor Microenvironment for Testing Breast Cancer Therapies. *Mol. Pharm.* **2014**, *11*, 2134–2142. [CrossRef] [PubMed]
203. Maharjan, S.; Cecen, B.; Zhang, Y.S. 3D Immunocompetent Organ-on-a-Chip Models. *Small Methods* **2020**, *4*, 1–21. [CrossRef] [PubMed]
204. Boussommier-Calleja, A.; Li, R.; Chen, M.B.; Wong, S.C.; Kamm, R.D. Microfluidics: A New Tool for Modeling Cancer–Immune Interactions. *Trends Cancer* **2016**, *2*, 6–19. [CrossRef] [PubMed]
205. Fedi, A.; Vitale, C.; Ponschin, G.; Ayehunie, S.; Fato, M.; Scaglione, S. In vitro models replicating the human intestinal epithelium for absorption and metabolism studies: A systematic review. *J. Control. Release* **2021**, *335*, 247–268. [CrossRef]
206. Adriani, G.; Pavesi, A.; Tan, A.T.; Bertoletti, A.; Thiery, J.P.; Kamm, R.D. Microfluidic models for adoptive cell-mediated cancer immunotherapies. *Drug Discov. Today* **2016**, *21*, 1472–1478. [CrossRef]
207. Pavesi, A.; Tan, A.T.; Koh, S.; Chia, A.; Colombo, M.; Antonecchia, E.; Miccolis, C.; Ceccarello, E.; Adriani, G.; Raimondi, M.T.; et al. A 3D microfluidic model for preclinical evaluation of TCR-engineered T cells against solid tumors. *JCI Insight* **2017**, *2*, e89762. [CrossRef]
208. Pavesi, A.; Tan, A.T.; Chen, M.B.; Adriani, G.; Bertoletti, A.; Kamm, R.D. Using microfluidics to investigate tumor cell extravasation and T-cell immunotherapies. In Proceedings of the 2015 37th Annual International Conference of the IEEE Engineering in Medicine and Biology Society, Milan, Italy, 25–29 August 2015; pp. 1853–1856. [CrossRef]
209. Ando, Y.; Siegler, E.L.; Ta, H.P.; Cinay, G.E.; Zhou, H.; Gorrell, K.A.; Au, H.; Jarvis, B.M.; Wang, P.; Shen, K. Evaluating CAR-T Cell Therapy in a Hypoxic 3D Tumor Model. *Adv. Healthc. Mater.* **2019**, *8*, 1900001. [CrossRef]
210. Ayuso, J.M.; Truttschel, R.; Gong, M.M.; Humayun, M.; Virumbrales-Munoz, M.; Vitek, R.; Felder, M.; Gillies, S.D.; Sondel, P.; Wisinski, K.B.; et al. Evaluating natural killer cell cytotoxicity against solid tumors using a microfluidic model. *Oncoimmunology* **2019**, *8*, 1–11. [CrossRef]
211. Park, D.; Son, K.; Hwang, Y.; Ko, J.; Lee, Y.; Doh, J.; Jeon, N.L. High-throughput microfluidic 3D cytotoxicity assay for cancer immunotherapy (CACI-ImpacT platform). *Front. Immunol.* **2019**, *10*, 1133. [CrossRef]
212. Lee, S.W.L.; Adriani, G.; Ceccarello, E.; Pavesi, A.; Tan, A.T.; Bertoletti, A.; Kamm, R.D.; Wong, S.C. Characterizing the role of monocytes in T cell cancer immunotherapy using a 3d microfluidic model. *Front. Immunol.* **2018**, *9*, 416. [CrossRef]
213. Bai, J.; Adriani, G.; Dang, T.M.; Tu, T.Y.; Penny, H.X.L.; Wong, S.C.; Kamm, R.D.R.D.; Thiery, J.P. Contact-dependent carcinoma aggregate dispersion by M2a macrophages via ICAM-1 and ß2 integrin interactions. *Oncotarget* **2015**, *6*, 25295–25307. [CrossRef] [PubMed]
214. Cui, X.; Morales, R.T.T.; Qian, W.; Wang, H.; Gagner, J.P.; Dolgalev, I.; Placantonakis, D.; Zagzag, D.; Cimmino, L.; Snuderl, M.; et al. Hacking macrophage-associated immunosuppression for regulating glioblastoma angiogenesis. *Biomaterials* **2018**, *161*, 164–178. [CrossRef]
215. Li, R.; Hebert, J.D.; Lee, T.A.; Xing, H.; Boussommier-Calleja, A.; Hynes, R.O.; Lauffenburger, D.A.; Kamm, R.D. Macrophage-secreted TNFα and TGFβ1 influence migration speed and persistence of cancer cells in 3D tissue culture via independent pathways. *Cancer Res.* **2017**, *77*, 279–290. [CrossRef] [PubMed]

216. Kim, H.; Chung, H.; Kim, J.; Choi, D.H.; Shin, Y.; Kang, Y.G.; Kim, B.M.; Seo, S.U.; Chung, S.; Seok, S.H. Macrophages-Triggered Sequential Remodeling of Endothelium-Interstitial Matrix to Form Pre-Metastatic Niche in Microfluidic Tumor Microenvironment. *Adv. Sci.* **2019**, *6*, 1900195. [CrossRef]
217. Mi, S.; Liu, Z.; Du, Z.; Yi, X.; Sun, W. Three-dimensional microfluidic tumor–macrophage system for breast cancer cell invasion. *Biotechnol. Bioeng.* **2019**, *116*, 1731–1741. [CrossRef]
218. Lee, S.W.L.; Seager, R.J.; Litvak, F.; Spill, F.; Sieow, J.L.; Leong, P.H.; Kumar, D.; Tan, A.S.M.; Wong, S.C.; Adriani, G.; et al. Integrated in silico and 3D in vitro model of macrophage migration in response to physical and chemical factors in the tumor microenvironment. *Integr. Biol.* **2020**, *12*, 90–108. [CrossRef]
219. Surendran, V.; Rutledge, D.; Colmon, R.; Chandrasekaran, A. A novel tumor-immune microenvironment TIME-on-chip mimics three dimensional neutorophil-tumor dynamics and neutrophil extracellular traps NETs-mediated collective tumor invasion. *Biofabrication* **2021**, *13*, 035029. [CrossRef]
220. Chen, M.B.; Hajal, C.; Benjamin, D.C.; Yu, C.; Azizgolshani, H.; Hynes, R.O.; Kamm, R.D. Inflamed neutrophils sequestered at entrapped tumor cells via chemotactic confinement promote tumor cell extravasation. *Proc. Natl. Acad. Sci. USA* **2018**, *115*, 7022–7027. [CrossRef]
221. Parlato, S.; De Ninno, A.; Molfetta, R.; Toschi, E.; Salerno, D.; Mencattini, A.; Romagnoli, G.; Fragale, A.; Roccazzello, L.; Buoncervello, M.; et al. 3D Microfluidic model for evaluating immunotherapy efficacy by tracking dendritic cell behaviour toward tumor cells. *Sci. Rep.* **2017**, *7*, 1093. [CrossRef]
222. Jenkins, R.W.; Aref, A.R.; Lizotte, P.H.; Ivanova, E.; Stinson, S.; Zhou, C.W.; Bowden, M.; Deng, J.; Liu, H.; Miao, D.; et al. Ex vivo profiling of PD-1 blockade using organotypic tumor spheroids. *Cancer Discov.* **2018**, *8*, 196–215. [CrossRef]
223. Nguyen, M.; De Ninno, A.; Mencattini, A.; Mermet-Meillon, F.; Fornabaio, G.; Evans, S.S.; Cossutta, M.; Khira, Y.; Han, W.; Sirven, P.; et al. Dissecting Effects of Anti-cancer Drugs and Cancer-Associated Fibroblasts by On-Chip Reconstitution of Immunocompetent Tumor Microenvironments. *Cell Rep.* **2018**, *25*, 3884–3893.e3. [CrossRef] [PubMed]
224. Marrella, A.; Fedi, A.; Varani, G.; Vaccari, I.; Fato, M.; Firpo, G.; Guida, P.; Aceto, N.; Scaglione, S. High blood flow shear stress values are associated with circulating tumor cells cluster disaggregation in a multi-channel microfluidic device. *PLoS ONE* **2021**, *16*, e0245536. [CrossRef] [PubMed]
225. Pulsoni, I.; Lubda, M.; Aiello, M.; Fedi, A.; Marzagalli, M.; von Hagen, J.; Scaglione, S. Comparison Between Franz Diffusion Cell and a novel Micro-physiological System for In Vitro Penetration Assay Using Different Skin Models. *SLAS Technol.* **2022**. [CrossRef] [PubMed]
226. Marrella, A.; Buratti, P.; Markus, J.; Firpo, G.; Pesenti, M.; Landry, T.; Ayehunie, S.; Scaglione, S.; Kandarova, H.; Aiello, M. In vitro demonstration of intestinal absorption mechanisms of different sugars using 3d organotypic tissues in a fluidic device. *ALTEX* **2020**, *37*, 255–264. [CrossRef]
227. Marrella, A. 3D fluid-dynamic ovarian cancer model resembling systemic drug administration for efficacy assay. *ALTEX* **2020**, *37*, 1–14. [CrossRef]
228. Haj-Shomaly, J.; Vorontsova, A.; Barenholz-Cohen, T.; Levi-Galibov, O.; Devarasetty, M.; Timaner, M.; Raviv, Z.; Cooper, T.J.; Soker, S.; Hasson, P.; et al. T Cells Promote Metastasis by Regulating Extracellular Matrix Remodeling following Chemotherapy. *Cancer Res.* **2022**, *82*, 278–291. [CrossRef]
229. Ferrari, E.; Palma, C.; Vesentini, S.; Occhetta, P.; Rasponi, M. Integrating Biosensors in Organs-on-Chip Devices: A Perspective on Current Strategies to Monitor Microphysiological Systems. *Biosensors* **2020**, *10*, 110. [CrossRef]
230. Kratz, S.R.A.; Höll, G.; Schuller, P.; Ertl, P.; Rothbauer, M. Latest trends in biosensing for microphysiological organs-on-a-chip and body-on-a-chip systems. *Biosensors* **2019**, *9*, 110. [CrossRef]

cancers

MDPI

Article

On the Evaluation of a Novel Hypoxic 3D Pancreatic Cancer Model as a Tool for Radiotherapy Treatment Screening

Gabrielle Wishart [1,2], Priyanka Gupta [1,3], Andrew Nisbet [4], Giuseppe Schettino [2,5] and Eirini Velliou [1,3,*]

1 Bioprocess and Biochemical Engineering Group (BioProChem), Department of Chemical and Process Engineering, University of Surrey, Guildford GU2 7XH, UK; g.wishart@surrey.ac.uk (G.W.); priyanka.g.gupta@ucl.ac.uk (P.G.)
2 Department of Physics, University of Surrey, Guildford GU2 7XH, UK; giuseppe.schettino@surrey.ac.uk
3 Centre for 3D Models of Health and Disease, Department of Targeted Intervention, Division of Surgery and Interventional Science, University College London (UCL), London W1W 7TY, UK
4 Department of Medical Physics and Biomedical Engineering, University College London (UCL), London WC1E 6BT, UK; andrew.nisbet@ucl.ac.uk
5 National Physical Laboratory, Teddington TW11 0LW, UK
* Correspondence: e.velliou@ucl.ac.uk

Simple Summary: Pancreatic cancer challenges global health with non-specific symptoms, devastatingly low survival rates, and high treatment resistance profiles. Tissue engineering is advancing to facilitate animal free tissue biomimicry, allowing the replication of tumour tissue specific hallmarks of pancreatic cancer that challenge modern treatments. Here, we report the development and characterisation of a low oxygen (hypoxic) 3D polyurethane scaffold system for long-term analysis of radiation responses. This finely tuned platform more accurately recapitulates bio-physical, biochemical, and structural-bio-mechanical in vivo tissue niches as well as tumour hypoxia. The latter is a treatment-limiting feature for radiotherapy, allowing the system to streamline the transition of clinical testing from bench to bedside.

Abstract: Tissue engineering is evolving to mimic intricate ecosystems of tumour microenvironments (TME) to more readily map realistic in vivo niches of cancerous tissues. Such advanced cancer tissue models enable more accurate preclinical assessment of treatment strategies. Pancreatic cancer is a dangerous disease with high treatment resistance that is directly associated with a highly complex TME. More specifically, the pancreatic cancer TME includes (i) complex structure and complex extracellular matrix (ECM) protein composition; (ii) diverse cell populations (e.g., stellate cells), cancer associated fibroblasts, endothelial cells, which interact with the cancer cells and promote resistance to treatment and metastasis; (iii) accumulation of high amounts of (ECM), which leads to the creation of a fibrotic/desmoplastic reaction around the tumour; and (iv) heterogeneous environmental gradients such as hypoxia, which result from vessel collapse and stiffness increase in the fibrotic/desmoplastic area of the TME. These unique hallmarks are not effectively recapitulated in traditional preclinical research despite radiotherapeutic resistance being largely connected to them. Herein, we investigate, for the first time, the impact of in vitro hypoxia (5% O_2) on the radiotherapy treatment response of pancreatic cancer cells (PANC-1) in a novel polymer (polyurethane) based highly macroporous scaffold that was surface modified with proteins (fibronectin) for ECM mimicry. More specifically, PANC-1 cells were seeded in fibronectin coated macroporous scaffolds and were cultured for four weeks in in vitro normoxia (21% O_2), followed by a two day exposure to either in vitro hypoxia (5% O_2) or maintenance in in vitro normoxia. Thereafter, in situ post-radiation monitoring (one day, three days, seven days post-irradiation) of the 3D cell cultures took place via quantification of (i) live/dead and apoptotic profiles and (ii) ECM (collagen-I) and HIF-1a secretion by the cancer cells. Our results showed increased post-radiation viability, reduced apoptosis, and increased collagen-I and HIF-1a secretion in in vitro hypoxia compared to normoxic cultures, revealing hypoxia-induced radioprotection. Overall, this study employed a low cost, animal free model enabling (i) the possibility of long-term in vitro hypoxic 3D cell culture for pancreatic cancer, and (ii) in vitro hypoxia associated PDAC radio-protection development. Our novel platform for

Citation: Wishart, G.; Gupta, P.; Nisbet, A.; Schettino, G.; Velliou, E. On the Evaluation of a Novel Hypoxic 3D Pancreatic Cancer Model as a Tool for Radiotherapy Treatment Screening. *Cancers* **2021**, *13*, 6080. https://doi.org/10.3390/cancers13236080

Academic Editor: Donatella Aldinucci

Received: 4 November 2021
Accepted: 30 November 2021
Published: 2 December 2021

radiation treatment screening can be used for long-term in vitro post-treatment observations as well as for fractionated radiotherapy treatment.

Keywords: pancreatic cancer; tissue engineering; tumour microenvironment (TME); treatment resistance; radiotherapy; radiation; radioprotection; hypoxia; polyurethane scaffolds; 3D cell culture; extracellular matrix (ECM); HIF-1a; PANC-1

1. Introduction

Pancreatic ductal adenocarcinoma (PDAC) is a cancer of the duct cells of the pancreas. Most commonly found in the head of the pancreas, this disease is the most common form of pancreatic cancer (96% of pancreatic cancers present as PDAC) as opposed to more rare pancreatic cancers (e.g., neoendocrine tumours) [1,2]. PDAC challenges global health with devastatingly low 5-year survival rates (10%) compared to other cancers (e.g., breast cancer (90%), prostate cancer (89%), and melanoma of the skin (93%)) [1,2]. Increasing incidence rates, non-specific symptoms, and late diagnosis of this disease elucidate high mortality rates [1–3]. At diagnosis, a small fraction of patients (20%) are eligible for curative surgery due to late detection and high metastatic occurrence [1]. Chemotherapeutics such as Gemcitabine, Capecitabine, and FOLFIRINOX are suggested for consideration for first line, adjuvant, or for metastatic PDAC treatment [3]. Radiotherapy for PDAC is suggested for consideration as an adjuvant therapy. However, this option has national variations in treatment recommendations due to limited data [2] and controversial European clinical trials [4,5]. A pooled analysis of 955 PDAC patients treated with adjuvant chemo-radiotherapy showed improved overall survival of patients compared to chemotherapy alone [6]. Thus, the role of radiotherapy for pancreatic cancer is largely debated and thought to be still evolving [2,4–8].

Pancreatic cancer is notoriously resistant to current chemotherapy and radiotherapy treatments due to an extremely complex tumour microenvironment [9–12]. The PDAC tumour microenvironment (TME) is a distinct in vivo milieu of cellular, biochemical, and biomechanical features that encompass distinct fingerprints that promote tumour cell survival, migration, and resistance to treatments [9,13–17]. These unique hallmarks include (i) complex structure and complex extracellular matrix (ECM) protein composition; (ii) diverse cell populations (e.g., stellate cells, cancer associated fibroblasts, endothelial cells, immune cells) that interact with the cancer cells and promote resistance to treatment and metastasis; (iii) accumulation of high amounts of ECM, which leads to the creation of a fibrotic/desmoplastic reaction around the tumour; and (iv) heterogeneous environmental gradients such as hypoxia, which result from vessel collapse and stiffness increase in the fibrotic/desmoplastic area of the TME. More specifically, chaotic cancer cell growth activates pancreatic stellate cell secretion of extracellular matrix (ECM) proteins, known as the desmoplastic reaction [9,14,15,18]. This large increase in ECM protein deposit (i.e., collagen, fibronectin, and laminin) and tumour stiffness, along with very high cancer cell growth, influences pro-survival characteristics and causes intra-tumoral blood vessel disruption and collapse, consequently impairing (chemotherapeutic) drug delivery (resulting in chemo-resistance) and causing heterogeneous expanses of low oxygen gradients (hypoxia) [9,13–18]. Tumour hypoxia influences instrumental changes to the TME micromilieu promoting cancer growth, metastasis, invasion, and resistance to radiotherapy, [17,19–22]. More specifically, over 50 years of research describes the reduced effectiveness of radiation (up to a factor of 2.5–3) in the absence of oxygen, as explained by the oxygenation fixation hypothesis (OFH) [17,22]. Attempts to target the hypoxic hallmark are emerging [23,24] with limited clinical progress. Unrealistic pre-clinical models that fail to recapitulate treatment-limiting hallmarks such as hypoxia are impeding treatment success [25]. Thus, the need to understand and recapitulate in vitro this complicated ecosystem of diverse cellular and non-cellular components that create the unique PDAC TME niche is a matter of clinical relevance for the optimisation of treatments such as radiotherapy for PDAC.

Preclinical treatment screening has traditionally utilised (a) 2D in vitro systems and (b) animal models. 2D cell culture systems are a fast cost-effective gold standard for therapy testing in vitro, however, this method does not incorporate realistic TME hallmarks such as microarchitecture, stiffness, spatial orientation, cell–cell and cell–extracellular matrix protein interactions, and environmental gradients of the TME that are associated with radiation response (i.e., hypoxia) [25–31]. In the literature, the exposure of 2D cell culture systems (tissue culture flasks or micro-plates) to hypoxia in gas controlled chambers has been utilised to identify hypoxia-associated genes [32], to evaluate hypoxic sensitisers [33], investigate damage repair pathways, and improve understanding of cell adaptions to hypoxia [34,35]. Typically 72 h of hypoxic exposure in 2D are described as chronic hypoxic conditions, a timeframe that is not in-line with heterogeneous oxygen expanses in vivo [34]. Animal models support more realistic microarchitecture, stiffness and cell–TME crosstalk compared to 2D cell culture systems [25,27,28,36]. These models are the most widely used for drug and radiotherapy pre-clinical testing, and can be utilised to study (i) the influence of hypoxia on metastatic disease progression [37]; (ii) hypoxia activated pro-drugs [38]; and (iii) radio-sensitisers [39]. However, animal models are expensive, complex, not always reproducible, challenging to use, and they also raise ethical concerns. Moreover, discrepancies in physiology, size and genetics limit the reproducibility of the studies and can result in translational errors in the clinic [25,27,28,36,40].

As a result, there is growing interest in the development and use of 3D cancer in vitro models for the replacement of 2D in vitro models and reduction in the use of animal models, the latter of which is an important objective in terms of the 3R framework (replace, reduction, and refinement of animals) [41,42]. 3D models for pre-clinical treatment testing are emerging to more realistically recapitulate TMEs, and more specifically TME hallmarks that are associated with treatment resistance. TME mimicry encourages the cells to employ behavioural and physiological characteristics that more similarly emulate realistic in vivo properties to allow for a more streamlined transition of clinical treatment testing from laboratory bench to patient bedside [25,27,28,43,44]. 3D PDAC models have emerged to mimic in vivo niches, these include (i) spheroids [45,46]; (ii) hydrogels [47]; and (iii) polymeric scaffolds [26–28,43]. With respect to radiotherapy in 3D PDAC in vitro models, spheroids and polymeric scaffolds as tools specifically for radiotherapy screening have started to emerge [25] and are required to test new modalities [48].

Spheroid models are simple cell clusters/aggregates in suspension, they have featured in research articles reporting radiation response studies and investigating radiosensitiser potential for pancreatic cancer. For example, Al-Ramadan et al. (2018) utilised spheroids to identify radiation dose (0–6 Gy) dependent sensitivity in the pancreatic neuroendocrine cell line BON-1 seven days post treatment via apoptosis induction [49]. Moreover, Al-Assar et al. (2014) reported that the co-culture of the pancreatic cancer cells (PANC-1) and stellate cells (PSC) enhanced radio-resistance (0–6 Gy) in sphere models (eight days) similar to their xenograft models (60 days) [50]. Furthermore, Hehlgans et al. (2009) identified Caveolin-1 and TAE226 as potential radiosensitisers (in radiotherapy doses of 0–6 Gy) for the pancreatic cancer cell line MiaPacCa2 in a spheroid system [51,52]. Moreover, PDAC spheroids are emerging as platforms to test new modalities such as proton therapy and boron neutron capture therapy [53,54]. Overall, spheroid models can be considered an advanced platform supporting more realistic 3D cellular interactions compared to traditional 2D cell cultures for radiation response studies, however, these models lack robust porosity, ECM controlled composition, and mechanical stability [25,27,28,36]. As a result, long-term radiation studies can be challenging.

Polymeric scaffolds have featured in very few research articles reporting radiation response studies for PDAC. These scaffolds are made of biocompatible polymers and have controlled stiffness and internal architecture (porosity or fibrous internal organisation). We have previously reported the long-term culture (35 days) of PDAC cells lines (PANC-1,

AsPC-1, BxPC-3) in highly macroporous polyurethane (PU) polymeric scaffolds. In our previous work, we surface modified our scaffolds with fibronectin. Fibronectin is one of the most abundant proteins present in the ECM of the pancreatic cancer tumour microenvironment [55]. Therefore, the scaffolds were surface modified with fibronectin to enhance ECM mimicry and improve cell–cell and cell–matrix interactions. Indeed fibronectin modified scaffolds have demonstrated dense spatial cellular masses, collagen-I production from the cancer cells, and environmental (hypoxic) gradients that followed an in vivo-like trends. In contrast, cells in non-coated scaffolds did not have a physiological behaviour [27]. Furthermore, we have performed chemotherapy (with 10 µm, 50 µm, 100 µm GEM), radiotherapy (with 0 Gy, 2 Gy, 6 Gy, 8 Gy), and chemoradiotherapy (with 10 µm GEM and radiotherapy of 6 Gy) screening on those fibronectin modified polyurethane (PU) scaffolds [28]. More specifically, we have reported dose dependent chemotherapy and radiotherapy viability drop and apoptosis induction after short-term (one day) and long-term (17 days) PANC-1 cell culture, with chemoradiotherapy being more efficient than single radiation doses. Similar trends were seen at both short- and long-term, with 17 days being the longest reported in vitro post-treatment timeframe [28]. This is the first long-term scaffold based platform employing radiation treatment screening for pancreatic cancer [28].

To the best of our knowledge, the combination of hypoxic exposure and radiation modalities have not been tested in 3D scaffold models for PDAC. Therefore, the aim of this work was to investigate, for the first time, the effect of hypoxia on the response of pancreatic cancer cells to radiotherapy, in our previously developed highly porous, fibronectin coated 3D polyurethane scaffold. Overall, we report (i) a novel long-term in vitro hypoxic 3D cell culture system for pancreatic cancer; and (ii) the development of in vitro hypoxia associated PDAC radio-protection. Our novel platform for radiation treatment screening can be used for long-term in vitro post-treatment observations as well as for fractionated radiotherapy treatment; the latter being standard practice in the clinic.

2. Materials and Methods

2.1. Scaffold Fabrication and Surface Modification

Polymeric scaffolds were fabricated via the thermally induced phase separation (TIPS) method as reported previously [27,28,56] (Appendix A Figure A1). More specifically, 3 g of polyurethane (PU) beads (Noveon, Oudergem, Belgium) was dissolved in 60 mL dioxane (5% w/v) (99.8% anhydrous pure, Sigma-Aldrich, Merck, Gillingham, UK) for 48 h before the solution was quenched at −80 °C for 3 h. The solvent was removed via freeze-drying in a polyethylene glycol (PEG) bath at −15 °C under 0.01 mbar vacuum pressure for 72 h. Scaffolds were snap frozen in liquid nitrogen following immediate cutting into $5 \times 5 \times 5$ mm^3 cubes. Thereafter, the scaffolds were sterilised via 70% ethanol submersion (3 h) and UV ray exposure (1 h). The average pore size of the scaffold was 100–150 µm, the porosity was 85–90%, and the elastic modulus 20 ± 2 kPa, with stiffness similar to ex vivo high stiffness diseased PDAC tissue, as previously reported [27,28,56–59]. Thereafter, the scaffolds were surface modified (adsorption) to enable coating with fibronectin (i.e., an ECM protein extensively present in the PDAC TME for ECM biomimicry). As mentioned in the introduction, we have previously reported physiological behaviour of PDAC cells in the presence of fibronectin (dense cell aggregates, collagen-I secretion by the cancer cells, and realistic environmental gradients) compared to sparser cell organisation and no collagen production in uncoated scaffolds [27]. Briefly, for surface modification with fibronectin, the scaffolds were centrifuged in phosphate buffered saline (PBS, Sigma-Aldrich, Merck, UK) for 10 min at 2500 rpm, then centrifuged in fibronectin solution (25 µg mL^{-1}) for 20 min at 2000 rpm, before finally being centrifuged in PBS for 10 min at 1500 rpm.

2.2. Cell Culture

The 3D cell culture (in the scaffolds) was accomplished as described previously [27]. More specifically, human pancreatic adenocarcinoma cells (PANC-1) (ATCC) were initially expanded in 2D flasks, in Dulbecco's Modified Eagle's Medium (DMEM) with high glucose

(Sigma-Aldrich, Merck, UK) supplemented with 10% foetal bovine serum (Fisher Scientific, Loughborough, UK), 1% penicillin/streptomycin (Fisher Scientific, UK), and 2 mM L-glutamine (Sigma-Aldrich, Merck UK) in 37 °C with 21% O_2 and 5% CO_2. PANC-1 cells were passaged when 80–90% confluency was reached, until the appropriate cell number for the 3D experiments was obtained. Thereafter, 0.5×10^6 PANC-1 cells were seeded per scaffold (re-suspended in 30 µL of cell culture media) and placed in 24-well plates. Thereafter, the scaffolds were placed in an incubator for 1 h to ensure adherence. Therein, 1.5 mL of cell culture media was added to each well, this was replaced every two days and each 24-well plate was replaced after one week to avoid cell egress from scaffolds and cell confluency on the bottom of each well. Incubation of all scaffolds took place in a humidified incubator at 37 °C with 21% O_2 and 5% CO_2 (in vitro normoxia) for 28 days (four weeks). Thereafter, half of the scaffolds were moved to in vitro hypoxic conditions at 37 °C with 5% O_2 and 5% CO_2 in a Ruskinn InvivO2 300 workstation (Baker Ruskinn, Ltd., Bridgend, UK) for a two day pre-treatment exposure to low oxygen. Post treatment analysis took place at one day, three days, and seven days post treatment in both (i) in vitro normoxic (21% O_2) and (ii) in vitro hypoxic (5% O_2) culture conditions.

2.3. Radiation Treatment

Radiation treatments were performed with a clinical 250 kVp X-ray irradiator, Xstrahl 300 (Xstrahl, Camberley, UK) at the Royal Surrey County Hospital NHS Foundation Trust (Calibrated against a previously published code of practice [60]). More specifically, at four weeks of culture of normoxic culture (21% O_2) and after two day pre-treatment exposure to either in vitro hypoxia (5% O_2) or normoxia maintenance, the scaffolds were irradiated with 6 Gy based on previous results and protocols for radiotherapy treatment in normoxia [28]. A square field applicator of 15 × 15 cm was placed 3 cm above the plate surfaced, which was placed on an epoxy resin water equivalent phantom (30 × 30 × 10 cm) in order to ensure a uniform radiation field with known radiation back scattering conditions. Control scaffolds were used to account for cellular stress during transportation from the University of Surrey to the Royal Surrey County Hospital. Thereafter, hypoxic scaffolds were placed back in 5% O_2, normoxic scaffolds were placed back at 21% O_2, and all scaffolds were analysed at one day, three days, and seven days post-treatment.

2.4. Live/Dead Analysis via Confocal Imaging

The distribution of live and dead cells within scaffolds for all conditions under study (treated and untreated for both normoxic and hypoxic scaffolds) was evaluated via the Live/Dead Viability/Cytotoxicity Kit (Molecular Probes, Thermo Scientific, Loughborough, UK). More specifically, scaffolds were snap frozen at time points of one day, three days, and seven days post-treatment in liquid nitrogen for 20 min and preserved thereafter at −80 °C as previously described [26–28,36]. Thereafter, the scaffolds were sectioned and stained with 2 µm of Calcien-AM (4 mM stock) and 4 µm of Ethidium Homodimer (2 mM stock) and incubated at 37 °C for 1 h. Thereafter, the samples were washed twice with PBS and visualised using a Nikon Ti-Eclipse inverted confocal microscope (Nikon Instruments, Amstelveen, The Netherlands).

2.5. Caspase 3/7 Analysis via Confocal Imaging

The distribution of apoptotic cells within scaffolds for all conditions under study was evaluated via Caspase 3/7 analysis. More specifically, scaffolds were snap frozen at time of points one day, three days, and seven days post-treatment in liquid nitrogen for 20 min and preserved thereafter at −80 °C. The scaffolds were sectioned and stained with Cell Event Caspase 3/7 green detection reagent (Fisher Scientific, UK) and DAPI (Fisher Scientific, UK) for 1 h at 37 °C. Thereafter, the samples were washed with PBS and visualised using a Nikon Ti-Eclipse inverted confocal microscope (Nikon Instruments, Amstelveen, The Netherlands).

2.6. *Immunofluorescence Assay*

Immunofluorescence staining took place to characterise collagen-I (Abcam, Cambridge, UK), and HIF-1a (Abcam, UK) spatial secretion in all conditions under study at various time points. More specifically, scaffolds were snap frozen at one day, three days, and seven days post-treatment in liquid nitrogen for 20 min and preserved thereafter at −80 °C prior to immunofluorescence staining. Scaffolds were sectioned and fixed for 4 h in 4% *w*/*v* paraformaldehyde (Sigma-Aldrich, Merck, UK). Thereafter, scaffold sections were permeabilised for 2 h with 0.1% Triton-X solution (Sigma-Aldrich, Merck, UK), followed by blocking with 10% donkey serum (Abcam, UK) for 3 h. Primary antibody staining was carried out overnight, followed by overnight secondary antibody staining including DAPI (Fisher Scientific, UK). Primary (Mouse HIF-1a Abcam, UK) and secondary (Alexa Flour 647 donkey anti-mouse IgG (H+L)) antibodies for HIF-1a were diluted in 1% *w*/*v* bovine serum albumin (Sigma-Aldrich, Merck, UK).

2.7. *Confocal Laser Scanning Microscopy*

Live/Dead, Caspase 3/7 analysis, and immunofluorescence (Sections 2.5–2.7) were imaged on a Nikon Ti-Eclipse inverted confocal microscope (Nikon Instruments, Amstelveen, The Netherlands) and processed with the NIS-Elements software, using 405, 488, 555, 604, 647 lasers for DAPI (blue), Calcein AM and Caspase 3/7 (green), collagen-I (yellow) and Ethidium Homodimer and HIF-1a (red) staining, respectively. Images were captured at a 10× objective and 10 μm Z-stack distance. Multiple scaffolds, scaffold sections, and scaffold areas were imaged for each condition under study to ensure reproducibility. The images presented here are representative images from each condition.

2.8. *Image Analysis*

ImageJ® software (Wayne Rasband, NIH, Bethesda, MD, USA) was utilised to quantify spatial characterisation of (i) Live/Dead; (ii) Caspase 3/7; (iii) collagen-I; (iv) HIF-1a; and (v) DAPI positive areas vs. negative areas. Multiple scaffolds (n = 3), scaffold sections (n = 3), and scaffold areas (n = 2) were analysed per condition to ensure reproducibility of results. The bars in each bar graph of the results represent averages of percentage areas of each fluorescence channel.

2.9. *Statistical Analysis and Data Analysis*

Graph Pad Prism® (GraphPad Software, San Diego, CA, USA) was utilised to determine statistical significance ($p < 0.05$). Analysis of variance (ANOVA), followed by Tukey's multiple comparison test were employed. Standard error of the mean was used to determine error bars in the bar graphs. Where data were normalised with respect to the control, the following equation was employed: % Caspase 3/7 area = ((treatment positive area)/(control positive area)) × 100.

3. Results

3.1. *Evaluation of the Effect of Radiation Treatment on PANC-1 Cells in 3D Scaffolds*

Following four weeks of culturing pancreatic cancer cells in 21% O_2 in our highly porous polymer fibronectin coated scaffolds (i.e., at a timeframe that we have previously shown) allowed for the formation of PDAC physiological features such as dense spatial cell aggregates, metabolic gradients, and secretion of collagen-I by the PDAC cells [24]. The 3D scaffolds were either maintained in 21% O_2 or were exposed to 5% O_2 for two days. Thereafter, radiation treatment (6 Gy) took place followed by post-treatment monitoring of (i) the spatial secretion of the hypoxic biomarker HIF-1a in the 3D scaffolds; (ii) the cell viability and cell apoptosis in the 3D scaffolds; and (iii) the secretion of collagen-I by the cancer cells in the 3D scaffolds.

3.2. Post-Radiation Monitoring of the Hypoxic Biomarker HIF-1a in the 3D Scaffolds

The biomarker hypoxia-inducible factor, HIF-1a was monitored for one, three, and seven days post-radiation in the 3D scaffolds, this hypoxic biomarker is hyper-expressed in PDAC in comparison to healthy tissue of the pancreas [61]. As one of the major hypoxia induced intracellular signalling pathways, the HIF-1a pathway is associated with pro-survival, metastasis, and resistance to therapy [17]. Figure 1 summarises confocal images of scaffold sections showing the HIF-1a spatial distribution along with a quantitative assessment of the percentage of HIF-1a positive areas in multiple images taken from multiple sections of multiple scaffolds. As can be seen in Figure 1, a significantly higher HIF-1a secretion in the 3D scaffolds was observed at 5% O_2 for both treated and untreated scaffolds compared to 21% O_2 for days 3 and 7 post-radiotherapy treatment, while no differences were observed between different oxygen conditions one day post-treatment.

3.3. Post-Radiation Monitoring of Cell Viability and Apoptosis in the 3D Scaffolds

The cell viability and the cell apoptosis in the 3D scaffolds were monitored for one, three, and seven days post-radiation with Live/Dead and Caspase 3/7 staining respectively.

Figures 2 and 3 summarise the confocal images of scaffold sections showing the spatial distribution of Live/Dead and apoptotic cell areas, respectively, along with a quantitative assessment of the Live/Dead and apoptotic areas in multiple images taken from multiple sections of multiple scaffolds. Our results show a systematic trend of hypoxia-associated radio-protection of PANC-1 cells in the 3D scaffolds. More specifically, a significantly lower cell viability in the 3D scaffolds was seen at 21% O_2 post-radiation treatment compared to 5% O_2, where no statistical difference was observed between the treated and untreated 3D scaffolds at one, three, and seven days post-treatment (Figure 2). Similar results/trends were observed for apoptosis. More specifically, significantly higher apoptosis responses of the 3D scaffolds were observed at 21% O_2 compared to 5% O_2 at one, three, and seven days post radiation treatment (Figure 3). It should be highlighted that the cell viability (Figure 2) and cell apoptosis (Figure 3) trends in the 3D scaffolds were in line with the increase in the HIF-1a hypoxic biomarker (Figure 1), which showed a hypoxia associated radioprotection in the 3D scaffolds.

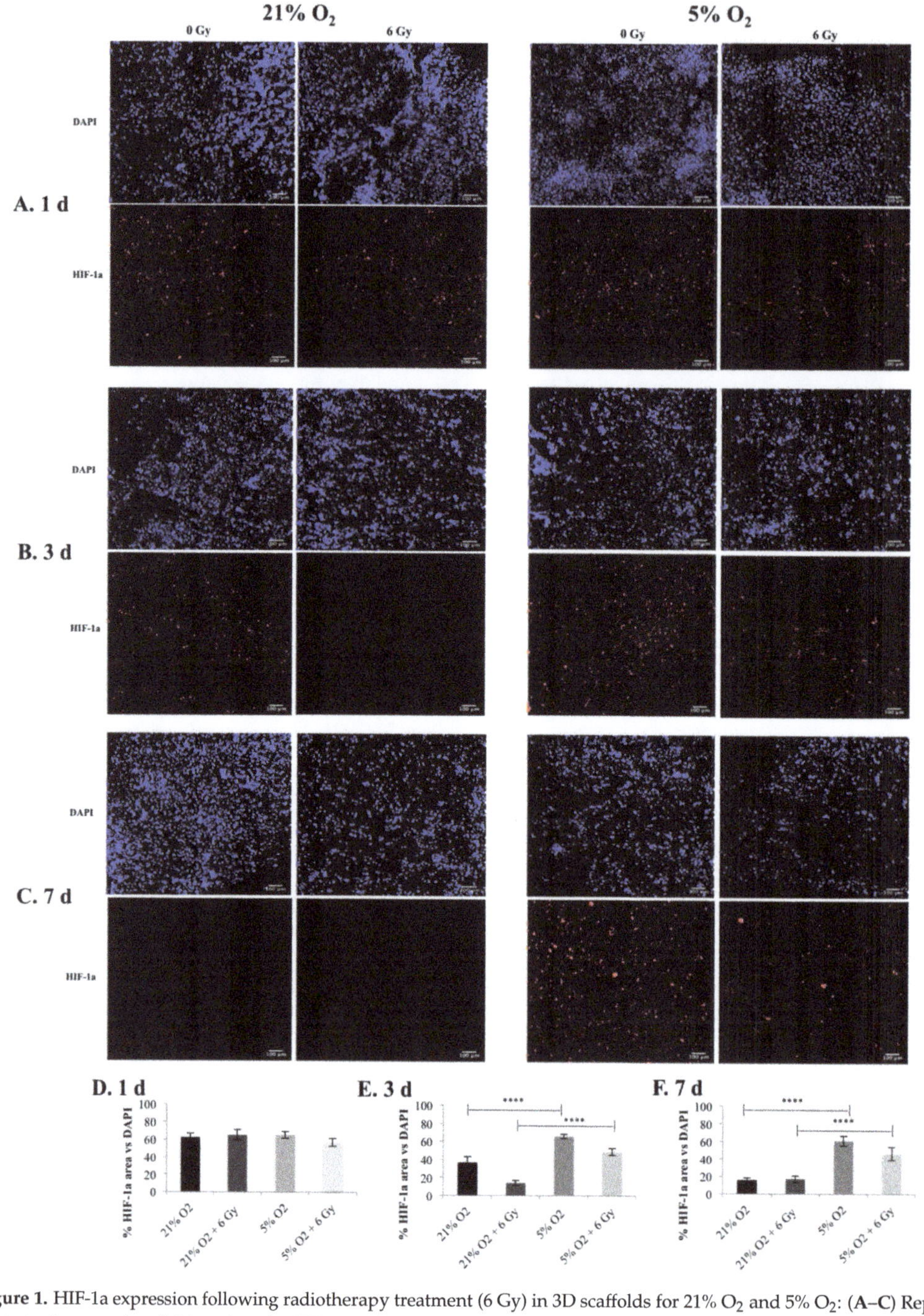

Figure 1. HIF-1a expression following radiotherapy treatment (6 Gy) in 3D scaffolds for 21% O_2 and 5% O_2: (**A–C**) Representative images of scaffold sections for HIF-1a (red) staining, counterstained with DAPI (blue) at 1 day, 3 days, and 7 days post radiation treatment. (**D–F**) Equivalent image analysis based quantification of the percentage of HIF-1a positive image areas over total cell (blue) image areas (DAPI) for 1 day (**D**), 3 days (**E**), and 7 days (**F**) post radiation treatment. Multiple scaffolds (3), scaffold sections (3), and images (2) were analysed, mean values are presented. (**** = $p < 0.0001$).

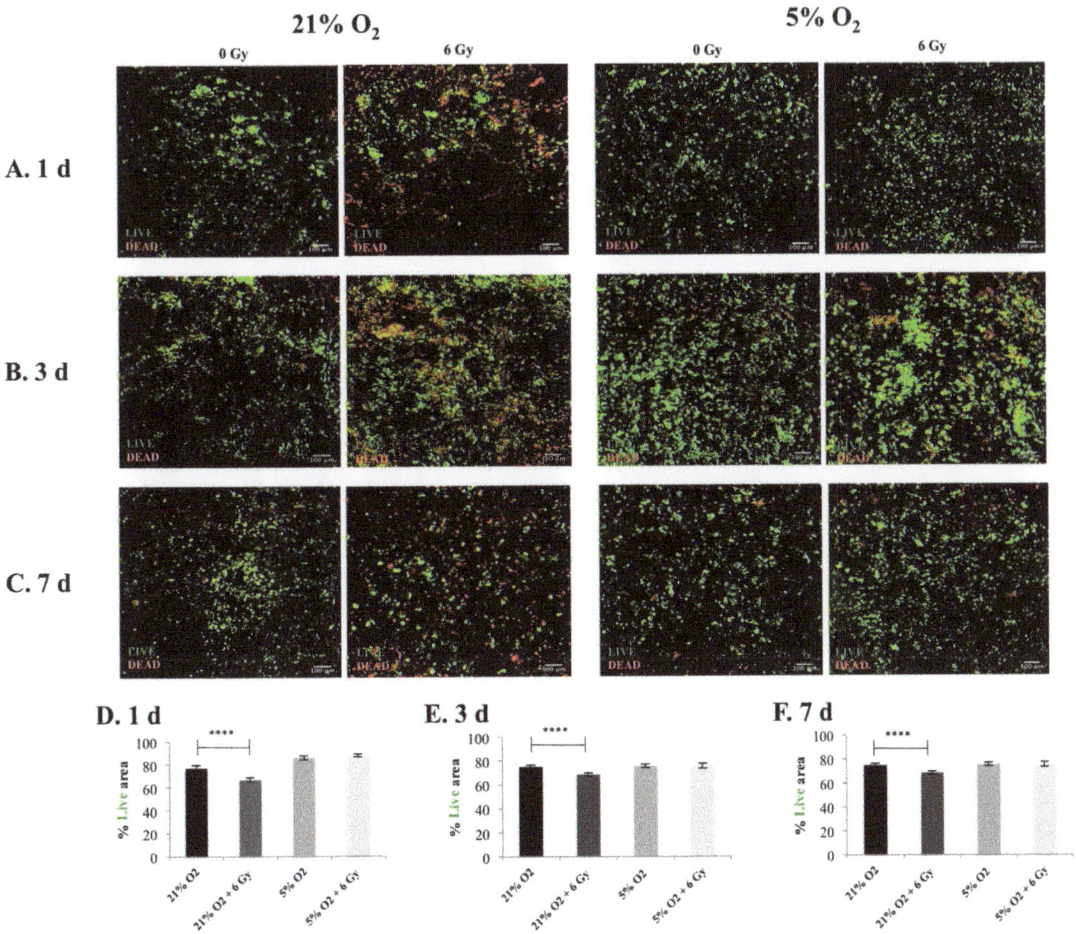

Figure 2. Cell viability (Live/Dead staining) following radiotherapy treatment (6 Gy) in 3D scaffolds for 21% O_2 and 5% O_2: (**A–C**) Representative images of scaffold sections for Live (green)/Dead (red) staining, 1 day, 3 days, and 7 days post-treatment and (**D–F**) equivalent image analysis based quantification of the percentage of Live (green) image areas for (**A–C**). Multiple scaffolds (3), scaffold sections (3), and images (2) were analysed, mean values are presented. (**** = $p < 0.0001$).

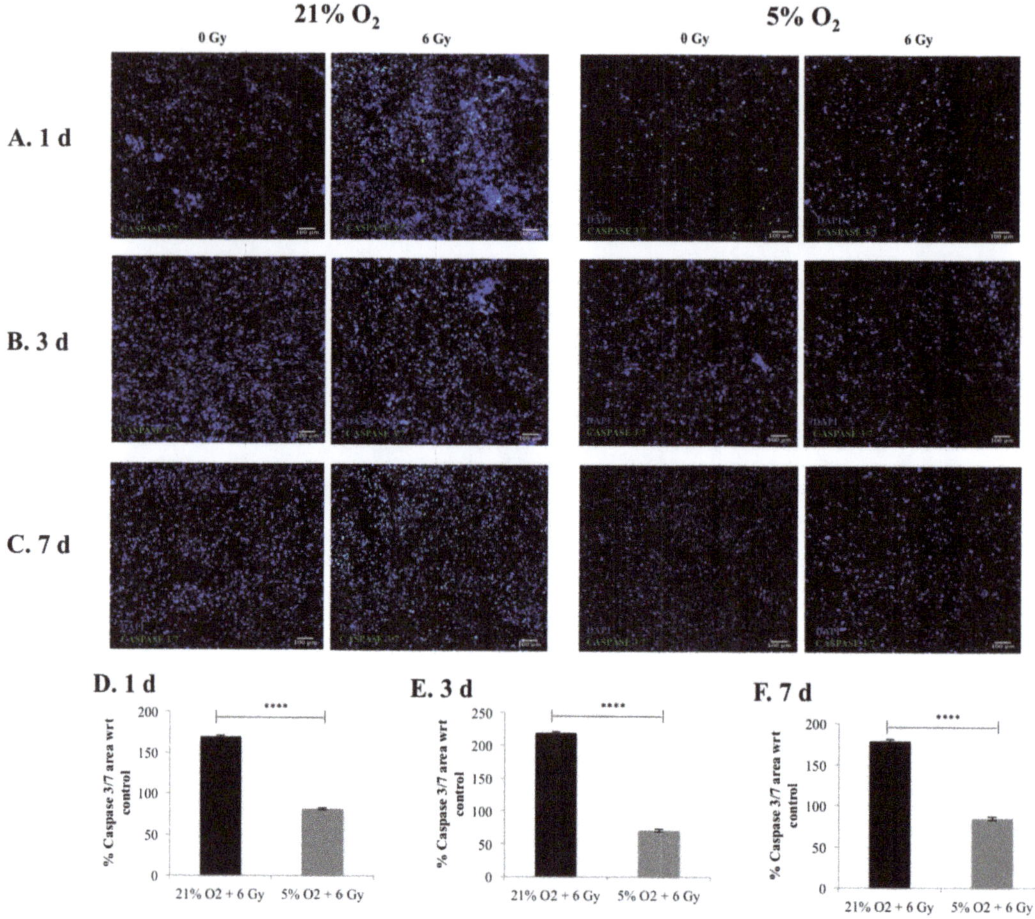

Figure 3. Apoptotic assay (Caspase 3/7) following radiotherapy treatment (6 Gy) in 3D scaffolds for 21% O_2 and 5% O_2: (**A–C**) Representative images of scaffold sections for Caspase 3/7 (green) and DAPI (blue) staining 1 day, 3 days, and 7 days post-treatment. (**D–F**) Equivalent image analysis based quantification of the percentage of Caspase 3/7 (green) image areas for (**A–C**) with respect to the control (untreated scaffolds). Multiple scaffolds (3), scaffold sections (3), and images (2) were analysed, mean values are presented. (**** = $p < 0.0001$).

3.4. Post-Radiotherapy Monitoring of Collagen-I Expression in the 3D Scaffolds

The expression of collagen-I (i.e., an ECM protein that is abundant in the PDAC tumour microenvironment and secreted by both stroma and cancer cells [55,62,63]) was monitored in the 3D scaffolds for all conditions under study. Figure 4 shows the immunofluorescence images of scaffold sections along with image quantification of the percentage of collagen-I positive areas for all conditions under study (Figure 4). Overall, we observed (i) higher collagen-I deposition in 5% O_2 culture compared to 21% O_2 culture for both the non-treated controls and the radiation treated samples at day 3 and 7, and (ii) the radiation treatment led to a substantially lower collagen-I secretion in 21% O_2 compared to the respective control (untreated sample in 21% O_2), while there was less collagen-I disruption in 5% O_2 compared to the respective control at day 3 and at day 7 post-radiation treatment (Figure 4).

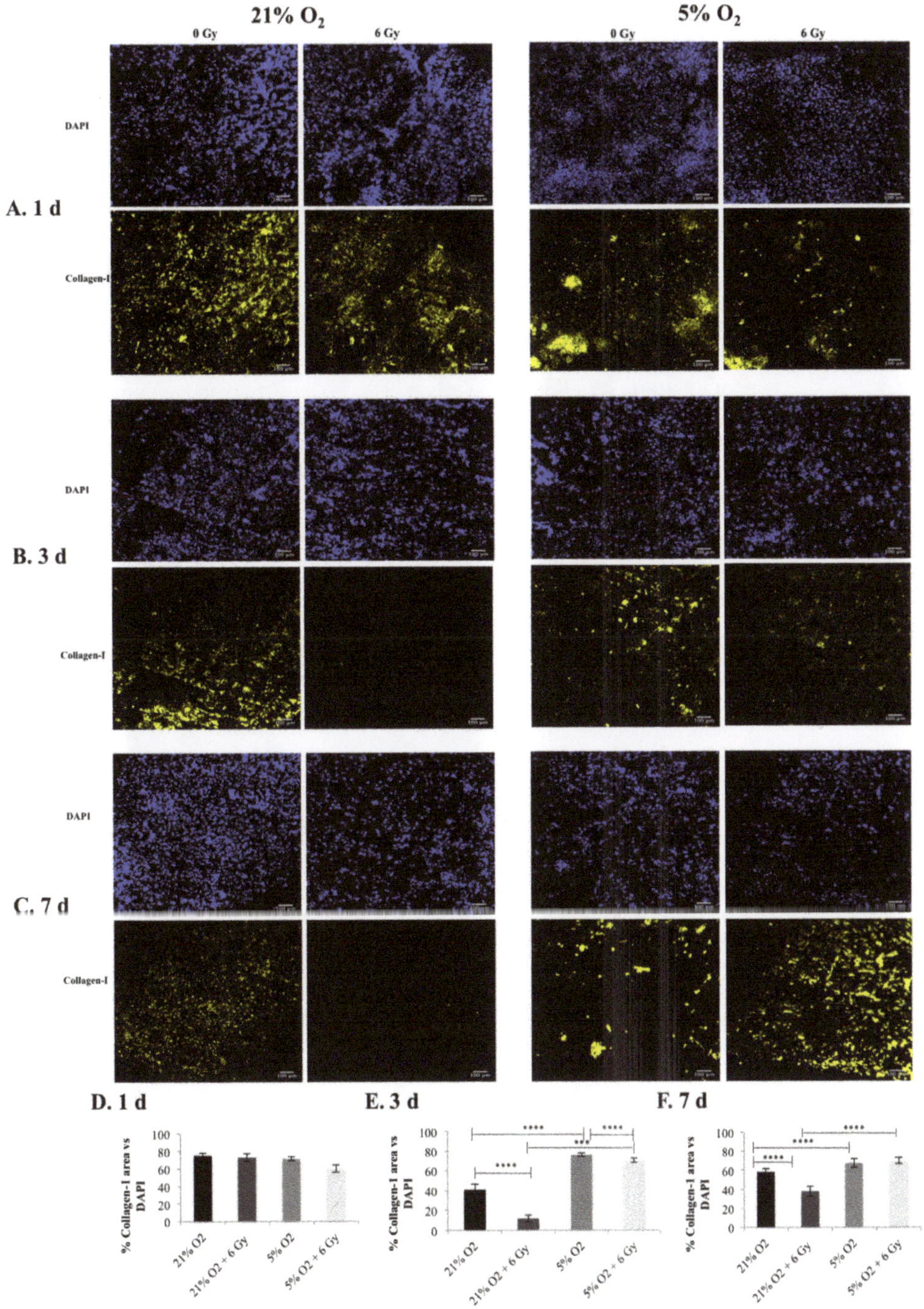

Figure 4. Collagen-I expression in 3D scaffolds, following radiotherapy treatment (6 Gy) for 21% O_2 and 5% O_2: (**A–C**) Representative images of scaffold sections for Collagen-I staining (yellow) and DAPI (blue) staining. (**D–F**) Equivalent image analysis based quantification of the percentage of collagen-I (yellow) areas for (**A–C**). Multiple scaffolds (3), scaffold sections (3), and images (2) were analysed, mean values were used. (**** = $p < 0.0001$). (*** = $p < 0.001$).

4. Discussion

In this work, we investigated, for the first time, the impact of in vitro hypoxia (5% O_2) on the radiation treatment response of pancreatic cancer cells (PANC-1) in our recently developed polymer (polyurethane) based highly macro-porous 3D scaffold, which is surface modified with proteins (fibronectin) for better ECM mimicry [27]. The scaffold supports the long-term (37 days) culture of PDAC cells with cancer cell proliferation distribution in the scaffold similar to the one reported for mouse models for this time-frame, without requiring cell re-suspension, which would affect the established cell–cell, cell–matrix interactions and metabolic gradients [27,64].

In the current work, PANC-1 cells were seeded in polymeric scaffolds and were cultured for four weeks in in vitro normoxia (21% O_2) followed by two days exposure to either in vitro hypoxia (5% O_2) or maintenance in in vitro normoxia. Thereafter, radiation treatment followed by in situ post-radiation monitoring (Appendix A Figure A1) (one day, three days, seven days post-treatment) of the 3D cell cultures took place via quantification of (i) HIF-1a secretion by the cancer cells; (ii) live/dead and apoptotic profiles; and (iii) ECM (collagen-I) secretion by the cancer cells. For radiation treatment, a dose of 6 Gy was selected based on our previously published radiotherapy screening experiments in in vitro normoxia (i.e., 6 Gy was the dose we have shown to cause a reduction in cell viability but not total death in 21% O_2 in our 3D scaffolds) [25]. In terms of the selection of the oxygen profile to create an in vitro hypoxic environment, 5% O_2 was selected, in line with the literature as most papers reported hypoxia studies at oxygen ranges from 0.1% to 10% [47]. We refer to 5% O_2 as in vitro hypoxia compared to 21% O_2, which is in vitro normoxia and the oxygen level for which the PANC-1 cells are established and authenticated.

We report a reduction in PANC-1 sensitivity to ionising radiation, associated with in vitro hypoxia (i.e., increased cell viability (live cell profiles) (Figure 2)) and decreased cell apoptosis (Caspase 3/7 profiles) (Figure 3) trends in the 3D scaffolds, in-line with the increase in the HIF-1a hypoxic biomarker (Figure 1). This indicates HIF-1a (hypoxia induced) associated radioprotection. Moreover, in line with increases in HIF-1a, we report (i) higher collagen-I deposition in hypoxia and (ii) less collagen-I disruption post-irradiation treatment in hypoxic cultures at three days and seven days post-treatment (Figure 4). To the best of our knowledge, this research is the first study to demonstrate the feasibility of using a complex biomaterial-based scaffold for radiation response studies of PDAC, specifically addressing the hypoxia factor. Furthermore, most in vitro treatment screening studies for PDAC have a maximum timeframe of one-week culture and generally have a hypoxic exposure time of up to 72 h [28,65], while in this study, the total culture frame was five weeks (i.e., four weeks prior to treatment followed by a week of post-treatment monitoring (up to nine days exposure to hypoxia)).

As previously mentioned, there are a limited number of studies investigating the treatment of radiation in 3D PDAC models. For example, Hehlgans et al. (2009) utilised a spheroid 3D model for the pancreatic cancer cell line MiaPacCa2 treated with 0–6 Gy of irradiation to identify Caveolin-1 and TAE226 as potential radiosensitisers using a two day post treatment timeframe. Similar to our work, this model facilitates a post treatment analysis, however, the 3D model is spatially and structurally different to our scaffold (i.e., it is a spheroid), and the timeframe is different to our model, where we report a more extended post-treatment timeframe of one day, three days, and seven days post treatment [51,52]. Similar to our study, Al-Assar et al. (2014) reported radio-resistance in their in vitro sphere model of pancreatic cancer, however, the radio-resistance reported was associated with the co-culture of pancreatic stellate cells with pancreatic cancer cells and the study did not evaluate the role of hypoxia. More specifically, sphere models of the pancreatic cancer cell line PANC-1 and pancreatic stellate cell line PSC were cultured for a total timeframe of eight days and treated with 0–6 Gy to identify co-culture enhancement of radio-resistance [50]. Furthermore, similar to our work, Al-Ramadan et al. (2018) employed a 7-day post treatment Caspase assay to pancreatic cancer spheroids (BON-1) (12 day total culture) to identify radiation dose (0–6 Gy) (treatment at five days after seeding) [49].

More recently, Görte et al. (2020) investigated 3D PDAC spheroids and found that proton (low-LET of 3.7 keV/µm) irradiation stimulated higher efficacy to tumouroid formation and greater phosphoproteome alterations compared to conventional photon (200-kVp X-ray) irradiation in MiaPacCa-2, Capan-1, Panc-1, Patu8902, and COLO357 based spheroids [53]. Moreover, these 3D PDAC models were treated with 2, 4 or 6 Gy of photons or protons 24 h after seeding, thereafter 7–13 days (cell dependent) of incubation was given before post radiation analysis [53]. This total culture time differed to the 37-day total scaffold culture time frame described here. Similar to Görte et al. (2020), Yu et al. (2021) investigated new modalities for PDAC, describing PANC-1 and BxPC-3 spheroids co-cultured with fibroblasts for the investigation of boron neutron capture therapy (1.2 MW for 28 min) [54]. This research reported lower survival rates and higher apoptosis rates in 3D spheroids compared to 2D cultures after thermal neutron irradiation treatment [54].

Gupta et al. (2019) were the first to investigate polymeric scaffolds as a treatment-screening platform for chemo-radiotherapy [28]. The studies in 3D models for PDAC radiation treatment described did not incorporate the treatment limiting hypoxic hallmark of PDAC, in fact, studies reporting the combination of hypoxia and radiation for 3D cancer models in general are very limited in the literature. For example, similar to our observations, Indovina et al. (2006) reported an over-expression of HIF-1a associated with inhibition of radiation (2 Gy, 5 Gy) effects for a shorter time period (48 h) for human osteosarcoma spheroids [66]. Similarly to this, Simon et al. (2016) report a lung cancer cell based Cells-in-Gels-in-Paper 3D model (in which polyvinyl chloride multilayer sheets simulate a poorly vascularised tumour), identifying a HIF-1a over-expression and reduced sensitivity to ionising radiation when the distance between cells and oxygenated medium was increased, suggesting low oxygen association with radioprotection [67]. Despite the very limited literature available on investigating the role of hypoxia in 3D PDAC and other (in vitro) cancer models, hypoxia related radio-protection has been extensively reported in clinic [68] as well as in both animal models [69–71] and 2D in vitro systems [69,70,72].

As previously mentioned, we also reported a correlation between increases in HIF-1a levels in our 3D scaffolds and (i) higher collagen-I deposition in hypoxic cultures and (ii) less collagen-I disruption post-radiation treatment in hypoxic cultures. Studies of tumour hypoxia have revealed not only the impact of hypoxia of pancreatic cancer progression and invasion, but the stimulation of fibrosis and angiogenesis including collagen-I expression [55,62,73–76]. As an abundant and important protein of ECM, the expression of collagen-I is also reported to be linked to pancreatic cancer cell survival and progression [62,73,74]. Moreover, enhanced radio resistance in pancreatic cancer cells has been identified in cells grown in the presence of ECM proteins (fibronectin) in 2D [75].

Generally, HIF-1a overexpression and collagen-I deposition correlation similar to that reported here, have mainly been described in 2D cell cultures, mouse models, and tissue samples in other cancer cell types. For example, HIF-1a activity is reported to promote ECM remodelling by inducing collagen-I expression in hypoxic fibroblast cultures [77]. Moreover, the study of liver fibrosis models revealed that HIF-1a deficient mouse models have shown reduced collagen-I and a-SMA levels (liver fibrosis) [78], suggesting hypoxia influence of collagen-I (ECM) secretion. Furthermore, collagen-I gene overexpression in line with lactate dehydrogenase (which aids glycolysis in hypoxic conditions) has been identified in radio-resistant cervical cancer tissue samples [79]. In recent years, there is a growing understanding of the importance of the ECM as a key player in treatment resistance and success, particularly in the complex pancreatic cancer TME (dense desmoplasia) [9,10,55,80,81]. Generally, the ability to map collagen-I deposition along with the response to in vitro hypoxia and radiation treatment is therefore an important aspect for the accuracy of a PDAC 3D model. Moreover, the biomimicry of cancerous tissues to develop models that accommodate the hypoxic and collagen rich environment to emulate physiological ecosystems is relevant for cancers that are renowned to be radio-resistant.

Overall, in this work, we describe a hypoxic PDAC 3D polymeric scaffold model for radiation treatment screening. We report increased levels of HIF-1a and collagen-I deposition in line with reduced PANC-1 sensitivity to ionising radiation in in vitro hypoxia. These cellular responses are characteristic of in vivo tumours. To the best of our knowledge, our work is the first to report long-term hypoxic PDAC culture and radiation treatment in biochemically and structurally complex 3D scaffolds. Generally, 3D hypoxic tissue research has been employed to understand the hypoxic behaviour of cells including cell migration and epithelial to mesenchymal transition [82–84]. However, there is a lack of 3D models investing the combination of hypoxia and radiotherapy treatment, despite the important role that hypoxia plays in radiotherapy treatment resistance. Therefore, developing structurally and biochemically complex models could accelerate our understanding of the link between radiotherapy treatment and spatial tissue characteristics, accelerating better therapies from bench to practice.

5. Conclusions

Overall, this work performed in vitro hypoxic radiation treatment screening on our recently published scaffold based PDAC model [27,28]. PANC-1 in vitro normoxia (21% O_2) scaffolds were cultured for four weeks and then exposed to in vitro hypoxia (5% O_2) or maintained in normoxic conditions (2 days), followed by radiation treatment (6 Gy). Thereafter, in situ post-radiotherapy monitoring (one day, three days, and seven days post-treatment) of the 3D cell cultures via quantification of (i) HIF-1a secretion by the cancer cells; (ii) Live/Dead and apoptotic profiles, and (iii) ECM (collagen-I) secretion by the cancer cells took place. Our analysis revealed increased levels of HIF-1a (Figure 1) in line with trends in increased cell viability (live cell profiles) (Figure 2) and decreases in cell apoptosis (Caspase 3/7 profiles) (Figure 3), which indicates HIF-1a (hypoxia) associated radioprotection. Moreover, we also report (in line with increases in HIF-1a) higher collagen-I deposition in in vitro hypoxic cultures and also less collagen-I disruption in in vitro hypoxic radiation treatment.

To the best of our knowledge, this is the first study to report an in vitro hypoxic PDAC long-term structurally and biochemically complex polymer based scaffold culture for radiation response studies and the first to correlate HIF-1a with increased collagen deposition post-radiation in a 3D PDAC model under hypoxia. Our system holds potential as an animal free alternative for predictive radiation research, providing more advanced spatial features compared to simple in vitro 2D models. Future work will focus on the investigation of the role of in vitro hypoxia on the PDAC cellular response to advanced radiotherapy modalities such as proton therapy and image guided radiotherapy using MR-Linacs in our complex multicellular scaffolds [36]. Furthermore, future research should also be performed in 3D systems with exposure to lower partial pressures of oxygen for even more physiologically relevant in vitro hypoxia.

Author Contributions: Conceptualization, E.V., G.W., P.G., G.S. and A.N.; Data collection, G.W.; Writing original draft preparation, G.W.; Writing and editing, G.W., E.V., P.G., G.S. and A.N.; Supervision, E.V. and G.S.; Funding acquisition, G.S., E.V. and A.N. All authors have read and agreed to the published version of the manuscript.

Funding: G.W. is funded by the Doctoral College of the University of Surrey and the National Physical Laboratory. P.G and E.V have received funding from 3DbioNet. E.V is grateful to the Royal Academy of Engineering for an Industrial Fellowship and to the Medical Research Council UK for a New Investigator Research Grant (MR/V028553/1), which also financially supports P.G.

Institutional Review Board Statement: Not applicable.

Informed Consent Statement: Not applicable.

Data Availability Statement: The datasets generated for this study are available on reasonable request to the corresponding author.

Acknowledgments: The authors thank the Royal Surrey County Hospital for radiation technical support.

Conflicts of Interest: The authors declare no conflict of interest.

Appendix A

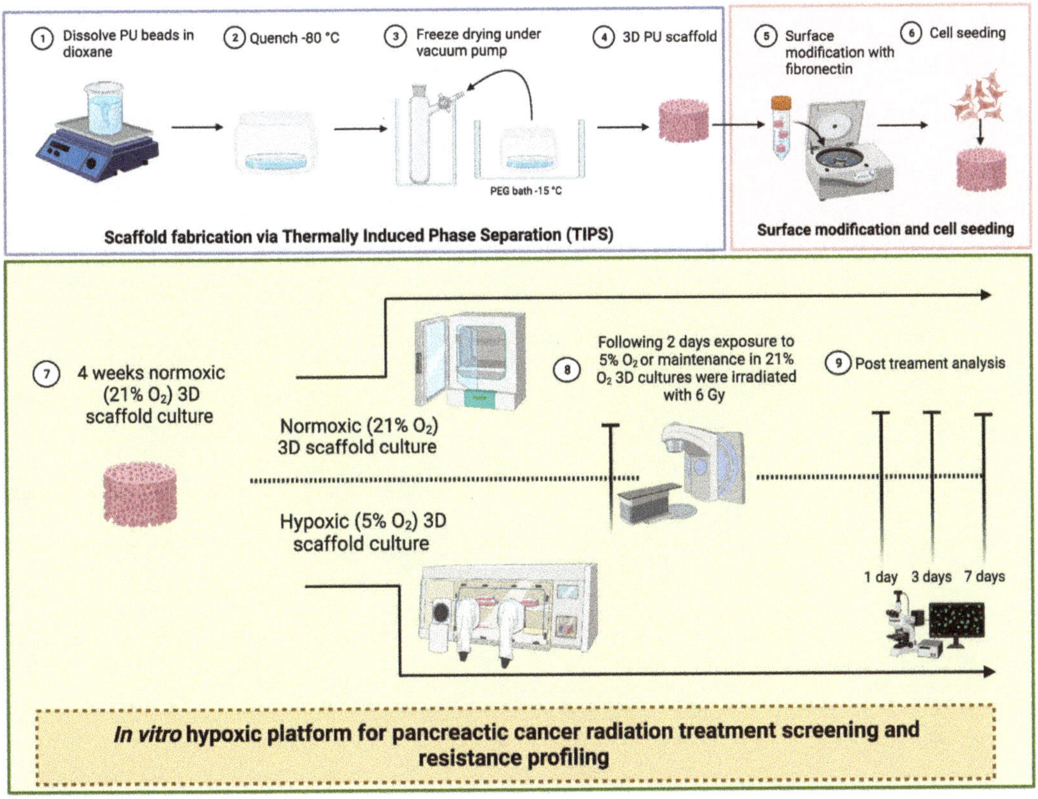

Figure A1. Experimental design. Including scaffold fabrication, surface modification, and cell seeding. Created with BioRender.com.

References

1. American Cancer Society. *American Cancer Society Cancer Facts and Figures*; American Cancer Society: Atlanta, GA, USA, 2021; pp. 1–4.
2. O'Reilly, D.; Fou, L.; Hasler, E.; Hawkins, J.; O'Connell, S.; Pelone, F.; Callaway, M.; Campbell, F.; Capel, M.; Charnley, R.; et al. Diagnosis and management of pancreatic cancer in adults: A summary of guidelines from the UK National Institute for Health and Care Excellence. *Pancreatology* **2018**, *18*, 962–970. [CrossRef]
3. Cascinu, S.; Falconi, M.; Valentini, V.; Jelic, S. Pancreatic cancer: ESMO Clinical Practice Guidelines for diagnosis, treatment and follow-up. *Ann. Oncol.* **2010**, *21*, v55–v58. [CrossRef]
4. Neoptolemos, J.P.; Stocken, D.D.; Friess, H.; Bassi, C.; Dunn, J.A.; Hickey, H.; Beger, H.; Fernandez-Cruz, L.; Dervenis, C.; Lacaine, F.; et al. A Randomized Trial of Chemoradiotherapy and Chemotherapy after Resection of Pancreatic Cancer. *N. Engl. J. Med.* **2004**, *350*, 1200–1210. [CrossRef]
5. Hammel, P.; Huguet, F.; van Laethem, J.-L.; Goldstein, D.; Glimelius, B.; Artru, P.; Borbath, I.; Bouché, O.; Shannon, J.; André, T.; et al. Effect of Chemoradiotherapy vs Chemotherapy on Survival in Patients with Locally Advanced Pancreatic Cancer Controlled after 4 Months of Gemcitabine with or without Erlotinib. *JAMA* **2016**, *315*, 1844. [CrossRef] [PubMed]
6. Morganti, A.G.; Falconi, M.; van Stiphout, R.G.P.M.; Mattiucci, G.-C.; Alfieri, S.; Calvo, F.A.; Dubois, J.-B.; Fastner, G.; Herman, J.M.; Maidment, B.W.; et al. Multi-institutional Pooled Analysis on Adjuvant Chemoradiation in Pancreatic Cancer. *Int. J. Radiat. Oncol.* **2014**, *90*, 911–917. [CrossRef]
7. Neoptolemos, J.P.; Palmer, D.H.; Ghaneh, P.; Psarelli, E.E.; Valle, J.W.; Halloran, C.M.; Faluyi, O.; O'Reilly, D.A.; Cunningham, D.; Wadsley, J.; et al. Comparison of adjuvant gemcitabine and capecitabine with gemcitabine monotherapy in patients with resected pancreatic cancer (ESPAC-4): A multicentre, open-label, randomised, phase 3 trial. *Lancet* **2017**, *389*, 1011–1024. [CrossRef]

8. Wishart, G.; Gupta, P.; Nisbet, A.; Velliou, E.; Schettino, G. Novel Anticancer and Treatment Sensitizing Compounds against Pancreatic Cancer. *Cancers* **2021**, *13*, 2940. [CrossRef] [PubMed]
9. Pickup, M.W.; Mouw, J.K.; Weaver, V.M. The extracellular matrix modulates the hallmarks of cancer. *EMBO Rep.* **2014**, *15*, 1243–1253. [CrossRef] [PubMed]
10. Yoshimura, M.; Itasaka, S.; Harada, H.; Hiraoka, M. Microenvironment and Radiation Therapy. *Biomed Res. Int.* **2013**, *2013*, 685308. [CrossRef]
11. Menon, H.; Ramapriyan, R.; Cushman, T.R.; Verma, V.; Kim, H.H.; Schoenhals, J.E.; Atalar, C.; Selek, U.; Chun, S.G.; Chang, J.Y.; et al. Role of radiation therapy in modulation of the tumor stroma and microenvironment. *Front. Immunol.* **2019**, *10*, 193. [CrossRef] [PubMed]
12. Mantoni, T.S.; Lunardi, S.; Al-Assar, O.; Masamune, A.; Brunner, T.B. Pancreatic Stellate Cells Radioprotect Pancreatic Cancer Cells through 1-Integrin Signaling. *Cancer Res.* **2011**, *71*, 3453–3458. [CrossRef] [PubMed]
13. Dauer, P.; Nomura, A.; Saluja, A.; Banerjee, S. Pancreatic Cancer: Neighborhood Matters. *Pancreatology* **2018**, *17*, 7–12. [CrossRef] [PubMed]
14. Melstrom, L.G.; Salazar, M.D.; Diamond, D.J. The pancreatic cancer microenvironment: A true double agent. *J. Surg. Oncol.* **2017**, *116*, 7–15. [CrossRef] [PubMed]
15. Xie, D.; Xie, K. Pancreatic cancer stromal biology and therapy. *Genes Dis.* **2015**, *2*, 133–143. [CrossRef]
16. Kamisawa, T.; Wood, L.D.; Itoi, T.; Takaori, K. Pancreatic cancer. *Lancet* **2016**, *388*, 73–85. [CrossRef]
17. Muz, B.; de la Puente, P.; Azab, F.; Azab, A.K. The role of hypoxia in cancer progression, angiogenesis, metastasis, and resistance to therapy. *Hypoxia* **2015**, *3*, 83–92. [CrossRef] [PubMed]
18. Li, D.; Xie, K.; Robert, W.; Abbruzzese, J.L. Pancreatic Cancer. *Lancet* **2004**, *363*, 10049–10057. [CrossRef]
19. Harris, A.L. Hypoxia—A key regulatory factor in tumour growth. *Nat. Rev. Cancer* **2002**, *2*, 38–47. [CrossRef]
20. Pouysségur, J.; Dayan, F.; Mazure, N.M. Hypoxia signalling in cancer and approaches to enforce tumour regression. *Nature* **2006**, *441*, 437–443. [CrossRef] [PubMed]
21. Gray, L.H.; Conger, A.D.; Ebert, M.; Hornsey, S.; Scott, O.C.A. The Concentration of Oxygen Dissolved in Tissues at the Time of Irradiation as a Factor in Radiotherapy. *Br. J. Radiol.* **1953**, *26*, 638–648. [CrossRef] [PubMed]
22. Grimes, D.R.; Partridge, M. A mechanistic investigation of the oxygen fixation hypothesis and oxygen enhancement ratio. *Biomed. Phys. Eng. Express* **2015**, *1*, 045209. [CrossRef]
23. Spiegelberg, L.; Houben, R.; Niemans, R.; de Ruysscher, D.; Yaromina, A.; Theys, J.; Guise, C.P.; Smaill, J.B.; Patterson, A.V.; Lambin, P.; et al. Hypoxia-activated prodrugs and (lack of) clinical progress: The need for hypoxia-based biomarker patient selection in phase III clinical trials. *Clin. Transl. Radiat. Oncol.* **2019**, *15*, 62–69. [CrossRef]
24. Zhou, M.; Xie, Y.; Xu, S.; Xin, J.; Wang, J.; Han, T.; Ting, R.; Zhang, J.; An, F. Hypoxia-activated nanomedicines for effective cancer therapy. *Eur. J. Med. Chem.* **2020**, *195*, 112274. [CrossRef] [PubMed]
25. Wishart, G.; Gupta, P.; Schettino, G.; Nisbet, A.; Velliou, E. 3d tissue models as tools for radiotherapy screening for pancreatic cancer. *Br. J. Radiol.* **2021**, *94*, 20201397. [CrossRef] [PubMed]
26. Totti, S.; Vernardis, S.I.; Meira, L.; Pérez-Mancera, P.A.; Costello, E.; Greenhalf, W.; Palmer, D.; Neoptolemos, J.; Mantalaris, A.; Velliou, E.G. Designing a bio-inspired biomimetic in vitro system for the optimization of ex vivo studies of pancreatic cancer. *Drug Discov. Today* **2017**, *22*, 690–701. [CrossRef] [PubMed]
27. Totti, S.; Allenby, M.C.; Dos Santos, S.B.; Mantalaris, A.; Velliou, E.G. A 3D bioinspired highly porous polymeric scaffolding system for in vitro simulation of pancreatic ductal adenocarcinoma. *RSC Adv.* **2018**, *8*, 20928–20940. [CrossRef]
28. Gupta, P.; Totti, S.; Pérez-Mancera, P.A.; Dyke, E.; Nisbet, A.; Schettino, G.; Webb, R.; Velliou, E.G. Chemoradiotherapy screening in a novel biomimetic polymer based pancreatic cancer model. *RSC Adv.* **2019**, *9*, 41649–41663. [CrossRef]
29. Little, J.B. Principal Cellular and Tissue Effects of Radiation. In *Holland-Frei Cancer Medicine*, 6th ed.; Kufe, D.W., Pollock, R.E., Weichselbaum, R.R., Bast, R.C., Jr., Gansler, T.S., Holland, J.F., Frei, E., III, Eds.; BC Decker: Hamilton, ON, Canada, 2003; p. 1305.
30. Malkinson, F.D. Some Principles of Radiobiology: A Selective Review. *J. Investig. Dermatol.* **1981**, *77*, 32–38. [CrossRef]
31. Whitmore, G.F.; Till, J.E. Quatitation of Cellular Radiobiological Responses 1.2. *Annu. Rev. Nucl. Sci.* **1964**, *14*, 347–374. [CrossRef] [PubMed]
32. Bobarykina, A.Y.; Minchenko, D.O.; Opentanova, I.L.; Moenner, M.; Caro, J.; Esumi, H.; Minchenko, O.H. Hypoxic regulation of PFKFB-3 and PFKFB-4 gene expression in gastric and pancreatic cancer cell lines and expression of PFKFB genes in gastric cancers. *Acta Biochim. Pol.* **2006**, *53*, 789–799. [CrossRef]
33. Shibamoto, Y.; Kubota, T.; Kishii, K.; Tsujitani, M. Radiosensitivity of human pancreatic cancer cells in vitro and in vivo, and the effect of a new hypoxic cell sensitizer, doranidazole. *Radiother. Oncol.* **2000**, *56*, 265–270. [CrossRef]
34. Chan, N.; Koritzinsky, M.; Zhao, H.; Bindra, R.; Glazer, P.M.; Powell, S.; Belmaaza, A.; Wouters, B.; Bristow, R.G. Chronic Hypoxia Decreases Synthesis of Homologous Recombination Proteins to Offset Chemoresistance and Radioresistance. *Cancer Res.* **2008**, *68*, 605–614. [CrossRef] [PubMed]
35. Riffle, S.; Hegde, R.S. Modeling tumor cell adaptations to hypoxia in multicellular tumor spheroids. *J. Exp. Clin. Cancer Res.* **2017**, *36*, 102. [CrossRef]
36. Gupta, P.; Pérez-Mancera, P.A.; Kocher, H.; Nisbet, A.; Schettino, G.; Velliou, E.G. A Novel Scaffold-Based Hybrid Multicellular Model for Pancreatic Ductal Adenocarcinoma—Toward a Better Mimicry of the in vivo Tumor Microenvironment. *Front. Bioeng. Biotechnol.* **2020**, *8*, 290. [CrossRef]

37. Büchler, P.; Reber, H.A.; Lavey, R.S.; Tomlinson, J.; Büchler, M.W.; Friess, H.; Hines, O.J. Tumor hypoxia correlates with metastatic tumor growth of pancreatic cancer in an orthotopic murine modell. *J. Surg. Res.* **2004**, *120*, 295–303. [CrossRef] [PubMed]
38. Hoang, N.T.H.; Kadonosono, T.; Kuchimaru, T.; Kizaka-Kondoh, S. Hypoxia-inducible factor-targeting prodrug TOP3 combined with gemcitabine or TS-1 improves pancreatic cancer survival in an orthotopic model. *Cancer Sci.* **2016**, *107*, 1151–1158. [CrossRef] [PubMed]
39. Doyen, J.; Parks, S.K.; Marcié, S.; Pouysségur, J.; Chiche, J. Knock-down of hypoxia-induced carbonic anhydrases IX and XII radiosensitizes tumor cells by increasing intracellular acidosis. *Front. Oncol.* **2013**, *2*, 199. [CrossRef] [PubMed]
40. Bailey, K.L.; Carlson, M.A. Porcine Models of Pancreatic Cancer. *Front. Oncol.* **2019**, *9*, 144. [CrossRef]
41. Kimlin, L.C.; Casagrande, G.; Virador, V.M. In vitro three-dimensional (3D) models in cancer research: An update. *Mol. Carcinog.* **2013**, *52*, 167–182. [CrossRef]
42. Pape, J.; Emberton, M.; Cheema, U. 3D Cancer Models: The Need for a Complex Stroma, Compartmentalization and Stiffness. *Front. Bioeng. Biotechnol.* **2021**, *9*, 660502. [CrossRef] [PubMed]
43. Ricci, C.; Danti, S. 3D Models of Pancreatic Ductal Adenocarcinoma via Tissue Engineering. In *Pancreatic Cancer: Methods in Molecular Biology*; Humana Press: New York, NY, USA, 2019; Volume 1882, pp. 81–95. [CrossRef]
44. Matta-Domjan, B.; King, A.; Totti, S.; Matta, C.; Dover, G.; Martinez, P.; Zakhidov, A.; La Ragione, R.; Macedo, H.; Jurewicz, I.; et al. Biophysical interactions between pancreatic cancer cells and pristine carbon nanotube substrates: Potential application for pancreatic cancer tissue engineering. *J. Biomed. Mater. Res. Part B Appl. Biomater.* **2018**, *106*, 1637–1644. [CrossRef]
45. Longati, P.; Jia, X.; Eimer, J.; Wagman, A.; Witt, M.-R.; Rehnmark, S.; Verbeke, C.; Toftgård, R.; Löhr, M.; Heuchel, R.L. 3D pancreatic carcinoma spheroids induce a matrix-rich, chemoresistant phenotype offering a better model for drug testing. *BMC Cancer* **2013**, *13*, 95. [CrossRef] [PubMed]
46. Lazzari, G.; Nicolas, V.; Matsusaki, M.; Akashi, M.; Couvreur, P.; Mura, S. Multicellular spheroid based on a triple co-culture: A novel 3D model to mimic pancreatic tumor complexity. *Acta Biomater.* **2018**, *78*, 296–307. [CrossRef] [PubMed]
47. Chiellini, F.; Puppi, D.; Piras, A.M.; Morelli, A.; Bartoli, C.; Migone, C. Modelling of pancreatic ductal adenocarcinoma in vitro with three-dimensional microstructured hydrogels. *RSC Adv.* **2016**, *6*, 54226–54235. [CrossRef]
48. Mohajer, J.K.; Nisbet, A.; Velliou, E.; Ajaz, M.; Schettino, G. Biological effects of static magnetic field exposure in the context of MR-guided radiotherapy. *Br. J. Radiol.* **2018**, *92*, 20180484. [CrossRef] [PubMed]
49. Al-Ramadan, A.; Mortensen, A.; Carlsson, J.; Nestor, M. Analysis of radiation effects in two irradiated tumor spheroid models. *Oncol. Lett.* **2017**, *15*, 3008–3016. [CrossRef]
50. Al-Assar, O.; Demiciorglu, F.; Lunardi, S.; Gaspar-Carvalho, M.M.; McKenna, W.G.; Muschel, R.M.; Brunner, T.B. Contextual regulation of pancreatic cancer stem cell phenotype and radioresistance by pancreatic stellate cells. *Radiother. Oncol.* **2014**, *111*, 243–251. [CrossRef] [PubMed]
51. Hehlgans, S.; Eke, I.; Storch, K.; Haase, M.; Baretton, G.B.; Cordes, N. Caveolin-1 mediated radioresistance of 3D grown pancreatic cancer cells. *Radiother. Oncol.* **2009**, *92*, 362–370. [CrossRef] [PubMed]
52. Hehlgans, S.; Lange, I.; Eke, I.; Cordes, N. 3D cell cultures of human head and neck squamous cell carcinoma cells are radiosensitized by the focal adhesion kinase inhibitor TAE226. *Radiother. Oncol.* **2009**, *92*, 371–378. [CrossRef]
53. Görte, J.; Beyreuther, E.; Danen, E.H.J.; Cordes, N. Comparative Proton and Photon Irradiation Combined with Pharmacological Inhibitors in 3D Pancreatic Cancer Cultures. *Cancers* **2020**, *12*, 3216. [CrossRef] [PubMed]
54. Yu, L.-S.; Jhunjhunwala, M.; Hong, S.-Y.; Yu, L.-Y.; Lin, W.-R.; Chen, C.-S. Tissue Architecture Influences the Biological Effectiveness of Boron Neutron Capture Therapy in In Vitro/In Silico Three-Dimensional Self-Assembly Cell Models of Pancreatic Cancers. *Cancers* **2021**, *13*, 4058. [CrossRef] [PubMed]
55. Perez, V.M.; Kearney, J.F.; Yeh, J.J. The PDAC Extracellular Matrix: A Review of the ECM Protein Composition, Tumor Cell Interaction, and Therapeutic Strategies. *Front. Oncol.* **2021**, *11*, 751311. [CrossRef] [PubMed]
56. Velliou, E.G.; Dos Santos, S.B.; Papathanasiou, M.M.; Fuentes-Gari, M.; Misener, R.; Panoskaltsis, N.; Pistikopoulos, E.N.; Mantalaris, A. Towards unravelling the kinetics of an acute myeloid leukaemia model system under oxidative and starvation stress: A comparison between two- and three-dimensional cultures. *Bioprocess Biosyst. Eng.* **2015**, *38*, 1589–1600. [CrossRef] [PubMed]
57. Pozzi, R.; Parzanese, I.; Baccarin, A.; Giunta, M.; Conti, C.B.; Cantù, P.; Casazza, G.; Tenca, A.; Rosa, R.; Gridavilla, D.; et al. Point shear-wave elastography in chronic pancreatitis: A promising tool for staging disease severity. *Pancreatology* **2017**, *17*, 905–910. [CrossRef]
58. Nabavizadeh, A.; Payen, T.; Iuga, A.C.; Sagalovskiy, I.R.; Desrouilleres, D.; Saharkhiz, N.; Palermo, C.F.; Sastra, S.A.; Oberstein, P.E.; Rosario, V.; et al. Noninvasive Young's modulus visualization of fibrosis progression and delineation of pancreatic ductal adenocarcinoma (PDAC) tumors using Harmonic Motion Elastography (HME) in vivo. *Theranostics* **2020**, *10*, 4614–4626. [CrossRef]
59. Chantarojanasiri, T.; Kongkam, P. Endoscopic ultrasound elastography for solid pancreatic lesions. *World J. Gastrointest. Endosc.* **2017**, *9*, 506–513. [CrossRef]
60. Aukett, R.J.; Burns, J.E.; Greener, A.G.; Harrison, R.M.; Moretti, C.; Nahum, A.E.; Rosser, K.E. Addendum to the IPEMB code of practice for the determination of absorbed dose for X-rays below 300 kV generating potential (0.035 mm Al–4 mm Cu HVL). *Phys. Med. Biol.* **2005**, *50*, 2739–2748. [CrossRef] [PubMed]
61. Welsh, S.; Williams, R.; Kirkpatrick, L.; Paine-Murrieta, G.; Powis, G. Antitumor activity and pharmacodynamic properties of PX-478, an inhibitor of hypoxia-inducible factor-1α. *Mol. Cancer Ther.* **2004**, *3*, 233–244.

62. Armstrong, T.; Packham, G.; Murphy, L.B.; Bateman, A.C.; Conti, J.A.; Fine, D.R.; Johnson, C.D.; Benyon, R.C.; Iredale, J.P. Type I Collagen Promotes the Malignant Phenotype of Pancreatic Ductal Adenocarcinoma. *Clin. Cancer Res.* **2004**, *10*, 7427–7437. [CrossRef]
63. Xu, S.; Xu, H.; Wang, W.; Li, S.; Li, H.; Li, T.; Zhang, W.; Yu, X.; Liu, L. The role of collagen in cancer: From bench to bedside. *J. Transl. Med.* **2019**, *17*, 309. [CrossRef]
64. Erstad, D.J.; Sojoodi, M.; Taylor, M.S.; Ghoshal, S.; Razavi, A.A.; Graham-O'Regan, K.A.; Bardeesy, N.; Ferrone, C.R.; Lanuti, M.; Caravan, P.; et al. Orthotopic and heterotopic murine models of pancreatic cancer and their different responses to FOLFIRINOX chemotherapy. *Dis. Model. Mech.* **2018**, *11*, dmm034793. [CrossRef] [PubMed]
65. Liu, Y.; Mohri, Z.; Alsheikh, W.; Cheema, U. The Role of Biomimetic Hypoxia on Cancer Cell Behaviour in 3D Models: A Systematic Review. *Cancers* **2021**, *13*, 1334. [CrossRef] [PubMed]
66. Indovina, P.; Ferrante, A.; Rainaldi, G.; Santini, M.T. Hypoxia and ionizing radiation: Changes in adhesive properties and cell adhesion molecule expression in MG-63 three-dimensional tumor spheroids. *Cell Commun. Adhes.* **2006**, *13*, 185–198. [CrossRef] [PubMed]
67. Simon, K.A.; Mosadegh, B.; Minn, K.T.; Lockett, M.R.; Mohammady, M.R.; Boucher, D.M.; Hall, A.B.; Hillier, S.M.; Udagawa, T.; Eustace, B.K.; et al. Metabolic response of lung cancer cells to radiation in a paper-based 3D cell culture system. *Biomaterials* **2016**, *95*, 47–59. [CrossRef]
68. Vaupel, P.; Mayer, A. Hypoxia in cancer: Significance and impact on clinical outcome. *Cancer Metastasis Rev.* **2007**, *26*, 225–239. [CrossRef] [PubMed]
69. Schwartz, D.L.; Bankson, J.A.; Lemos, R.; Lai, S.Y.; Thittai, A.K.; He, Y.; Hostetter, G.; Demeure, M.J.; Von Hoff, D.D.; Powis, G. Radiosensitization and stromal imaging response correlates for the HIF-1 inhibitor PX-478 given with or without chemotherapy in pancreatic cancer. *Mol. Cancer Ther.* **2010**, *9*, 2057–2067. [CrossRef]
70. Schwartz, D.L.; Powis, G.; Thitai-Kumar, A.; He, Y.; Bankson, J.; Williams, R.; Lemos, R.; Oh, J.; Volgin, A.; Soghomonyan, S.; et al. The selective hypoxia inducible factor-1 inhibitor PX-478 provides in vivo radiosensitization through tumor stromal effects. *Mol. Cancer Ther.* **2009**, *8*, 947–958. [CrossRef] [PubMed]
71. Hajj, C.; Russell, J.; Hart, C.P.; Goodman, K.A.; Lowery, M.A.; Haimovitz-Friedman, A.; Deasy, J.O.; Humm, J.L. A Combination of Radiation and the Hypoxia-Activated Prodrug Evofosfamide (TH-302) is Efficacious against a Human Orthotopic Pancreatic Tumor Model. *Transl. Oncol.* **2017**, *10*, 760–765. [CrossRef] [PubMed]
72. Liu, J.; Zhang, J.; Wang, X.; Li, Y.; Chen, Y.; Li, K.; Zhang, J.; Yao, L.; Guo, G. HIF-1 and NDRG2 contribute to hypoxia-induced radioresistance of cervical cancer Hela cells. *Exp. Cell Res.* **2010**, *316*, 1985–1993. [CrossRef] [PubMed]
73. Olivares, O.; Mayers, J.R.; Gouirand, V.; Torrence, M.E.; Gicquel, T.; Borge, L.; Lac, S.; Roques, J.; Lavaut, M.-N.; Berthezène, P.; et al. Collagen-derived proline promotes pancreatic ductal adenocarcinoma cell survival under nutrient limited conditions. *Nat. Commun.* **2017**, *8*, 16031. [CrossRef]
74. Aguilera, K.Y.; Rivera, L.B.; Hur, H.; Carbon, J.G.; Toombs, J.E.; Goldstein, C.D.; Dellinger, M.T.; Castrillon, D.H.; Brekken, R.A. Collagen Signaling Enhances Tumor Progression after Anti-VEGF Therapy in a Murine Model of Pancreatic Ductal Adenocarcinoma. *Cancer Res.* **2014**, *74*, 1032–1044. [CrossRef] [PubMed]
75. Cordes, N.; Meineke, V. Extracellular Matrix-Dependent Improvement of Cell Survival in Human Tumor and Normal Cells In Vitro. *Strahlenther. Onkol.* **2003**, *179*, 337–344. [CrossRef]
76. Masamune, A.; Kikuta, K.; Watanabe, T.; Satoh, K.; Hirota, M.; Shimosegawa, T. Hypoxia stimulates pancreatic stellate cells to induce fibrosis and angiogenesis in pancreatic cancer. *Am. J. Physiol. Liver Physiol.* **2008**, *295*, G709–G717. [CrossRef]
77. Gilkes, D.M.; Bajpai, S.; Chaturvedi, P.; Wirtz, D.; Semenza, G.L. Hypoxia-inducible factor 1 (HIF-1) promotes extracellular matrix remodeling under hypoxic conditions by inducing P4HA1, P4HA2, and PLOD2 expression in fibroblasts. *J. Biol. Chem.* **2013**, *288*, 10819–10829. [CrossRef] [PubMed]
78. Moon, J.-O.; Welch, T.P.; Gonzalez, F.J.; Copple, B.L. Reduced liver fibrosis in hypoxia-inducible factor-1α-deficient mice. *Am. J. Physiol. Liver Physiol.* **2009**, *296*, G582–G592. [CrossRef]
79. Kitahara, O.; Katagiri, T.; Tsunoda, T.; Harima, Y.; Nakamura, Y. Classification of Sensitivity or Resistance of Cervical Cancers to Ionizing Radiation According to Expression Profiles of 62 Genes Selected by cDNA Microarray Analysis. *Neoplasia* **2002**, *4*, 295–303. [CrossRef] [PubMed]
80. Winkler, J.; Abisoye-Ogunniyan, A.; Metcalf, K.J.; Werb, Z. Concepts of extracellular matrix remodelling in tumour progression and metastasis. *Nat. Commun.* **2020**, *11*, 5120. [CrossRef]
81. Carter, E.P.; Roozitalab, R.; Gibson, S.V.; Grose, R.P. Tumour microenvironment 3D-modelling: Simplicity to complexity and back again. *Trends Cancer* **2021**, *7*, 1033–1046. [CrossRef] [PubMed]
82. Puls, T.J.; Tan, X.; Whittington, C.F.; Voytik-Harbin, S.L. 3D collagen fibrillar microstructure guides pancreatic cancer cell phenotype and serves as a critical design parameter for phenotypic models of EMT. *PLoS ONE* **2017**, *12*, e0188870. [CrossRef] [PubMed]
83. Liverani, C.; De Vita, A.; Minardi, S.; Kang, Y.; Mercatali, L.; Amadori, D.; Bongiovanni, A.; La Manna, F.; Ibrahim, T.; Tasciotti, E. A biomimetic 3D model of hypoxia-driven cancer progression. *Sci. Rep.* **2019**, *9*, 12263. [CrossRef]
84. Wang, Y.; Mirza, S.; Wu, S.; Zeng, J.; Shi, W.; Band, H.; Band, V.; Duan, B. 3D hydrogel breast cancer models for studying the effects of hypoxia on epithelial to mesenchymal transition. *Oncotarget* **2018**, *9*, 32191–32203. [CrossRef] [PubMed]

cancers

MDPI

Commentary

Combined Application of Patient-Derived Cells and Biomaterials as 3D In Vitro Tumor Models

Asbiel Hasbum [1,†], Ozan Karabulut [2,3,†], Ruben Edgar Reyes [2,4,†], Claudio Ricci [5], Alessandro Franchi [6], Serena Danti [5,*] and Sue Anne Chew [2,*]

1 School of Medicine, University of Texas Rio Grande Valley, Edinburg, TX 78520, USA; asbiel.hasbum01@utrgv.edu
2 Department of Health and Biomedical Sciences, University of Texas Rio Grande Valley, Brownsville, TX 78520, USA; ozan_karabulut1@baylor.edu (O.K.); ruben.reyes@edu.uag.mx (R.E.R.)
3 Department of Chemistry and Biochemistry, Baylor University, Waco, TX 76706, USA
4 School of Medicine, Universidad Autónoma de Guadalajara, Zapopan 45129, Mexico
5 Department of Civil and Industrial Engineering, University of Pisa, 56122 Pisa, Italy; claudio.ricci@unipi.it
6 Department of Translational Research and New Technology in Medicine and Surgery, University of Pisa, 56126 Pisa, Italy; alessandro.franchi@unipi.it
* Correspondence: serena.danti@unipi.it (S.D.); sueanne.chew@utrgv.edu (S.A.C.)
† These authors contributed equally to this work.

Simple Summary: For years, cancer has remained the second leading cause of death in U.S. and Europe even though cancer mortality has decreased, as new advances in medical treatment have made this decrease possible. Chemotherapy has remained the gold standard and "one-size-fits-all" treatment for cancer, yet this approach has lacked precision and, at times, failed. Recent studies attempt to mimic the spatial microenvironment of cancer tissue to better study chemotherapy agents by combining patient-derived cells and three-dimensional (3D) scaffold, bioprinting, spheroid, and hydrogel culturing. This commentary aims to collect and discuss recent findings concerning the combined application of biomaterials with patient-derived cancer cells to better study and test therapies in vitro, that will further personalize and facilitate the treatment of various cancers, and also address the limitation and challenges in developing these 3D models.

Abstract: Although advances have been made in cancer therapy, cancer remains the second leading cause of death in the U.S. and Europe, and thus efforts to continue to study and discover better treatment methods are ongoing. Three-dimensional (3D) tumor models have shown advantages over bi-dimensional (2D) cultures in evaluating the efficacy of chemotherapy. This commentary aims to highlight the potential of combined application of biomaterials with patient-derived cancer cells as a 3D in vitro model for the study and treatment of cancer patients. Five studies were discussed which demonstrate and provided early evidence to create 3D models with accurate microenvironments that are comparable to in vivo tumors. To date, the use of patient-derived cells for a more personalized approach to healthcare in combination with biomaterials to create a 3D tumor is still relatively new and uncommon for application in clinics. Although highly promising, it is important to acknowledge the current limitations and challenges of developing these innovative in vitro models, including the need for biologists and laboratory technicians to become familiar with biomaterial scaffolds, and the effort for bioengineers to create easy-to-handle scaffolds for routine assessment.

Keywords: personalized therapy; scaffold; tissue engineering; primary cancer cells; experimental models; screening; 3Rs

Citation: Hasbum, A.; Karabulut, O.; Reyes, R.E.; Ricci, C.; Franchi, A.; Danti, S.; Chew, S.A. Combined Application of Patient-Derived Cells and Biomaterials as 3D In Vitro Tumor Models. *Cancers* **2022**, *14*, 2503. https://doi.org/10.3390/cancers14102503

Academic Editor: Sudip Das

Received: 6 April 2022
Accepted: 16 May 2022
Published: 19 May 2022

1. Introduction

Although current trends show that overall cancer death rates have decreased for men, women and children, cancer remains the second leading cause of mortality in the U.S.

behind cardiovascular disease and is responsible for millions of deaths worldwide [1,2]. In 2020, 1,806,590 new cases and 606,520 cancer-related deaths have been estimated. The prevalence rate for all cancers combined in the U.S. since 2018 is approximately 5% of the population. Cancer incidence rates for all ages (per 100,000 people) between 2014 and 2018 were 450.5. Between 2015 and 2019, the average mortality rate for men and women combined was 152.4 (per 100,000 men and women). Although there has been a steep decline in the death rates for melanoma and lung cancers, which can be attributed to advances in treatment such as immune checkpoint inhibitors, targeted drug therapy, and a decrease in cancer risk factors, lung cancer mortality remains the leading cause of cancer death among men and women [3]. Despite advances in novel, targeted interventions and therapeutics, chemotherapeutic drugs remain the gold standard treatment and employ a "one-size-fits-most" approach, which lack precision and result in significant variations in patient response to therapy. Recent studies have attempted to mimic the spatial microenvironment of cancer tissue to better study chemotherapy agents through various techniques such as three-dimensional (3D) scaffold, bioprinting, spheroid and hydrogel culturing 3D tumor models, which have been shown to have advantages over bidimensional (2D) cultures in evaluating the efficacy of chemotherapeutics due to their heterogeneity and simulating the tumor microenvironment [4]. Aside from evaluating the efficacy and pharmacodynamics, 3D tissue models have been used to determine toxicity and drug resistance to chemotherapeutic agents simultaneously across different cells [5]. Several research papers have been published in the last decade using cancer cell lines to build up 3D in vitro tumor models, with the promise of delivering a useful tool for personalized therapy to the patients. This commentary aims at collecting and discussing the up-to-date findings concerning the use of biomaterials with patient-derived cancer cells for a near application in the clinics.

2. 3D In Vitro Models for Therapy Screening: The Role of Patient-Derived Cancer Cells and Biomaterials

While considerable progress has been made in 3D bioprinting, many obstacles remain in creating tumor models that provide physiological relevance and reliable data for the development of personalized treatment. The ability to replicate tumor microenvironments and establish vasculature for appropriate oxygen and nutrient distribution to specific areas within the 3D culture are challenges that need to be addressed [4,6,7]. Despite the advantages and increased use and acceptance of 3D tumor models, they are still more expensive and time intensive than their 2D counterparts [4]. Consequently, conventional 2D models are still widely used by pharmaceutical companies for drug development even though they do not accurately represent the tumor microenvironment which limits their use for anticancer drug screening [6]. 2D cell cultures have demonstrated minimal drug resistance compared to 3D cultures which has contributed to the high failure rates in drug discovery [6]. The physiological features of tumor tissue including oxygen and nutrient delivery, gene expression, and cell proliferation are better recapitulated in 3D tissue models [4]. Factors such as immune cells, inflammatory mediators, and vasculature add complexity and significantly influence the tumor microenvironment. Thus, 3D tumor models based on patient derived cancer tissue will more closely resemble the in vivo microenvironment and have better predictive value when compared to traditional models [6].

Conventional preclinical cancer models have generally used tumor cell lines as their source of cell derivation. Immortalized cell lines have been a preference for in vitro and in vivo preclinical models because of their ease of acquisition, production, reproducibility, and proliferation rates compared to primary cells [8,9]. Cell lines are commonly used for in vivo xenograft models and for in vitro scaffolds or organoid/spheroid formation models. Santoro et al. reported differences in signaling transduction pathway by applying shear stress to 3D constructs made of Ewing sarcoma TC71 cells and electrospun scaffolds, which highlighted the importance of recapitulating not only the compositional but also the mechanical features of the tissue microenvironment in 3D bone cancer models [10].

Silk scaffolded 786-O cell models developed by Abbott et al. showed increased lipid drop development, a significant upregulation of genes that signal cytokines and immune checkpoint inhibition therapy markers as well as downregulation of the genes CXCLS, ACACA, FASN, and CD10 comparatively between 2D and 3D scaffolded 786-O cells [11]. Persson et al. reported distinct differences of proteins between 2D and 3D tumor model as well as larger variations and more diversity between the secretome of all 3D cultures [12]. Additionally, they found correlation of the proteins secreted and clinical parameters of the original breast cancer [12]. Sun et al. utilized purchased HepG2 cells to 3D print an effective tumor model and compare it to a conventional 2D model using the same HepG2 cells. The 3D model showed significantly higher levels of mRNA related to liver function, higher expression of liver-associated proteins, a differing gene expression profile, and large differences in drug resistance [13].

Despite their upsides, immortalized cell lines have a handful of drawbacks that significantly affect their overall effectiveness as tumor modeling agents. Their biggest drawback occurs in their production; because cells derived from immortalized cell lines must have uncontrolled tumor-like growth in ex vivo conditions, they may suffer from altered genetic material, differing biological or tumorigenic properties compared to primary cells. Pan et al. discovered proteomic differences of the hepatoma cell line Hepa1-6 compared to primary hepatocytes [9]. Other tumor cell lines, like those of ovarian, breast, and neck cancer, also showed higher rates of mutations in comparison to their primary cell counterparts [14–16]. Because of the inevitable differences of cell lines to primary cells, in addition to the inherent individuality of every tumor, using cell lines as a preclinical tumor modeling tool may not be accurate. Alternatively, using patient derived cells as tumor models may guarantee increased similarity in biological and tumorigenic properties.

Patient-derived cells for use in 3D models offer a unique and relatively new approach to tumors. Unlike tumor cell lines that have garnered phenotypic and functional changes throughout their use, patient-derived cell models allow for the retention of most biochemical and physiological features from the in vivo tissue [17–19]. Furthermore, various studies have shown the possibility of using these patient-derived cells to determine drug combinations and resistance and identify the most effective treatment [19,20]. Patient-derived cells open the door to more targeted treatment, however, still carry some challenges due to the limited availability of these cells and difficulty in proliferating them as a result of tumor cell senescence [21,22]. The comparison of cell lines and patient-derived cells is summarized in Table 1.

Advancements in tissue engineering have slowly allowed biomaterials to play a bigger role in creating these 3D cancer models than using only cells (e.g., spheroids). Not only do biomaterials allow for a more realistic 3D structure, but they offer more realistic cell-to-cell interactions and microenvironments as opposed to 2D models [23–25]. Additionally, Rao et al. demonstrated that biomaterials can be used to monitor tumor progression and metastases [26]. Biomaterials can be natural materials, such as alginate, hyaluronic acid, gelatin, which tend to be highly biocompatible and can be degraded enzymatically by the body; however, they are often associated with immunogenicity or homogeneity [27]. Alginate is a naturally occurring anionic polysaccharide that can easily gel by ionic crosslinking using divalent metal ions, such as calcium. The ability to gel allows cancer cells to be encapsulated in alginate microparticles or scaffolds. DelNero et al. utilized alginate-based 3D scaffolds that can control oxygen concentration, resulting in homogeneous oxygen levels in the scaffold which permitted them to better study tumor hypoxia and angiogenesis [28]. Chitosan is a natural cationic polysaccharide, which is obtained from chitin, present in arthropod exoskeletons and some mushrooms, after deacetylation. It can be used to produce films, fibers or porous scaffolds and thus, is a very versatile biomaterial [29,30]. Dhiman et al. employed chitosan scaffolds for the culture of breast cancer cell lines and determined that chitosan polymer with high degree of deacetylation favored adsorption and cell growth [31]. Collagen is an important protein in tissues for physical support [32]. Duarte Campos et al. applied collagen as bioprintable bioinks and demonstrated that, since the printed bioink was stable enough, cells seeded in the bioprinted models maintained their ability to proliferate. They suggest that this biomaterial may be promising to be used with patient-derived primary tumor cells for precision medicine therapy [33]. Gelatin is another natural polymer that is widely used as a biomaterial. It is the denatured form of collagen, and unlike collagen which has low water solubility, it is a water-soluble biomaterial. Nii et al. utilized gelatin to fabricate microparticles as a 3D cell culture system combined with drug delivery, as a cancer invasion model [27]. Hyaluronic acid is a key ECM component and thus, is a suitable biomaterial for 3D cell culture. Engel et al. fabricated a multi-layered hyaluronic acid hydrogel to coculture cancer and stromal cells and demonstrated that it can improve drug screening predictability compared to 2D cultures [34].

Table 1. Comparison of cell lines vs. patient-derived cells.

	Cell Lines	Patient-Derived Cells
Accessibility	Easily accessible	Difficult to access, limited availability of cells/tissues
Cost	Low cost to obtain and culture	Increased cost to obtain and culture
Proliferation	Proliferates rapidly and indefinitely	More difficult to proliferate due to tumor cell senescence, limited amount of pasages
Ease in culturing	Robust and easy to work with and maintain	More fragile and difficult to work with and maintain
Reproducibility	Pure population of cells therefore reproducible data can be obtained	Heterogenous population of cells therefore data can differ between cell populations
Ability to mimic TME and clinical response	Lack of complexity to mimic tumor environment and clinical response	Can better mimic the TME and clinical response
Ethical issues and research compliance	No ethical concerns and no need for institutional review board approval to obtain and use	Need to obtain institutional review board approval to obtain and use

Tumor microenvironment (TME).

Among synthetic polymers, aliphatic polyesters are another class of biomaterials that is usually biodegraded through hydrolysis and their characteristics can be controlled easily [27]. One of the synthetic polymer that is widely used for 3D biomaterial models is poly(lactic-co-glycolic acid) (PLGA), whose degradation rate can be altered by controlling the lactic and glycolic acid ratio, the polymer molecular weight, as well as its end caps. Luo et al. demonstrated that PLGA electrospun scaffold incorporating hydroxyapatite (HA), which is often used for bone regeneration, can better support cancer cells compared to PLGA scaffold alone as the HA induce cell growth, DNA synthesis and cell division [35]. A biodegradable polymer widely used as a biomaterial is polycaprolactone (PCL), which is also a polyester like PLGA and thus, degrades by hydrolysis. Chen et al. utilized PCL to 3D print scaffolds and co-cultured colorectal cancer cells, cancer-associated fibroblasts and tumor-associated endothelial cells on them to develop a 3D model of tumor tissue for colorectal cancer [36].

Natural polymers, due to their water solubility, usually result in hydrogels, which are common types of biomaterials for 3D cancer models due to their ease of encapsulating the cells of interest. Synthetic polymers, which are usually more hydrophobic in nature are thermoplastic and soluble in non-water solvents. Therefore, they often result in scaffolds fabricated either a sponge or fiber form, obtained via emulsion, compression molding, 3D printing or electrospinning, among other methods. Such prefabricated structures can offer microenvironments for 3D cancer models, in which geometrical features, porosity, mechanical properties and roughness can be all tuned. Just like hydrogels, they provide the vital cell–cell and cell-ECM interactions that would mimic real tumors. Additionally, these scaffolds account for controlled mechanical properties, thus can be stiffer and more stable and can better withstand shear stress, as opposed to hydrogels [37].

3. Combined Application of Biomaterials and Patient-Derived Cancer Cells

Personalized healthcare is a much more effective form of treatment for very complex diseases that have been, for the longest time, receiving broad and comparatively generic treatments. Tumors and their complete environment in a body are incredibly complex and individualized and the utilization of patient-derived cells with the combination of biomaterials to create a 3D model may be a promising method to address this challenge. Cell lines are most often used to create a 3D in vitro model, while patient derived cells are most often used as patient derived xenographs (PDX) models in vivo models. Comparatively, very few studies use both patient derived cells in tandem with biomaterials to create an in vitro 3D models of tumors. The flow chart explaining the process to build those models for personalized therapy is shown in Figure 1.

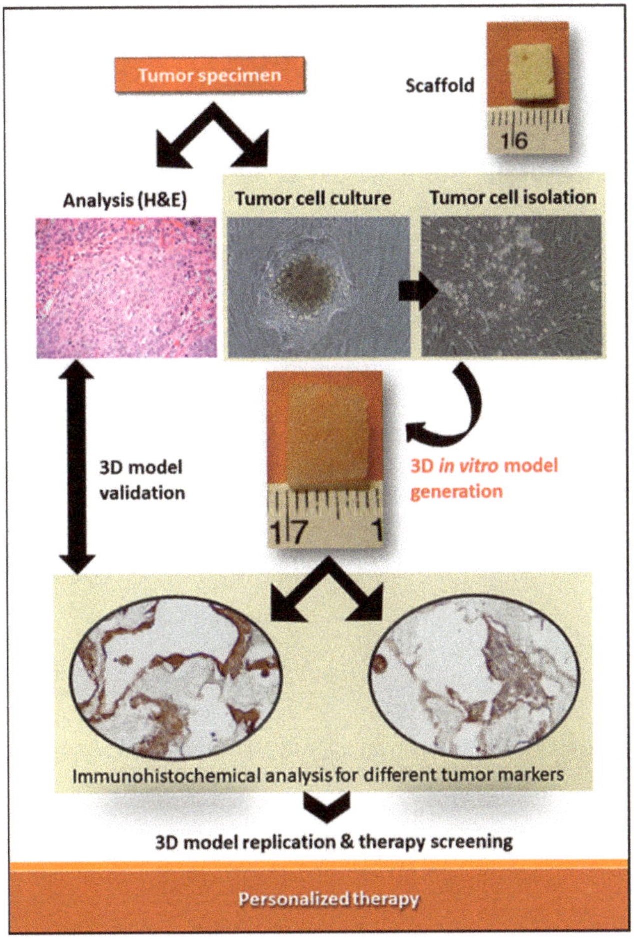

Figure 1. Schematic depicting the combined application of patient's larynx tumor-derived cells (TCCR3) and polyvinyl alcohol (PVA) spongy scaffold to create 3D in vitro tumor models. Immunohistochemical analysis shows immunopositivity (in brown) for Integrin α5 (on the left) and Smad4 (on the right). All figures in the schematic are original unpublished material of the authors.

There are many applications of biomaterials as 3D in vitro tumor models; however, they are usually coupled with cell lines instead of patient-derived cells. For example, Sun et al. utilized bioprinted scaffolds seeded with hepatocellular carcinoma HepG2 cell line [13] and Abbot et al. utilized silk scaffold seeded with renal cell carcinoma 786-O cell line [11]. Patient-derived tissues have also been utilized as 3D tumor models by creating patient-derived scaffolds (PDS) that are obtained via decellularization of surgically resected tumors and then used as a substrate for cell line culture [38]. Parkinson et al. cultured colon cancer cell line HT29 in PDSs [38]. They demonstrated that the PDSs can result in induced transcriptomic and proteomic responses that align with patient-specific clinical disease information and thus, can be a potential tumor model for predicting the effectiveness of cancer therapies. Pearson et al. recellularized PDSs with breast cancer cell lines, MCF7 and MDA-MB-231 [12], whereas Gustafsson et al. used PDS to culture breast cancer cell lines MCF7 and T47D [39], as in vitro 3D tumor models. Besides PDX and PDS, fresh tumor samples, employed directly without separating the cells by enzymatic or mechanical digestion from the tumor specimens, have also been applied as ex vivo 3D tumor models [40]. These 3D tumor models conserved the original tissue architecture and cell heterogeneity better than other 3D models where the cells are removed from the original tumor [41,42]; however, they cannot be used for longer cultures and passages (i.e., usually can be culture for a week) [40]. Therefore, such tumor explants although able to mimic the TME very well, may not be as cost effective and easily accessible models.

To the best of our knowledge, there are only a few studies reporting the combination of both patient-derived cells and biomaterials as 3D in vitro models, which are summarized in Table 2. Below, we discuss each of these applications.

Table 2. 3D in vitro models that utilize patient-derived cells and biomaterials.

Cancer Type	Type of Cells	Biomaterial/Scaffold	Main Outcome	Ref.
Pancreatic ductal adenocarcinoma (PDAC)	Cells obtained from PDAC pieces/explants	(PVA/G) blend sponges; (PEOT/PBT) copolymer compression molded scaffolds and electrospun fibers meshes.	PDAC cells demonstrated various behaviors when exposed to different scaffold types. Sponge-like pores allowed for cellular clustering resembling the native cancer morphostructure. In PVA/G sponges the active MMP-2 enzyme was the highest.	[43]
Intrahepatic cholangiocarcinoma (ICC)	Cells obtained from a male patient diagnosed with ICC	Gelatin-alginate-Matrigel™ hydrogel bioink	Printed ICC cells showed colony forming capacity, high survival rate, active proliferation, invasive and metastatic phenotype and other characteristics of ICC cells, e.g., expression levels of tumor markers and cancer stem cell markers.	[44]
Pediatric brain tumors	Cells obtained from 11 pediatric tumor cases, consisting of three medulloblastoma (MB) patients, three ependymoma (EPN) patients, one glioblastoma (GBM) patient, and four juvenile pilocytic astrocytoma (Ast) patients.	Bombyxmoricocoons silk porous scaffold infused with liquid ECM gel	The 3D scaffold silk base structure had a vital role in supporting tumor spheroids, giving structural stability to gels, and maintaining tumor stem cells in 3D.	[45]

Table 2. *Cont.*

Cancer Type	Type of Cells	Biomaterial/Scaffold	Main Outcome	Ref.
PDAC	Cells obtained from PDAC patient	Co-assembly of PAs with custom ECM components (PA-ECM)	This model of PDAC was able to sustain patient-specific transcriptional profiles and demonstrated high cancer stem cell functionality.	[46]
Acute myeloid leukemia (AML) and myeloproliferative neoplasms (MPN)	Cells obtained from malignant human malignant hematopoietic stem and progenitor cells (HSPCs)	HA scaffold with perfusion bioreactor	The 3D model provided an environment that could sustain $CD34^+$ cells from acute myeloid leukemia (AML) and myeloproliferative neoplasm patients for up to 3 weeks.	[47]

Abbreviated as follows: Poly(vinyl alcohol)/gelatin (PVA/G); poly(ethylene oxide terephthalate)/poly(butylene terephthalate) (PEOT/PBT); extracellular matrix (ECM); peptide amphiphiles (PAs); Hydroxyapatite (HA).

3.1. Polymeric Scaffolds and Primary Pancreatic Ductal Adenocarcinoma Cells

In their study, Ricci et al. isolated a pancreatic ductal adenocarcinoma (PDAC) cells from an explant of a patient, created different scaffold architectures, and used the scaffold as well as the PDAC cells in conjunction to explore biomaterial-based 3D tumor models [43]. Three different polymeric scaffolds architectures were used in the creation of the 3D in vitro tumor models of the patient-derived PDAC [43]. The tumor samples were obtained from surgical procedures and cleaned thoroughly before use. The study used polyvinyl alcohol (PVA)/gelatin (G) at 80/20 (*w*/*w*)% and a poly(ethylene oxide terephthalate)/poly(butylene terephthalate) (PEOT/PBT) copolymer as their structure materials [43]. PDAC cells preferred to aggregate in sponge-like material rather than nanofiber structures and preferred the PVA/G sponge compared to PEOT/PBT sponge and PEOT/PBT fiber mesh [43]. Matrix metalloproteinases (MMPs) are enzymes that are important in protein degradation and have been directly correlated with cancer development and invasion [48]. MMP-2 and MMP-9 in particularly, because of their relationship with the tumor suppressor Smad4, which is downregulated in PDAC progression [49,50], are important factors to study PDAC. As shown in section A2 in Figure 2, PDAC cells in PVA/G sponges had higher levels of active MMP-2 production and protein synthesis [43]. Results in section A3–D3 of Figure 2 demonstrate that MMP-9 expression was strongly positive in all three cell/scaffold constructs. Nonetheless, PDAC cell MMP-2 and MMP-9 expression in the PEOT/PBT copolymer structure is still reduced in the production of active MMP-9 [43].

In conclusion, this study showed the possibility of using various biomaterial scaffolds with patient-derived cells to find a compatible pair that could work to model real pancreatic tumors. Among them, spongy scaffolds, like those obtained via PVA/G emulsion and freeze-drying, were the most suitable. They showed volume swelling ratio higher than 200% and were mechanically soft, with material stiffness increasing with G content $\geq$20%, due to enhanced sites of G crosslinked by glutaraldehyde [51].

The comparative analyses demonstrated that PVA/G 70/30 and PVA/G 80/20 (*w*/*w*)% were similar in terms of morphology, swelling behavior, water stability, physico-chemical and viscoelastic mechanical properties, with an apparent compressive modulus of about 7 kPa at strain rates of 0.005 s^{-1} [51]. In dry conditions, PVA/G 80/20 (*w*/*w*)% sponges showed high volume porosity (i.e., 84.43%) and pore interconnectivity (97.44%), the latter under pore-pore openings $\geq$51.2 μm [52]. It is possible that this scaffold could mimic the morphological and mechanical features of the pancreas, whose stiffness increased from 7.72 to 10.97 kPa under shear wave velocity measurements, from normal to fibrotic organs [53].

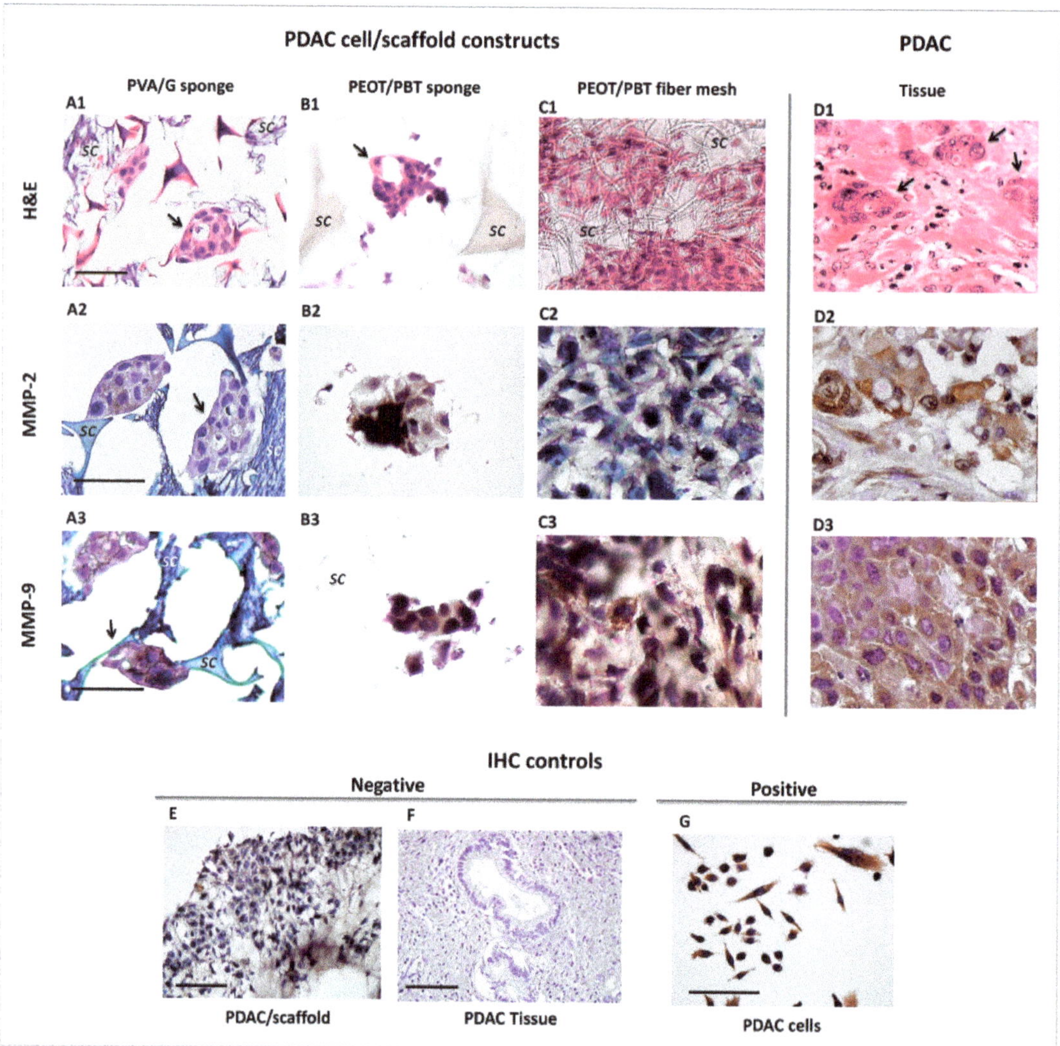

Figure 2. Histological micrographs of (**A–C**) pancreatic ductal adenocarcinoma (PDAC) cell/scaffold constructs, and (**D**) tumor tissue: (**A**) PVA/G spongy scaffold prepared via emulsion and freeze-drying, (**B**) PEOT/PBT spongy scaffold prepared via compression molding and salt leaching, and (**C**) PEOT/PBT fiber mesh prepared via electrospinning. (**A1–D1**) Hematoxylin and eosin staining, and (**A2–D2**, **A3–D3**) immunohistochemistry for metalloproteinases (MMPs) MMP-2 and MMP-9. Arrows indicate some organized clusters of cells with duct formation; "sc" indicates the scaffold material. (**E–G**) Controls of immunohistochemical reactions. Scale bar is 50 μm. (Reprinted from Taylor & Francis, Ricci et al., Biomatter, 2014 (Ricci, 2014 [43]).

The easy procedure leading to the fabrication of such scaffolds and their usefulness in pancreatic cancer in vitro modeling was described by Ricci et al. [54]. Among other interesting characteristics, such as durability (i.e., non-biodegradability, thus suitability even for long term cell cultures, since mechanical and pore properties remain consistent over time), PVA/G sponges were fully processable via routine histology processing, which made them interesting scaffolds for in-hospital research. In fact, it is possible to directly compare the generated 3D model with the patient's tumor under histology; therefore, the morphological features of the cells and the newly formed in vitro tumors can support the suitability of the 3D model for a possible therapeutic screening. s [51,54]. This same principle could be exploited to model other types of cancers throughout their various phases and screen for drug susceptibility.

3.2. Bioprinting of Patient-Derived Intrahepatic Cholangiocarcinoma

In this study, Mao et al. used a bioprinting process to create a 3D model of intrahepatic cholangiocarcinoma. Bioprinting offers uniqueness to creating 3D tumor models; while other modeling procedures have random arrangements and densities of molecules, cells, and biomaterials, bioprinting offers control over their density, arrangement, and structural design [44]. This study also used patient-derived tumor specimens acquired through a resection surgery on a single man diagnosed with intrahepatic cholangiocarcinoma and a composite hydrogel system containing gelatin-alginate-Matrigel to print their tumor model [44]. The 3D printed culture exhibited a more uniform distribution of cell clusters, as well as a faster aggregation process, in comparison to the sandwich culture [44]. Tumor markers CA19-9 and CEA, cancer stem cell markers CD133 and EpCAM, relative gene expression, liver function markers, as well as pathological markers of ICC cells were significantly higher in the 3D tumor model microenvironment [44]. Additionally, the MMP and fibrosis makers of the 3D model were significantly higher and showed much better drug resistance than the sandwich model [44]. This drug resistance opens the possibility of using these models as personalized therapy due to displaying stem-like properties. Overall, this study demonstrated the possibility of retaining cell viability while bioprinting, the vital up-regulation of tumorigenic phenotypes in 3D models when compared to 2D models, and the potential of using these models to study drug resistance for a more personalized treatment.

3.3. Porous Scaffold Composed from Bombyxmoricocoons Silk Was Infused with Liquid ECM Gel for Pediatric Brain Tumors

Porous scaffold composed from Bombyxmoricocoons silk was infused with liquid ECM gel (Figure 3) and employed for a variety of pediatric brain tumors, namely, 11 pediatric tumor cases, consisting of three medulloblastoma (MB), three ependymoma (EPN), one glioblastoma (GBM), and four juvenile pilocytic astrocytoma (Ast) patients [45]. Tang-Schomer et al. found that the 3D scaffold alone supported cell heterogeneity and had the ability to form tumor type-dependent spheroids, which were not possible in 2D or gel-only control cultures. They concluded that the 3D scaffold silk-based structure had a vital role in supporting tumor spheroids, giving structural stability to gels, and maintaining tumor stem cells in 3D.

3.4. Co-Assembly of Peptide Amphiphiles (PAs) with Custom ECM Components (PA-ECM) with Pancreatic Ductal Adenocarcinoma (PDAC)

Osuna de la Pena et al. cocultured patient-derived cells obtained from PDAC with a 3D co-assembly of peptide amphiphiles (PAs) with custom ECM components (PA-ECM) for ex vivo tissue modeling with increased adaptability [46]. This model of PDAC was able to sustain patient-specific transcriptional profiles and demonstrated high cancer stem cell functionality. These peptides provided a reductionist approach to bioengineering complicated microenvironments by regulating nanoscale geometries and epitope presentation to selectively signal cells.

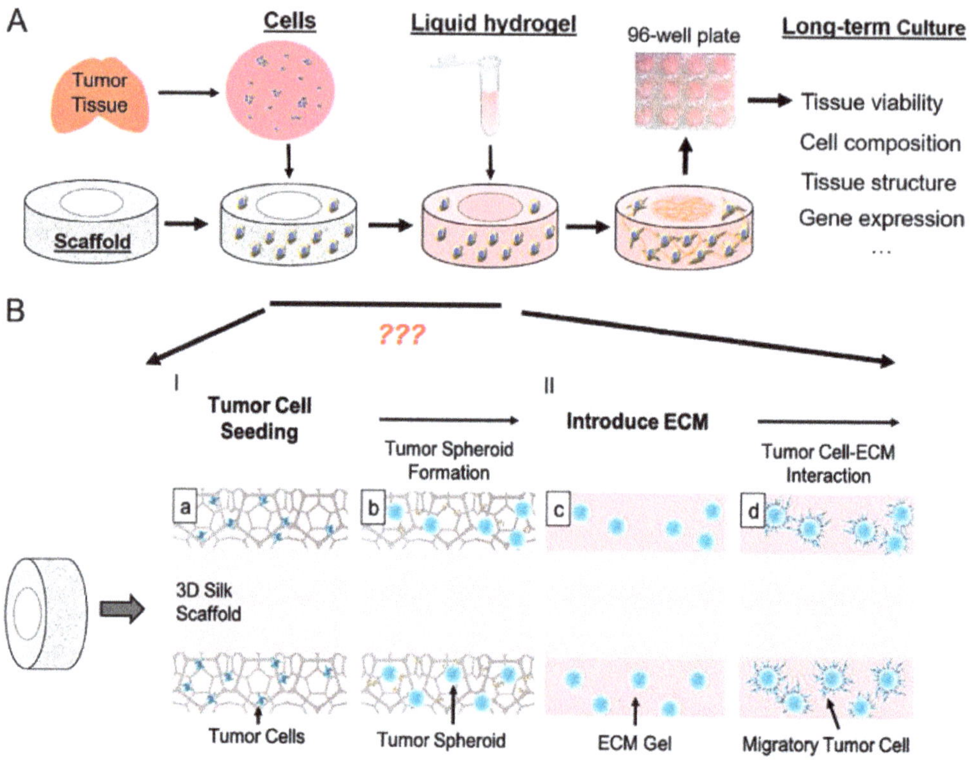

Figure 3. Schematics of the 3D modeling process: (**A**) Schematics of the 3D brain tissue engineering process. (**B**) To adapt the process for brain tumor model, questions regarding media conditions, ECM and timing for the change of culture conditions need to be addressed. Dissociated tumor cells are seeded onto a donut-shaped 3D silk-based porous scaffold, from which tumor spheroid develops. ECM gels are introduced to the scaffold filling the pores and the center-hole (CH) region, providing a permissive environment for the migrating tumor cells and cell–cell interaction. (**a**) Tumore cell seeding, (**b**) tumor spheroid formation, (**c**) introduce ECM, (**d**) tumor cell-ECM interaction. (Reprinted from Elsevier, Tang-Schomer et al., Translational Oncology, 2022 [45]).

3.5. Patient-Derived Malignant Hematopoietic Stem Cells in HA Scaffold Developed with a Bioreactor System for Acute Myeloid Leukemia and Myeloproliferative Neoplasms

Andrés García-García et al. demonstrated how cellular niches may be built in a 3D hydroxyapatite scaffold and perfusion flow-based bioreactor system and used to maintain, expand, and regulate the phenotypic and functional properties of patient-derived human malignant hematopoietic stem and progenitor cells (HSPCs) ex vivo [47]. The fully humanized model was used to study human leukemogenesis in the presence of tailored niche components (e.g., osteoblastic vs. stromal-vascular elements) and to assess chemotherapeutic responsiveness. Human osteoblastic bone marrow niches were produced by culturing mesenchymal stromal cells in the scaffolds under perfusion flow in a bioreactor system. They demonstrated that the 3D model provided an environment that could sustain $CD34^+$ cells from acute myeloid leukemia (AML) and myeloproliferative neoplasm (MPN) patients for up to 3 weeks.

4. Conclusions

The use of patient-derived cells for a more personalized approach to healthcare and biomaterials to create 3D tumor are relatively new and uncommon. Although highly promising due to their more accurate cellular environments that mimics that of the patient, it is important to acknowledge the current limitations of these innovative in vitro models. From having to mimic the molecular biology, physiology and genetic makeup of these cancer cells and ultimately reproduce the heterogenous nature of cancer cells, these models face multiple obstacles. A key challenge is developing a way to create a network of vasculatures in tumors. Without these capillary networks, the 3D tumor growth would be difficult and possibly impossible to occur. Nonetheless, studies have shown that if these bioprinted models are able to account for cellular and molecular factors, it is possible to transform stem cells into endothelial cells, thus promoting angiogenesis [55,56]. Moreover, recreating these complex microenvironments has deemed difficult as various studies have shown a multitude of cancers interacting with surrounding stromal cells and environmental factors to further their growth. This sets up a challenge as it is not as simple as combining these cells and factors with each other and expecting the same system to exist [57–59]. Cell viability has played a factor in the creation of these 3D models as the shear forces of bioprinting often damages the cells ultimately deeming them unusable. Current technology has made it possible for significant progress to occur; nonetheless, it is still very limited due to the expensive and time-consuming process to create these personalized and accurate 3D tumor microenvironments.

As technology and knowledge about how these cancers rapidly progresses, current conventional cancer models still carry several limitations. In future studies, it will be vital to minimize the cost and focus on the development and fabrication of less time-consuming models, while accounting the importance of every cellular and molecular factors when using patient-derived cells and designing these 3D tumor constructs. Furthermore, keeping cell viability and replicating the complex environment of tumors, such as vascular network and surrounding stromal cell to cell interactions, are areas of research that ongoing and future studies are and should focus on. A recent study by Contessi Negrini et al. has suggested the use of human mesenchymal stromal cells to pregenerate bone ECM on a 3D printed polyurethane scaffold, to be used, after cell lysis, as a bioactive and biomimetic environment for osteosarcoma cell growth [60], which demonstrates the possibility of integrating a synthetic biomaterial with biomolecules produced by patient-derived healthy cells to replicate the some complexity of the TME. Overall, the incredible effectiveness of biomaterial-based scaffolding and 3D printing has shown remarkable promise, hence, allowing a more personalized healthcare approach; however, much work is needed to continue to progress in the development of these in vitro systems.

Author Contributions: Conceptualization, S.D. and S.A.C.; methodology, S.D. and S.A.C.; software, S.A.C.; validation, S.A.C.; formal analysis, A.H., O.K. and R.E.R.; investigation, A.H., O.K. and R.E.R.; resources, A.F., S.D. and S.A.C.; data curation, C.R. and A.F.; writing—original draft preparation, A.H., O.K. and R.E.R.; writing—review and editing, C.R., A.F., S.D. and S.A.C.; visualization, C.R., A.F. and S.D.; supervision, S.D. and S.A.C.; project administration, S.D. and S.A.C.; funding acquisition, A.F., S.D. and S.A.C. All authors have read and agreed to the published version of the manuscript.

Funding: This research was funded by Tuscany Region under Health Program 2018, project name "Sinonasal cancer: In depth genetic analysis of patients for personalized treatment and disease monitoring (acronym: ADAPTA)", grant number CUP I58D20000540002.

Institutional Review Board Statement: Not applicable.

Informed Consent Statement: Not applicable.

Data Availability Statement: Not applicable.

Acknowledgments: The 3R Center (Pisa, Italy) is acknowledged for supporting research on the application of 3D in vitro models, according to the 3R's principles. C.R., A.F. and S.D. acknowledge Delfo D'Alessandro (University of Pisa) for his vital technical contribution to 3D in vitro models.

Conflicts of Interest: The authors declare no conflict of interest.

References

1. Ahmad, F.B.; Cisewski, J.A.; Minino, A.; Anderson, R.N. Provisional Mortality Data—United States, 2020. *Morb. Mortal. Wkly. Rep.* **2021**, *70*, 519–522. [CrossRef] [PubMed]
2. Ferlay, J.; Colombet, M.; Soerjomataram, I.; Parkin, D.M.; Pineros, M.; Znaor, A.; Bray, F. Cancer statistics for the year 2020: An overview. *Int. J. Cancer* **2021**, *70*, 519–522. [CrossRef] [PubMed]
3. Islami, F.; Ward, E.M.; Sung, H.; Cronin, K.A.; Tangka, F.K.L.; Sherman, R.L.; Zhao, J.; Anderson, R.N.; Henley, S.J.; Yabroff, K.R.; et al. Annual Report to the Nation on the Status of Cancer, Part 1: National Cancer Statistics. *J. Natl. Cancer Inst.* **2021**, *113*, 1648–1669. [CrossRef] [PubMed]
4. Rodrigues, J.; Heinrich, M.A.; Teixeira, L.M.; Prakash, J. 3D In Vitro Model (R)evolution: Unveiling Tumor-Stroma Interactions. *Trends Cancer* **2021**, *7*, 249–264. [CrossRef]
5. King, S.M.; Presnell, S.C.; Nguyen, D.G. Abstract 2034: Development of 3D bioprinted human breast cancer for in vitro drug screening. *Cancer Res.* **2014**, *74*, 2034. [CrossRef]
6. Augustine, R.; Kalva, S.N.; Ahmad, R.; Zahid, A.A.; Hasan, S.; Nayeem, A.; McClements, L.; Hasan, A. 3D Bioprinted cancer models: Revolutionizing personalized cancer therapy. *Transl. Oncol.* **2021**, *14*, 101015. [CrossRef]
7. Mehrotra, S.; Moses, J.C.; Bandyopadhyay, A.; Mandal, B.B. 3D Printing/Bioprinting Based Tailoring of in Vitro Tissue Models: Recent Advances and Challenges. *ACS Appl. Bio Mater.* **2019**, *2*, 1385–1405. [CrossRef]
8. Goodspeed, A.; Heiser, L.M.; Gray, J.W.; Costello, J.C. Tumor-Derived Cell Lines as Molecular Models of Cancer Pharmacogenomics. *Mol. Cancer Res.* **2016**, *14*, 3–13. [CrossRef]
9. Pan, C.; Kumar, C.; Bohl, S.; Klingmueller, U.; Mann, M. Comparative proteomic phenotyping of cell lines and primary cells to assess preservation of cell type-specific functions. *Mol. Cell Proteom.* **2009**, *8*, 443–450. [CrossRef]
10. Santoro, M.; Lamhamedi-Cherradi, S.E.; Menegaz, B.A.; Ludwig, J.A.; Mikos, A.G. Flow perfusion effects on three-dimensional culture and drug sensitivity of Ewing sarcoma. *Proc. Natl. Acad. Sci. USA* **2015**, *112*, 10304–10309. [CrossRef]
11. Abbott, A.; Bond, K.; Chiba, T.; Sims-Lucas, S.; Oxburgh, L.; Coburn, J.M. Development of a mechanically matched silk scaffolded 3D clear cell renal cell carcinoma model. *Mater. Sci. Eng. C Mater. Biol. Appl.* **2021**, *126*, 112141. [CrossRef] [PubMed]
12. Persson, E.; Gregersson, P.; Gustafsson, A.; Fitzpatrick, P.; Rhost, S.; Stahlberg, A.; Landberg, G. Patient-derived scaffolds influence secretion profiles in cancer cells mirroring clinical features and breast cancer subtypes. *Cell Commun. Signal.* **2021**, *19*, 66. [CrossRef] [PubMed]
13. Sun, L.; Yang, H.; Wang, Y.; Zhang, X.; Jin, B.; Xie, F.; Jin, Y.; Pang, Y.; Zhao, H.; Lu, X.; et al. Application of a 3D Bioprinted Hepatocellular Carcinoma Cell Model in Antitumor Drug Research. *Front. Oncol.* **2020**, *10*, 878. [CrossRef] [PubMed]
14. Domcke, S.; Sinha, R.; Levine, D.A.; Sander, C.; Schultz, N. Evaluating cell lines as tumour models by comparison of genomic profiles. *Nat. Commun.* **2013**, *4*, 2126. [CrossRef]
15. Li, H.; Wawrose, J.S.; Gooding, W.E.; Garraway, L.A.; Lui, V.W.; Peyser, N.D.; Grandis, J.R. Genomic analysis of head and neck squamous cell carcinoma cell lines and human tumors: A rational approach to preclinical model selection. *Mol. Cancer Res.* **2014**, *12*, 571–582. [CrossRef]
16. Neve, R.M.; Chin, K.; Fridlyand, J.; Yeh, J.; Baehner, F.L.; Fevr, T.; Clark, L.; Bayani, N.; Coppe, J.P.; Tong, F.; et al. A collection of breast cancer cell lines for the study of functionally distinct cancer subtypes. *Cancer Cell* **2006**, *10*, 515–527. [CrossRef]
17. Czekanska, E.M.; Stoddart, M.J.; Ralphs, J.R.; Richards, R.G.; Hayes, J.S. A phenotypic comparison of osteoblast cell lines versus human primary osteoblasts for biomaterials testing. *J. Biomed. Mater. Res. A* **2014**, *102*, 2636–2643. [CrossRef]
18. Gillet, J.P.; Calcagno, A.M.; Varma, S.; Marino, M.; Green, L.J.; Vora, M.I.; Patel, C.; Orina, J.N.; Eliseeva, T.A.; Singal, V.; et al. Redefining the relevance of established cancer cell lines to the study of mechanisms of clinical anti-cancer drug resistance. *Proc. Natl. Acad. Sci. USA* **2011**, *108*, 18708–18713. [CrossRef]
19. Yi, H.G.; Jeong, Y.H.; Kim, Y.; Choi, Y.J.; Moon, H.E.; Park, S.H.; Kang, K.S.; Bae, M.; Jang, J.; Youn, H.; et al. A bioprinted human-glioblastoma-on-a-chip for the identification of patient-specific responses to chemoradiotherapy. *Nat. Biomed. Eng.* **2019**, *3*, 509–519. [CrossRef]
20. Crystal, A.S.; Shaw, A.T.; Sequist, L.V.; Friboulet, L.; Niederst, M.J.; Lockerman, E.L.; Frias, R.L.; Gainor, J.F.; Amzallag, A.; Greninger, P.; et al. Patient-derived models of acquired resistance can identify effective drug combinations for cancer. *Science* **2014**, *346*, 1480–1486. [CrossRef]
21. Liu, X.; Ory, V.; Chapman, S.; Yuan, H.; Albanese, C.; Kallakury, B.; Timofeeva, O.A.; Nealon, C.; Dakic, A.; Simic, V.; et al. ROCK inhibitor and feeder cells induce the conditional reprogramming of epithelial cells. *Am. J. Pathol.* **2012**, *180*, 599–607. [CrossRef] [PubMed]
22. Brower, M.; Carney, D.N.; Oie, H.K.; Gazdar, A.F.; Minna, J.D. Growth of cell lines and clinical specimens of human non-small cell lung cancer in a serum-free defined medium. *Cancer Res.* **1986**, *46*, 798–806. [PubMed]
23. Ricci, C.; Moroni, L.; Danti, S. Cancer tissue engineering—New perspectives in understanding the biology of solid tumors—A critical review. *OA Tissue Eng.* **2013**, *1*, 1–7. [CrossRef]
24. Wang, F.; Weaver, V.M.; Petersen, O.W.; Larabell, C.A.; Dedhar, S.; Briand, P.; Lupu, R.; Bissell, M.J. Reciprocal interactions between beta1-integrin and epidermal growth factor receptor in three-dimensional basement membrane breast cultures: A different perspective in epithelial biology. *Proc. Natl. Acad. Sci. USA* **1998**, *95*, 14821–14826. [CrossRef] [PubMed]
25. Gu, L.; Mooney, D.J. Biomaterials and emerging anticancer therapeutics: Engineering the microenvironment. *Nat. Rev. Cancer* **2016**, *16*, 56–66. [CrossRef]

26. Rao, S.S.; Bushnell, G.G.; Azarin, S.M.; Spicer, G.; Aguado, B.A.; Stoehr, J.R.; Jiang, E.J.; Backman, V.; Shea, L.D.; Jeruss, J.S. Enhanced Survival with Implantable Scaffolds That Capture Metastatic Breast Cancer Cells In Vivo. *Cancer Res.* **2016**, *76*, 5209–5218. [CrossRef]
27. Nii, T.; Makino, K.; Tabata, Y. Three-Dimensional Culture System of Cancer Cells Combined with Biomaterials for Drug Screening. *Cancers* **2020**, *12*, 2754. [CrossRef]
28. DelNero, P.; Lane, M.; Verbridge, S.S.; Kwee, B.; Kermani, P.; Hempstead, B.; Stroock, A.; Fischbach, C. 3D culture broadly regulates tumor cell hypoxia response and angiogenesis via pro-inflammatory pathways. *Biomaterials* **2015**, *55*, 110–118. [CrossRef]
29. Healy, K.E.; Thomas, C.H.; Rezania, A.; Kim, J.E.; McKeown, P.J.; Lom, B.; Hockberger, P.E. Kinetics of bone cell organization and mineralization on materials with patterned surface chemistry. *Biomaterials* **1996**, *17*, 195–208. [CrossRef]
30. Minuth, W.W.; Sittinger, M.; Kloth, S. Tissue engineering: Generation of differentiated artificial tissues for biomedical applications. *Cell Tissue Res.* **1998**, *291*, 1–11. [CrossRef]
31. Dhiman, H.K.; Ray, A.R.; Panda, A.K. Characterization and evaluation of chitosan matrix for in vitro growth of MCF-7 breast cancer cell lines. *Biomaterials* **2004**, *25*, 5147–5154. [CrossRef] [PubMed]
32. Gelse, K.; Poschl, E.; Aigner, T. Collagens—Structure, function, and biosynthesis. *Adv. Drug Deliv. Rev.* **2003**, *55*, 1531–1546. [CrossRef] [PubMed]
33. Duarte Campos, D.F.; Bonnin Marquez, A.; O'Seanain, C.; Fischer, H.; Blaeser, A.; Vogt, M.; Corallo, D.; Aveic, S. Exploring Cancer Cell Behavior In Vitro in Three-Dimensional Multicellular Bioprintable Collagen-Based Hydrogels. *Cancers* **2019**, *11*, 180. [CrossRef] [PubMed]
34. Engel, B.J.; Constantinou, P.E.; Sablatura, L.K.; Doty, N.J.; Carson, D.D.; Farach-Carson, M.C.; Harrington, D.A.; Zarembinski, T.I. Multilayered, Hyaluronic Acid-Based Hydrogel Formulations Suitable for Automated 3D High Throughput Drug Screening of Cancer-Stromal Cell Cocultures. *Adv. Healthc. Mater.* **2015**, *4*, 1664–1674. [CrossRef]
35. Luo, H.L.; Zhang, Y.; Gan, D.Q.; Yang, Z.W.; Ao, H.Y.; Zhang, Q.C.; Yao, F.L.; Wan, Y.Z. Incorporation of hydroxyapatite into nanofibrous PLGA scaffold towards improved breast cancer cell behavior. *Mater. Chem. Phys.* **2019**, *226*, 177–183. [CrossRef]
36. Chen, H.; Cheng, Y.; Wang, X.; Wang, J.; Shi, X.; Li, X.; Tan, W.; Tan, Z. 3D printed in vitro tumor tissue model of colorectal cancer. *Theranostics* **2020**, *10*, 12127–12143. [CrossRef]
37. Velliou, E.; Gupta, P.; Ricci, C.; Danti, S. Chapter 11—Biomaterial-based in vitro models for pancreatic cancer. In *Biomaterials for 3D Tumor Modeling*; Kundu, S.C., Reis, R.L., Eds.; Elsevier: Amsterdam, The Netherlands, 2020; pp. 235–249. [CrossRef]
38. Parkinson, G.T.; Salerno, S.; Ranji, P.; Hakansson, J.; Bogestal, Y.; Wettergren, Y.; Stahlberg, A.; Lindskog, E.B.; Landberg, G. Patient-derived scaffolds as a model of colorectal cancer. *Cancer Med.* **2021**, *10*, 867–882. [CrossRef]
39. Gustafsson, A.; Garre, E.; Leiva, M.C.; Salerno, S.; Stahlberg, A.; Landberg, G. Patient-derived scaffolds as a drug-testing platform for endocrine therapies in breast cancer. *Sci. Rep.* **2021**, *11*, 13334. [CrossRef]
40. Boucherit, N.; Gorvel, L.; Olive, D. 3D Tumor Models and Their Use for the Testing of Immunotherapies. *Front. Immunol.* **2020**, *11*, 603640. [CrossRef]
41. Kenerson, H.L.; Sullivan, K.M.; Seo, Y.D.; Stadeli, K.M.; Ussakli, C.; Yan, X.W.; Lausted, C.; Pillarisetty, V.G.; Park, J.O.; Riehle, K.J.; et al. Tumor slice culture as a biologic surrogate of human cancer. *Ann. Transl. Med.* **2020**, *8*, 114. [CrossRef]
42. Weiswald, L.B.; Bellet, D.; Dangles-Marie, V. Spherical cancer models in tumor biology. *Neoplasia* **2015**, *17*, 1–15. [CrossRef] [PubMed]
43. Ricci, C.; Mota, C.; Moscato, S.; D'Alessandro, D.; Ugel, S.; Sartoris, S.; Bronte, V.; Boggi, U.; Campani, D.; Funel, N.; et al. Interfacing polymeric scaffolds with primary pancreatic ductal adenocarcinoma cells to develop 3D cancer models. *Biomatter* **2014**, *4*, e955386. [CrossRef] [PubMed]
44. Mao, S.; He, J.; Zhao, Y.; Liu, T.; Xie, F.; Yang, H.; Mao, Y.; Pang, Y.; Sun, W. Bioprinting of patient-derived in vitro intrahepatic cholangiocarcinoma tumor model: Establishment, evaluation and anti-cancer drug testing. *Biofabrication* **2020**, *12*, 045014. [CrossRef] [PubMed]
45. Tang-Schomer, M.D.; Chandok, H.; Wu, W.B.; Lau, C.C.; Bookland, M.J.; George, J. 3D patient-derived tumor models to recapitulate pediatric brain tumors In Vitro. *Transl. Oncol.* **2022**, *20*, 101407. [CrossRef]
46. Osuna de la Pena, D.; Trabulo, S.M.D.; Collin, E.; Liu, Y.; Sharma, S.; Tatari, M.; Behrens, D.; Erkan, M.; Lawlor, R.T.; Scarpa, A.; et al. Bioengineered 3D models of human pancreatic cancer recapitulate in vivo tumour biology. *Nat. Commun.* **2021**, *12*, 5623. [CrossRef]
47. Garcia-Garcia, A.; Klein, T.; Born, G.; Hilpert, M.; Scherberich, A.; Lengerke, C.; Skoda, R.C.; Bourgine, P.E.; Martin, I. Culturing patient-derived malignant hematopoietic stem cells in engineered and fully humanized 3D niches. *Proc. Natl. Acad. Sci. USA* **2021**, *118*, e2114227118. [CrossRef]
48. Coussens, L.M.; Fingleton, B.; Matrisian, L.M. Matrix metalloproteinase inhibitors and cancer: Trials and tribulations. *Science* **2002**, *295*, 2387–2392. [CrossRef]
49. Wiercinska, E.; Naber, H.P.; Pardali, E.; van der Pluijm, G.; van Dam, H.; ten Dijke, P. The TGF-beta/Smad pathway induces breast cancer cell invasion through the up-regulation of matrix metalloproteinase 2 and 9 in a spheroid invasion model system. *Breast Cancer Res. Treat.* **2011**, *128*, 657–666. [CrossRef]
50. Bera, A.; Zhao, S.; Cao, L.; Chiao, P.J.; Freeman, J.W. Oncogenic K-Ras and loss of Smad4 mediate invasion by activating an EGFR/NF-kappaB Axis that induces expression of MMP9 and uPA in human pancreas progenitor cells. *PLoS ONE* **2013**, *8*, e82282. [CrossRef]

51. Feula, M.; Milazzo, M.; Giannone, G.; Azimi, B.; Trombi, L.; Cacopardo, L.; Moscato, S.; Lazzeri, A.; Ahluwalia, A.; Berrettini, S.; et al. Bioartificial Sponges for Auricular Cartilage Engineering. In Proceedings of the Advances in Bionanomaterials II, Cham, Switzerland, 29 September–3 October 2020; pp. 191–209.
52. D'Alessandro, D.; Moscato, S.; Fusco, A.; De la Ossa, J.G.; D'Acunto, M.; Trombi, L.; Feula, M.; Serino, L.P.; Donnarumma, G.; Petrini, M.; et al. Poly(vinyl alcohol)/Gelatin Scaffolds Allow Regeneration of Nasal Tissues. *Appl. Sci.* **2021**, *11*, 3651. [CrossRef]
53. Sezgin, O.; Yaras, S.; Ozdogan, O. The course and prognostic value of increased pancreas stiffness detected by ultrasound elastography during acute pancreatitis. *Pancreatology* **2021**, *21*, 1285–1290. [CrossRef] [PubMed]
54. Ricci, C.; Danti, S. 3D Models of Pancreatic Ductal Adenocarcinoma via Tissue Engineering. *Methods Mol. Biol.* **2019**, *1882*, 81–95. [CrossRef] [PubMed]
55. Ghosh, S.; Joshi, M.B.; Ivanov, D.; Feder-Mengus, C.; Spagnoli, G.C.; Martin, I.; Erne, P.; Resink, T.J. Use of multicellular tumor spheroids to dissect endothelial cell-tumor cell interactions: A role for T-cadherin in tumor angiogenesis. *FEBS Lett.* **2007**, *581*, 4523–4528. [CrossRef] [PubMed]
56. Wang, R.; Chadalavada, K.; Wilshire, J.; Kowalik, U.; Hovinga, K.E.; Geber, A.; Fligelman, B.; Leversha, M.; Brennan, C.; Tabar, V. Glioblastoma stem-like cells give rise to tumour endothelium. *Nature* **2010**, *468*, 829–833. [CrossRef]
57. Dominiak, A.; Chelstowska, B.; Olejarz, W.; Nowicka, G. Communication in the Cancer Microenvironment as a Target for Therapeutic Interventions. *Cancers* **2020**, *12*, 1232. [CrossRef]
58. Dumas, J.F.; Brisson, L. Interaction between adipose tissue and cancer cells: Role for cancer progression. *Cancer Metastasis Rev.* **2021**, *40*, 31–46. [CrossRef]
59. Calorini, L.; Bianchini, F. Environmental control of invasiveness and metastatic dissemination of tumor cells: The role of tumor cell-host cell interactions. *Cell Commun. Signal.* **2010**, *8*, 24. [CrossRef]
60. Contessi Negrini, N.; Ricci, C.; Bongiorni, F.; Trombi, L.; D'Alessandro, D.; Danti, S.; Farè, S. An Osteosarcoma Model by 3D Printed Polyurethane Scaffold and In Vitro Generated Bone Extracellular Matrix. *Cancers* **2022**, *14*, 2003. [CrossRef]

cancers MDPI

Article

Longitudinal Monitoring of Intra-Tumoural Heterogeneity Using Optical Barcoding of Patient-Derived Colorectal Tumour Models

Carolyn Shembrey [1,2], Jai Smith [1,2], Mélodie Grandin [1,2], Nathalia Williams [1,2], Hyun-Jung Cho [3], Christina Mølck [1,2], Corina Behrenbruch [1,2,4,5], Benjamin NJ. Thomson [5,6], Alexander G. Heriot [4,7,8], Delphine Merino [9,10,11,12] and Frédéric Hollande [1,2,*]

1 Department of Clinical Pathology, University of Melbourne, Melbourne, VIC 3000, Australia; Carolyn.Shembrey@petermac.org (C.S.); jai.smith@unimelb.edu.au (J.S.); melodie.grandin@unimelb.edu.au (M.G.); Nathalia.Williams@unige.ch (N.W.); christina.moelck@gmail.com (C.M.); Cori.Behrenbruch@petermac.org (C.B.)

2 Victorian Comprehensive Cancer Centre, University of Melbourne Centre for Cancer Research, Melbourne, VIC 3000, Australia

3 Biological Optical Microscopy Platform, University of Melbourne, Melbourne, VIC 3010, Australia; hcho@unimelb.edu.au

4 The Sir Peter MacCallum Department of Oncology, University of Melbourne, Melbourne, VIC 3010, Australia; Alexander.Heriot@petermac.org

5 Department of General Surgical Specialties, The Royal Melbourne Hospital, University of Melbourne, Melbourne, VIC 3050, Australia; Benjamin.Thomson@mh.org.au

6 Department of Surgery, the Royal Melbourne Hospital, University of Melbourne, Melbourne, VIC 3050, Australia

7 Department of Cancer Surgery, Peter MacCallum Cancer Centre, Melbourne, VIC 3000, Australia

8 Department of Surgery, St Vincent's Hospital, Melbourne, VIC 3065, Australia

9 Olivia Newton-John Cancer Research Institute, Heidelberg, VIC 3084, Australia; delphine.merino@onjcri.org.au

10 School of Cancer Medicine, La Trobe University, Melbourne, VIC 3086, Australia

11 Immunology Division, The Walter and Eliza Hall Institute of Medical Research, Melbourne, VIC 3052, Australia

12 Department of Medical Biology, The Faculty of Medicine, Dentistry and Health Science, University of Melbourne, Melbourne, VIC 3010, Australia

* Correspondence: frederic.hollande@unimelb.edu.au

Citation: Shembrey, C.; Smith, J.; Grandin, M.; Williams, N.; Cho, H.-J.; Mølck, C.; Behrenbruch, C.; Thomson, B.N.; Heriot, A.G.; Merino, D.; et al. Longitudinal Monitoring of Intra-Tumoural Heterogeneity Using Optical Barcoding of Patient-Derived Colorectal Tumour Models. *Cancers* **2022**, *14*, 581. https://doi.org/10.3390/cancers14030581

Academic Editors: Serena Danti, Nicola Contessi Negrini and Alessandro Franchi

Received: 22 December 2021
Accepted: 19 January 2022
Published: 24 January 2022

Simple Summary: Colorectal cancer (CRC) is the second most common cancer worldwide. Despite improvements in the clinical management of CRC, outcomes for those with metastatic disease remain extremely poor. One reason for this is tumour heterogeneity, which refers to the observation that each cell within complex tumour cell populations displays different genetic features and biological behaviours. Such tumour heterogeneity is known to impact treatment efficacy and promote tumour recurrence. Here, we present a multi-colour barcoding methodology that allows for different lineages of colorectal cancer cells to be identified and monitored, thus allowing for tumour heterogeneity to be quantified in real-time. We show that discrete cell lineages can be quantified by both fluorescence microscopy and flow cytometry. Using this approach, we show that the cell culture models that are traditionally used in cancer research display limited heterogeneity, whereas patient-derived organoids—which are generated from fresh tumour resections—more faithfully represent the heterogeneity observed in cancer patients.

Abstract: Geno- and phenotypic heterogeneity amongst cancer cell subpopulations are established drivers of treatment resistance and tumour recurrence. However, due to the technical difficulty associated with studying such intra-tumoural heterogeneity, this phenomenon is seldom interrogated in conventional cell culture models. Here, we employ a fluorescent lineage technique termed "optical barcoding" (OBC) to perform simultaneous longitudinal tracking of spatio-temporal fate in 64 patient-derived colorectal cancer subclones. To do so, patient-derived cancer cell lines and organoids were labelled with discrete combinations of reporter constructs, stably integrated into the genome and thus passed on from the founder cell to all its clonal descendants. This strategy enables the longitudinal monitoring of individual cell lineages based upon their unique optical barcodes. By designing a novel panel of six fluorescent proteins, the maximum theoretical subpopulation resolution of 64 discriminable subpopulations was achieved, greatly improving throughput compared with previous studies. We demonstrate that all subpopulations can be purified from complex clonal mixtures via flow cytometry, permitting the downstream isolation and analysis of any lineages of interest. Moreover, we outline an optimized imaging protocol that can be used to image optical barcodes in real-time, allowing for clonal dynamics to be resolved in live cells. In contrast with the limited intra-tumour heterogeneity observed in conventional 2D cell lines, the OBC technique was successfully used to quantify dynamic clonal expansions and contractions in 3D patient-derived organoids, which were previously demonstrated to better recapitulate the heterogeneity of their parental tumour material. In summary, we present OBC as a user-friendly, inexpensive, and high-throughput technique for monitoring intra-tumoural heterogeneity in in vitro cell culture models.

Keywords: organoids; tumour heterogeneity; colorectal neoplasms; clonal evolution; longitudinal imaging; neoplasm recurrence; metastasis; cell lineage; self-renewal; cell culture techniques

1. Introduction

Cancer is a highly heterogeneous disease whereby individual tumours and tumour cell subpopulations can display significant differences in their genetic, histopathologic, metabolic and immunologic profiles [1,2]. This inter- and intra-tumour heterogeneity is observed between tumours arising from different tissue or cell types, across patients with the same subtype of tumour, between primary and metastatic tumours within the same patient, and amongst individual cells from a single tumour. Indeed, tumours with a higher degree of clonal diversity have been shown to exhibit an enhanced capacity for metastatic progression [3–7] and treatment resistance [8,9], thereby conferring worse overall survival. This is notably the case in colorectal cancer (CRC) [10], the study model of the present work.

Lineage tracing is an approach used to assess intra-tumoural heterogeneity, and such clonal tracking can be adapted for both in vitro and in vivo settings. This suite of technologies (reviewed in [11]), in which the integration of a unique fluorescent or genetic barcode into individual cells enables their identification amongst heterogeneous cell pools, are powerful tools for addressing biologically relevant questions regarding clonal interactions, developmental trajectories, and the regulation of tissue homeostasis [12–14]. Lineage tracing is highly versatile and can be applied to virtually all cell types provided that three fundamental requirements are met: the lineage tracer should be passed on from founder cells to all clonal descendants, should be retained over time, and should not be transferable to unrelated or neighbouring cells. The major benefit of this approach is that it does not require prior definition of cellular markers of interest, allowing researchers to simultaneously evaluate various subpopulations without phenotypic bias. It was successfully used to monitor heterogeneity during the growth of CRC xenografts following transplantation in immunocompromised mice [15].

Previous genetic barcoding studies have revealed the enormous scope of intra-tumour heterogeneity, with reports of up to 1700 clones with distinct phenotypes and growth rates within a single tumour [16]. However, assessments of clonal contributions using this approach rely entirely upon the quantification of sequence reads, and the fidelity between barcode abundance and the resulting read counts has recently been questioned [17]. Furthermore, all sequencing methods face certain drawbacks, including errors associated with the amplification of small starting materials, challenging rates of false hits and a requirement for complex bioinformatics analysis. Most importantly, genomic data alone is unable to report on the functional importance of the cells interrogated, as endpoint assays preclude the continued monitoring of cells in culture. This caveat is particularly important when studying cellular interactions at the subpopulation level, as it limits the ability of researchers to determine whether the features which support the outgrowth of dominant clones are pre-existing or acquired de novo in response to selective pressures. For this reason, methods allowing for the longitudinal examination of clonal interactions in live tumour cell populations are desirable.

As compared with genetic barcoding, fluorescent barcoding approaches use multiple fluorescent proteins (FPs) to generate stable fluorescent signatures in recipient cells which can be visualized in real-time. Perhaps the most renowned example is the in vivo mapping of neuronal networks using transgenic Cre/lox *Brainbow* mice, in which site-specific recombination results in the stochastic expression of two, three or four fluorescent proteins [18,19]. *Brainbow* has been adapted for use in other model organisms, including drosophila melanogaster [20,21] and zebrafish [22,23], and it is possible to use the very similar Clonal Labelling of Neural Progenies (CLoNe) tool in all vertebrate species [24]. Moreover, to enable investigation into non-neural systems, the *Brainbow 2.1* cassette was modulated to produce the *R26R-Confetti* construct, which has been used for in vivo lineage tracing in various cell types, including glomerular [25], corneal [26] and Lgr5+ intestinal stem cells [27]. However, as the probes used in each of these methods exist on a continuum of fluorescence intensities, it is difficult to reproducibly purify these subpopulations using FACS, and thus, such approaches are primarily qualitative in nature and preclude the downstream analysis of any subpopulations of interest. There is, therefore, an unmet need for novel techniques that allow for both the in situ monitoring of individually tagged subpopulations as well as their isolation from complex polyclonal mixtures.

A novel approach to study heterogeneity is through optical barcoding (OBC), achieved by the simultaneous introduction of multiple reporter constructs per cell that each encode a different fluorescent protein (FP). The combinatorial expression of these FPs allows for multiple cell lineages to be traced based on their unique fluorescent signatures. Given the stable integration of these constructs into the genome of the target cell, these fluorescent signatures are passed on from the founder cell to all its clonal descendants, thus enabling the longitudinal monitoring of discrete subpopulations (hereby referred to as a "cluster") over time. Importantly, as the number of FPs employed for barcoding is user-defined, and the throughput of this technique can be upscaled depending on the application of interest.

Early iterations of the OBC technique were based upon the additive red-green-blue (RGB) additive colour principle, in which the combinatorial expression of red, green and blue fluorescent proteins (mCherry, Venus and Cerulean, respectively) in heterogeneous cell populations can theoretically produce cellular clones with fluorescence signatures corresponding to all hues of the visible colour spectrum [28]. RGB marking has been implemented to monitor the clonality of liver regeneration following transplantation of RGB-barcoded primary hepatocytes [29], as well as to study the heterogeneous sensitivity of CRC cells to stress signals [30]. However, like the aforementioned techniques, these studies define individual subpopulations based on the fluorescent intensities of each vector and therefore make FACS-based quantitation difficult.

Recently, the OBC technique has been expanded from three to six fluorescent vectors [31], greatly expanding the number of subpopulations that can be unambiguously identified and traced within a single experiment. The major benefit of upscaling the number of fluorescent constructs used within an OBC panel is that the subpopulations can be defined in a binary manner (i.e., "blue" or "not blue"), thus circumventing the issues created by fluctuating fluorescent intensities.

Using this binary approach, recent studies have been able to trace clonal heterogeneity in breast cancer [32] and malignant glioma [31]. However, until now, optical barcoding has been limited to the quantitation of up to 41 subpopulations, using flow cytometry (FC) only. Imaging was used on smaller numbers of subclones (up to 31), limiting the spatial analysis of large sub-clonal interactions. In this study, we report a novel panel of FPs that allows for all theoretical subpopulations to be unambiguously discriminated by flow cytometry, thereby enhancing the resolution of the OBC technique. This panel is used in patient-derived colorectal cancer (CRC) cells grown in 2D as well as organoids grown in 3D. We also demonstrate for the first time that this optimized FP panel can expand the OBC approach to allow for live-cell imaging studies, thus permitting real-time monitoring of cellular responses in situ. Finally, by combining these approaches, we demonstrate that OBC can be used to trace and quantify dynamic clonal interactions in 3D patient-derived CRC organoids.

2. Results

2.1. Validation of OBC Conditions in Patient-Derived CRC Cells

OBC is achieved via the simultaneous transduction of wild-type cells with several fluorescent constructs, thus generating a multi-coloured cell pool. As fluorescent signatures are passed on from the parental cells to all their clonal descendants, individual cellular subpopulations are identifiable based on their distinct fluorescent signature (Figure 1A,B). In our study, the OBC panel is composed of six different FPs (EBFP, blue; tSapphire, green; Cerulean, cyan; Venus, yellow; mOrange, orange; dKatushka, red) with unique excitation and emission spectra (Figure 1C). Thus, using a binomial coefficient, the concurrent expression of up to six of these colours should theoretically generate 64 clusters discriminable by their unique fluorescent signatures (Figure 1D). The proportion of each cluster is defined by the transduction efficiency for each of the six colours. At a transduction efficiency of 50% per colour, 64 subpopulations of equal size (each representing 1.56% of the total) should be generated. We observed that actual values deviate from this prediction, depending on the actual TE values obtained experimentally. In this study, the submaximal transduction efficiency of 40–70% was accepted, as this facilitates optimal colour diversity. Lower percentages result in a large proportion of indistinguishable, non-barcoded cells, whilst higher percentages increase the proportion of five and six vector positive cells, decreasing cluster diversity.

It is well established that various cellular events such as cell division can modulate FP intensity in vitro. Although fluctuations in fluorescence can be exploited for quantitative imaging of both RNA biogenesis and protein–protein interactions [33], this property would be a hindrance in the context of lineage tracing as it may cause cells to "shift" from one lineage assignment to another based on their functional state. Thus, rather than enumerate fluorescent intensity for each probe, the OBC method quantifies the expression of each of the six FPs is annotated in a binary manner (i.e., "blue" or "not blue") which, when combined, generates a fluorescent barcode that is less sensitive to transient changes in fluorescent signal. This feature ensures that clusters can be unambiguously identified and purified via FC, facilitating the cross-comparison of specific clusters throughout the course of a single experiment as well as across independent experiments. When expression levels are scored in this manner, fluorescent signatures can be described using a barcode nomenclature where solid and empty bars denote positivity and negativity for each FP, respectively (Figure 1E).

2.2. OBC Does Not Alter Biological Properties of Wildtype Cells

Although lentiviral transduction is a common tool employed in biomedical research, there is a small possibility that the introduced constructs may lead to insertional mutagenesis, resulting in genetic dysregulation at the site of viral integration. Such insertional mutagenesis can influence the composition of polyclonal populations, promoting the outgrowth of certain clones whilst suppressing others [34,35]. Although the LeGO vectors employed in this study are rendered self-inactivating (SIN) via a deletion in the 3′ LTR, which is known to lower the likelihood of insertional mutagenesis events, multiplexing six simultaneous transductions may conceivably result in off-target effects. Thus, to validate that the optically barcoded cells are representative of the parental cell lines from which they were derived, we tested the impact of optical barcoding on the proliferation, self-renewal properties and drug sensitivity of labelled CRC cell lines.

Three patient-derived CRC cell lines [36,37] were optically barcoded and their biological properties compared with that of their parental cell lines. The CPP14 and CPP35 lines were derived in our laboratory from treatment-naïve primary CRCs, whereas the

CPP19 was generated from the liver metastasis of a patient treated with neo-adjuvant chemotherapy.

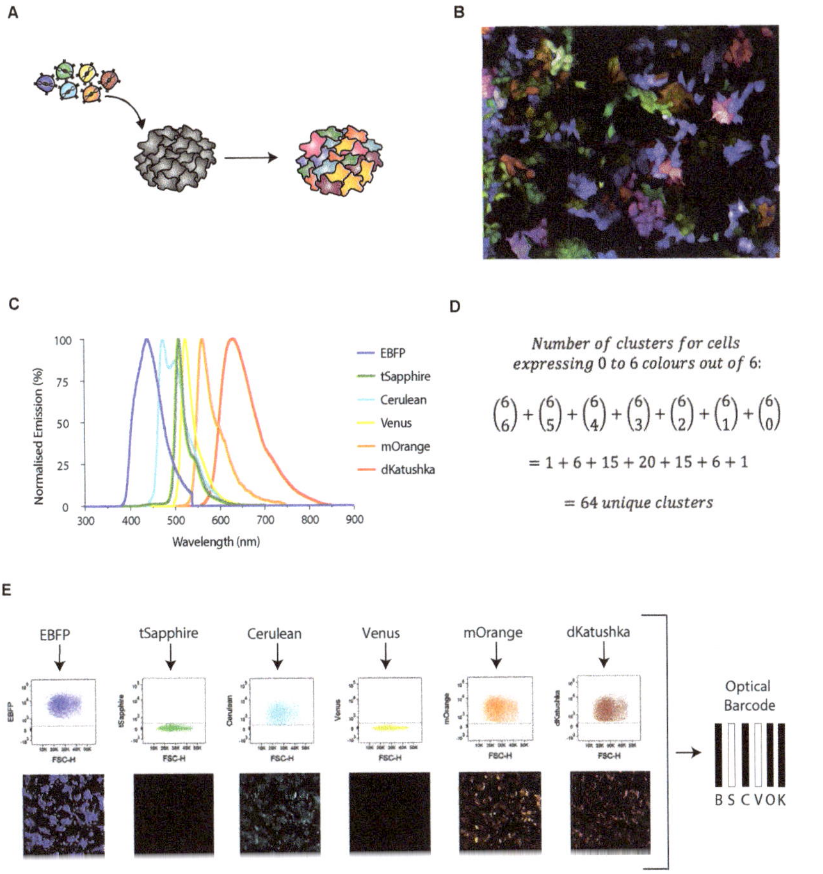

Figure 1. Optical barcoding as a tool to monitor intra-tumoural heterogeneity. (**A**) Optical barcoding is facilitated by the simultaneous transduction of heterogeneous cell populations with six different FPs, generating a multicolour pool of cells with unique fluorescent signatures. (**B**) Representative image of an optically barcoded cell line. (**C**) The six FPs employed in the optical barcoding panel are EBFP (B, blue), tSapphire (S, green), Cerulean (C, cyan), Venus (V, yellow), mOrange (O, Orange) and dKatushka (K, red). Curves show emission spectra only. (**D**) Using a binomial coefficient, 64 distinct barcodes will be generated when cells are transduced with 6 constructs and allowed to express 0 to 6 colours simultaneously (1 six-colour combination, 6 five-colour combinations, 15 four-colour combinations, etc.). (**E**) Expression levels for each FP are scored in a binary manner, generating a 'barcode' nomenclature that summarizes the fluorescent signature of each cluster. Solid bars denote FP-positivity, whereas blank bars indicate FP-negativity.

Proliferation assays conducted over a 96 h period confirmed that there was no significant difference in growth kinetics nor doubling time between the parental (WT) and optically barcoded (L6) cells (Figure 2A; Table S1). Similarly, as a defining property of stem-like cancer cells is an increased capacity for self-renewal, we used Extreme Limiting Dilution Analysis (ELDA) [38] to infer the stem cell frequency (SCF) of the wild-type cell lines and their optically barcoded variants. This sphere-forming assay is designed for the enrichment of tumour-initiating cells and follows a single-hit hypothesis which postulates that only single cells endowed with self-renewal capacity should be able to generate spheres. In all three cell lines, there was no significant difference in SCF observed between WT and L6 pairings (Figure 2B). Lastly, to evaluate whether the OBC approach modulates therapeutic response, we compared the dose-response profiles of our overall WT and L6 cell populations after exposure to two targeted inhibitors of the MAPK pathway, cobimetinib and selumetinib. As the 95% confidence intervals for LogIC50 values were overlapping for all WT/L6 pairings, with LogIC50 fold changes ranging between 0.94 and 1.34, indicating that OBC did not significantly modify sensitivity to these compounds (Figure 2C; Table S2). Overall, these results indicated that the transduction of CRC cell lines with LeGO vectors didn't affect their proliferation, self-renewal capacity and sensitivity to targeted inhibitors.

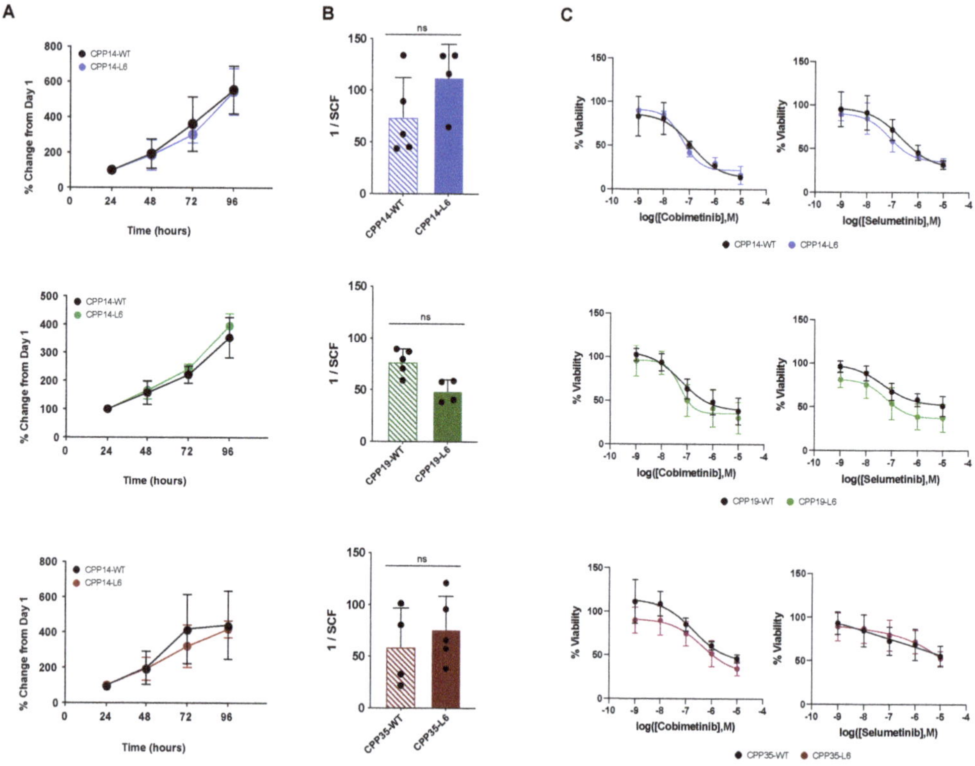

Figure 2. Optical barcoding does not perturb the biological properties of wild-type cells. Three patient-derived cell lines (CPP14, blue; CPP19, green; and CPP35, red) were optically barcoded (denoted by "L6" variants). (**A**) Proliferation was assessed via a resazurin metabolic assay performed at 24 h intervals for a period of 96 h (n = 4 independent experiments, mean ± SD). Doubling time (hours) ± 95% CI for the WT and L6 pairs were calculated by fitting an exponential growth equation (ns = non-significant, see Table S1). (**B**) ELDA performed to determine the SCF of the WT vs L6 cell pairs (n = 4 independent experiments, mean ± SD). The presence or absence of colonospheres was assessed 10 days after seeding at densities of 1000, 1000, 10 or 1 cells per well and is reported as the mean SCF ± SD (P = ns, Student's t-test). (**C**) WT and L6 cells were treated with escalating doses of the MEK inhibitors cobimetinib and Selumetinib for 72 h, at which point cell viability was assessed via resazurin assay. Data is normalized to the vehicle control for each compound and reported as mean % viability ± SD (n > 4). LogIC50 ± 95% CI values were calculated by interpolating sigmoidal dose-response curves (see Table S2). WT, wild-type; ELDA, Extreme limiting dilution analysis; SCF, stem cell frequency.

2.3. A Novel FP Panel Allows for Enhanced OBC Resolution and Precise Barcode Quantification Using Flow Cytometry

Next, we confirmed that our OBC panel performed robustly when analyzed by flow cytometry. Compensation matrices derived from this new panel demonstrates that all possible combinations of the six LeGO FPs can be clearly separated using FC (Figure 3A), including spectrally adjacent FPs (such as tSapphire/Cerulean). We next confirmed that the spectral barcodes assigned to different cell subpopulations were stable over time. Subpopulations with specific barcodes were sorted via FC, serially passaged in culture, and their fluorescence profiles were analyzed after every second passage. Over the course of one month and eight passages, no changes were observed in the barcode profiles of the barcoded population (Figure 3B).

We next asked whether a novel gating strategy that would allow for the purification of all 64 theoretical barcoded subpopulations could be designed. This would significantly increase the resolution of the OBC technique, which was previously restricted to the analysis of only 41 subpopulations. To this end, a hierarchical gating strategy was established (Figure 3C); here, the cells of interest are sequentially passed through a combination of univariate (i.e., Cerulean+ vs Cerulean-) and bivariate (i.e., EBFP+/mOrange-, EBFP+/mOrange+, EBFP-/mOrange+, EBFP-/mOrange-) gates encompassing each of the 6 FPs utilized in the OBC panel. As the near-infrared (near-IR) region of the visible spectrum is unoccupied, assessments of cell viability and/or other antibody applications can be performed upstream of this 6-colour deconvolution.

To test the fidelity of this gating strategy, heterogeneous mixtures of cells were purified into 64 uni-barcoded populations. Reanalysis of the sorted cells confirmed that the sorted populations maintained the expected fluorescence profile with no spectral leakage (Figure 3D), validating the precision of the gating and compensation settings.

2.4. Spectral Imaging with Linear Unmixing Allows for Real-Time Tracing of Optically Barcoded Subpopulations

Although flow cytometry analysis is a powerful tool for the enumeration and purification of optically barcoded cells, it does mandate that the cells of interest be detached from cultureware or excised from recipient mice prior to analysis. As the in situ analysis of barcoded subpopulations would allow for additional morphological and spatial information to be acquired, we next asked whether live-cell imaging could be used for barcode quantification.

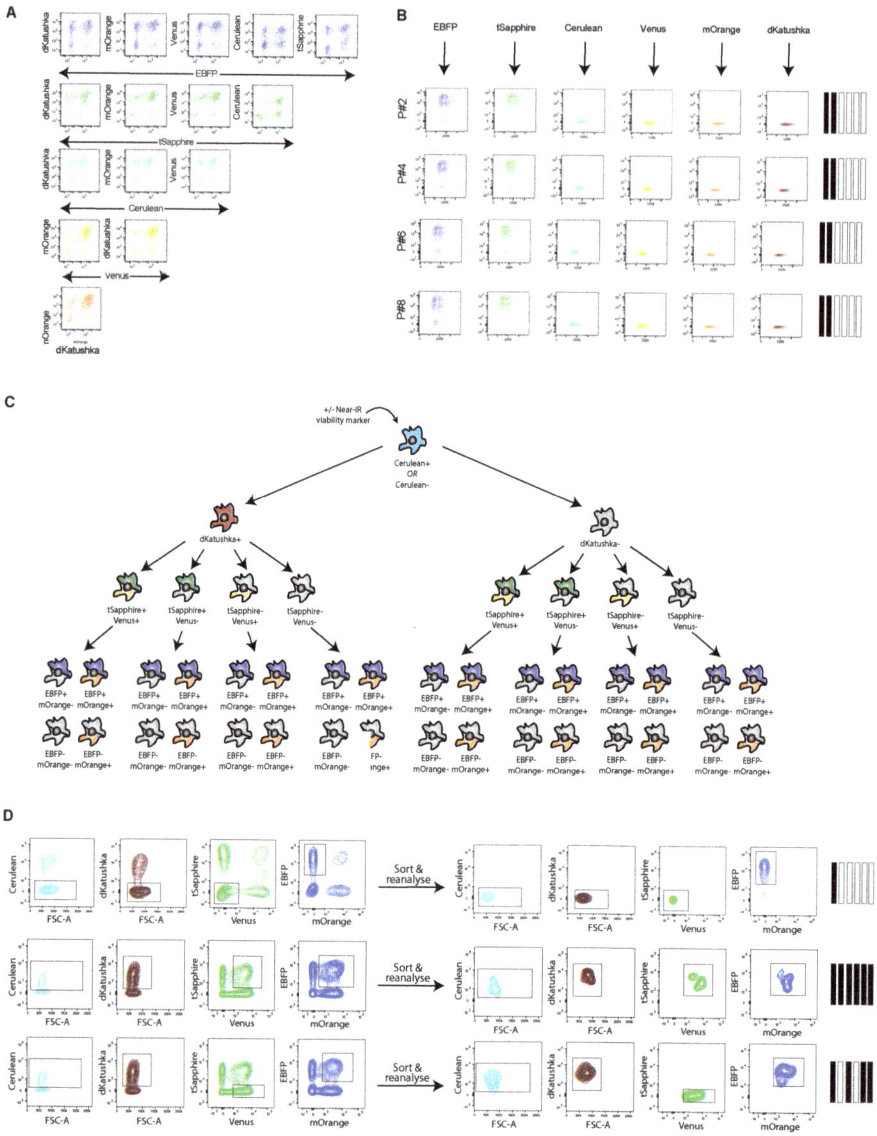

Figure 3. An optimized OBC panel allows for robust cluster separation via flow cytometry. **(A)** Bivariate compensation matrix demonstrates that all combinations of the six OBC constructs can be clearly distinguished using flow cytometry. **(B)** Barcoded subpopulations were isolated from the CPP35-L6 cell line and maintained in culture. Fluorescence signatures were analyzed by flow cytometry at each passage. Representative plots demonstrate the fluorescence stability of a single barcoded subpopulation for a duration of 8 passages (~4 weeks in culture). **(C)** A hierarchical gating strategy that combines sequential uni- and bivariate gates allow for each of the 64 barcoded subpopulations to be isolated via flow cytometry. Viability markers or antibodies are compatible with this strategy, provided they are conjugated to near-IR FPs. **(D)** Representative plots demonstrating that single barcoded populations can be purified from heterogeneous mixtures. Reanalysis of the sorted population confirms the accuracy of the gating strategy. IR, infra-red.

To image optical barcodes in live cells, cells expressing known fluorescence barcodes were purified by FACS and subsequently imaged using spectral imaging with linear unmixing (SILU) protocol (Figure 4A; see Methods for further details). Irrespective of the number of FPs present in the barcoded population (zero through six out of six), the fluorescence profiles obtained by SILU analysis were consistent with the ground-truth profiles obtained by flow cytometry (Figure 4B). Additionally, proof-of-concept staining with an AlexaFluor 647-conjugated antibody against E-cadherin revealed that the far-red region of the visible spectrum was compatible with spectral deconvolution using SILU (Figure 4C). Like the near-IR marker used for FC analysis, the addition of this 7th far-red probe allows for additional antibody-based studies to be multiplexed with the OBC panel, with the added benefit of being able to determine the cellular localization of the target of interest.

2.5. Applying OBC to Monitor Clonal Outgrowth in Heterogenous Tumour Cell Populations

Phylogenetic lineage-tracing studies have demonstrated that, over time, tumour cell populations may evolve in a manner that either supports the concomitant growth of multiple minor clones or promotes the emergence of a single, dominant clone [39]. Thus, having validated the technical aspects of the OBC system, we sought to determine whether OBC could be used to monitor differential growth kinetics at the subpopulation level. Preliminary analyses using conventional adherent CRC cell lines revealed limited shift in clonal composition over time (Figure S1), consistent with previous reports that the clonal diversity of cell lines grown as 2D culture models declines with increased passaging [40]. Subsequently, we elected to profile cluster dynamics in primary CRC cells grown as 3D structures termed patient-derived organoids (PDOs). By embedding patient-derived cells in a Matrigel matrix supplemented with physiologically relevant growth factors, PDOs adopt a 3D morphology that faithfully recapitulates the architecture of the parental tissue. The provision of ECM components like collagen and laminin not only supports 3D growth of these organoids but additionally mimics the physiological barriers which influence compound delivery and immune cell infiltration in in vivo lesions. Molecular and functional studies of large organoid biobanks have confirmed that organoids retain patient-specific mutation profiles and mimic clinical responses to a variety of anti-cancer compounds [41,42].

We utilized a panel of PDOs obtained from the liver metastases of three Stage IV CRC patients, recently developed and characterized in our group [43], including one treatment-naïve sample, P275_LT, and two neoadjuvant FOLFOX-treated samples, yP315_LT and yP295_LT. Each PDO was optically barcoded (P275_LT-L6, yP315_LT L6 and yP295_LT L6; Figure 5A) and subjected to serial flow cytometric analysis to enumerate cluster distribution after every second passage (approximately 2–3 week intervals, depending on the growth rate of each PDO).

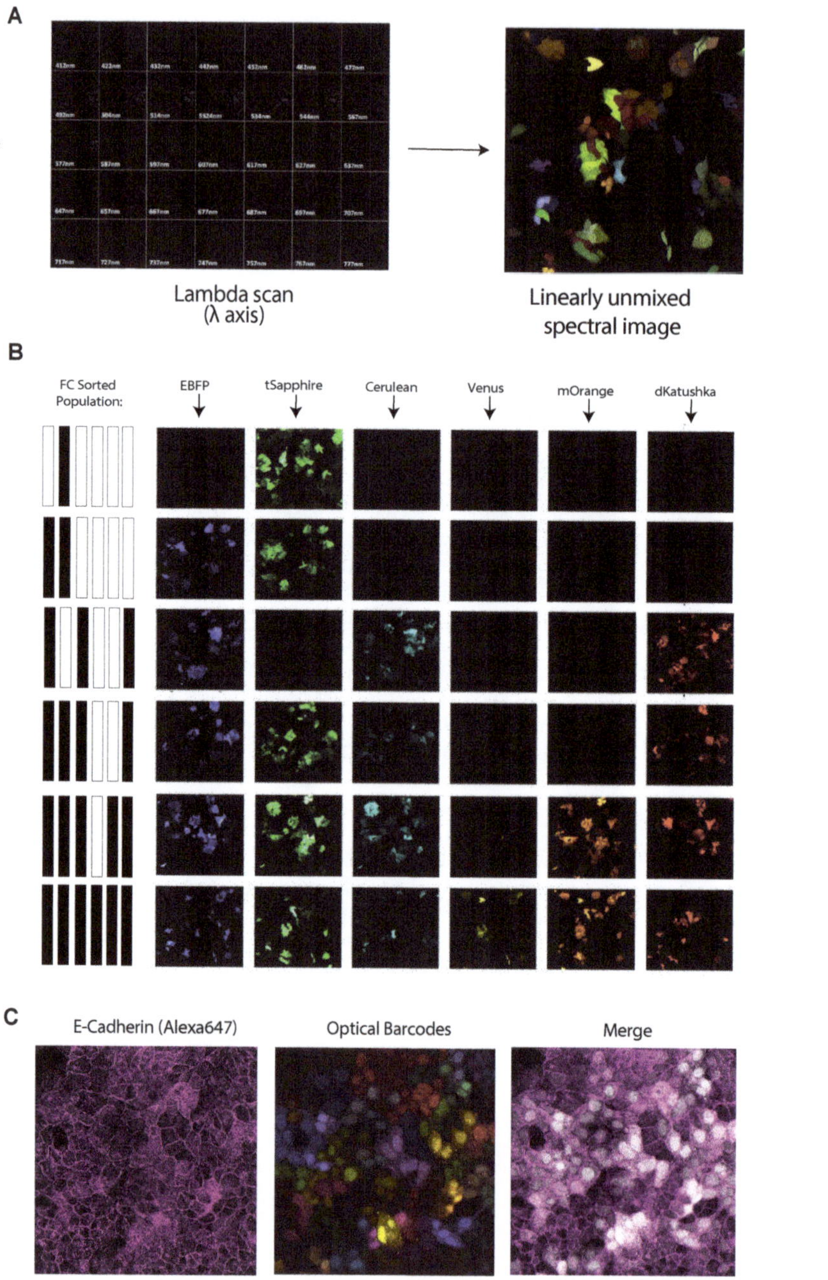

Figure 4. Spectral imaging with linear unmixing allows for real-time identification of optically barcoded subpopulations. (**A**) Schematic of the SILU analysis pipeline used to image optical barcodes in live cells. (**B**) Subpopulations with barcodes indicated at left were sorted, and their fluorescence signatures were corroborated via SILU. (**C**) CPP35-L6 cells were stained for E-cadherin to demonstrate the compatibility of far-red fluorochromes with the OBC panel, enabling seven-colour applications. In merged panel, OBC channels have been converted to greyscale. SILU, spectral imaging with linear unmixing.

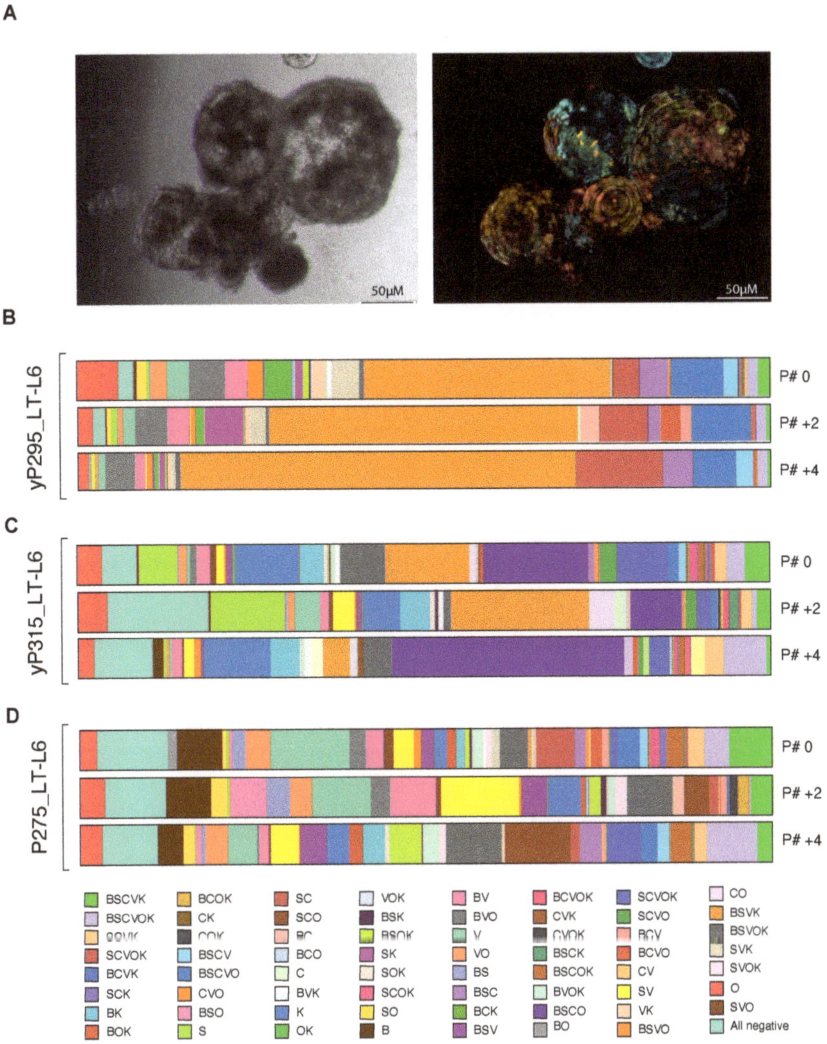

Figure 5. Optical barcoding of patient-derived organoids allows for the real-time estimation of subpopulation growth kinetics. (**A**) Representative brightfield (left) and fluorescence (right) images of optically barcoded CRC PDOs. (**B–D**) Bar-charts illustrating cluster frequencies following serial flow cytometric analysis of three optically barcoded PDO lines derived from primary colorectal liver metastasis samples. (**B**) yP295_LT-L6 (**C**) yP315_LT-L6 and (**D**) P275_LT-L6. Each of the 64 colours represents a single optically barcoded subpopulation, analyzed at baseline (P#0) and after 2 (P# +2) and 4 (P# +4) passages (~4 weeks total duration).

The BSVK+ population (orange) was the most dominant subpopulation in the yP295_LT-L6 sample (Figure 5B), steadily expanding from 35.6% of the total population to at P#0 to 56.8% by P#+4. Similarly, the SC+ population (salmon) steadily increased throughout the analysis period, expanding ~3.5-fold from 3.6% at P#0 to 12.5% by P#+4. Collectively, these results suggest that serial passaging of the yP295_LT-L6 line leads to profound contraction of most subpopulations in favour of the steady outgrowth of the major clusters.

For yP315_LT-L6 (Figure 5C), whilst the BSVK+ cluster (orange) was dominant at P#0 (12.2% of total cells), it had contracted to 3.8% by P#+4. Of note, this contraction was accompanied by the outgrowth of the other major subpopulation in this line, BSCO (purple), from 15.3% at P#0 to 33.7% at P#+4, suggestive of clonal competition between these 2 subpopulations.

Conversely, P275_LT-L6 (Figure 5D) demonstrated a comparative lack of inter-clonal competition. Few dominant clusters were observed, with the largest subpopulation at any time point, V+ (seafoam green), comprising only 11.2% of the total cell population at P#0 (later contracting to = 4.2% by P# +4). Moreover, only 3 subpopulations (V+, seafoam green; BSVOK, grey; and the all-negative, teal) had populations above 5% of the total cell number at 2 out of 3 timepoints.

Collectively, these results demonstrate that the OBC technique can be used to monitor the growth kinetics of individual cellular lineages as well as the complex interplay between these subpopulations in real-time.

3. Discussion

The isolation and cloning of green fluorescent protein (GFP) from *Aequorea victoria* revolutionized biomedical research, and since then, the repertoire of FPs with unique absorption and emission spectra has greatly expanded [44]. Multiplexing of these FPs has spawned several colour-based fluorescent barcoding technologies, such as OBC, which allow for cellular lineage tracing in real-time. Here, individual cells and all their clonal descendants within the barcoded population express a unique fluorescent signature, facilitating the spatio-temporal monitoring of unique subpopulations and assessment of their divergent behaviours. A major benefit of the OBC approach is the unbiased nature of cluster labelling, which does not require prior knowledge of specific markers of interest. Additionally, as compared with other endpoint analyses such as genetic barcoding, the ability to monitor fluorescent barcodes in real-time allows for longitudinal, live-cell analyses to be conducted, as well as enabling the prospective identification and isolation of different clusters of interest for subsequent studies.

In this study, we present a novel panel of FPs which, using a purpose-built multiplexed FC gating strategy as well as a novel SILU image analysis pipeline, enabled the unambiguous identification and purification of all 64 possible OBC subpopulations. This greatly improves the resolution of OBC as first proposed by Mohme and colleagues [31], where only 41 subpopulations were discriminable via FC, and simultaneously allows for the spatial distribution of the various subpopulations to be visualized. Indeed, amenability to FC analysis is a significant bottleneck that frequently reduces the throughput of fluorescent reporter technologies. For example, a novel transgenic mouse model has recently been introduced whereby stochastic recombination of a Cre-reporter construct can generate cells with over 100 unique multicolour signatures [45]. Although this, in theory, greatly extends the colour repertoire reported in previous techniques such as the *Brainbow* [18] and *Confetti* [46] mice, due to technical limitations associated with FC analysis, the quantification was restricted to a maximum of 15 colours per animal. Similarly, Maetzig and colleagues [47] recently described a method of multiplexed fluorescent barcoding that can generate up to 26 different colours; however, only 12 of these were traced in vivo. Although the SILU-based analyzes may need to be revised for ex vivo or intravital colour deconvolution, the FC pipeline presented herein is suitable for the analysis of optically barcoded cells cultured in vitro as well as dissociated primary tumour material recovered from animal models. Thus, it is expected that the novel OBC panel presented in this work may facilitate higher-throughput lineage tracing studies both in vitro and in vivo.

Optically barcoding a panel of PDOs derived from colorectal liver metastases allowed for distinct subpopulation interaction dynamics to be observed. For two of the three PDOs profiled, yP295_LT-L6 and yP315_LT-L6, serial passaging led to the emergence of a small number of dominant subpopulations. This is consistent with previous studies investigating the clonal composition of metastases. Studies using genetic barcoding in PDX models of triple-negative breast cancer (TNBC) demonstrated that a small proportion of subclones present in the primary tumour were the major contributors of metastases [48,49].

Interestingly, these results were not mimicked in the 2D setting, where limited dynamic clonal interactions were observed.

The OBC technique presented herein could also extend beyond monitoring growth kinetics at steady state. For example, as the presence of drug-resistant cells is known to undermine treatment efficacy [50–52], future studies could determine whether the OBC technique could be implemented to identify, monitor, and characterize drug-resistant subpopulations with temporal and spatial resolution. In this respect, it is noteworthy that P275_LT-L6, the only chemo-naïve sample included in this study, showed a relatively stable clonal distribution pattern, whereas yP295_LT-L6 and yP315_LT-L6, which were both previously exposed to neoadjuvant chemotherapy, demonstrated profound clonal expansion and contraction upon serial passaging.

4. Methods

4.1. Cell Lines and Tissue Culture

The CPP14, CPP19 and CPP35 patient-derived colorectal cancer cells (CPPs) were prepared from fresh colorectal cancer biopsies under Human ethics agreement #2011-A01141-100 40 (Nimes University Hospital, France), as described in [36,37]. Cells were maintained in Dulbecco's Modified Eagle's Medium (DMEM; Lonza, #12-614F) supplemented with 10% fetal bovine serum (FBS) (Bovogen Biologicals), Glutamax (2 mM; Life Technologies, #35050-061) and penicillin-streptomycin (10.00 U/mL; Life Technologies, #15140-122) at 37 °C in a humidified atmosphere with 5% CO_2.

4.2. Patient-Derived Organoid Culture

Colorectal liver metastases were collected from patients undergoing treatment at Peter MacCallum Cancer Centre and St Vincent's Hospital, Melbourne, Australia. This study was approved by the Peter MacCallum Cancer Centre Human Research Ethics Committee (HREC/15/PMCC/112, project #15/169), and written consent was obtained from each patient according to National Health and Medical Research Council guidelines.

Tumour samples of approximately 0.25–1 cm^3 were collected directly from the operating theatre and used for the establishment of PDOs, which were maintained in Advanced DMEM/F12 (#2634010, Thermo Fisher Scientific) supplemented with Glutamax (2 mM; Life Technologies, #35050-061), penicillin-streptomycin (10.00 U/mL; Life Technologies, #15140-122), recombinant human EGF (50 ng/mL, Miltenti Biotec, #130-097-749), Leu [15] human Gastrin I (1 µg/mL, Merck, # G9145), N-acetyl cysteine (1 mM, Merck, #A9165), B27 (2X; Life Technologies, #17504044), A83-01 (ALK inhibitor; 500 nM, Merck, #SML0788), SB202190 (p38/MAPK inhibitor; 10 nM, Miltenyi Biotec, #130-106-275) and YP-27632 (ROCK inhibitor; 10 µM, Abcam, #120129) and at 37 °C in a humidified atmosphere with 5% CO_2.

Organoids were cultivated in 24-well plates and typically passaged every 5–7 days at a split ratio of 1:2–1:4. Media was aspirated, and each well was washed with ice-cold PBS. Growth-factor reduced Matrigel matrix (Corning, #356231) was detached from the cultureware by scratching with the tip of a p1000 pipette and contents of each well were transferred to a falcon tube containing 10 mL of ice-cold PBS. Following centrifugation at 1200 rpm for 5 min, pelleted organoids were dissociated by mechanical trituration such that large aggregates were broken into small (<100 um) fragments. Fragments were resuspended in the appropriate amount of Matrigel (as determined by split ratio) and plated by adding 40 µL of cell suspension per well. When single-cell suspensions were required for flow, pelleted organoids were resuspended in 1 mL of pre-warmed TrypLE Express and incubated for up to 1 h at 37 °C in a temperature-controlled orbital shaker. To correct for variation in protein concentration between Matrigel batches, all Matrigel was pre-diluted to a concentration of 8.7 mg/mL in ice-cold basal DMEM/-F12. Matrigel domes were incubated at 37 °C and allowed to solidify for 30–60 min. Following incubation, 500 µL of complete organoid media was added to each well.

4.3. Production of Viruses Used for Optical Barcoding

To generate viral particles for optical barcoding, HEK-293T packaging cells were transiently transfected using $CaCl_2$. Packaging cells were seeded in complete DMEM in 10 cm dishes at 80% confluency and incubated overnight. Four hours prior to transfection, complete media was refreshed. A transfection cocktail composed of packaging vectors (pMDLg/pRRE [10 mg; Addgene #12251], pRSV-Rev [5 mg; Addgene #12253] and pMD2.G/VSVg [2 mg; Addgene #12259]) and target DNA (10 mg per LeGO vector; Table 1) was combined with $CaCl_2$ (2 M). Whilst air was bubbled through the solution, the DNA-$CaCl_2$ mix was then added to an equal volume of 2 ×HEPES buffered saline (HEPES 0.05 M, NaCl 0.28 M, Na_2HPO_4 1.5 mM [pH 7]). The solution was allowed to incubate for 20 min at RT, mixed well, then added gently to packaging cells and incubated overnight at 37 °C and 5% CO_2. The following day, DMEM media was refreshed, and for the next 48 h, lentivirus-containing supernatants were harvested once daily. The harvested media was passed through a 0.4 mm filter to exclude cells and cellular debris. Viral titers were determined via titration of the 6 supernatants (one per vector) on each target cell line, analyzed 5 days later by flow cytometry on a FACSAria Fusion (Becton Dickinson) with a custom optical configuration (described in Section 2.5).

Table 1. Constructs used for optical barcoding.

Fluorescent Protein	Plasmid Name	Plasmid ID (Addgene #)	Excitation (nm)	Emission (nm)
EBFP	LeGO-EBFP2	85213	360–400	410–480
tSapphire	LeGO-S2	85211	360–400	500–550
Cerulean	LeGO-Cer2	27338	410–430	460–500
Venus	LeGO-V2	27340	490–510	500–550
mOrange	LeGO-mOrange2	85211	520–550	560–630
dKatushka	LeGO-dKatushka2	85214	600–630	640–680

4.4. Optical Barcoding of Primary Cell Lines and Organoids

Six third-generation HIV-1-derived self-inactivating lentiviral gene ontology (LeGO) vectors (Table 1) were used for stable transduction of target cells as described previously [51]. Briefly, target cell lines (CPP14, CPP19, CPP35) were plated in 1 mL complete DMEM in a 24-well plate at a density of 50,000 cells per well. The following day, various amounts (as determined by the viral titres) of all 6 LeGO viruses were combined with 1 mg/mL polybrene, and the resulting mixture was added dropwise to the target cells. After overnight incubation at 37 °C and 5% CO_2, complete DMEM media was refreshed. To optically barcode patient-derived organoids, the above procedure was performed; however, ULA plates were used to prevent cell adherence, and, the morning after transduction, organoids were harvested and replated in Matrigel domes (as described in Section 4.2). For both cell lines and organoids, the transduction efficiencies for each fluorescent protein were analyzed 2 passages later by flow cytometry on a FACSAria Fusion (Becton Dickinson) with raw cytometry data processed using FlowJo software (v 10.4, Tree Star Incorporated). Samples were accepted for further experimentation if the 6 fluorescent proteins were each expressed by a submaximal proportion of the barcoded cells (40–70%).

4.5. Resazurin Reduction Assay

Resazurin sodium salt working solution (Sigma-Aldrich, #R7017-1G) was prepared by resuspension in PBS (pH 7.4) to a final concentration of 0.15 mg/mL. Working solutions were filter sterilized and stored at 4 °C in light-protected tubes. Adherent cells were harvested using trypsin and resuspended at a concentration of 50,000 cells/mL. Cells were seeded into clear 96-well plates at a density of 5000 cells per well (100 µL of cell suspension) and allowed to adhere for 24 h. For proliferation assays, 20 µL of resazurin working solution was added to a set of quintuplicate wells every 24 h for a duration of 96 h. After incubation at 37 °C for 2 h, the resazurin reduction outcome measure was obtained by subtracting background readings from the absorbance measurement at 560 nM using the EnSpire multimode plate reader (Perkin Elmer, #6055400). For cytotoxicity assays, cells were treated, in triplicate, with the MEK inhibitors cobimetinib (Selleckchem, #S8041) or selumetinib (Selleckchem, #S1008) at various concentrations. After 72 h, 20 µL of resazurin working solution was added to each well and allowed to incubate at 37 °C for 2 h prior to absorbance measurement. Concentration-response curves were generated using non-linear regression analysis to determine drug potency (logIC_{50}). Pooled graphs were generated from the means of data, normalized to the vehicle controls, from each individual experiment.

4.6. Extreme Limiting Dilution Analysis (ELDA)

ELDA was used to evaluate the self-renewal capacity of individual tumour cell lines as described previously. Briefly, adherent cells were harvested using Accumax Cell Dissociating Solution (Innovative Cell Technologies) for 15 min at 37 °C and mechanically pipetted until a single-cell suspension was obtained. Cells were plated in 96-well ultra-low attachment (ULA) plates (Corning Life Sciences, #CLS3474) in 100 µL Advanced DMEM/F-12 (Gibco, #12634010) supplemented with Glutamax (2 mM, Life Technologies, #35050-061) D-glucose (10% (*v*/*v*); Sigma-Aldrich, #G7021), N-2 Supplement (1% (*v*/*v*); Thermo Fisher Scientific, #17502048), human recombinant insulin (0.2% (*v*/*v*); Sigma-Aldrich, #I0908), human recombinant EGF (0.02% (*v*/*v*); Miltenyi Biotec, #130-097-749) and human recombinant FGF (0.01% (*v*/*v*); Miltenyi Biotec, #130-093-841) at cell densities of 1000, 100, 10 and 1 cell(s) per well, reflecting the dynamic range of the sphere-forming frequency of adherent cell lines. Twelve replicates for each dilution were obtained for each of the four treatment groups. 50 µL Media was replenished after 7 days in culture to prevent wells from drying out.

After 11 days in culture, sphere formation was evaluated by scoring each well for the presence (+) or absence (−) of a sphere(s) in a binary manner. The fraction of responding cells obtained at each density was then analyzed using the ELDA Webtool made available by the Bioinformatics Division of the Walter and Eliza Hall Institute of Medical Research (http://bioinf.wehi.edu.au/software/elda/, accessed on 18 January 2022) to derive the stem cell frequency (SCF) with corresponding 95% confidence intervals (CIs) for all treatments groups tested. For individual experiments, pairwise differences in SCFs across treatment groups were assessed using a χ^2 test. For collated data, statistical analysis was performed using one-way, unpaired ANOVA followed by the Bonferroni correction for multiple comparisons.

4.7. Flow Cytometry

For analysis of optically barcoded cells, a custom optical configuration was designed (Table 2). To ensure accurate spectral deconvolution of optically barcoded cells, single-colour positive control cells were used for instrument compensation for each independent experiment. When viability staining was required, cells were incubated with LIVE/DEAD™ Fixable Near-IR Dead Cell Stain (Thermo Fisher Scientific, #L34975) for 20 min on ice, protected from light.

Table 2. Optical configuration of BD FACSAria Fusion used for flow cytometric analysis 1.

Fluorochrome	Laser Line	Long-Pass Filter	Band-Pass Filter	Detector
EBFP	405 nm	N/A	430/25	C
tSapphire	405 nm	505	510/21	A
Cerulean	405 nm	450	485/22	B
Venus	488 nm	525	543/23	A
mOrange	561 nm	570	582/15	B
dKatushka	561 nm	750	780/60	A

For all flow cytometry analysis and cell sorting, single-cell suspensions were re-suspended in FACS Buffer (2% FBS, 0.2% EDTA in PBS) and passed through a 35 μm nylon mesh cell strainer to exclude large aggregates. All cytometry was performed on a FACSAria™ Fusion cytometer (Becton Dickinson). Raw cytometry data were analyzed using FlowJo software (Tree Star Incorporated, v 10.4)

4.8. Spectral Imaging with Linear Unmixing

Imaging of optically barcoded cells was performed using the Olympus FluoView™ FV3000 (Tokyo, Japan) confocal microscope using a 20× air objective (numerical aperture of 1.2). To generate reference spectra for each FP, the confocal lambda-scanning mode (xyλ) was used to acquire emitted light in 10 nm bandwidths (2 nm step size) of the visible spectrum from CPP35 cells transduced with a single fluorescent construct. These spectral traces were recorded using the 3 laser excitations wavelengths, which correspond to those used for flow cytometric analyzes (405 nm, 488 nm and 561 nm) to harmonize the 2 techniques. Spectral deconvolution and image processing was performed using the normal unmixing function with background correction.

4.9. Statistical Analyses

All statistical analyses were performed using GraphPad Prism software (GraphPad, v8.2.1). Data are expressed as mean ± standard deviation (SD), pooled from a minimum of 3 independent experiments. The minimum threshold for rejecting the null hypothesis was $p < 0.05$. For results where statistics are shown, Student's t-test was used (unless otherwise indicated) and significance is denoted as: * = $p < 0.05$; ** = $p < 0.01$; *** = $p < 0.001$.

5. Conclusions

Collectively, this study demonstrates that the user-friendly and high-throughput technique of OBC can be used as a tool for monitoring intra-tumoural heterogeneity in 2D and 3D cell culture models. We extend previous works by combining both FC and live-cell SILU-based analyses, allowing for quantitative readouts of subpopulation distributions to be complemented with spatial and temporal resolution. We envision that this optimized OBC panel will be useful not only in studying intra-tumoural heterogeneity but for lineage tracing experiments in other biological contexts.

Supplementary Materials: The following are available online at https://www.mdpi.com/article/10.3390/cancers14030581/s1, Figure S1: Optical barcoding of 2D CRC cell lines show limited change in cluster distrubtion upon serial passaging; Table S1: Doubling time estimation of wild-type and optically barcoded cell lines; Table S2: Profiling of MEK-inhibitor responses in wild-type and optically barcoded cell lines.

Author Contributions: Conceptualization, C.S. and F.H.; methodology, C.S., H.-J.C., C.M., D.M. and F.H.; validation, C.S., M.G. and C.B.; formal analysis, C.S., J.S., M.G. and N.W.; investigation, C.S., J.S., M.G., N.W. and C.B.; resources, A.G.H., B.N.J.T. and C.B.; writing—original draft preparation, C.S.; writing—review and editing, F.H., D.M. and C.M.; visualization, C.S., H.-J.C. and C.B.; supervision, F.H.; project administration, F.H.; funding acquisition, F.H. All authors have read and agreed to the published version of the manuscript.

Funding: This research was funded by The National Health and Medical Research Council of Australia (project grant 1164081, F.H.) and by the Tour de Cure Foundation (Senior research grant, F.H.). D.M. and F.H. were supported by the Australian NBCF (Investigator Initiated Research Grant IIRS-19-082). D.M. is supported by Susan G. Komen and Cancer Australia (CCR19606878).

Institutional Review Board Statement: The study was conducted according to the guidelines of the Declaration of Helsinki and approved by the Peter MacCallum Cancer Centre Human Research Ethics Committee (HREC/15/PMCC/112, project #15/169, for human PDOs) and by the Nimes University Hospital institutional board (#2011-A01141-100 40, for CPP lines).

Informed Consent Statement: Informed consent was obtained from all subjects involved in the study.

Data Availability Statement: The data presented in this study are available on request from the corresponding author.

Acknowledgments: The authors wish to acknowledge the support of Ralph Rossi and his team at the Peter MacCallum Cancer Centre (PMCC) flow cytometry facility, as well as the Centre for Advanced Histology & Microscopy (PMCC) and Biological Optical Microscopy Platform (University of Melbourne).

Conflicts of Interest: The authors declare no conflict of interest. The funders had no role in the design of the study; in the collection, analyses, or interpretation of data; in the writing of the manuscript, or in the decision to publish the results.

References

1. McGranahan, N.; Swanton, C. Clonal Heterogeneity and Tumor Evolution: Past, Present, and the Future. *Cell* **2017**, *168*, 613–628. [CrossRef] [PubMed]
2. Dagogo-Jack, I.; Shaw, A.T. Tumour heterogeneity and resistance to cancer therapies. *Nat. Rev. Clin. Oncol.* **2018**, *15*, 81–94. [CrossRef] [PubMed]
3. Bochtler, T.; Stölzel, F.; Heilig, C.E.; Kunz, C.; Mohr, B.; Jauch, A.; Janssen, J.W.G.; Kramer, M.; Benner, A.; Bornhaüser, M.; et al. Clonal heterogeneity as detected by metaphase karyotyping is an indicator of poor prognosis in acute myeloid leukemia. *J. Clin. Oncol.* **2013**, *31*, 3898–3905. [CrossRef] [PubMed]
4. Morris, L.G.T.; Riaz, N.; Desrichard, A.; Senbabaoglu, Y.; Ari Hakimi, A.; Makarov, V.; Reis-Filho, J.S.; Chan, T.A. Pan-cancer analysis of intratumor heterogeneity as a prognostic determinant of survival. *Oncotarget* **2016**, *7*, 10051–10063. [CrossRef] [PubMed]
5. Mroz, E.A.; Tward, A.D.; Pickering, C.R.; Myers, J.N.; Ferris, R.L.; Rocco, J.W. High intratumor genetic heterogeneity is related to worse outcome in patients with head and neck squamous cell carcinoma. *Cancer* **2013**, *119*, 3034–3042. [CrossRef]
6. Oh, B.Y.; Shin, H.T.; Yun, J.W.; Kim, K.T.; Kim, J.; Bae, J.S.; Cho, Y.B.; Lee, W.Y.; Yun, S.H.; Park, Y.A.; et al. Intratumor heterogeneity inferred from targeted deep sequencing as a prognostic indicator. *Sci. Rep.* **2019**, *9*, 4542. [CrossRef]
7. Suguro, M.; Yoshida, N.; Umino, A.; Kato, H.; Tagawa, H.; Nakagawa, M.; Fukuhara, N.; Karnan, S.; Takeuchi, I.; Hocking, T.D.; et al. Clonal heterogeneity of lymphoid malignancies correlates with poor prognosis. *Cancer Sci.* **2014**, *105*, 897–904. [CrossRef]
8. Chabon, J.J.; Simmons, A.D.; Lovejoy, A.F.; Esfahani, M.S.; Newman, A.M.; Haringsma, H.J.; Kurtz, D.M.; Stehr, H.; Scherer, F.; Karlovich, C.A.; et al. Circulating tumour DNA profiling reveals heterogeneity of EGFR inhibitor resistance mechanisms in lung cancer patients. *Nat. Commun.* **2016**, *7*, 11815. [CrossRef]
9. Lee, H.J.; Seo, A.N.; Kim, E.J.; Jang, M.H.; Suh, K.J.; Ryu, H.S.; Kim, Y.J.; Kim, J.H.; Im, S.-A.; Gong, G.; et al. HER2 Heterogeneity Affects Trastuzumab Responses and Survival in Patients with HER2-Positive Metastatic Breast Cancer. *Am. J. Clin. Pathol.* **2014**, *142*, 755–766. [CrossRef]
10. Sagaert, X.; Vanstapel, A.; Verbeek, S. Tumor Heterogeneity in Colorectal Cancer: What Do We Know So Far? *Pathobiol. J. Immunopathol. Mol. Cell. Biol.* **2018**, *85*, 72–84. [CrossRef]
11. Gui, P.; Bivona, T.G. (2021). Evolution of metastasis: New tools and insights. *Trends Cancer.* **2022**, *8*, 98–109. [CrossRef] [PubMed]
12. Gerrits, A.; Dykstra, B.; Kalmykowa, O.J.; Klauke, K.; Verovskaya, E.; Broekhuis, M.J.C.; de Haan, G.; Bystrykh, L.v. Cellular barcoding tool for clonal analysis in the hematopoietic system. *Blood* **2010**, *115*, 2610–2618. [CrossRef] [PubMed]
13. Schepers, K.; Swart, E.; van Heijst, J.W.J.; Gerlach, C.; Castrucci, M.; Sie, D.; Heimerikx, M.; Velds, A.; Kerkhoven, R.M.; Arens, R.; et al. Dissecting T cell lineage relationships by cellular barcoding. *J. Exp. Med.* **2008**, *205*, 2309–2318. [CrossRef] [PubMed]
14. Verovskaya, E.; Broekhuis, M.J.C.; Zwart, E.; Ritsema, M.; van Os, R.; de Haan, G.; Bystrykh, L.V. Heterogeneity of young and aged murine hematopoietic stem cells revealed by quantitative clonal analysis using cellular barcoding. *Blood* **2013**, *122*, 523–532. [CrossRef]
15. van der Heijden, M.; Miedema, D.M.; Waclaw, B.; Veenstra, V.L.; Lecca, M.C.; Nijman, L.E.; van Dijk, E.; van Neerven, S.M.; Lodestijn, S.C.; Lenos, K.J.; et al. Spatiotemporal regulation of clonogenicity in colorectal cancer xenografts. *Proc. Natl. Acad. Sci. USA* **2019**, *116*, 6140–6145. [CrossRef]
16. Nguyen, L.v.; Pellacani, D.; Lefort, S.; Kannan, N.; Osako, T.; Makarem, M.; Cox, C.L.; Kennedy, W.; Beer, P.; Carles, A.; et al. Barcoding reveals complex clonal dynamics of de novo transformed human mammary cells. *Nature* **2015**, *28*, 267–271. [CrossRef]
17. Thielecke, L.; Aranyossy, T.; Dahl, A.; Tiwari, R.; Roeder, I.; Geiger, H.; Fehse, B.; Glauche, I.; Cornils, K. Limitations and challenges of genetic barcode quantification. *Sci. Rep.* **2017**, *7*, 43249. [CrossRef]
18. Cai, D.; Cohen, K.B.; Luo, T.; Lichtman, J.W.; Sanes, J.R. Improved tools for the Brainbow toolbox. *Nat. Methods* **2013**, *10*, 540–547. [CrossRef]
19. Livet, J.; Weissman, T.A.; Kang, H.; Draft, R.W.; Lu, J.; Bennis, R.A.; Sanes, J.R.; Lichtman, J.W. Transgenic strategies for combinatorial expression of fluorescent proteins in the nervous system. *Nature* **2007**, *450*, 56–62. [CrossRef]
20. Hadjieconomou, D.; Rotkopf, S.; Alexandre, C.; Bell, D.M.; Dickson, B.J.; Salecker, I. Flybow: Genetic multicolor cell labeling for neural circuit analysis in Drosophila melanogaster. *Nat. Methods* **2011**, *8*, 260–266. [CrossRef]
21. Hampel, S.; Chung, P.; McKellar, C.E.; Hall, D.; Looger, L.L.; Simpson, J.H. Drosophila Brainbow: A recombinase-based fluorescence labeling technique to subdivide neural expression patterns. *Nat. Methods* **2011**, *8*, 253–259. [CrossRef] [PubMed]

22. Pan, Y.A.; Livet, J.; Sanes, J.R.; Lichtman, J.W.; Schier, A.F. Multicolor brainbow imaging in Zebrafish. *Cold Spring Harb. Protoc.* **2011**, *6*. [CrossRef] [PubMed]
23. Pan, Y.A.; Freundlich, T.; Weissman, T.A.; Schoppik, D.; Wang, X.C.; Zimmerman, S.; Ciruna, B.; Sanes, J.R.; Lichtman, J.W.; Schier, A.F. Zebrabow: Multispectral cell labeling for cell tracing and lineage analysis in zebrafish. *Development* **2013**, *140*, 2835–2846. [CrossRef] [PubMed]
24. Garcia-Moreno, F.; Vasistha, N.A.; Begbie, J.; Molnar, Z. CLoNe is a new method to target single progenitors and study their progeny in mouse and chick. *Development* **2014**, *141*, 1589–1598. [CrossRef]
25. Hackl, M.J.; Burford, J.L.; Villanueva, K.; Lam, L.; Suszták, K.; Schermer, B.; Benzing, T.; Peti-Peterdi, J. Tracking the fate of glomerular epithelial cells in vivo using serial multiphoton imaging in new mouse models with fluorescent lineage tags. *Nat. Med.* **2013**, *19*, 1661–1666. [CrossRef]
26. Amitai-Lange, A.; Altshuler, A.; Bubley, J.; Dbayat, N.; Tiosano, B.; Shalom-Feuerstein, R. Lineage Tracing of Stem and Progenitor Cells of the Murine Corneal Epithelium. *Stem Cells* **2015**, *33*, 230–239. [CrossRef]
27. Snippert, H.J.; Schepers, A.G.; van Es, J.H.; Simons, B.D.; Clevers, H. Biased competition between Lgr5 intestinal stem cells driven by oncogenic mutation induces clonal expansion. *EMBO Rep.* **2014**, *15*, 62–69. [CrossRef]
28. Weber, K.; Bartsch, U.; Stocking, C.; Fehse, B. A multicolor panel of novel lentiviral "gene ontology" (LeGO) vectors for functional gene analysis. *Mol. Ther.* **2008**, *16*, 698–706. [CrossRef]
29. Weber, K.; Thomaschewski, M.; Warlich, M.; Volz, T.; Cornils, K.; Niebuhr, B.; Täger, M.; Lütgehetmann, M.; Pollok, J.M.; Stocking, C.; et al. RGB marking facilitates multicolor clonal cell tracking. *Nat. Medicine.* **2011**, *17*, 504–509. [CrossRef]
30. Behrenbruch, C.; Shembrey, C.; Paquet-Fifield, S.; Mølck, C.; Cho, H.J.; Michael, M.; Thomson, B.N.J.; Heriot, A.G.; Hollande, F. Surgical stress response and promotion of metastasis in colorectal cancer: A complex and heterogeneous process. *Clin. Exp. Metastasis* **2018**, *35*, 333–345. [CrossRef]
31. Mohme, M.; Maire, C.L.; Riecken, K.; Zapf, S.; Aranyossy, T.; Westphal, M.; Lamszus, K.; Fehse, B. Optical Barcoding for Single-Clone Tracking to Study Tumor Heterogeneity. *Mol. Ther.* **2017**, *25*, 621–633. [CrossRef] [PubMed]
32. Berthelet, J.; Wimmer, V.C.; Whitfield, H.J.; Serrano, A.; Boudier, T.; Mangiola, S.; Merdas, M.; El-Saafin, F.; Baloyan, D.; Wilcox, J.; et al. The site of breast cancer metastases dictates their clonal composition and reversible transcriptomic profile. *Sci. Adv* **2021**, *7*, eabf4408. [CrossRef] [PubMed]
33. Priest, D.G.; Solano, A.; Lou, J.; Hinde, E. Fluorescence fluctuation spectroscopy: An invaluable microscopy tool for uncovering the biophysical rules for navigating the nuclear landscape. In *Biochemical Society Transactions*; Portland Press Ltd.: London, UK, 2019; Volume 47, pp. 1117–1129. [CrossRef]
34. Kustikova, O.; Fehse, B.; Modlich, U.; Yang, M.; Düllmann, J.; Kamino, K.; von Neuhoff, N.; Schlegelberger, B.; Li, Z.; Baum, C. Clonal dominance of hematopoietic stem cells triggered by retroviral gene marking. *Science* **2005**, *308*, 1171–1174. [CrossRef]
35. Ott, M.G.; Schmidt, M.; Schwarzwaelder, K.; Stein, S.; Siler, U.; Koehl, U.; Glimm, H.; Kühlcke, K.; Schilz, A.; Kunkel, H.; et al. Correction of X-linked chronic granulomatous disease by gene therapy, augmented by insertional activation of MDS1-EVI1, PRDM16 or SETBP1. *Nat. Med.* **2006**, *12*, 401–409. [CrossRef]
36. Giraud, J.; Failla, L.M.; Pascussi, J.M.; Lagerqvist, E.L.; Ollier, J.; Finetti, P.; Bertucci, F.; Ya, C.; Gasmi, I.; Bourgaux, J.F.; et al. Autocrine Secretion of Progastrin Promotes the Survival and Self-Renewal of Colon Cancer Stem-like Cells. *Cancer Res.* **2016**, *76*, 3618–3628. [CrossRef]
37. Grillet, F.; Bayet, E.; Villeronce, O.; Zappia, L.; Lagerqvist, E.L.; Lunke, S.; Charafe-Jauffret, E.; Pham, K.; Molck, C.; Rolland, N.; et al. Circulating tumour cells from patients with colorectal cancer have cancer stem cell hallmarks in ex vivo culture. *Gut* **2017**, *66*, 1802–1810. [CrossRef] [PubMed]
38. Hu, Y.; Smyth, G.K. ELDA: Extreme limiting dilution analysis for comparing depleted and enriched populations in stem cell and other assays. *J. Immunol. Methods* **2009**, *347*, 70–78. [CrossRef]
39. Sottoriva, A.; Barnes, C.P.; Graham, T.A. Catch my drift? Making sense of genomic intra-tumour heterogeneity. *Biochim. Biophys. Acta (BBA) Rev. Cancer* **2017**, *1876*, 95–100. [CrossRef]
40. Hollmann, J.; Brecht, J.; Goetzke, R.; Franzen, J.; Selich, A.; Schmidt, M.; Eipel, M.; Ostrowska, A.; Hapala, J.; Fernandez-Rebollo, E.; et al. Genetic barcoding reveals clonal dominance in iPSC-derived mesenchymal stromal cells. *Stem Cell Res. Ther* **2020**, *11*, 105. [CrossRef]
41. Vlachogiannis, G.; Hedayat, S.; Vatsiou, A.; Jamin, Y.; Fernández-Mateos, J.; Khan, K.; Lampis, A.; Eason, K.; Huntingford, I.; Burke, R.; et al. Patient-derived organoids model treatment response of metastatic gastrointestinal cancers. *Science* **2018**, *359*, 920–926. [CrossRef]
42. Yao, Y.; Xu, X.; Yang, L.; Zhu, J.; Wan, J.; Shen, L.; Xia, F.; Fu, G.; Deng, Y.; Pan, M.; et al. Patient-Derived Organoids Predict Chemoradiation Responses of Locally Advanced Rectal Cancer. *Cell Stem Cell* **2020**, *26*, 17–26.e6. [CrossRef] [PubMed]
43. Behrenbruch, C.; Foroutan, M.; Lind, P.; Smith, J.; Grandin, M.; Cooper, B.; Shembrey, C.; Ramm, S.; Cowley, K.; Nikolic, I.; et al. Targeting of TP53-independent cell cycle checkpoints overcomes FOLFOX resistance in Metastatic Colorectal Cancer. *BioRxiv* **2021**. [CrossRef]
44. Prasher, D.C.; Eckenrode, V.K.; Ward, W.W.; Prendergast, F.G.; Cormier, M.J. Primary structure of the Aequorea victoria green-fluorescent protein. *Gene* **1992**, *111*, 229–233. [CrossRef]

45. Yu, V.W.C.; Yusuf, R.Z.; Oki, T.; Wu, J.; Saez, B.; Wang, X.; Cook, C.; Baryawno, N.; Ziller, M.J.; Lee, E.; et al. Epigenetic Memory Underlies Cell-Autonomous Heterogeneous Behavior of Hematopoietic Stem Cells. In *Cell;* Cell Press: Cambridge, MA, USA, 2016; Volume 167, pp. 1310–1322. [CrossRef]
46. Snippert, H.J.; van der Flier, L.G.; Sato, T.; van Es, J.H.; van den Born, M.; Kroon-Veenboer, C.; Barker, N.; Klein, A.M.; van Rheenen, J.; Simons, B.D.; et al. Intestinal crypt homeostasis results from neutral competition between symmetrically dividing Lgr5 stem cells. *Cell* **2010**, *143*, 134–144. [CrossRef] [PubMed]
47. Maetzig, T.; Ruschmann, J.; Sanchez Milde, L.; Lai, C.K.; von Krosigk, N.; Humphries, R.K. Lentiviral Fluorescent Genetic Barcoding for Multiplex Fate Tracking of Leukemic Cells. *Mol. Ther. Methods Clin. Dev.* **2017**, *6*, 54–65. [CrossRef]
48. Echeverria, G.v.; Powell, E.; Seth, S.; Ge, Z.; Carugo, A.; Bristow, C.; Peoples, M.; Robinson, F.; Qiu, H.; Shao, J.; et al. High-resolution clonal mapping of multi-organ metastasis in triple negative breast cancer. *Nat. Commun.* **2018**, *9*, 5079. [CrossRef]
49. Merino, D.; Weber, T.S.; Serrano, A.; Vaillant, F.; Liu, K.; Pal, B.; di Stefano, L.; Schreuder, J.; Lin, D.; Chen, Y.; et al. Barcoding reveals complex clonal behavior in patient-derived xenografts of metastatic triple negative breast cancer. *Nat. Commun.* **2019**, *10*, 766. [CrossRef]
50. Brady, S.W.; McQuerry, J.A.; Qiao, Y.; Piccolo, S.R.; Shrestha, G.; Jenkins, D.F.; Layer, R.M.; Pedersen, B.S.; Miller, R.H.; Esch, A.; et al. Combating subclonal evolution of resistant cancer phenotypes. *Nat. Commun.* **2017**, *8*, 1231. [CrossRef]
51. Kreso, A.; O'Brien, C.A.; van Galen, P.; Gan, O.I.; Notta, F.; Brown, A.M.K.; Ng, K.; Jing, M.; Wienholds, E.; Dunant, C.; et al. Variable clonal repopulation dynamics influence chemotherapy response in colorectal cancer. *Science* **2013**, *339*, 543–548. [CrossRef]
52. Saeed, K.; Ojamies, P.; Pellinen, T.; Eldfors, S.; Turkki, R.; Lundin, J.; Järvinen, P.; Nisen, H.; Taari, K.; af Hällström, T.M.; et al. Clonal heterogeneity influences drug responsiveness in renal cancer assessed by ex vivo drug testing of multiple patient-derived cancer cells. *Int. J. Cancer* **2019**, *144*, 1356–1366. [CrossRef]

cancers

MDPI

Systematic Review

The Development of a Three-Dimensional Platform for Patient-Derived Ovarian Cancer Tissue Models: A Systematic Literature Review

Lusine Sevinyan [1,2], Priyanka Gupta [3,4], Eirini Velliou [3,4] and Thumuluru Kavitha Madhuri [1,2,*]

1 Department of Gynaecological Oncology, Royal Surrey NHS Foundation Trust, Guildford GU2 7XX, UK
2 Cancer Research, School of Applied Sciences, University of Brighton, Brighton BN2 4HQ, UK
3 Centre for 3D Models of Health and Disease, Division of Surgery and Interventional Science, University College London, London WC1E 6BT, UK
4 Bioprocess and Biochemical Engineering Group (BioProChem), Department of Chemical and Process Engineering, University of Surrey, Guildford GU2 7XH, UK
* Correspondence: docmadhuri231@doctors.org.uk

Simple Summary: The aim of this systematic review is to provide an overview of the state of the art on in vitro models for ovarian cancer studies, with focus on patient derived studies, which provides a personalised approach to treatment of patients with ovarian cancer.

Abstract: There is an unmet biomedical need for ex vivo tumour models that would predict drug responses and in turn help determine treatment regimens and potentially predict resistance before clinical studies. Research has shown that three dimensional models of ovarian cancer (OvCa) are more realistic than two dimensional in vitro systems as they are able to capture patient in vivo conditions in more accurate manner. The vast majority of studies aiming to recapitulate the ovarian tumour morphology, behaviors, and study chemotherapy responses have been using ovarian cancer cell lines. However, despite the advantages of utilising cancer cell lines to set up a platform, they are not as informative as systems applying patient derived cells, as cell lines are not able to recapitulate differences between each individual patient characteristics. In this review we discussed the most recent advances in the creation of 3D ovarian cancer models that have used patient derived material, the challenges to overcome and future applications.

Keywords: scaffolds; 3D; three-dimensional; model; ovarian; cancer; patient-derived; personalised

Citation: Sevinyan, L.; Gupta, P.; Velliou, E.; Madhuri, T.K. The Development of a Three-Dimensional Platform for Patient-Derived Ovarian Cancer Tissue Models: A Systematic Literature Review. *Cancers* **2022**, *14*, 5628. https://doi.org/10.3390/cancers14225628

Academic Editor: Donatella Aldinucci

Received: 25 October 2022
Accepted: 9 November 2022
Published: 16 November 2022

1. Introduction

Worldwide there were nearly 300,000 cases of OvCa diagnosed in 2018 with the 5-year survival between 30–50% despite the advances in diagnostics and treatment. In the UK alone, there were 7300 new cases reported in 2017 with 4200 deaths in 2018, and the numbers continue to steadily increase [1]. The most common type of ovarian malignancy is epithelial ovarian cancer (EOC) most of which is represented by the serous subtype [2].

The current standard for treatment of EOC is cytoreductive surgery combined with platinum-based chemotherapy [3]. In patients diagnosed with EOC where upfront surgery is medically contraindicated (e.g., comorbidities or poor performance status), or where complete cytoreduction cannot be achieved, neoadjuvant chemotherapy prior to interval debulking surgery, and adjuvant chemotherapy is an alternative therapeutic option [4]. However approximately 80% of patients will relapse following first-line chemotherapy [3]. Moreover, despite treatment, most patients appear to develop a recurrent disease which is platinum resistant, partially due to the complex and heterogeneous tumour microenvironment. Many chemotherapy agents then fail to have sustained efficacy in clinical practice, As a result, there is an unmet biomedical need for ex vivo tumour models that would predict drug responses and in turn help determine treatment regimens and possibly predict resistance before clinical studies [5].

There have been attempts to create patient-derived OvCa models since the first report by Griffon et al., in 1995 [6]. In the last decades, a number of studies [7–41] have been conducted aiming to create in vitro tumour models, which could be a platform for OvCa research and drug testing for specific patients prior to treatment, thereby moving towards a more personalised/individualised treatment. For the latter to be achieved efficiently, accurate in vitro models with high predicting capability are required. Such models should simulate ex vivo the OvCa evolution and response to treatment maintaining each patient's phenotypic and genetic characteristics. The aim of this systematic review is to provide an overview of the state of the art on in vitro models for OvCa studies, with focus on patient derived studies.

1.1. Ovarian Cancer Tumour Characteristics

In order to create a model of OvCa that would represent the in vivo environment and reflect the processes happening in the body it is important to understand the general morphology and physiology of OvCa tumours and the way it spreads.

It is well established that ovarian tumours consist of mixture of epithelial, stromal, immune, and endothelial cells which form the complex tumour microenvironment [42–45]. In their recent review Horst et al., have discussed in detail the impact and the role of heterogenous cellular components on the extracellular matrix [46]. And as outlined by Lengyel et al., this heterogeneity of cell types likely impacts tumour histology, growth potential and ability to evade chemotherapy [42]. Hence it is important to acknowledge this when creating the ovarian tumour ex vivo model.

The primary microenvironment for the ovarian carcinoma cell is the mesothelium [43,47]. An intact mesothelial cell layer can very efficiently inhibit the invasion of OvCa cells, suggesting that mesothelial cells can delay OvCa attachment and invasion [48]. There is also a pool of evidence suggesting that adipocytes play a crucial role in creating tumour microenvironment and promoting metastasis [49,50].

Fibronectin as well as integrins play an important role in spheroid growth and the abundant presence of several isoforms of fibronectin in malignant ascites suggests the importance of the microenvironment in OvCa metastasis [43]. Furuya outlines the importance of extracellular matrix, stroma and omental adipose tissue in the development of the OvCa [51]. Other sources highlight the role of fibroblasts in the OvCa growth, adhesion and invasiveness [52].

The unique anatomical location of the ovary renders OvCa cells the ability to easily metastasise in comparison to other cancers. Once the cancer cells have detached as single cells or clusters from the primary tumour, it is thought that they metastasise through a passive mechanism, carried by the physiological movement of peritoneal fluid to the peritoneum and omentum.

It is not entirely clear whether single cells detach and then aggregate to form spheroids, or if the cells detach as cell clumps that stay together while floating in ascites [43]. The other major mechanism of metastasis of the OvCa is via haematological route, which requires intra- and extravasation of the cancer cells [53].

It is also necessary to consider when creating a tumour model that in order for the tumour to grow over a certain size (>1–2 mm) a process of neovascularisation has to occur [54]. However the role of angiogenesis in OvCa development remains unclear and as mentioned by Duncan et al. [55] there are contradictory studies with regards to the influence of microvessel density in OvCa prognosis [56–58]. Worzfeld et al., also outline the possible role of extracellular microvesicles in invasion and metastasis and their contribution to the drug-resistance in the patients with OvCa [47]. From the therapeutic perspective, it has been demonstrated that poly(ADP-ribose) polymerase (PARP) inhibition decreases angiogenesis whereas, hypoxic state and vascular endothelial growth factor receptor 3 (VEGFR3) inhibitors induce down-regulation of BRCA1/2 and RAD51, which potentiate

PARP inhibitors sensitivity. However, hypoxia is also associated with hypoxia inducible factor 1 alpha (HIF1α) up-regulation, and therefore resistance to angiogenesis inhibitors. Though, PARP1 is involved in HIF1α stabilization and consequently, inhibition of PARP may prevent HIF1α accumulation that leads to targeted hypoxic-induced apoptosis [59].

It seems evident from the up-to-date scientific reports that there are certain conditions to be met for an ideal OvCa tumour model to be achieved, and these key requirements are summarised in Table 1 below.

Table 1. Key characteristics of an ideal model of ovarian tumour.

Key Element/Characteristic of Ovarian Tumour Model	In Vivo Function/Repercussion
Complex microenvironment (cellular and architectural) [42–45]	• Reflects tumour histology • Tumour growth • Resistance to chemotherapeutic agents
Mesothelial cells [43,47]	• Attachment and invasion of cancer cells
Fibronectin/Integrins [43]	• Spheroidal structure growth
Fibroblasts [52]	• Tumour growth, adhesion and invasiveness
Adipocytes [49–51]	• Tumour growth and metastasis promotion
Extracellular matrix and stroma [51]	• Tumour growth, adhesion
Extracellular microvesicles [47]	• Invasion and methastasis • Drug-resistance
Angiogenesis (PARP/VEGFR3)/Neovascularisation [54,55,59]	• Ability to grow over a certain size • Invasion and metastasis • Drug-resistance
Ability to self-organise in 3D structures [43]	• Invasion and metastasis • Drug-resistance

1.2. Current Models

Animal models, three-dimensional (3D) models and two-dimensional (2D) cell culture are currently the most widely used methods for creating various tissue models including tumour tissues. 2D cell cultures (in T-flasks or microplates) have dominated the in vitro landscape for the last decade [60,61]. Researchers have widely used the monolayer cell model due to its easy accessibility, relative ease of use and low cost. However, this model has several limitations. Carvalho et al., stressed the inability of 2D models to recapitulate the complex nature of tumours and the influence of the surrounding tumour microenvironment (TME) [60]. Moreover as the monolayer model has a different morphology from the in vivo model considering the lack of interactions with surrounding cells and surrounding matrix, molecular differences and inability to mimic the complex TME, it would not be the primary choice for researchers to use as a drug-testing platform as this non-physiological screening approach often results in poor predictive power for drug efficacy in patients [48,61,62]. In fact, many studies have shown that 2D models have different cellular responses to the environment [63] and drug response patterns as compared to 3D systems and to in vivo studies [64–73]. Pinto et al., note in their recent review that generally most of published work around tumour cancer models is done exclusively utilising cancer cells without any other cell types, which can not closely represent the processes occurring in the complex tumour microenvironment [74].

In order to replicate genuine tumour morphology and microenvironment animal models and 3D models have been created. Currently, animal models are considered to be gold standard in pre-clinical studies. They are more realistic than 2D in vitro systems as they are able to capture patient in vivo conditions in a more accurate manner. Currently, both small and large animal models are available for OvCa. They are mostly murine models (mouse and rat), but also include hens and SCID pigs [26,72,73,75–79]. Although animal studies are informative, high cost, complexity of reproduction and length of development time as well as several deficiencies, that make extrapolation to human tumour biology problematic [48], makes them less attractive for researchers [61]. As a result of limitations of 2D and the animal studies it has encouraged development of three-dimensional models of tumours, including OvCa [60].

Three-dimensional models offer the potential for ex vivo research of cell to cell interactions thus making the study of the nature of tumours easier, especially where extracellular matrix cover and stromal cells are added [58,60,62,63,80–90]. They also can allow the real-life oxygen, nutrient and temperature distribution and more realistic drug resistance studies [61,63,91]. Additionally, it is also feasible to culture patient samples directly in 3D models allowing for a much more realistic assessment of various therapeutic methods and leading to a personalised medicine-based approach towards cancer treatment [15].

3D models of OvCa have become highly interesting for investigation as according to recent oncogenesis theories the progression of OvCa involves detachment of cancer cells from the in situ carcinoma into cell aggregates/spheroids, and further attachment to mesothelial-lined surfaces. [92] For instance, Bapat et al., were one of the first who were able to isolate spheroid aggregates from malignant ascites of a patient [93]. Thus, a 3D model of OvCa cells could morphologically resemble multicellular aggregates in cancerous ascites [22]. It has also been shown that spheroid aggregates of malignant ascites compared to 2D models are more resistant to chemotherapy treatments, including cisplatin and taxol [94,95]. This has made the 3D model of OvCa highly useful for research.

Most OvCa in vitro remodelling approaches in 3D involve, like in other diseases [63,65,66,68,96–99], (i) spheroids, (ii) hydrogel type scaffolds, (iii) synthetic highly porous polymer-based scaffolds and synthetic matrices (Figure 1).

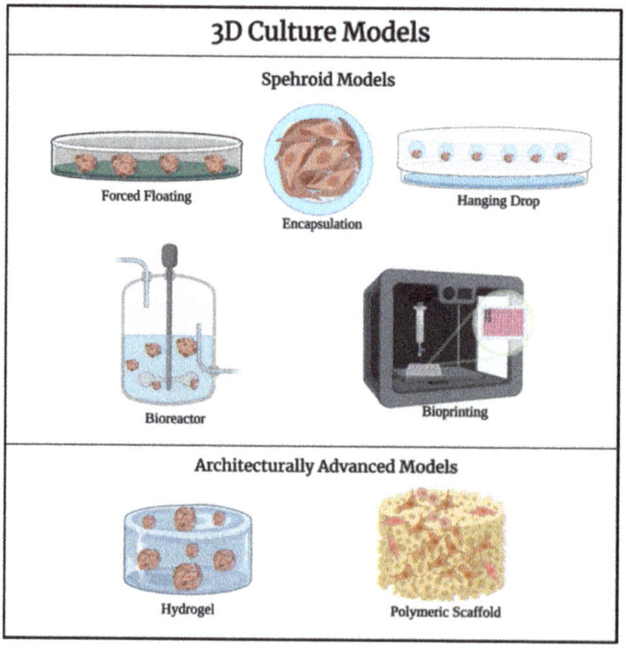

Figure 1. A graphical schematic of key techniques used for three-dimensional cell culture (created on BioRender.com (accessed on 14 March 2022)).

1.3. *Spheroid Models for Ovarian Cancer*

Spheroid systems are one of the oldest and most widely used 3D cell culture set up in the field of tumour model development, drug discovery, therapeutic assessments etc. They are cell aggregates/clusters, which are formed as a result of cell-cell adhesion without attachment to culture vessel surfaces. In vitro 3D approaches have been established till date for spheroid formation like, forced floating method, hanging drop method, encapsulation-based method, 3D bioprinting and agitation method to name a few [61,100–102]. The close cell-cell interactions in spheroid systems assist in production of various ECM (extra cellular matrix) proteins by the cells, allowing them to form their own niche/microenvironment.

Similar to other cancer models, spheroids are also one of the most widely used system for in vitro models of OvCa and its metastasis. Ishiguro et al., have done a comprehensive overview of the representative methods for spheroid cultures of cancer cells, which include organotypic multicellular spheroids, multicellular tumour spheroids, tumour-derived organoids and tumour-derived spheroids [69]. For example, low cell number spheroids (as few as 10 cells) were created, and significant chemoresistance compared to 2D model was demonstrated by Raghavan et al. [64] Zietarska et al., were able to form spheroids up to 500 µm in size at 4 day of culture, generated from EOC cell line in hanging droplets, which allowed the researchers to have a better understanding of the cancer biology as compared to the 2D model of OvCa [21]. According to Grun et al., spheroids grown using Rotary cell Culture Systems could grow for longer periods and reach significantly higher volumes [24]. Increased chemotherapy resistance in 3D compared to 2D culture was also exhibited by Lee et al. [73] and Moraya et al. [103].

However even with extensive research of OvCa using spheroid models, they have certain inherent disadvantages. Due to their spatial characteristics, high diffusion gradient in terms of nutrients and oxygen is formed within them. This in turn results in the formation of necrotic cores at the center and decreasing cellular proliferation over time like it has been shown for other types of cancer models [61,104,105]. Spheroid systems are also difficult to maintain over long time periods (weeks or months) without re-suspending the cells to form fresh cellular aggregates. Depending on the method it is also difficult to control spheroid size and shape of the spheroids [106]. These disadvantages cannot be neglected as they can lead to differences in experimental results obtained. For instance, Lal-Nag et al., showed on modified HEY A8 cell line that spheroid size is an important consideration when comparing chemotherapeutic responses because cell metabolism, proliferation and survival vary within spheroids as they grow larger, and nutrient and oxygen gradients become pronounced [5].

1.4. Ovarian Cancer Models Using Biomaterials with Advances Structural Complexity

Although relatively easy to use, the disadvantages associated with spheroids models have resulted in the development of 3D cancer models with more structural complexity and stability using natural and artificial biomaterials in the form of hydrogels and polymeric scaffolds. Hydrogels are cross linked polymeric networks consisting of a high amount of water and are able to simulate the native tissues in terms of architectural and spatial characteristics, biocompatibility and also allows nutrient and oxygen diffusion. Hydrogels can be formed of natural molecules like collagen, Matrigel and other extra cellular matrix (ECM) proteins or of synthetic materials like poly–ethylene glycol (PEG), agarose, alginate etc. Similar to other cancers, OvCa 3D models using hydrogels have been established by various groups [65,66,68,72,96,107,108]. The length of the various studies varies between 7–28 days highlighting the feasibility of relatively long-term culture in hydrogel systems. It has also been reported that hydrogel-based tumour models of OvCa shows higher resistance to chemotherapeutic agents like paclitaxel in comparison to 2D culture systems [65,66,68,96]. However, despite its multiple advantages, depending on the material properties and structural configurations (porosity, pore size, pore interconnectivity) hydrogels might not provide consistent cell distribution within them resulting in different densities of cells within them and reducing the chances of consistent spheroid formation [104]. In addition, hydrogels due to their high water content also lack mechanical strength resulting in handling difficulty [61,98,109,110]. Moreover, for natural polymers, long term culture is a difficult proposition due to high batch-to-batch variations, undefined matrix composition and restricted modification possibilities [109]. A lot of studies have been conducted to compare different types of hydrogels [65,108,109,111–115], however each of them were found to have certain limitations. Li et al., have done a comprehensive overview of the main types of hydrogel microenvironments discussing their pros and cons and their application for cancer research and drug screening in different types of cancer cells, and in their review they agreed that further research is necessary to focus on improvement of the modelling of tumour microenvironment and confirmation of in vitro results, and that patient-derived cells are more desirable than well established cell lines [116].

To overcome the limitations of the above methods there have been a number of developments in tissue engineering to construct polymeric scaffolds-based tumour models. Those include reports describing scaffolding systems for OvCa [16,81,117], lung cancer [117,118], pancreatic cancer [80,119–124], breast cancer [117,125], prostate cancer, melanoma [117] and others [117]. Amongst the advantages of the polymeric scaffolds, authors highlight their ability to be 'customised' for any type of tumour and the potential of researchers to endue scaffolds with the desired extracellular matrix properties [126]. Polymeric scaffolds are especially very promising for future research as, they incorporate the advantages of hydrogel and spheroid scaffolds, such as provision of structure, realistic spatial arrangement, realistic cell-cell and cell-ECM interactions, and also possess such qualities as good mass transfer, porosity, architectural tuneability as well as tuneable mechanical properties. However, polymeric scaffolds are more complex in terms of synthesis procedure in comparison to spheroids and hydrogels and cell retrieval can be difficult in scaffolds depending on the material used.

Graphic summary and comparison of different types of platforms used to replicate and research OvCa models can be found in the Appendix A (Figure A1).

1.5. Cell Sources in Available Ovarian Cancer In Vitro 3D Studies

The vast majority of studies aimed to research OvCa have used cancer cell lines with very few utilising patient derived samples. Despite the advantages of using cancer cell lines to set up a platform/system, i.e., availability, consistency/reproducibility, cancer cell lines are not as informative as patient derived cells, as they would not able to recapitulate differences between each individual patient characteristics, including differences in drug sensitivities due to changes in gene expression following culture and passage [127]. Furthermore, they have been optimised to grow in a 2D environment and they cannot account for patient and tumour heterogeneity, therefore making personalised medicine impossible. Létourneau et al., developed new cancer cell lines in their study and had shown that there were differences in spheroid formation between cell lines derived before and after chemotherapy treatment [25]. This would be an important consideration when using cancer cell lines as the sensitivities shown in preclinical studies could be misinterpreted.

The focus of this review is to explore the existing knowledge base and evidence with regards patient derived samples of OvCa cells for fabrication of a 3D model to study the properties in vitro.

2. Materials and Methods

This review subscribes to the new PRISMA (Preferred Reporting Items for Systematic reviews and Meta-Analyses) guidelines [128].

2.1. Search Strategy

The eligibility criteria were studies published in English language and there were no other limitations to the search. Thorough literature search was performed using electronic databases of MEDLINE, EMBASE, Library, Information Science & Technology Abstracts, British Library Document Supply Centre Inside Serials & Conference Proceedings, ScienceDirect, Oxford Handbooks Online, Academic Search Index, Supplemental Index, Complementary Index, Directory of Open Access Journals, British Library EthOS, Digital Access to Scholarship at Harvard (DASH), University Press Scholarship Online, BioOne Complete, Center for Research Libraries, Research Starters, Oxford Medicine Online, Cochrane Database of Systematic Reviews, Oxford Bibliographies, Gale OneFile: Health and Medicine, Kent, Surrey and Sussex NHS Libraries, Springer Protocols, McGraw-Hill

Medical, VleBooks, Kortext eBook Catalogue, BMJ Best Practice, ClinicalTrials.gov, Emerald InsightNCI (National Cancer Institute at the National Institutes of Health) (Bethesda, MD, USA)), Directory of Open Access Books by two reviewers to identify relevant studies. Bibliographies of the relevant literature were also screened for any additional studies that were missed through the electronic search. Additional searching of the grey literature has also been conducted.

The exact syntax of search terms included 'scaffold', 'three-dimensional', '3D' and 'spheroid' each of which were combined with term 'ovarian cancer'. These terms were used as historically there has been interchangeability between the terms of "3D models" and "scaffolds". All search terms were expanded, and all sub-categories were included. Thesaurus search was also used to identify additional terms. All duplicates were removed. The search was independently run by two of the authors. Databases last accessed 5 October 2022. The search protocol available on demand.

2.2. Selection Process

Initial screening of article headings was performed identifying the potential studies that could be used in current review. The selected studies were further screened using the abstracts and the irrelevant ones were excluded. The full text of these potentially eligible studies was retrieved and assessed for eligibility. Any discrepancies over the eligibility of particular studies were resolved through discussion with two other reviewers.

2.3. Study Selection

The studies included in the current review meet the following criteria:

- The type of cells used for the 3D model were exclusively patient derived OvCa cells or newly established cell lines derived directly from primary OvCa cells.
- Only multicellular tumour spheroid and tumour-derived spheroid models were reviewed in the current study.
- The main focus of the research was to build a three-dimensional model of OvCa cells regardless of method used to accomplish it.

Studies that used established OvCa cell lines or other types of cancer cells; organotypic multicellular spheroid and tumour derived organoid models and studies that used 3D modelling in their research but did not mention methods of construction were not included in the current review.

2.4. Data Extraction

Data was extracted from the selected papers using pre-designed data collection forms. Attempts were made to contact authors where data was missing from the papers.

3. Results

Following the electronic database search 3839 articles were identified and 2844 articles remained after removal of duplicates. Title screening resulted in rejection of 2435 papers as they did not meet the inclusion criteria. Out of the remaining 409 articles 326 were excluded based on abstract review.

Full manuscripts of all the remaining papers were reviewed. Following the screening of the bibliographies of the full manuscripts two additional papers were also reviewed.

The exclusion of the papers after abstract and full manuscript review was based on the following criteria:

- established OvCa cell lines were used for creating tumour spheroids opposed to patient derived samples;
- not OvCa cells used;
- animal models studied;
- only conference abstract was available which did not contain enough information for detailed analysis;
- review article;

- 3D model construction was described in the previous study of the same group, sensitivities were evaluated in the article;
- in vivo formed spheroids were the focus of the research;
- topic was found irrelevant.

Twenty-five full papers and eleven conference abstracts were included in the review. Figure A2 in the Appendix A shows the full summary of the search and the process of the paper selection.

4. Discussion

The vast majority of the studies on OvCa models use in their research well-established cell lines, which in itself becomes a significant limitation as we discussed above. In this review we would like to focus on the papers that have used patient-derived samples to create OvCa models.

The relevant reports on 3D models for OvCa have been overviewed in Tables 2–7 and will be further discussed herein.

Table 2 summarised the papers where fresh OvCa specimens were used to create 3D OvCa models along with the conference abstracts (Table 7), where all the authors used fresh specimens. Various samples have been used—solid tumours, biopsy specimens, ascites and pleural effusions. In some cases, the patients were chosen specifically to be chemotherapy naïve [8,17]. As evident from Tables 2 and 7, in most of the cases if fresh tumour specimens were obtained, mincing and enzyme digestion was used to isolate cells for further cultivation.

To our knowledge the first study to utilise human tissue for construction of 3D ovarian tumour model was reported by Griffon et al., who were able to demonstrate spheroid growth in situ in eight cell suspensions out of eighteen, harvested from eight solid tumours, nine ascitic fluids and one pleural effusion. They noted that effusions were more productive in spheroid formation, than solid tumours and that samples obtained from mucinous ovarian adenocarcinomas did not form spheroids at all [6].

Other studies have been able to establish new cell lines derived directly from fresh tumour specimens to further study the spheroid formation (Table 3). Again, surgical specimens and ascites of OvCa patients were used, however other methods of obtaining the material were described, like scrape method [23,25] and Grun et al., used cytobrushings of squamous ovarian carcinoma to establish a cell line [24]. In their study Puiffe et al., described spheroid formation using OV-90 cell line; as well as the effects of acellular fraction of patient-derived ascites on the spheroid formation, growth and invasion [22].

It becomes evident from the analysis of Tables 2, 3 and 7 that more recently there is an ongoing trend towards utilising patient derived material to study tumour microenvironment and its molecular and biochemical features [9–11,28–30].

Table 2. Summary of papers, which used primary patient specimens to develop spheroid models.

Paper	Number of Patients	Type of Specimens Collected
Griffon et al. [6]	18	8 solid tumours finely chopped and enzymatically disaggregated, 9 ascitic fluids and 1 pleural effusion
Zhang et al. [7]	5	Tumour specimens of stage III serous adenocarcinomas—minced and enzymatically disaggregated
Kryczek et al. [8]	25	Cells and tissues obtained from ascites and tumours of chemotherapy naïve patients with EOC
He et al. [12]	6	Tumour specimens from OvCa patients mechanically dissociated and enzymatically disaggregated within 30 min of surgery
Martinez-Serrano et al. [13]	10	Ovarian tumour mass from chemotherapy naïve patients with papillary serous EOC processed using enzymatic cell tissue dissociation

Table 2. *Cont.*

Paper	Number of Patients	Type of Specimens Collected
Rafehi et al. [14]	At least 4 independent patient samples and at least 3 experimental replicates	Ascites fluid obtained from OvCa patients at the time of debulking surgery or paracentesis
Raghavan et al. [15]	3	Primary patient ascites cells (centrifuged) from tumour bank with confirmed OvCa origin
Loessner et al. [16]	n/a	Primary OvCa cells isolated from patients with high grade serous OvCa
Shuford et al. [17]	92	Fresh tissue from either a primary debulking surgery (n = 76) or laparoscopic biopsy (n = 16) of chemotherapy naïve patients
Maru et al. [18]	15	Tissue fragments of approximately 500–1000 mm^3 obtained from ovarian tumours immediately after tumour resection. Non-necrotic lesions with solid or papillary growth selected. Tissue fragments cut into 2–3 mm pieces and enzymatically disaggregated.
Nelson et al. [19]	12	Primary patient ascites cells (centrifuged) from tumour bank and solid tumour samples processed using a tumour dissociation kit
Park et al. [9]	3	Fresh tumours minced and dissociated with collagenase
Huang et al. [10]	7	Fresh tumours minced and dissociated with collagenase and hyaluronidase
Hedemann et al. [11]	n/a	Fresh tumour cleared, fragmentated and enzymatically disaggregated

Table 3. Summary of papers, which used cell lines derived directly from primary patient specimens to develop spheroid models.

Paper	Number of Patients	Type of Specimens Collected
Sonoda et al. [20]	n/a	OVMG-1 and OVMG-2 serous adenocarcinoma cell lines from surgical specimens
Zietarska et al. [21]	n/a	TOV-21G and TOV-112I cell lines from primary ovarian malignant tumours; OV-90 cell line from ovarian malignant ascites from chemotherapy naïve patients
Puiffe et al. [22]	OV-90 cell line derived from 1 patient; ascites of 54 EOC patients	OV-90 cell line derived from cellular fraction of ascites from a chemotherapy-naïve patient
Ouellet et al. [23]	2	TOV-1946 (scape method used) and TOV-2223G (collagenase method used) cell lines derived from solid tumours and OV-1946 cell lines from a mass of cells from ascites (micro-dissection into small pieces) of chemotherapy naïve patients with grade 3 serous papillary cystadenoma at stage IIIC
Grun et al. [24]	n/a	OV-TRL12B cell line established from cytobrushing of a squamous ovarian carcinoma
Létourneau et al. [25]	3	TOV cell lines (n = 4) derived from solid ovarian tumour (scape method) and OV cell lines (n = 5) established from the cellular fraction of ascites (centrifugation)
Liao et al. [26]	30	Primary EOC cell lines obtained from tumour specimens (finely minced) and ascitic fluid (centrifugation) obtained from patients undergoing tumour debulking surgery for EOC
Fleury et al. [27]	6	Solid ovarian tumour (TOV) derived cell lines (TOV2978G, TOV3041G, TOV3291G) (scrape method). The OV cell lines (OV866(2), OV4453, OV4485) established from the cellular fraction of ascites (centrifugation).
Noguchi et al. [28]	1	NCC-cOV1-C1 cell line derived from cellular fraction of ascites of a patient with clear cell carcinoma
Silva et al. [29]	1	IPO43 cell line established from the ascitic fluid of a patient with a diagnosis of high-grade serous carcinoma (HGSC) of the ovary, previously treated with chemotherapy
Parashar et al. [30]	1	Ovarian tumour samples minced and enzymatically disaggregated. Cells strained and centrifuged.

It is also evident from the literature that spheroid models are the most widely used systems for 3D in vitro models of OvCa, although few groups have assessed complex systems like hydrogels. Amongst the spheroid models, forced floating technique using commercially available specialised spheroid forming ultra-low attachment dishes is the most common method [5–8,10–12,14,17,20,22], followed by hanging drop method [15,21,23,25,27], all in static culture. A key difficulty of static 3D culture is their long-term maintenance, which is evident from the literature wherein most spheroid models are maintained for 24 h–7 days. Very few studies have been able to investigate long term culture of spheroid models. For example, some authors maintained their model for 2 weeks [7,12,13,16] while Kryczek et al., went further and investigated the spheroids in the culture for 6 weeks [8].

Grun et al., established spheroids in Rotary Cell Culture System and highlighted that the used method had allowed spheroids to grow for longer periods and reach significantly greater 'tumour' volumes than when using hanging droplet method. Although in their study authors were able to reach a maximum diameter of 4 mm, extensive areas of necrosis were present [24]. Loessner et al., described an encapsulation-based spheroid formation using a hydrogel system, which allowed them to create a 3D culture showing cell proliferation and aggregation similar to in vivo. The method and the 3D system were optimised using cell line OVMZ-6 and then used for primary patient samples to establish that their method worked on primary samples [16]. Finally, Maru et al., used hydrogel-based sandwich method with incubation time of 5 days to create the 3D model. Their 3D hydrogel-based model of patient samples was able to maintain original tumour characteristics. However, they found a significant limitation of this method for gynaecological tumours, as a result of which insufficient number of cells were able to attach to hydrogel platform [18] The time of incubation ranged from minimum of 24 h [17] to 6 weeks [8] with most of the studies validating the findings in vivo.

Table 4. Summary of methods used to originate 3D structures from patient derived cells.

Hanging Drop	Forced Floating	Bioreactor	Others
Zietarska et al. (4 days)	Griffon et al. (4–5 days)	Grun et al. (3–4 weeks)	Loessner et al. (hydrogel system; 2 weeks)
Ouellet et al. (3 days)	Sonoda et al. (7 days)		Maru et al. (hydrogel-based sandwich method; 5 days)
Létourneau et al. (4 days)	Puiffe et al. (4 days)		
Fleury et al. (5–7 days)	Zhang et al. (11–14 days)		
Raghavan et al. (7 days)	Kryczek et al. (1–6 weeks)		
	Liao et al. (3 weeks early culture- > dissociation and replating fornightly)		
	Rafehi et al. (72 h)		
	He et al. (14 days)		
	Martinez-Serrano et al. (average 28 days)		
	Shuford et al. (24–72 h)		
	Nelson et al. (2–4 days)		
	Vader et al.		
	Basten et al.		
	Mikkonen et al.		
	Nanki et al.		
	Park et al. (7 days)		
	Noguchi et al. (4 days)		
	Hedemann et al. (4 days)		
	Huang et al.		
	Silva et al. (72 h)		
	Parashar et al. (7 days)		

The biggest size of spheroids was demonstrated by Grun et al., who used the Rotary Cell Culture System to culture them [24], whereas others grew spheroids with a maximum size up to 500 μm [21,22] and the smallest of 50–100 μm by Zhang et al. [7]. Zietarska et al., reported an absence of hypoxic or necrotic cores, which they related to relatively short culture time (4 days) as well as small size of spheroids [23].

One of the key findings of Kryczek et al., was that deletion of ALDH+ cells or CD133+ cells dramatically reduced the quantity and size of spheres formed, whereas their simultaneous deletion drastically reduced sphere formation. Furthermore, they were able to show that the expression of ALDH+ and CD133+ gradually reduced following prolonged in vitro cycles [8]. Other studies have gone down the path of comparing the 2D and 3D systems [20,21,24]. Sonoda et al., looked at the expression of VEGF, IL-8 etc., which was compared between monolayer, spheroids and animal model [20]; Zietarska et al., following cluster analysis of gene expression suggested differences amongst the three types of models, such as the expression of *THBS1, PECAM1* genes and others [21]; Grun et al., carried-out comparison between 2D and 3D culture in terms of proteomic profiling and observed differences between the two for various markers including those of proliferation, apoptosis (CA125, BCL2, proliferation marker, Mib-1, p53, CK7 and others) [24]. Another study looking at the molecular effects of spheroids, reported that spheroid formation promoted/induced epithelial to mesenchymal transition (EMT) for EOC cells which was decreased on re-attachment of the spheroids. They also observed increased expression of TGFβ 1 in EOC spheroids, which they hypothesis modulates EMT in spheroid cultures [14].

A lot of the studies aimed to look at the cellular characteristics of the established spheroids. For example, it was shown by Puiffe et al., that compact spheroids were able to form by the cell line culture in the presence of patient derived acellular fraction of ascites suggesting the importance and subtleties of ascites in modulating the tumour microenvironment. They also studied the effects of ascites on invasion, proliferation and gene expression in the developed spheroids [22]. The same authors developed cell lines from patient samples and characterised different phenotypic characteristics including their ability to form 3D tumour models as spheroids. For instance, the authors showed that only TOV-112D cell line was able to form compact spheroids when hanging droplet method used, whereas other cell lines either formed cell clusters or did not show any aggregation properties at all [23]. Similar to these group, other authors also developed new cell lines and tested the ability of those to form 3D structures [25–27]. Liao et al., were able to isolate spheroids that were tumorigenic in vivo and had higher proliferation and migration in comparison to their parent non spheroid cells; they also observed a difference in expression of various stem cell markers, like Notch1, Nanog, D34 etc., higher than non-spheroid cells under same growth conditions [26]. Fleury et al., in their study had also shown that different cell lines derived from patient samples had different mutations seen in EOC cells (TP53, BRCA1, BRCA2 etc.) [27]. From the studies using patient derived material, Loessner et al., established the importance of co-culture system involving OvCa and mesothelial cells [16]; Zhang et al., described the process of isolation and characterisation of highly tumorigenic subpopulation of cells (malignant progenitors) [7]; whereas Maru et al., showed that 3D hydrogel-based model of patient samples was able to maintain original tumour characteristics. They also inferred that spheroid-based models are better for assessment of treatments in comparison to hydrogel-based 3D in vitro models [18].

All of the above studies confirm the benefits of 3D models in the investigation of the tumour characteristics, microenvironment and its 'behavior' and as it was highlighted by Maru et al., the choice of 3D system should be dependent on the end goal/objectives [18]. A key takeaway point from the various publications involving 3D models of OvCa is their similarity to in vivo conditions. For example, Sonoda et al., showed that VEGF expression was enhanced in spheroid models in comparison to 2D monolayer of the same cell lines. [20] Gene expression analysis carried out by Zietarska et al., identified genes in 3D model which mimic in vivo tumour gene expression in contrast to 2D culture wherein such genes were not expressed [21]. It has also been reported by Raghavan et al., that spheroid models were able to mimic A2780 and OVCAR3′s in vivo characteristic of resistance to cisplatin, which is not seen in 2D in vitro culture [15]. These aspects once again highlight the need for the research community to move towards a 3D model-based approach in comparison to 2D systems.

The similarity to in vivo conditions of the patient derived 3D culture has a major advantage of helping in treatment studies for cancer patients. The first study to utilise this was by Griffon et al., who assessed the radiosensitivity (0–8 Gy) of primary OvCa cells from various locations using tumour spheroid model. They observed extensive variation between 3D spheroids from different patients in terms of their response to radiotherapy, highlighting the need for personalised treatment protocol for patients in a clinical setting [6]. Similar studies were conducted to assess chemosensitivity of OvCa cells in 3D spheroid models by Zhang et al. They reported higher resistance to cisplatin and paclitaxel for 3D spheroids with stem cell like properties in comparison to differentiated ones, suggesting a plausible reason behind OvCa's high recurrence rate [7]. Liao et al., along with He et al., who had similar findings, showed that spheroid forming cells that maintained stem cell like properties were more resistant to cisplatin in comparison to their parent cells [12,26]. More recent studies have been testing patterns of chemoresistance [15] and therapy response predictions [17] on in vitro models. Raghavan et al., used cells recovered from primary patient malignant ascites by centrifugation and demonstrated differences in therapeutic response between patient-derived samples as well as showed correlation with in vivo drug studies in xenografts [15], whereas Shuford et al., conducted a large study including 92 samples of fresh tissue from either a primary debulking surgery or laparoscopic biopsy of chemotherapy naïve patients [17]. Looking at the abstracts included in this review it also becomes evident how important it is becoming in the recent years to create a 3D model for treatment prediction as one of the key highlights of 3D systems is that they mimic the effect of therapy better than 2D. Multiple study reports show differences in sensitivities of drug agents used for treatment of OvCa, such as carboplatin [17,34,41], cisplatin [40,127], paclixatel [33,40,41,127] and others [34–38]. Details of each of the studies included in this review could be seen in Tables 5–7.

These reports have shown the most recent advances in personalised approach to treatment prediction in OvCa patients, however it is still evident from this analysis that further research is required of patient derived studies in 3D due to their obvious advantages for investigation of tumour characteristics and more importantly better and personalised drug screening.

The vast majority of the 3D in vitro models developed for OvCa studies are spheroid type models and hydrogel types of models. Despite advantages of spheroids (especially compared to 2D cultures), such as simplicity of fabrication, achievement of heterogeneity in phenotype and gene expression and altered cell metabolism they do have several limitations. Specifically, there is an exceedingly high variability of their aggregate forming densities, which is also supported in the most recent research papers [11,30], they are susceptible to dissociation during handling and experimentation, crucial tumour microenvironment (TME) conditions (ECM, cell-matrix interactions, stiffness and mechanical properties) cannot be controlled, they cannot recapitulate in vivo mass transfer limitations, vascularisation does not occur, and they cannot be cultured long term [10]. For instance Hedemann et al., had shown, that the spheroids grown from primary cancer cells were not able to exhibit the same degree of growth as cell-lines in the same "environment" conditions [11]. Hydrogels, based on their chemistry are more advanced than spheroids, offering some level of structure, ECM mimicry and porosity, however, they have low mechanical strength and artificially high-water content (some of them up to 95%), which leads to an unrealistic microenvironment for the cells. Moreover, although generally hydrogels depending on their porosity and pore inter-connectivity could mimic various densities of different tissues, they lack other components present in connective tissue; whereas while Matrigel resembles the laminin/collagen IV-rich basement membrane extracellular environment, it does not accurately mimic the basement membranes [22].

Finally, as it is evident from the review there are currently no studies that have been able to create OvCa polymeric scaffold 3D model utilising patient-derived cells, which would significantly improve patient care by predicting the efficacy of potential chemotherapy treatment and be a further step forward into investigating the microenvironment and biology of EOC.

Table 5. Summary of papers, which used primary patient specimens to develop spheroid models—part 2.

Paper	Construct Development Method	Time of Incubation	Size of Spheroids	General Comment
Griffon et al. [6]	6-well plates coated with 1 mL of 0.5% agarose (forced floating/aggregation)	10 days in vitro	Mean of 198 (±7.7) µm	Radiosensitivities of spheroids obtained from human ovarian carcinoma cells tested.
Zhang et al. [7]	Ultra Low Attachment plates (forced floating/aggregation)	11–14 days in vitro Validation in vivo	50–100 µm	Isolation and characterisation of highly tumorigenic subpopulation of cells (malignant progenitors) described.
Kryczek et al. [8]	Ultralow attachment plates (forced floating/aggregation)	Spheres counted for 1–6 weeks Validation in vivo	>50 µm	Expression of multiple cancer stem cell markers in fresh OvCa and established primary OvCa cell lines investigated and the stem cell properties of potential OvCa stem cells in vitro and in vivo examined.
He et al. [12]	96-well ultra-low attachment plates (forced floating/aggregation)	2 weeks in vitro Validation in vivo	n/a	Subpopulation of stem cell-like cells that form spheroids and possess self-renewal capacity, strong tumour-initiating ability, and higher resistance to chemotherapy derived from high grade serous carcinoma studied.
Martinez-Serrano et al. [13]	Corning Ultra-Low attachment surface T25 flask (forced floating/aggregation)	Median period of 28 days in vitro cultivation	n/a	Specificity of cell surface markers to discriminate the tumour initiating cells (isolated from EOC) from somatic stem cells (isolated from healthy women) investigated.
Rafehi et al. [14]	Ultralow attachment plates (forced floating/aggregation)	3 days in vitro	n/a	Findings demonstrating that intact TGFβ signalling is required to control epithelial-mesenchymal transition in EOC ascites-derived cell spheroids, and it promotes the malignant characteristics of these structures.
Raghavan et al. [15]	Spheroids formed by hanging drop method	7 days in vitro Validation in vivo	n/a	The responses to varying drug treatments were different in patient-derived samples and correlated with in vivo drug studies in xenografts.
Loessner et al. [16]	Encapsulation based spheroid formation	2 weeks in vitro	n/a	3D culture showed cell proliferation profile and aggregation similar to in vivo. Expression of integrins, MMP enhanced in 3D culture in comparison to 2D. Spheroids showed higher chemoresistance in comparison to 2D for paclitaxel.
Shuford et al. [17]	84-well spheroid microplates (forced floating/aggregation)	24 h in vitro	n/a	Analytical and prospective clinical validation of a new test that utilizes primary patient tissue in 3D cell culture to make patient specific response predictions prior to initiation of treatment in the clinic presented.
Maru et al. [18]	Hydrogel based sandwich method	5 days in vitro	n/a	3D hydrogel-based model of patient samples was able to maintain original tumour characteristics. Spheroid based models are better for assessment of treatments in comparison to hydrogel-based 3D in vitro models.
Nelson et al. [19]	Matrigel in 24-well plate (forced floating/aggregation)	2–4 days in vitro	n/a	Ex vivo cultures from patient biopsies used to provide models that support interrogation of chromosome instability mechanisms.
Park et al. [9]	Ultra-attachment 6-well culture plates (forced floating/aggregation)	7 days in vitro	n/a	mRNA expression of transcription factors and miRNA expression of spheroids derived from primary ovarian cancers to identify factors regulating ovarian cancer stem cells.

Table 5. *Cont.*

Paper	Construct Development Method	Time of Incubation	Size of Spheroids	General Comment
Huang et al. [10]	6-well ultra-low attachment plates. (forced floating/aggregation)	7–10 days in vitro Validation in vivo	>50 μm	Cell lines and primary tissue used to grow spheroids, which were tested against platinum-chemotherapy agents, correlated with in vivo drug studies in xenografts.
Hedemann et al. [11]	Ultralow attachment plates (forced floating/aggregation)	4 days in vitro	~150–300 μm	A combination of ADAM17 inhibitor with cisplatin tested in 2D and 3D culture of cells derived from cell lines and primary ovarian tumor- and ascites-derived cells.

Table 6. Summary of papers, which used cell lines derived directly from primary patient specimens to develop spheroid models—part 2.

Paper	Construct Development Method	Time of Incubation	Size of Spheroids	General Comment
Sonoda et al. [20]	24-well culture plate coated with 1% agarose (forced floating/aggregation)	7 days in vitro Validation in vivo	n/a	Angiogenesis factors expression measured and compared in 2D, 3D and xenografts.
Zietarska et al. [21]	Spheroids formed by hanging drop method	10 days in vitro (spheroids formed by day 4) Validation in vivo	Maximum size of 500 μm	Molecular comparison of spheroid model versus 2D and xenograft model described.
Puiffe et al. [22]	Spheroids formed by modified hanging drop method	4 days in vitro	Small in the absence And ~500 μm in the presence of ascites	Effect of the acellular fraction of ascites on OV-90 addressed.
Ouellet et al. [23]	Spheroids formed by modified hanging drop method	4 days in vitro Validation in vivo	TOV-1946—aggregate OV-1946—semi-compact TOV-2223—none	New serous EOC cell lines from both solid tumours and ascites of the same patient were derived and characterised.
Grun et al. [24]	Rotary Cell Culture System	3–4 weeks in vitro (spheroids formed in 1 week)	Maximum diameter of 4 mm	3D culture established, biological features (morphological characteristics, expression of tumour markers, proteomic profiles). compared between 2D, 3D and primary tumours.
Létourneau et al. [25]	Spheroids formed by hanging drop method	4 days in vitro Validation in vivo	n/a	New OvCa cell lines described.
Liao et al. [26]	Ultralow attachment plates (forced floating/aggregation)	Cultivation period n/a Validation in vivo	n/a	To study EOC pathogenesis, EOC primary cells under stem cell selective conditions were cultured and generated anchorage-independent, self-renewing spheroids morphologically similar to spheroids isolated from patient ascites.
Fleury et al. [27]	Spheroids formed by hanging droplet method	5–7 days in vitro Validation in vivo	TOV2978G, OV4453—aggregate OTOV3291G—semi compact TOV3041G—compact	Six new EOC cell lines spontaneously derived from high grade serous tumours or ascites established and described.

Table 6. *Cont.*

Paper	Construct Development Method	Time of Incubation	Size of Spheroids	General Comment
Noguchi et al. [28]	96-well culture plates (forced floating/aggregation)	4 days in vitro	n/a	NCC-cOV1-C1 cell line established and characterised. Anticancer drug screening conducted.
Silva et al. [29]	Stirred-tank culture system placed on a magnetic stirrer	3 days of in vitro	n/a	IPO43 cell line established and characterised.
Parashar et al. [30]	24-well, growth factor reduced Matrigel-coated non-adherent plates (forced floating/aggregation)	7 days in vitro	~30–100 μm	MCW-OV-SL-3 endometrioid subtype of ovarian cancer cell line established, chemoresistance mechanisms studied.

Table 7. Summary of conference abstracts.

Abstract	Year	Number of Patients, Specimen and Method	General Comment
Sun et al. [31]	2012	Fresh specimens of OvCa minced, enzymatically digested, rinsed, incubated in DMEM (monolayer cultures) Mammosphere media (spheroids). Validation in vivo.	Spheroids are enriched for expression of markers including CD133, CD44, NANOG and OCT4, suggesting that spheroid formation enhances stem cell-like markers. Increased expression of miR-26b in spheroids compared to monolayer culture.
Shuford et al. [32]	2014	OvCa samples—standard mincing & digestion.	Ex vivo 3D (EV3D™) culture and testing of primary human OvCa was described. Carboplatin & taxane based combination therapy was used in most cases.
Ishiguro et al. [33]	2014	OvCa cells from surgical specimen. Validation in vivo.	Differentiation of spheroid cells associated with the downregulation of the stem cell-specific regulators Nanog, Sox2, and ALDH1A1 and the up-regulation of cytokeratin and it is associated with increased paclitaxel resistance. The changes are reversible.
Desrochers et al. [34]	2015	OvCa samples (newly diagnosed, treatment naïve and relapsed) standard mincing & digestion. 3D spheroids were developed and 3D perfused Ovarian Microtumours were cultured using the 3DKUBETM.	Carboplatin, gemcitabine, erlotinib and afatanib responses tested.
Vader et al. [35] Basten et al. [36] Dijkmans et al. [37]	2017 2018 2018	3D cultures embedded in a protein-rich hydrogel (384 well plates) are generated from tumour biopsies (endometrial, cervical, and OvCa patients–fresh and cryopreserved material).	3D cultures exposed to standard-of-care therapies, targeted therapies and drug combinations.

Table 7. *Cont.*

Abstract	Year	Number of Patients, Specimen and Method	General Comment
Mikkonen et al. [38]	2018	Processed fresh cancer tissue (ovarian)—cells cultivated in Matrigel or in cellulose-based hydrogel, GrowDex.	Genetic profiling and image-based phenotyping, phenomics done. Drug responses (52 agents) tested in 2D and 3D, significant differences in sensitivity to several drugs observed.
Nanki et al. [39]	2018	Intraoperative ascites and tissue samples from primary ovarian, peritoneal, and fallopian tube cancer patients. 3D culture obtained using 96-well plates—14 days.	Spheroids-like structures were formed in 30% (1/3) of ascites samples and 50% (4/8) of tissue samples. The tumorigenicity and invasiveness of the cells were demonstrated using new 3D model cultured in vitro by NanoCulture Plate LH96.
Tanaka et al. [40]	2018	13 primary ovarian tumour surgical samples (8—OvCa, 2—borderline, 3—benign) and 1 malignant effusion (ascitic and pleural) of OvCa patient. Matrigel-based organoid culture, or spheroid culture.	Long-term 3D cultures established from 4 samples. Drug responses tested for 2 cultures (cisplatin and paclitaxel).
Ito et al. [41]	2018	OvCa cells from patient tumours (61 cancer tissue-originated spheroid (CTOS) method).	Sensitivity assay for paclitaxel and carboplatin conducted and compared to clinical outcome.

5. Conclusions

There is a clear need for the development of an accurate, robust, 3D system which will enable the culture and drug screening of patient derived ovarian tumours. Such a system will allow screening of drugs as well as genetic analysis of the cancer of a specific individual, therefore, optimising/tailoring the treatment towards that individual. Furthermore, such an in vitro 3D system which would account for the tumour microenvironment heterogeneity, would help elucidate developmental and evolutionary aspects of the disease. Finally, for the development of such system with a tangible clinical outcome a systematic rigorous experimentation with patient derived tumours (and not with cell lines) is essential.

While this review covers studies of 3D OvCa models utilising patient derived samples, we have limited ourselves to multicellular tumour spheroid and tumour-derived spheroid models; and the review focused mainly on the models of EOC.

Author Contributions: Conceptualization, T.K.M. and E.V.; methodology, T.K.M. and L.S.; validation, T.K.M., L.S., P.G. and E.V.; data curation, L.S. and T.K.M., writing—original draft preparation, L.S., P.G., E.V. and T.K.M.; writing—review and editing, L.S. and P.G.; supervision, E.V. and T.K.M.; project administration, T.K.M. and L.S.; funding acquisition, E.V. All authors have read and agreed to the published version of the manuscript.

Funding: P.G. and E.V. have received funding from the 3DbioNet (MR/R025762/1). E.V. is grateful to the Royal Academy of Engineering for an Industrial Fellowship and to the Medical Research Council UK for a New Investigator Research Grant (MR/V028553/1), which also financially supports P.G.

Institutional Review Board Statement: Not applicable.

Informed Consent Statement: Not applicable.

Data Availability Statement: Data supporting reported results can be found on: https://search.ebscohost.com/login.aspx?direct=true&AuthType=sso&bquery=AB+scaffold+AND+TI+(+ovarian+cancer+or+ovarian+neoplasms+)+OR+AB+3d+AND+TI+(+ovarian+cancer+or+ovarian+neoplasms+)+OR+AB+three+dimensional+AND+TI+(+ovarian+cancer+or+ovarian+neoplasms+)+OR+AB+spheroid+AND+TI+(+ovarian+cancer+or+ovarian+neoplasms+)&type=1&searchMode=And&site=eds-live&ssl=y&custid=ns124426 (accessed on 25 October 2022).

Acknowledgments: E.V. is grateful to the Royal Academy of Engineering for an Industrial Fellowship and to the Medical Research Council UK for a New Investigator Research Grant (MR/V028553/1), which also financially supports P.G. T.K.M. is grateful to GRACE Charity for research funding that supports L.S. Doctoral studies at Brighton.

Conflicts of Interest: The authors declare no conflict of interest.

Abbreviations

OvCa	ovarian cancer
EOC	epithelial ovarian cancer
3D	three dimensional
2D	two dimensional
TME	tumour microenvironment
ECM	extracellular matrix
PEG	poly-ethylene glycol
EMT	epithelial to mesenchymal transition
MMP	matrix metalloproteinases
CTOS	cancer tissue originated spheroids
TGFβ	transforming growth factor β
VEGF	vascular endothelial growth factor

Appendix A

PLATFORM	ADVANTAGES	DISADVANTAGES
 2D CULTURE	Easy to use and replicate Low cost and high throughput Uniform distribution of nutrients Compatible with currently available assay systems	Unable to mimic complex *in vivo* tissue microenvironment Lack of topologically relevant cell-cell and cell-ECM interaction Unable to support long term culture without continuous passaging
 SPHEROID 3D CULTURE	Presence of topologically relevant cell-cell and cell- ECM interaction Relatively easy to develop, use and replicate Low cost and relatively high throughput Can mimic *in vivo* complexities to some extent	Presence of artificial diffusion gradient for oxygen, essential nutrients and waste materials Moderate control over spheroid shape and size Difficulty in maintaining long term culture without regular resuspension
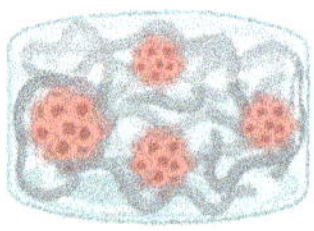 HYDROGEL 3D CULTURE	Presence of topologically relevant cell-cell and cell- ECM interaction Can mimic *in vivo* complexities Allows nutrient and oxygen diffusion Can maintain long term *in vitro* culture without resuspension	Difficulty in achieving uniform cell distribution Lack of mechanical strength due to high water content Low structural and architectural tunability Presence of material batch-to-batch variation Expensive and more difficult to reproduce
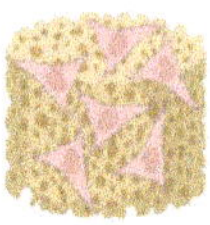 POLYMERIC SCAFFOLD 3D CULTURE	Presence of topologically relevant cell-cell and cell- ECM interaction Can mimic *in vivo* complexities Allows nutrient and oxygen diffusion and control Can maintain long term *in vitro* culture without resuspension Mechanically stable and allows for structural and mechanical tunability	Expensive and more complex synthesis process Difficulty in cell retrieval and observation while in culture

Figure A1. A graphical schematic of key techniques used for creation of OvCa tumour models (created on BioRender.com (accessed on 26 July 2022)).

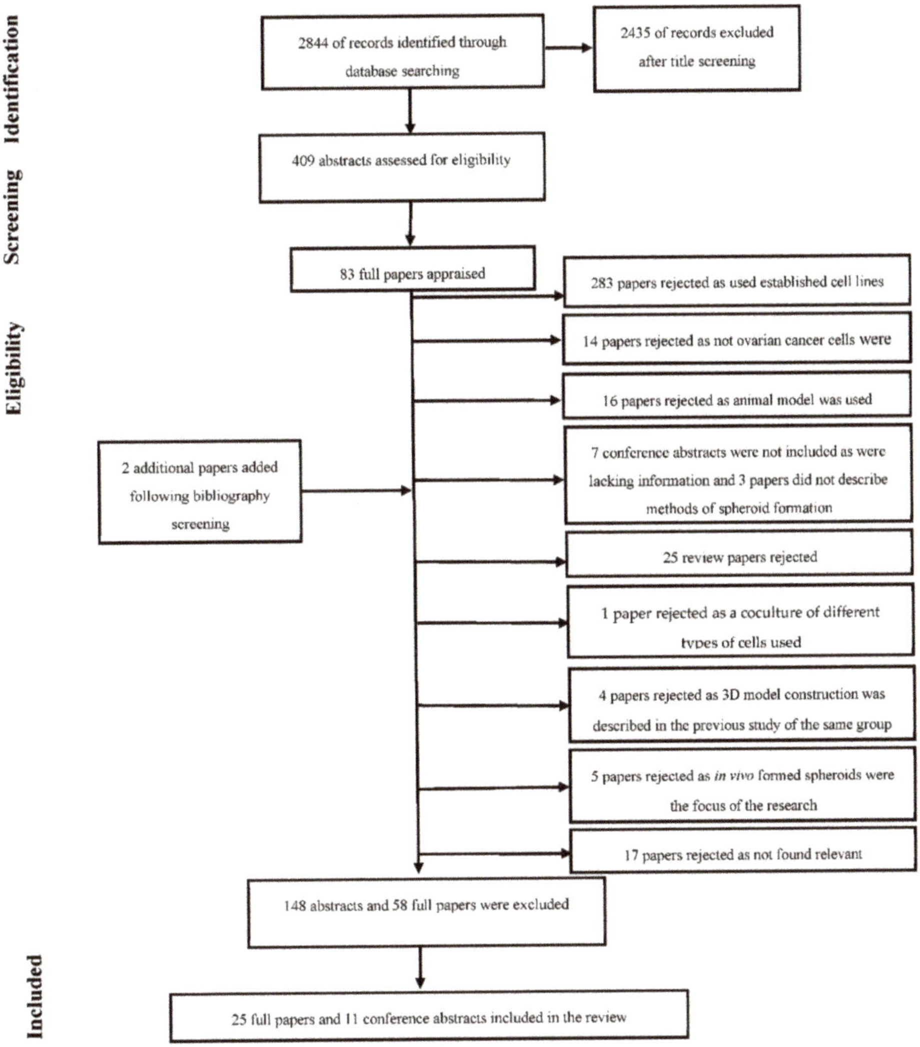

Figure A2. PRISMA flow chart.

References

1. Cancer Research, UK. Ovarian Cancer Statistics. Available online: https://www.cancerresearchuk.org/health-professional/cancer-statistics/statistics-by-cancer-type/ovarian-cancer (accessed on 27 January 2020).
2. Torre, L.A.; Trabert, B.; Desantis, C.E.; Miller, K.D.; Samimi, G.; Runowicz, C.D.; Gaudet, M.M.; Jemal, A.; Siegel, R.L. Ovarian Cancer Statistics, 2018. *CA Cancer J. Clin.* **2018**, *68*, 284–296. [CrossRef] [PubMed]
3. PDQ Adult Treatment Editorial Board. *Ovarian Epithelial, Fallopian Tube, and Primary Peritoneal Cancer Treatment (PDQ®): Health Professional Version*; National Cancer Instutute: Bethesda, MD, USA, 2002.
4. Moschetta, M.; Boussios, S.; Rassy, E.; Samartzis, E.P.; Funingana, G.; Uccello, M. Neoadjuvant treatment for newly diagnosed advanced ovarian cancer: Where do we stand and where are we going? *Ann. Transl. Med.* **2020**, *8*, 1710. [CrossRef] [PubMed]
5. Lal-Nag, M.; McGee, L.; Titus, S.A.; Brimacombe, K.; Michael, S.; Sittampalam, G.; Ferrer, M. Exploring Drug Dosing Regimens In Vitro Using Real-Time 3D Spheroid Tumor Growth Assays. *SLAS Discov.* **2017**, *22*, 537–546. [CrossRef] [PubMed]
6. Griffon, G.; Marchal, C.; Merlin, J.L.; Marchal, S.; Parache, R.M.; Bey, P. Radiosensitivity of multicellular tumour spheroids obtained from human ovarian cancers. *Eur. J. Cancer* **1995**, *31A*, 85–91. [CrossRef]

7. Zhang, S.; Balch, C.; Chan, M.W.; Lai, H.C.; Matei, D.; Schilder, J.M.; Yan, P.S.; Huang, T.H.M.; Nephew, K.P. Identification and characterization of ovarian cancer-initiating cells from primary human tumors. *Cancer Res.* **2008**, *68*, 4311–4320. [CrossRef]
8. Kryczek, I.; Liu, S.; Roh, M.; Vatan, L.; Szeliga, W.; Wei, S.; Banerjee, M.; Mao, Y.; Kotarski, J.; Wicha, M.S.; et al. Expression of aldehyde dehydrogenase and CD133 defines ovarian cancer stem cells. *Int. J. Cancer* **2012**, *130*, 29–39. [CrossRef] [PubMed]
9. Park, H.; Hwang, S.; Jeong, J.Y.; Jung, S.G.; Choi, M.C.; Joo, W.D.; Song, S.H.; Lee, C.; An, H.J. Integrative analysis of transcription factors and microRNAs in ovarian cancer cell spheroids. *J. Ovarian Res.* **2020**, *13*, 16. [CrossRef]
10. Huang, Z.; Kondoh, E.; Visco, Z.R.; Baba, T.; Matsumura, N.; Dolan, E.; Whitaker, R.S.; Konishi, I.; Fujii, S.; Berchuck, A.; et al. Targeting dormant ovarian cancer cells in vitro and in an in vivo mouse model of platinum resistance. *Mol. Cancer Ther.* **2021**, *20*, 85–95. [CrossRef]
11. Hedemann, N.; Herz, A.; Schiepanski, J.H.; Dittrich, J.; Sebens, S.; Dempfle, A.; Feuerborn, J.; Rogmans, C.; Tribian, N.; Flörkemeier, I.; et al. Adam17 inhibition increases the impact of cisplatin treatment in ovarian cancer spheroids. *Cancers* **2021**, *13*, 2039. [CrossRef]
12. He, Q.Z.; Luo, X.Z.; Wang, K.; Zhou, Q.; Ao, H.; Yang, Y.; Li, S.X.; Li, Y.; Zhu, H.T.; Duan, T. Isolation and characterization of cancer stem cells from high-grade serous ovarian carcinomas. *Cell. Physiol. Biochem.* **2014**, *33*, 173–184. [CrossRef]
13. Martínez-Serrano, M.J.; Caballero-Baños, M.; Vilella, R.; Vidal, L.; Pahisa, J.; Martínez-Roman, S. Is sphere assay useful for the identification of cancer initiating cells of the ovary? *Int. J. Gynecol. Cancer* **2015**, *25*, 12–17. [CrossRef] [PubMed]
14. Rafehi, S.; Valdes, Y.R.; Bertrand, M.; McGee, J.; Préfontaine, M.; Sugimoto, A.; Dimattia, G.E.; Shepherd, T.G. TGFβ signaling regulates Epithelial-mesenchymal plasticity in ovarian cancer ascites-derived spheroids. *Endocr. Relat. Cancer* **2016**, *23*, 147–159. [CrossRef] [PubMed]
15. Raghavan, S.; Mehta, P.; Ward, M.R.; Bregenzer, M.E.; Fleck, E.M.A.; Tan, L.; McLean, K.; Buckanovich, R.J.; Mehta, G. Personalized medicine–based approach to model patterns of chemoresistance and tumor recurrence using ovarian cancer stem cell spheroids. *Clin. Cancer Res.* **2017**, *23*, 6934–6945. [CrossRef] [PubMed]
16. Loessner, D.; Rockstroh, A.; Shokoohmand, A.; Holzapfel, B.M.; Wagner, F.; Baldwin, J.; Boxberg, M.; Schmalfeldt, B.; Lengyel, E.; Clements, J.A.; et al. Biomaterials A 3D tumor microenvironment regulates cell proliferation, peritoneal growth and expression patterns. *Biomaterials* **2019**, *190–191*, 63–75. [CrossRef]
17. Shuford, S.; Wilhelm, C.; Rayner, M.; Elrod, A.; Millard, M.; Mattingly, C.; Lotstein, A.; Smith, A.M.; Guo, Q.J.; O'Donnell, L.; et al. Prospective Validation of an Ex Vivo, Patient-Derived 3D Spheroid Model for Response Predictions in Newly Diagnosed Ovarian Cancer. *Sci. Rep.* **2019**, *9*, 11153. [CrossRef]
18. Maru, Y.; Tanaka, N.; Itami, M.; Hippo, Y. Efficient use of patient-derived organoids as a preclinical model for gynecologic tumors. *Gynecol. Oncol.* **2019**, *154*, 189–198. [CrossRef]
19. Nelson, L.; Tighe, A.; Golder, A.; Littler, S.; Bakker, B.; Moralli, D.; Murtuza Baker, S.; Donaldson, I.J.; Spierings, D.C.J.; Wardenaar, R.; et al. A living biobank of ovarian cancer ex vivo models reveals profound mitotic heterogeneity. *Nat. Commun.* **2020**, *11*, 822. [CrossRef]
20. Sonoda, T.; Kobayashi, H.; Kaku, T.; Hirakawa, T.; Nakano, H. Expression of angiogenesis factors in monolayer culture, multicellular spheroid and in vivo transplanted tumor by human ovarian cancer cell lines. *Cancer Lett.* **2003**, *196*, 229–237. [CrossRef]
21. Zietarska, M.; Maugard, C.M.; Filali-Mouhim, A.; Alam-Fahmy, M.; Tonin, P.N.; Provencher, D.M.; Mes-Masson, A.M. Molecular description of a 3D in vitro model for the study of epithelial ovarian cancer (EOC). *Mol. Carcinog.* **2007**, *46*, 872–885. [CrossRef]
22. Puiffe, M.L.; Le Page, C.; Filali-Mouhim, A.; Zietarska, M.; Ouellet, V.; Tonin, P.N.; Chevrette, M.; Provencher, D.M.; Mes-Masson, A.M. Characterization of ovarian cancer ascites on cell invasion, proliferation, spheroid formation, and gene expression in an in vitro model of epithelial ovarian cancer. *Neoplasia* **2007**, *9*, 820–829. [CrossRef]
23. Ouellet, V.; Zietarska, M.; Portelance, L.; Lafontaine, J.; Madore, J.; Puiffe, M.L.; Arcand, S.L.; Shen, Z.; Hébert, J.; Tonin, P.N.; et al. Characterization of three new serous epithelial ovarian cancer cell lines. *BMC Cancer* **2008**, *8*, 152. [CrossRef] [PubMed]
24. Grun, B.; Benjamin, E.; Sinclair, J.; Timms, J.F.; Jacobs, I.J.; Gayther, S.A.; Dafou, D. Three-dimensional in vitro cell biology models of ovarian and endometrial cancer. *Cell Prolif.* **2009**, *42*, 219–228. [CrossRef] [PubMed]
25. Létourneau, I.J.; Quinn, M.C.J.; Wang, L.L.; Portelance, L.; Caceres, K.Y.; Cyr, L.; Delvoye, N.; Meunier, L.; de Ladurantaye, M.; Shen, Z.; et al. Derivation and characterization of matched cell lines from primary and recurrent serous ovarian cancer. *BMC Cancer* **2012**, *12*, 379. [CrossRef] [PubMed]
26. Liao, J.; Qian, F.; Tchabo, N.; Mhawech-Fauceglia, P.; Beck, A.; Qian, Z.; Wang, X.; Huss, W.J.; Lele, S.B.; Morrison, C.D.; et al. Ovarian cancer spheroid cells with stem cell-like properties contribute to tumor generation, metastasis and chemotherapy resistance through hypoxia-resistant metabolism. *PLoS ONE* **2014**, *9*, e84941. [CrossRef]
27. Fleury, H.; Communal, L.; Carmona, E.; Portelance, L.; Arcand, S.L.; Rahimi, K.; Tonin, P.N.; Provencher, D.; Mes-Masson, A.M. Novel high-grade serous epithelial ovarian cancer cell lines that reflect the molecular diversity of both the sporadic and hereditary disease. *Genes Cancer* **2015**, *6*, 378–398. [CrossRef]
28. Noguchi, R.; Yoshimatsu, Y.; Sei, A.; Yoshida, H.; Katou, T.; Kondo, T. Drug screening and kinase activity profiling of a novel patient-derived cell line of clear cell ovarian carcinoma. *J. Electrophor.* **2021**, *65*, 23–31. [CrossRef]
29. Silva, F.; Coelho, F.; Peixoto, A.; Pinto, P.; Martins, C.; Frombach, A.S.; Santo, V.E.; Brito, C.; Guimarães, A.; Félix, A. Establishment and characterization of a novel ovarian high-grade serous carcinoma cell line—IPO43. *Cancer Cell Int.* **2022**, *22*, 175. [CrossRef]

30. Parashar, D.; Geethadevi, A.; Mittal, S.; McAlarnen, L.A.; George, J.; Kadamberi, I.P.; Gupta, P.; Uyar, D.S.; Hopp, E.E.; Drendel, H.; et al. Correction to: Parashar et al. Patient-Derived Ovarian Cancer Spheroids Rely on PI3K-AKT Signaling Addiction for Cancer Stemness and Chemoresistance. *Cancers* **2022**, *14*, 958. [CrossRef]
31. Sun, G.; Meng, F.; Zhong, M.; Yu, Y.; Shan, W.; Anderson, M.; Brewer, M.A. Abstract 1406: A new ovarian cancer metastasis model using multicellular spheroids generated from human ovarian cancers tissues. *Cancer Res.* **2012**, *72*, 1406. [CrossRef]
32. Shuford, S.; Widener, R.; Cheluvaraju, C.; Desrochers, T.; Mattingly, C.; Puls, L.; Gevaert, M.; Orr, D.; Crosswell, H.E. Abstract LB-36: Chemotherapy testing of primary human ovarian cancers in an ex vivo 3D culture platform: A novel method of phenotypic profiling for clinical trial selection and personalized medicine. *Cancer Res.* **2014**, *74*, LB-36. [CrossRef]
33. Ishiguro, T.; Ohata, H.; Nakagama, H.; Okamoto, K.; Tanaka, K.; Enomoto, T. Abstract 3056: Human ovarian cancer stem cells: In vitro cultivation and characterization. *Cancer Res.* **2014**, *74*, 3056. [CrossRef]
34. Desrochers, T.; Shuford, S.; Mattingly, C.; Holmes, L.; Gevaert, M.; Elder, J.; Orr, D.; Corless, C.; Puls, L.; Crosswell, H.E. Abstract LB-282: Ex vivo 3d drug response profiling of primary human ovarian cancer differentiates treatment-naive and relapsed patients and molecular subtypes. *Cancer Res.* **2015**, *75*, LB-282. [CrossRef]
35. Vader, W.; Price, L.; Herpers, B.; Basten, S. 3D cultured tumour from patients to predict treatment response. *Ann. Oncol.* **2017**, *28*, V451. [CrossRef]
36. Basten, S.; Herpers, B.; Yan, K.; Vader, W.; Price, L. Abstract LB-A09: Predicting PARPi sensitivity in patient derived ex vivo 3D tumor cultures. *Mol. Cancer Ther.* **2018**, *17*, LB-A09. [CrossRef]
37. Dijkmans, T.; Basten, S.; Herpers, B.; Yan, K.; Giesemann, T.; Schueler, J.; Vader, W.; Price, L. Abstract 4644: Patient-derived 3D tumor cultures for clinical diagnostics and pre-clinical drug development. *Cancer Res.* **2018**, *78*, 4644. [CrossRef]
38. Mikkonen, P.; Turunen, L.; Paasonen, L.; Potdar, S.; Paavolainen, L.; Murumägi, A.; Kallioniemi, O.; Pietiäinen, V.M. Abstract 5029: Precision cancer medicine based on 3D drug profiling of patient-derived cancer cell spheroid models. *Cancer Res.* **2018**, *78*, 5029. [CrossRef]
39. Nanki, Y.; Hirasawa, A.; Nomura, H.; Okubo, A.; Itoh, M.; Akahane, T.; Chiyoda, T.; Kataoka, F.; Tominaga, E.; Aoki, D. Abstract A61: Ascites-derived and tissue-derived ovarian cancer cell primary 3D cultures aimed for personalized medicine. *Clin. Cancer Res.* **2018**, *24*, A61. [CrossRef]
40. Tanaka, N.; Suzuka, K.; Maru, Y.; Hippo, Y.; Itami, M. Development of three dimensional culture method for ovarian cancer toward clinical application. *Int. J. Gynecol. Cancer* **2018**, *28*, 209. [CrossRef]
41. Ito, Y.; Endo, H.; Kondo, J.; Matsuzaki, S.; Ueda, Y.; Kimura, T.; Yoshino, K. Ex vivo chemosensitivity assay using patient-derived spheroids of epithelial ovarian cancer. *Cancer Sci.* **2018**, *109*, 603. [CrossRef]
42. Lengyel, E.; Burdette, J.E.; Kenny, H.A.; Matei, D.; Pilrose, J.; Haluska, P.; Hales, D.B.; Stack, M.S. Epithelial Ovarian Cancer Experimental Models. *Oncogene* **2014**, *33*, 3619–3633. [CrossRef]
43. Lengyel, E. Ovarian cancer development and metastasis. *Am. J. Pathol.* **2010**, *177*, 1053–1064. [CrossRef] [PubMed]
44. Landen, C.N.; Birrer, M.J.; Sood, A.K. Early events in the pathogenesis of epithelial ovarian cancer. *J. Clin. Oncol.* **2008**, *26*, 995–1005. [CrossRef] [PubMed]
45. Pearce, O.M.T.; Delaine-Smith, R.; Maniati, E.; Nichols, S.; Wang, J.; Böhm, S.; Rajeeve, V.; Ullah, D.; Chakravarty, P.; Jones, R.R.; et al. Deconstruction of a metastatic tumor microenvironment reveals a common matrix response in human cancers Europe PMC Funders Group. *Cancer Discov.* **2018**, *8*, 304–319. [CrossRef] [PubMed]
46. Horst, E.N.; Bregenzer, M.E.; Mehta, P.; Snyder, C.S.; Repetto, T.; Yang-Hartwich, Y.; Mehta, G. Personalized models of heterogeneous 3D epithelial tumor microenvironments: Ovarian cancer as a model. *Acta Biomater.* **2021**, *132*, 401–420. [CrossRef]
47. Worzfeld, T.; von Strandmann, E.P.; Huber, M.; Adhikary, T.; Wagner, U.; Reinartz, S.; Müller, R. The unique molecular and cellular microenvironment of Ovarian cancer. *Front. Oncol.* **2017**, *7*, 24. [CrossRef]
48. Kenny, H.A.; Dogan, S.; Zillhardt, M.; Anirban Mitra, S.D.Y.; Krausz, T.; Lengyel, E. *Organotypic Models of Metastasis: A 3 Dimensional Culture Mimicking the Human Peritoneum and Omentum for the Study of the Early Steps of Ovarian Cancer Metastasis*; Springer: New York, NY, USA, 2013; pp. 3–6. [CrossRef]
49. Mukherjee, A.; Chiang, C.Y.; Daifotis, H.A.; Nieman, K.M.; Fahrmann, J.F.; Lastra, R.R.; Romero, I.L.; Fiehn, O.; Lengyel, E. Adipocyte-induced FABP4 expression in ovarian cancer cells promotes metastasis and mediates carboplatin resistance. *Cancer Res.* **2020**, *80*, 1748–1761. [CrossRef]
50. Nieman, K.M.; Kenny, H.A.; Penicka, C.V.; Ladanyi, A.; Buell-Gutbrod, R.; Zillhardt, M.R.; Romero, I.L.; Carey, M.S.; Mills, G.B.; Hotamisligil, G.S.; et al. Adipocytes promote ovarian cancer metastasis and provide energy for rapid tumor growth. *Nat. Med.* **2011**, *17*, 1498–1503. [CrossRef]
51. Furuya, M. Ovarian cancer stroma: Pathophysiology and the roles in cancer development. *Cancers* **2012**, *4*, 701–724. [CrossRef]
52. Cai, J.; Tang, H.; Xu, L.; Wang, X.; Yang, C.; Ruan, S.; Guo, J.; Hu, S.; Wang, Z. Fibroblasts in omentum activated by tumor cells promote ovarian cancer growth, adhesion and invasiveness. *Carcinogenesis* **2012**, *33*, 20–29. [CrossRef]
53. Yeung, T.L.; Leung, C.S.; Yip, K.P.; Yeung, C.L.A.; Wong, S.T.C.; Mok, S.C. Cellular and molecular processes in ovarian cancer metastasis. A review in the theme: Cell and molecular processes in cancer metastasis. *Am. J. Physiol.—Cell Physiol.* **2015**, *309*, C444–C456. [CrossRef]
54. Nwani, N.G.; Sima, L.E.; Nieves-Neira, W.; Matei, D. Targeting the microenvironment in high grade serous ovarian cancer. *Cancers* **2018**, *10*, 266. [CrossRef] [PubMed]

55. Duncan, T.J.; Al-Attar, A.; Rolland, P.; Scott, I.V.; Deen, S.; Liu, D.T.Y.; Spendlove, I.; Durrant, L.G. Vascular endothelial growth factor expression in ovarian cancer: A model for targeted use of novel therapies? *Clin. Cancer Res.* **2008**, *14*, 3030–3035. [CrossRef] [PubMed]
56. Hollingsworth, H.C.; Kohn, E.C.; Steinberg, S.M.; Rothenberg, M.L.; Merino, M.J. Tumor angiogenesis in advanced stage ovarian carcinoma. *Am. J. Pathol.* **1995**, *147*, 33–41.
57. Gasparini, G.; Bonoldi, E.; Viale, G.; Verderio, P.; Boracchi, P.; Panizzoni, G.A.; Radaelli, U.; Di Bacco, A.; Guglielmi, R.B.; Bevilacqua, P. Prognostic and predictive value of tumour angiogenesis in ovarian carcinomas. *Int. J. Cancer* **1996**, *69*, 205–211. [CrossRef]
58. Orre, M.; Lotfi-Miri, M.; Mamers, P.; Rogers, P.A.W. Increased microvessel density in mucinous compared with malignant serous and benign tumours of the ovary. *Br. J. Cancer* **1998**, *77*, 2204–2209. [CrossRef]
59. Boussios, S.; Karathanasi, A.; Cooke, D.; Neille, C.; Sadauskaite, A.; Moschetta, M.; Zakynthinakis-Kyriakou, N.; Pavlidis, N. PARP inhibitors in ovarian cancer: The route to "ITHacA". *Diagnostics* **2019**, *9*, 55. [CrossRef]
60. Carvalho, M.R.; Lima, D.; Reis, R.L.; Oliveira, J.M.; Correlo, V.M. Anti-Cancer Drug Validation: The Contribution of Tissue Engineered Models. *Stem Cell Rev. Rep.* **2017**, *13*, 347–363. [CrossRef]
61. Totti, S.; Vernardis, S.I.; Meira, L.; Pérez-Mancera, P.A.; Costello, E.; Greenhalf, W.; Palmer, D.; Neoptolemos, J.; Mantalaris, A.; Velliou, E.G. Designing a bio-inspired biomimetic in vitro system for the optimization of ex vivo studies of pancreatic cancer. *Drug Discov. Today* **2017**, *22*, 690–701. [CrossRef]
62. Luvero, D.; Milani, A.; Ledermann, J.A. Treatment options in recurrent ovarian cancer: Latest evidence and clinical potential. *Ther. Adv. Med. Oncol.* **2014**, *6*, 229–239. [CrossRef]
63. Velliou, E.G.; Dos Santos, S.B.; Papathanasiou, M.M.; Fuentes-Gari, M.; Misener, R.; Panoskaltsis, N.; Pistikopoulos, E.N.; Mantalaris, A. Towards unravelling the kinetics of an acute myeloid leukaemia model system under oxidative and starvation stress: A comparison between two- and three-dimensional cultures. *Bioprocess Biosyst. Eng.* **2015**, *38*, 1589–1600. [CrossRef]
64. Raghavan, S.; Ward, M.R.; Rowley, K.R.; Wold, R.M.; Buckanovich, R.J.; Mehta, G.; Arbor, A.; Arbor, A.; Arbor, A.; Arbor, A. Formation of stable small cell number three-dimensional ovarian cancer spheroids using hanging drop arrays for preclinical drug sensitivity assays. *Gynecol. Oncol.* **2016**, *138*, 181–189. [CrossRef] [PubMed]
65. Loessner, D.; Stok, K.S.; Lutolf, M.P.; Hutmacher, D.W.; Clements, J.A.; Rizzi, S.C. Bioengineered 3D platform to explore cell-ECM interactions and drug resistance of epithelial ovarian cancer cells. *Biomaterials* **2010**, *31*, 8494–8506. [CrossRef] [PubMed]
66. Liu, M.; Zhang, X.; Long, C.; Xu, H.; Cheng, X.; Chang, J.; Zhang, C.; Zhang, C.; Wang, X. Collagen-based three-dimensional culture microenvironment promotes epithelial to mesenchymal transition and drug resistance of human ovarian cancer: In vitro. *RSC Adv.* **2018**, *8*, 8910–8919. [CrossRef] [PubMed]
67. Xu, G.; Yin, F.; Wu, H.; Hu, X.; Zheng, L.; Zhao, J. In vitro ovarian cancer model based on three-dimensional agarose hydrogel. *J. Tissue Eng.* **2014**, *5*, 2041731413520438. [CrossRef] [PubMed]
68. Yang, Z.; Zhao, X. A 3D model of ovarian cancer cell lines on peptide nanofiber scaffold to explore the cell-scaffold interaction and chemotherapeutic resistance of anticancer drugs. *Int. J. Nanomed.* **2011**, *6*, 303–310. [CrossRef]
69. Ishiguro, T.; Ohata, H.; Sato, A.; Yamawaki, K.; Enomoto, T.; Okamoto, K. Tumor-derived spheroids: Relevance to cancer stem cells and clinical applications. *Cancer Sci.* **2017**, *108*, 283–289. [CrossRef]
70. Lal-Nag, M.; McGee, L.; Guha, R.; Lengyel, E.; Kenny, H.A.; Ferrer, M. A high throughput screening model of the tumor microenvironment for ovarian cancer cell growth. *SLAS Discov.* **2017**, *22*, 494–506. [CrossRef]
71. Heredia-Soto, V.; Redondo, A.; Berjón, A.; Martín, M.M.; Díaz, E.; Crespo, R.; Hernández, A.; Yébenes, L.; Gallego, A.; Feliu, J.; et al. High-throughput 3-dimensional culture of epithelial ovarian cancer cells as preclinical model of disease. *Oncotarget* **2018**, *9*, 21893–21903. [CrossRef]
72. Chen, J.; Wang, J.; Chen, D.; Yang, J.; Yang, C.; Zhang, Y.; Zhang, H.; Dou, J. Evaluation of characteristics of CD44+CD117+ ovarian cancer stem cells in three dimensional basement membrane extract scaffold versus two dimensional monocultures. *BMC Cell Biol.* **2013**, *14*, 7. [CrossRef]
73. Myungjin Lee, J.; Mhawech-Fauceglia, P.; Lee, N.; Cristina Parsanian, L.; Gail Lin, Y.; Andrew Gayther, S.; Lawrenson, K. A three-dimensional microenvironment alters protein expression and chemosensitivity of epithelial ovarian cancer cells in vitro. *Lab. Investig.* **2013**, *93*, 528–542. [CrossRef]
74. Pinto, B.; Henriques, A.C.; Silva, P.M.A.; Bousbaa, H. Three-Dimensional Spheroids as In Vitro Preclinical Models for Cancer Research. *Pharmaceutics* **2020**, *12*, 1186. [CrossRef] [PubMed]
75. Fu, X.; Hoffman, R.M. Human ovarian carcinoma metastatic models constructed in nude mice by orthoptic transplantation of histologically-intact patient specimens. *Anticancer Res.* **1993**, *57*, 283–286.
76. Ricci, F.; Bizzaro, F.; Cesca, M.; Guffanti, F.; Ganzinelli, M.; Decio, A.; Ghilardi, C.; Perego, P.; Fruscio, R.; Buda, A.; et al. Patient-derived ovarian tumor xenografts recapitulate human clinicopathology and genetic alterations. *Cancer Res.* **2014**, *74*, 6980–6990. [CrossRef] [PubMed]
77. Heo, E.J.; Cho, Y.J.; Cho, W.C.; Hong, J.E.; Jeon, H.K.; Oh, D.Y.; Choi, Y.L.; Song, S.Y.; Choi, J.J.; Bae, D.S.; et al. Patient-derived xenograft models of epithelial ovarian cancer for preclinical studies. *Cancer Res. Treat.* **2017**, *49*, 915–926. [CrossRef]
78. Wang, M.; Yao, L.C.; Cheng, M.; Cai, D.; Martinek, J.; Pan, C.X.; Shi, W.; Ma, A.H.; De Vere White, R.W.; Airhart, S.; et al. Humanized mice in studying efficacy and mechanisms of PD-1-targeted cancer immunotherapy. *FASEB J.* **2018**, *32*, 1537–1549. [CrossRef]

79. Magnottia, E.; Marascoa, W.A. The latest animal models of ovarian cancer for novel drug discovery. *Expert Opin.* **2018**, *13*, 249–257. [CrossRef]
80. Totti, S.; Allenby, M.C.; Dos Santos, S.B.; Mantalaris, A.; Velliou, E.G. A 3D bioinspired highly porous polymeri system for: In vitro simulation of pancreatic ductal adenocarcinoma. *RSC Adv.* **2018**, *8*, 20928–20940. [CrossRef]
81. De Jaeghere, E.; De Vlieghere, E.; Van Hoorick, J.; Van Vlierberghe, S.; Wagemans, G.; Pieters, L.; Melsens, E.; P Dorpe, J.; Boone, M.N.; et al. Heterocellular 3D scaffolds as biomimetic to recapitulate the tumor microenvironment metastases in vitro and in vivo. *Biomaterials* **2018**, *158*, 95–105. [CrossRef]
82. Brooks, E.A.; Gencoglu, M.F.; Corbett, D.C.; Stevens, K.R.; Peyton, S.R. An omentum-inspired 3D PEG hydrogel fo ECM-drivers of drug resistant ovarian cancer. *APL Bioeng.* **2019**, *3*, 026106. [CrossRef]
83. Johnson, P.A.; Giles, J.R. The hen as a model of ovarian cancer. *Nat. Rev. Cancer* **2013**, *13*, 432–436. [CrossRef]
84. Bolland, D.E.; McLean, K. Preclinical models in ovarian cancer. In *Animal Models in Cancer Drug Discovery*; Elsevier The Netherlands, 2019; pp. 31–57. ISBN 9780128147054.
85. Boettcher, A.N.; Kiupel, M.; Adur, M.K.; Cocco, E.; Santin, A.D.; Bellone, S.; Charley, S.E.; Blanco-Fernandez, B.; Ross, J.W.; et al. Human ovarian cancer tumor formation in severe combined immunodeficient (SCID) pigs. *Front.* 9. [CrossRef] [PubMed]
86. Kenny, H.A.; Krausz, T.; Yamada, S.D.; Lengyel, E. Use of a novel 3D culture model to elucidate the role of mes fibroblasts and extra-cellular matrices on adhesion and invasion of ovarian cancer cells to the omentum. *Int. J. Can* 1463–1472. [CrossRef] [PubMed]
87. Masiello, T.; Dhall, A.; Hemachandra, L.; Tokranova, N.; Melendez, J.; Castracane, J. A Dynamic Culture Metho Ovarian Cancer Spheroids under Physiologically-Relevant Shear Stress. *Cells* **2018**, *7*, 277. [CrossRef]
88. Totti, S.; Ng, K.W.; Dale, L.; Lian, G.; Chen, T.; Velliou, E.G. A novel versatile animal-free 3D tool for rapid low-cost immunodiagnostic microneedles. *Sens. Actuators B Chem.* **2019**, *296*, 126652. [CrossRef]
89. Fuentes-Garí, M.; Velliou, E.; Misener, R.; Pefani, E.; Rende, M.; Panoskaltsis, N.; Mantalaris, A.; Pistikopoulos, E.N. framework for the design, simulation and optimization of personalized healthcare: Making and healing blood. *Eng.* **2015**, *81*, 80–93. [CrossRef]
90. Velliou, E.; Fuentes-Garí, M.; Misener, R.; Pefani, E.; Rende, M.; Panoskaltsis, N.; Mantalaris, A.; Pistikopoulos, E.N. for the design, modeling and optimization of biomedical systems. In *Computer Aided Chemical Engineering*; Elsevier The Netherlands, 2014; Volume 34, pp. 225–236.
91. Zanoni, M.; Piccinini, F.; Arienti, C.; Zamagni, A.; Santi, S.; Polico, R.; Bevilacqua, A.; Tesei, A. 3D tumor sphero in vitro therapeutic screening: A systematic approach to enhance the biological relevance of data obtained. *Sci* 19103. [CrossRef] [PubMed]
92. Watters, K.M.; Bajwa, P.; Kenny, H.A. Organotypic 3D models of the ovarian cancer tumor microenvironment. *Ca* 265. [CrossRef] [PubMed]
93. Bapat, S.A.; Mali, A.M.; Koppikar, C.B.; Kurrey, N.K. Stem and progenitor-like cells contribute to the aggressiv human epithelial ovarian cancer. *Cancer Res.* **2005**, *65*, 3025–3029. [CrossRef]
94. Desoize, B.; Jardillier, J.C. Multicellular resistance: A paradigm for clinical resistance? *Crit. Rev. Oncol. Hematol.* **200** [CrossRef]
95. Yoshida, Y.; Kurokawa, T.; Nishikawa, Y.; Orisa, M.; Kleinman, H.K.; Kotsuji, F. Laminin-1-derived scrambled p disaggregates laminin-1-induced ovarian cancer cell spheroids and improves the efficacy of cisplatin. *Int. J. O* 673–681. [CrossRef]
96. Wan, X.; Ball, S.; Willenbrock, F.; Yeh, S.; Vlahov, N.; Koennig, D.; Green, M.; Brown, G.; Jeyaretna, S.; Li, Z.; e Three-dimensional Organotypic Culture of Human Cancer Cells for Therapeutic Evaluation. *Sci. Rep.* **2017**, *7*, 94 [PubMed]
97. Jong, B.K.; Stein, R.; O'Hare, M.J. Three-dimensional in vitro tissue culture models of breast cancer—A review. *Bre Treat.* **2004**, *85*, 281–291. [CrossRef]
98. Nyga, A.; Cheema, U.; Loizidou, M. 3D tumour models: Novel in vitro approaches to cancer studies. *J. Cell Commu* 5, 239–248. [CrossRef] [PubMed]
99. Gomez-Roman, N.; Stevenson, K.; Gilmour, L.; Hamilton, G.; Chalmers, A.J. A novel 3D human glioblastoma cell for modeling drug and radiation responses. *Neuro. Oncol.* **2017**, *19*, 229–241. [CrossRef]
100. Xu, F.; Celli, J.; Rizvi, I.; Moon, S.; Hasan, T.; Demirci, U. A three-dimensional in vitro ovarian cancer coculture high-throughput cell patterning platform. *Biotechnol. J.* **2011**, *6*, 204–212. [CrossRef]
101. Cui, X.; Hartanto, Y.; Zhang, H. Advances in multicellular spheroids formation. *J. R. Soc. Interface* **2017**, *14*, 877. [C
102. Weiswald, L.B.; Bellet, D.; Dangles-Marie, V. Spherical Cancer Models in Tumor Biology. *Neoplasia* **2015**, *17*, 4. [Cr
103. Moraya, A.I.; Ali, J.L.; Samadder, P.; Liang, L.; Morrison, L.C.; Werbowetski-Ogilvie, T.E.; Ogunsina, M.; Schweizer, Nachtigal, M.W. Novel glycolipid agents for killing cisplatin-resistant human epithelial ovarian cancer cells. *J. Ex Res.* **2017**, *36*, 67. [CrossRef]
104. Burdett, E.; Kasper, F.K.; Mikos, A.G.; Ludwig, J.A. Engineering Tumors: A Tissue Engineering Perspective in C *Tissue Eng. Part B Rev.* **2010**, *16*, 351–359. [CrossRef]

Knight, E.; Przyborski, S. Advances in 3D cell culture technologies enabling tissue-like structures to be created in vitro. *J. Anat.* **2015**, *227*, 746–756. [CrossRef]

Sodek, K.L.; Ringuette, M.J.; Brown, T.J. Compact spheroid formation by ovarian cancer cells is associated with contractile behavior and an invasive phenotype. *Int. J. Cancer* **2009**, *124*, 2060–2070. [CrossRef] [PubMed]

Loessner, D.; Rizzi, S.C.; Stok, K.S.; Fuehrmann, T.; Hollier, B.; Magdolen, V.; Hutmacher, D.W.; Clements, J.A. A bioengineered 3D ovarian cancer model for the assessment ofpeptidase-mediated enhancement of spheroid growth andintraperitoneal spread. *Biomaterials* **2013**, *34*, 7389–7400. [CrossRef] [PubMed]

Zheng, L.; Hu, X.; Huang, Y.; Xu, G.; Yang, J.; Li, L. In vivo bioengineered ovarian tumors based on collagen, matrigel, alginate and agarose hydrogels: A comparative study. *Biomed. Mater.* **2015**, *10*, 15016. [CrossRef] [PubMed]

Kaemmerer, E.; Melchels, F.P.W.; Holzapfel, B.M.; Meckel, T.; Hutmacher, D.W.; Loessner, D. Gelatine methacrylamide-based hydrogels: An alternative three-dimensional cancer cell culture system. *Acta Biomater.* **2014**, *10*, 2551–2562. [CrossRef] [PubMed]

Hutmacher, D.W. Scaffold design and fabrication technologies for engineering tissues—State of the art and future perspectives. *J. Biomater. Sci. Polym. Ed.* **2001**, *12*, 107–124. [CrossRef]

Sodek, K.L.; Brown, T.J.; Ringuette, M.J. Collagen I but not Matrigel matrices provide an MMP-dependent barrier to ovarian cancer cell penetration. *BMC Cancer* **2008**, *8*, 223. [CrossRef]

Zhou, N.; Hu, K.; Guo, Z.; Zhang, Q.; Chen, J.; Zhang, T.; Gu, N. Thermo-Sensitive PLGA-PEG-PLGA Tri-Block Copolymer Hydrogel as Three-Dimensional Cell Culture Matrix for Ovarian Cancer Cells. *J. Nanosci. Nanotechnol.* **2018**, *18*, 5252–5255. [CrossRef]

Zhang, T.; Zhang, Q.; Chen, J.; Fang, K.; Dou, J.; Gu, N. The controllable preparation of porous PLGA microspheres by the oil/water emulsion method and its application in 3D culture of ovarian cancer cells. *Colloids Surfaces A Physicochem. Eng. Asp.* **2014**, *452*, 115–124. [CrossRef]

Sun, B.; Taha, M.S.; Ramsey, B.; Torregrosa-Allen, S.; Elzey, B.D.; Yeo, Y. Intraperitoneal chemotherapy of ovarian cancer by hydrogel depot of paclitaxel nanocrystals. *J. Control. Release* **2016**, *235*, 91–98. [CrossRef]

McKenzie, A.J.; Hicks, S.R.; Svec, K.V.; Naughton, H.; Edmunds, Z.L.; Howe, A.K. The mechanical microenvironment regulates ovarian cancer cell morphology, migration, and spheroid disaggregation. *Sci. Rep.* **2018**, *8*, 7228. [CrossRef]

Li, Y.; Kumacheva, E. Hydrogel microenvironments for cancer spheroid growth and drug screening. *Sci. Adv.* **2018**, *4*, 8998. [CrossRef] [PubMed]

Girard, Y.K.; Wang, C.; Ravi, S.; Howell, M.C.; Mallela, J.; Alibrahim, M.; Green, R.; Hellermann, G.; Mohapatra, S.S.; Mohapatra, S. A 3D Fibrous Scaffold Inducing Tumoroids: A Platform for Anticancer Drug Development. *PLoS ONE* **2013**, *8*, 75345. [CrossRef] [PubMed]

Godugu, C.; Patel, A.R.; Desai, U.; Andey, T.; Sams, A.; Singh, M. AlgiMatrix^{TM} Based 3D Cell Culture System as an In-Vitro Tumor Model for Anticancer Studies. *PLoS ONE* **2013**, *8*, 53708. [CrossRef] [PubMed]

Matta-Domjan, B.; King, A.; Totti, S.; Matta, C.; Dover, G.; Martinez, P.; Zakhidov, A.; La Ragione, R.; Macedo, H.; Jurewicz, I.; et al. Biophysical interactions between pancreatic cancer cells and pristine carbon nanotube substrates: Potential application for pancreatic cancer tissue engineering. *J. Biomed. Mater. Res. Part B Appl. Biomater.* **2018**, *106*, 1637–1644. [CrossRef] [PubMed]

Wang, X.; Zhang, X.; Fu, Z.; Yin, H. A bioengineered metastatic pancreatic tumor model for mechanistic investigation of chemotherapeutic drugs. *J. Biotechnol.* **2013**, *166*, 166–173. [CrossRef] [PubMed]

He, Q.; Wang, X.; Zhang, X.; Han, H.; Han, B.; Xu, J.; Tang, K.; Fu, Z.; Yin, H. A tissue engineered subcutaneous pancreatic cancer model for antitumor drug evaluation. *Int. J. Nanomed.* **2013**, *8*, 1167–1176. [CrossRef]

Ricci, C.; Mota, C.; Moscato, S.; D'Alessandro, D.; Ugel, S.; Sartoris, S.; Bronte, V.; Boggi, U.; Campani, D.; Funel, N.; et al. Interfacing polymeric scaffolds with primary pancreatic ductal adenocarcinoma cells to develop 3D cancer models. *Biomatter* **2014**, *4*, e955386. [CrossRef]

Gupta, P.; Totti, S.; Pérez-Mancera, P.A.; Dyke, E.; Nisbet, A.; Schettino, G.; Webb, R.; Velliou, E.G. Chemoradiotherapy screening in a novel biomimetic polymer based pancreatic cancer model. *RSC Adv.* **2019**, *9*, 41649–41663. [CrossRef]

Gupta, P.; Perez-Mancera, P.; Kocher, H.; Nisbet, A.; Schettino, G.; Velliou, E. A novel scaffold based hybrid multicellular model for pancreatic ductal adenocarcinoma—Towards a better mimicry of the in vivo tumour microenvironment. *Front. Bioeng. Biotechnol.* **2020**, *8*, 290. [CrossRef]

Rijal, G.; Bathula, C.; Li, W. Application of Synthetic Polymeric Scaffolds in Breast Cancer 3D Tissue Cultures and Animal Tumor Models. *Int. J. Biomater.* **2017**, *2017*, 8074890. [CrossRef]

Liu, Z.; Vunjak-Novakovic, G. Modeling tumor microenvironments using custom-designed biomaterial scaffolds. *Curr. Opin. Chem. Eng.* **2016**, *11*, 94–105. [CrossRef] [PubMed]

Birgersdotter, A.; Sandberg, R.; Ernberg, I. Gene expression perturbation in vitro—A growing case for three-dimensional (3D) culture systems. *Semin. Cancer Biol.* **2005**, *15*, 405–412. [CrossRef] [PubMed]

Moher, D.; Liberati, A.; Tetzlaff, J.; Altman, D.G.; Altman, D.; Antes, G.; Atkins, D.; Barbour, V.; Barrowman, N.; Berlin, J.A.; et al. Preferred reporting items for systematic reviews and meta-analyses: The PRISMA statement. *PLoS Med.* **2009**, *6*, 1000097. [CrossRef] [PubMed]

MDPI
St. Alban-Anlage 66
4052 Basel
Switzerland
Tel. +41 61 683 77 34
Fax +41 61 302 89 18
www.mdpi.com

Cancers Editorial Office
E-mail: cancers@mdpi.com
www.mdpi.com/journal/cancers

www.ingramcontent.com/pod-product-compliance
Lightning Source LLC
Chambersburg PA
CBHW061800210926
51850CB00006B/604
9783036568423